Physics 1

Nick England
Carol Davenport
Jeremy Pollard
Nicky Thomas

Approval message from AQA

This textbook has been approved by AQA for use with our qualification. This means that we have checked that it broadly covers the specification and we are satisfied with the overall quality. Full details of our approval process can be found on our website.

We approve textbooks because we know how important it is for teachers and students to have the right resources to support their teaching and learning. However, the publisher is ultimately responsible for the editorial control and quality of this book.

Please note that when teaching the ***AQA A-level Physics*** course, you must refer to AQA's specification as your definitive source of information. While this book has been written to match the specification, it does not provide complete coverage of every aspect of the course.

A wide range of other useful resources can be found on the relevant subject pages of our website: www.aqa.org.uk.

Photo Credits

p. 1 © Graham J. Hills/Science Photo Library; **p. 2** t © Peter Tuffy, University Of Edinburgh/Science Photo Library, l © Graham J. Hills/Science Photo Library; **p. 3** t © Athol Pictures / Alamy, l © egorxfi – Fotolia; **p. 8** tl C.T.R. Wilson/Science Photo Library, *tr* © C. T. R. Wilson (1912) Proc. Roy. Soc., A, vol 87, © The Royal Society, cl © N. Feather/Science Photo Library, cr © Science Photo Library; **p. 11** © v.poth – Fotolia; **p. 18** © dkidpix - Fotolia; **p. 23** © sciencephotos / Alamy; **p. 24** © Klaus Guldbrandsen/ Science Photo Library; **p. 42** l © Carlos Clarivan/Science Photo Library; **p. 43** l © Carlos Clarivan/ Science Photo Library; **p. 48** t © Jim Elder, Ottawa, Canada, lb © Jim Elder, Ottawa, Canada; **p. 57** © E. R. Degginger/Science Photo Library; **p. 64** l © Omikron/Science Photo Library, r © E. R. Degginger/ Science Photo Library; **p. 69** t Download for free at http://cnx.org/contents/d45b204f-4b3a-4c7a-9018-d091e9270579@5/Probability:_The_Heisenberg_Un; c Download for free at http://cnx.org/contents/ d45b204f-4b3a-4c7a-9018-d091e9270579@5/Probability:_The_Heisenberg_Un; b Image courtesy of A. Weis and T.L. Dimitrova; **p. 70** © roblan – Fotolia; **p. 71** t © Science and Society / SuperStock, b © roblan – Fotolia; **p. 81** c © adrian davies / Alamy; **p. 92** c © Colin Reid / Colin Reid Glass; **p. 95** © Science Photo Library / Andrew Lambert; **p. 96** © Comstock/Getty Images; **p. 108** © Science Photo Library / Andrew Lambert; **p. 116** l © eyewave – Fotolia, c © dpa picture alliance archive / Alamy, r © eyewave – Fotolia; **p. 133** © ParagonSpaceDevelopment/Splash/Splash News/Corbis; **p. 134** © ParagonSpaceDevelopment/ Splash/Splash News/Corbis; **p. 151** © Perry van Munster / Alamy; **p. 152** © Perry van Munster / Alamy; **p. 168** © NASA/Science Photo Library; **p. 183** © xavier gallego morel - Fotolia; **p. 184** © Delphimages - Fotolia; **p. 186** © Loren Winters, Visuals Unlimited /Science Photo Library; **p. 187** © culture-images GmbH / Alamy; **p. 199** © Science Photo Library; **p. 200** l © NREL/US Department of Energy/Science Photo Library, r © NASA/JPL/Science Photo Library; **p. 201** © D P P I/REX; **p. 202** t © Carol Davenport, c © NREL/US Department of Energy/Science Photo Library, b © NASA/JPL; **p. 221** © Vendigo – Fotolia; **p. 222** © Vendigo – Fotolia; **p. 226** © Daniel L. Osborne, University of Alaska/ Detlev Van Ravenswaay/ Science Photo Library; **p. 228** © Science Photo Library; **p. 229** t © David Parker/Imi/University of Birmingham High TC Consortium/Science Photo Library, b © Bernd Mellmann / Alamy; **p. 244** © khunaspix - Fotolia; **p. 246** © Science Photo Library; **p. 250** © Lucie Lang / Alamy; **p. 253** © Science Photo Library; **p. 259** © Science Photo Library; **p. 274** © Coloures-pic - Fotolia; **p. 289** l © NASA; **p. 289** r © NASA; **p. 298** l © ptnphotof - Fotolia; **p. 299** © Carol Davenport;

t = top, *b* = bottom, *l* = left, *c* = centre, *r* = right

Special thanks to Robin Hughes and The British Physics Olympiad for permission to use a selection of their questions within this textbook.

Figures 4.1, 4.2, 4.3, 4.10, and 13.3 are based on artworks from School Physics, http://www.schoolphysics.co.uk/, and are reproduced with special thanks to Keith Gibbs – © Keith Gibbs 2015.

Orders: please contact Bookpoint Ltd, 130 Milton Park, Abingdon, Oxon OX14 4SB. Telephone: +44 (0)1235 827720. Fax: +44 (0)1235 400454. Lines are open 9.00a.m.–5.00p.m., Monday to Saturday, with a 24-hour message answering service. Visit our website at www.hoddereducation.co.uk

First published in 2015 by
Hodder Education,
An Hachette UK Company
Carmelite House
50 Victoria Embankment
London EC4Y 0DZ

Impression number 10 9 8 7 6 5 4 3 2

Year 2019 2018 2017 2016 2015

Cover photo © Kesu – Fotolia
Illustrations by Aptara
Typeset in 11/13 pt ITC Berkeley Oldstyle by Aptara, Inc.
Printed in Italy

A catalogue record for this title is available from the British Library.

ISBN 978 1471 807732

Contents

Get the most from this book iv

1 Particles and nuclides 1

2 Fundamental particles 18

3 Electrons and energy levels 42

4 Particles of light 57

5 Waves 70

6 Combining waves 95

7 Introduction to mechanics 116

8 Motion and its measurement 133

9 Newton's laws of motion 151

10 Work, energy and power 168

11 Momentum 183

12 Properties of materials 200

13 Current electricity 221

14 Electrical circuits 244

15 Maths in physics 274

16 Developing practical skills in physics 289

17 Preparing for written assessments 298

Index 302

Free online resources 305

Get the most from this book

Welcome to the **AQA A-level Physics Year 1 Student's Book**. This book covers Year 1 of the AQA A-level Physics specification and all content for the AQA AS Physics specification.

The following features have been included to help you get the most from this book.

Prior knowledge

This is a short list of topics that you should be familiar with before starting a chapter. The questions will help to test your understanding.

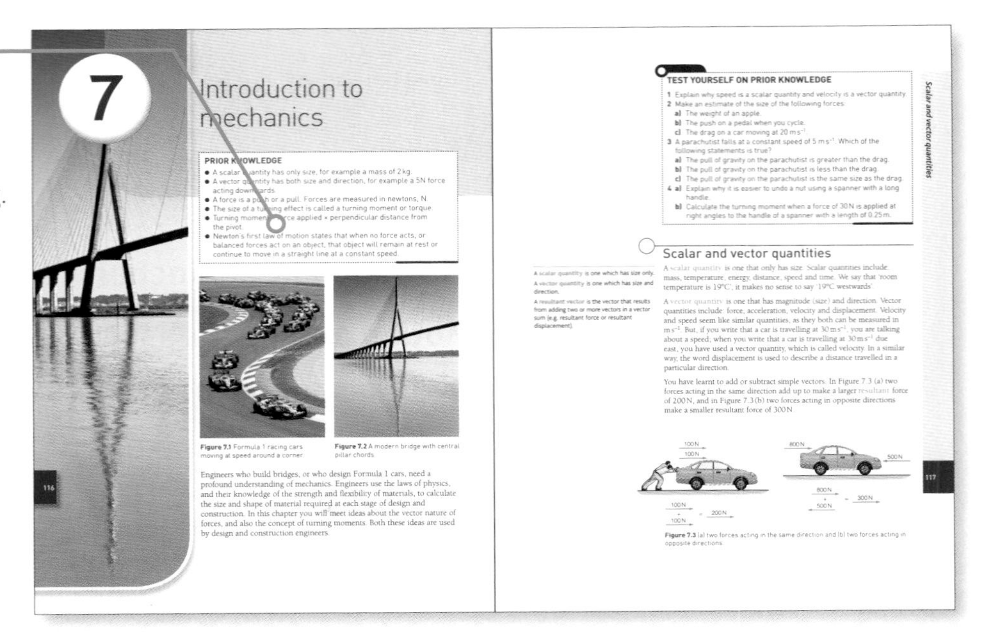

Activities and Required practicals

These practical-based activities will help consolidate your learning and test your practical skills. AQA's required practicals are clearly highlighted.

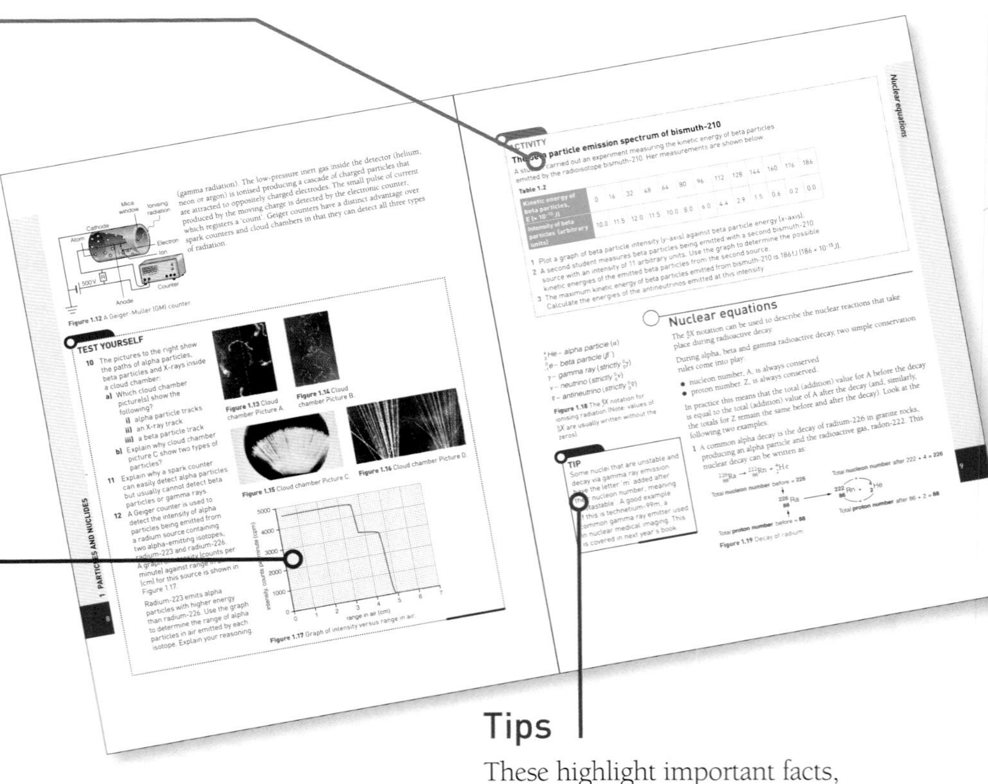

Test yourself questions

These short questions, found throughout each chapter, are useful for checking your understanding as you progress through a topic.

Tips

These highlight important facts, common misconceptions and signpost you towards other relevant topics.

Practice questions

You will find Practice questions at the end of every chapter. These follow the style of the different types of questions you might see in your examination, including multiple-choice questions, and are colour coded to highlight the level of difficulty. Test your understanding even further with Stretch and challenge questions.

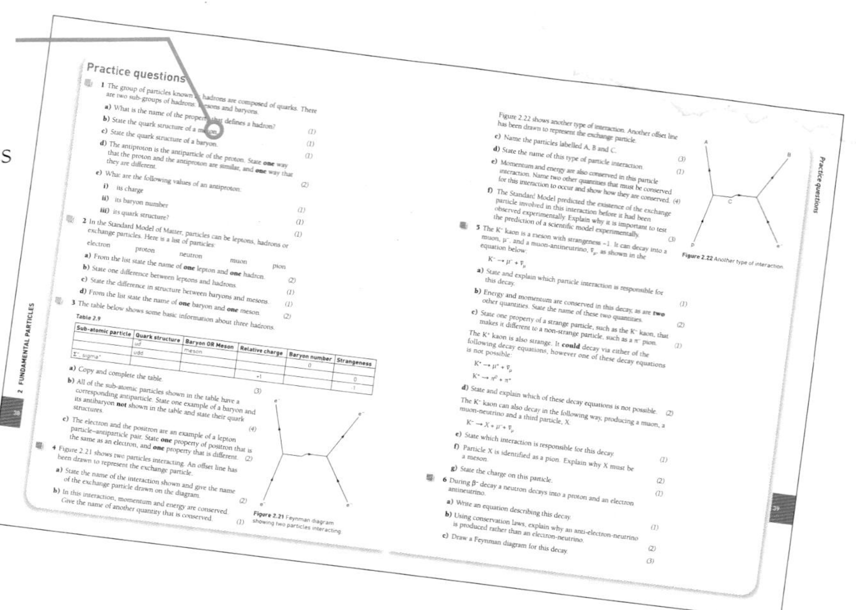

Questions are colour-coded, to help target your practice:

- Green – Basic questions that everyone should be able to answer without difficulty.
- Orange – Questions that are a regular feature of exams and that all competent candidates should be able to handle.
- Purple – More demanding questions which the best candidates should be able to do.
- Stretch and challenge – Questions for the most able candidates to test their full understanding and sometimes their ability to use ideas in a novel situation.

Key terms and formulae

These are highlighted in the text and definitions are given in the margin to help you pick out and learn these important concepts.

Examples

Examples of questions and calculations feature full workings and sample answers.

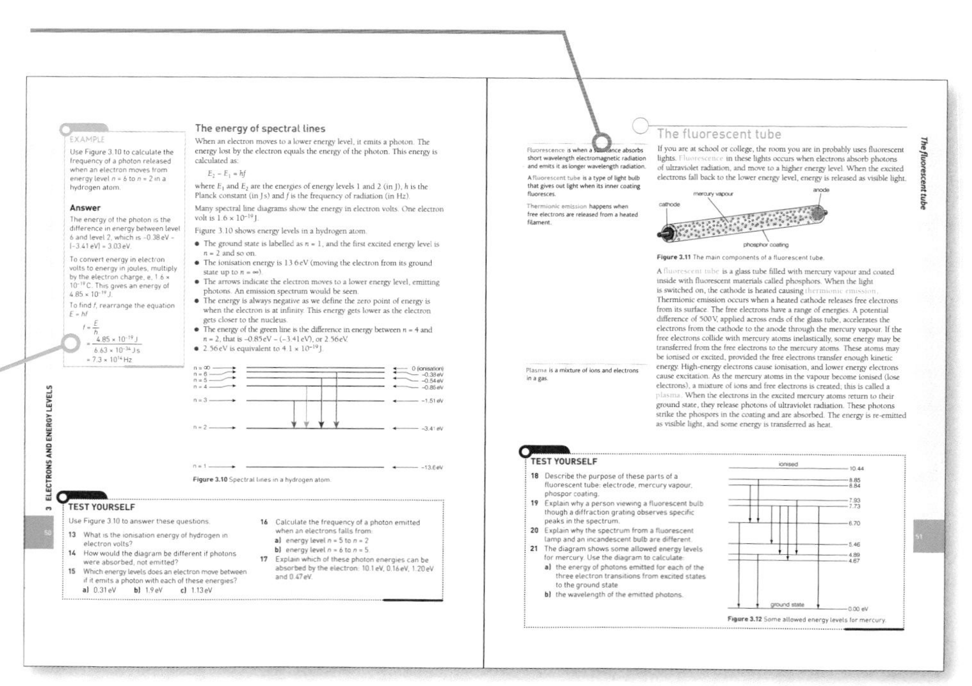

Dedicated chapters for developing your **Maths** and **Practical skills** and **Preparing for your exam** can be found at the back of this book.

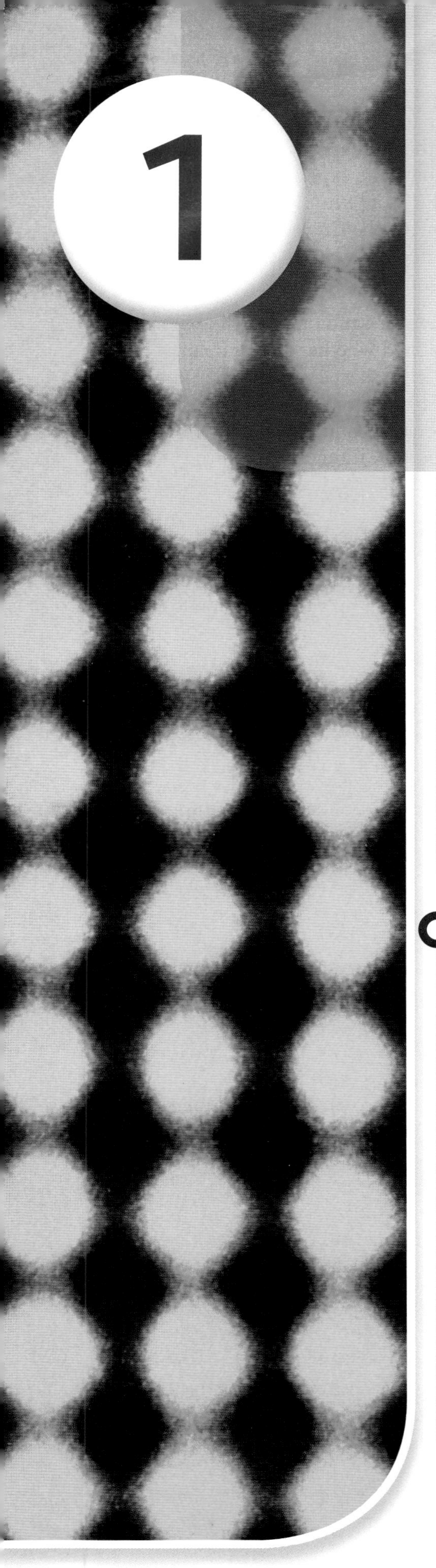

Particles and nuclides

PRIOR KNOWLEDGE

- Matter is made of atoms. Atoms are made up of a very small central nucleus containing particles called protons and neutrons, surrounded by orbiting electrons. The number of protons contained in the nucleus of an atom is called the proton or atomic number and the total number of protons and neutrons is called the nucleon or mass number.
- Protons are positively charged particles with a relative charge of +1; electrons are negatively charged particles with a relative charge of −1; neutrons are electrically neutral with a relative charge of zero.
- Atoms are electrically neutral overall, which means that they have the same number of protons and electrons. When atoms lose or gain electrons they become ions. An excess of electrons produces a negatively charged ion whereas a shortage of electrons produces a positively charged ion.
- The diameter of an atomic nucleus is of the order of 10 000 times smaller than the diameter of an atom, but contains the vast majority of the atomic mass. Protons and neutrons have a relative mass of 1, with electrons about 1800 times less massive.
- Elements are made up of atoms with the same proton number. Atoms can have the same number of protons but different numbers of neutrons: these are called isotopes.
- The relative atomic mass of an element compares the mass of atoms of the element with the carbon-12 isotope.

TEST YOURSELF ON PRIOR KNOWLEDGE

1 Naturally occurring hydrogen has three isotopes: ('normal') hydrogen, deuterium and tritium. What three properties do atoms of each of these isotopes have in common? How do the isotopes differ?

2 What is the relative charge of a copper atom that has lost two electrons?

3 Chlorine has two naturally occurring isotopes that exist in an almost 75 : 25 abundance ratio. The average relative atomic mass of chlorine is 35.5. What are the relative atomic masses of these two isotopes? Find out their proton and nucleon numbers.

4 As at spring 2015, the element with the highest atomic number so far discovered is ununoctium, (Uuo, atomic number 118), 'discovered' in 2005, of which only three or four atoms have been observed, each with a relative atomic mass of 294. Determine its:

a) proton number
b) neutron number
c) electron number
d) proton : electron mass ratio.

Figure 1.1 Peter Higgs and the Higgs Boson Theory.

Figure 1.2 A more recent image taken by a transmission electron microscope showing individual gold atoms each separated from its neighbour by a distance of 2.3 nm (2.3×10^{-9} m).

On 8th October 2013 the Nobel Prize in Physics was awarded jointly to François Englert and Peter Higgs for the theoretical discovery of the particle known as the Higgs Boson. (This followed its experimental discovery by the Large Hadron Collider team at CERN on 4th July 2012). This discovery completes our current best model of what the Universe is built from, called the Standard Model, a model that started development nearly two and a half thousand years ago by a Greek philosopher called Democritus.

It is said that Democritus observed that the sand on a beach was once part of the rocks of the cliffs and he questioned whether the sand could be cut into ever smaller pieces by a succession of sharper and smaller knives. It was Democritus who first coined the word 'atom', meaning the smallest indivisible piece that the sand could be cut up into. Nearly 2500 years later, JJ Thomson was able to extend Democritus' thought experiment when he discovered the electron, splitting up atoms and discovering the first sub-atomic particle.

In the early 1980s, researchers at IBM in Zurich produced a machine called an atomic force microscope that was able to image individual atoms. Figure 1.2 shows a more recent image taken by a transmission electron microscope showing individual gold atoms. Each individual gold atom is separated from its neighbour by a distance of 2.3 nm (2.3×10^{-9} m).

What are the building blocks of the Universe?

What is the material of the Universe made of? Like so many things in physics, it depends on which model you are using to explain things. Physics is a series of evermore complex, layered models designed to explain how the Universe behaves as we see it now. Over the years, as our observations of the Universe – particularly on the sub-atomic and cosmological scales – have become more and more detailed, so the physical models have had to adapt to the new observations and measurements. It is rare now for a theory to be developed as the result of a 'thought' experiment such as Democritus'.

Models in physics

In science, and in physics in particular, we rely on models to explain how the Universe around us works. Models make complex, often invisible things or processes easier to visualise. Models take many forms. Some, like the models of atoms and nuclei used in this chapter, are visual models used as analogies of the real thing. The Rutherford-Bohr model used in the next section is a good example of this type of model, where the nucleus is modelled as a small ball containing neutrons and protons, with electrons whizzing round the outside. Atoms don't actually 'look' like this, but it is a convenient, easy first model to use because it uses the analogy of everyday objects that we are very familiar with.

Other types of model in physics are more mathematically based. A good example of this is the kinetic theory of gases, where simple physical rules expressed mathematically are applied to a model of gas molecules behaving as hard, bouncy elastic balls. These rules allow the model to predict the behaviour of real gases.

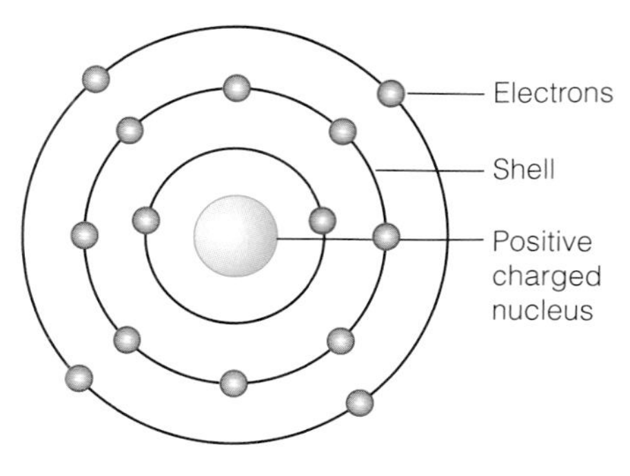

Figure 1.3 The Rutherford-Bohr model of the atom.

Some models are very good analogies and others not so. Models and analogies have their limits. The best physical models are those that are adaptable to take in new experimental discoveries. Good physicists always state and explain the limitations of the models that they are using.

JJ Thomson was the first person to create a model of the structure of the atom, which he called the plum pudding model. Thomson produced this model just after he had discovered the electron in 1897, and it was named after a popular steamed dessert. His model consisted of a sphere of positive charge with the electrons embedded throughout the positive charge, rather like the small pieces of plum (or currants) inside the steamed pudding. The electrons were allowed to move through the structure in ringed orbits. Thomson's model, however, was short-lived. The nucleus of the atom was discovered by Rutherford, Geiger and Marsden in 1909.

The Rutherford-Bohr atom

The Rutherford-Bohr model of the atom consists of a tiny, central, positively charged nucleus, containing protons and neutrons, surrounded by orbiting negatively charged electrons (analogous to the Solar System). This model is particularly useful for chemists as it can be used to explain much of the chemical behaviour of atoms. The Rutherford-Bohr model is a useful model for visualising the atom, and is used a great deal by the popular media to describe atoms.

Figure 1.4 If the atom was the size of the Wembley Stadium complex, then the nucleus would be the size of a pea on the centre spot.

Rutherford scattering

Experiments carried out by Ernest Rutherford and his research assistants Hans Geiger and Ernest Marsden in 1911 proved that the nuclear radius is about 5000 times smaller than the atomic radius. Geiger and Marsden performed alpha-particle scattering experiments on thin films of gold, which has an atomic radius of 134 pm (picometres) – where 1 pm = 1×10^{-12} m. They found that the radius of the nucleus of gold is about 27 fm (femtometres or fermi) – where 1 fm = 1×10^{-15} m. However, modern measurements of the radius of a gold nucleus produce a value of about 5 fm. So a more accurate ratio of the radius of a gold atom to the radius of a gold nucleus is about 25 000.

As a way of visualising the scale of the atom and the nucleus, if the atom had a radius equal to that of the Wembley Stadium complex, then the nucleus would be about the size of a small pea on the centre spot.

Charges, masses and specific charges

The charges and masses of the proton, neutron and electron are shown in Table 1.1 below:

Table 1.1

Sub-atomic particle	Charge/C	Relative charge	Mass/kg
Proton	$+1.60 \times 10^{-19}$	+1	1.673×10^{-27}
Neutron	0	0	1.675×10^{-27}
Electron	-1.60×10^{-19}	−1	9.11×10^{-31}

TIP
Next year you will meet the concept of binding energy, where some of the mass of the particles involved with forming nuclei is transferred into nuclear energy holding the nuclei together. So the calculations in the examples are approximations based on the separate masses of the protons and neutrons.

The specific charge of a particle is defined as the charge per unit mass, and its units are $\mathrm{C\,kg^{-1}}$. Specific charge is calculated using the formula:

$$\text{specific charge of a particle} = \frac{\text{charge of the particle}}{\text{mass of the particle}} = \frac{Q}{m}$$

So the specific charge of the proton is calculated by:

$$\text{specific charge of a proton} = \frac{+1.60 \times 10^{-19}}{1.673 \times 10^{-27}}$$
$$= 9.56 \times 10^{7}\ \mathrm{C\,kg^{-1}}$$

The specific charges of the neutron and the electron are 0 and $-1.76 \times 10^{11}\ \mathrm{C\,kg^{-1}}$ respectively.

EXAMPLE

a) Calculate the specific charge of a beryllium-9 nucleus, containing 5 neutrons and 4 protons.

b) Calculate the specific charge of a lithium-7 (+1) ion, containing 4 neutrons, 3 protons and 2 electrons.

Answer

a) The total charge, Q, of the nucleus is $4 \times (+1.60 \times 10^{-19})$

$= +6.40 \times 10^{-19}\ \mathrm{C}$

The total mass, m, of the nucleus is (mass of protons) + (mass of neutrons)

$m = (4 \times 1.673 \times 10^{-27}) + (5 \times 1.675 \times 10^{-27})$

$= 1.507 \times 10^{-26}\ \mathrm{kg}$

The specific charge of the nucleus is:

$$\text{specific charge} = \frac{Q}{m} = \frac{+6.40 \times 10^{-19}}{1.507 \times 10^{-26}} = +4.25 \times 10^{7}\ \mathrm{C\,kg^{-1}}$$

b) The relative total charge, Q, of the ion is +1 (3 protons and 2 electrons) so the charge is

$1 \times (+1.60 \times 10^{-19})$

$= +1.60 \times 10^{-19}\ \mathrm{C}$

The total mass, m, of the ion is (mass of protons) + (mass of neutrons) + (mass of electrons)

$m = (3 \times 1.673 \times 10^{-27}) + (4 \times 1.675 \times 10^{-27}) + (2 \times 9.11 \times 10^{-31})$

$= 1.17 \times 10^{-26}\ \mathrm{kg}$

The specific charge of the ion is:

$$\text{specific charge} = \frac{Q}{m} = \frac{+1.60 \times 10^{-19}}{1.17 \times 10^{-26}} = +1.37 \times 10^{7}\ \mathrm{C\,kg^{-1}}$$

Nucleon is the word used to describe protons or neutrons. These are particles that exist in the nucleus.

The **proton number** of a nucleus is the number of protons in the nucleus.

The **nucleon number** of a nucleus is the number of protons plus the number of neutrons.

Describing nuclei and isotopes

The number of protons in any given nucleus is called the proton number, Z. The total number of protons and neutrons in the nucleus is called the nucleon number, A. (Protons and neutrons are both found in the nucleus and are therefore called nucleons). These two numbers completely describe any nucleus, as the number of neutrons in a nucleus can be calculated by subtracting the proton number from the nucleon number. The two nuclear numbers and the chemical symbol are used together as a shorthand way of describing any nucleus. This is called the $^{A}_{Z}X$ **notation.**

The nucleon number, A, is always written as a superscripted prefix to the chemical symbol, and the proton number, Z, is always a subscripted prefix.

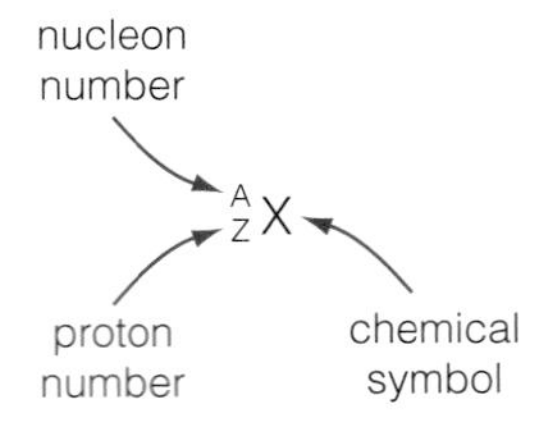

Figure 1.5 The $^{A}_{Z}X$ notation.

Isotopes are nuclei with the same number of protons, but different numbers of neutrons.

TIP

Note: chemists use different names for A and Z. They call A the mass number and Z the atomic number; this is because chemists generally use this notation for describing **atoms** on the Periodic Table whereas, as physicists, we are using the notation to very specifically describe **nuclei**.

This notation allows all nuclei to be described uniquely, including isotopes, which are nuclei (and atoms) of the same element having the same proton number, Z, and therefore chemical symbol, X, but different numbers of neutrons and hence different nucleon numbers, A.

A good example of how the $^{A}_{Z}X$ notation is used is when describing the two main isotopes of uranium used in nuclear fission reactors. About 99.3% of all naturally occurring uranium atoms (and therefore nuclei) are uranium-238, meaning 238 nucleons (the nucleon number, A), comprising of 92 protons (the proton number, Z) and 146 neutrons (238 − 92 = 146). Only about 0.7% of naturally occurring uranium is the more useful uranium-235, which can be used in nuclear reactors as nuclear fuel. Uranium-235 has the same proton number, 92, but only 143 neutrons (235 − 92 = 143). Using the $^{A}_{Z}X$ notation, the two nuclei are written as:

Figure 1.6 The two most abundant isotopes of uranium.

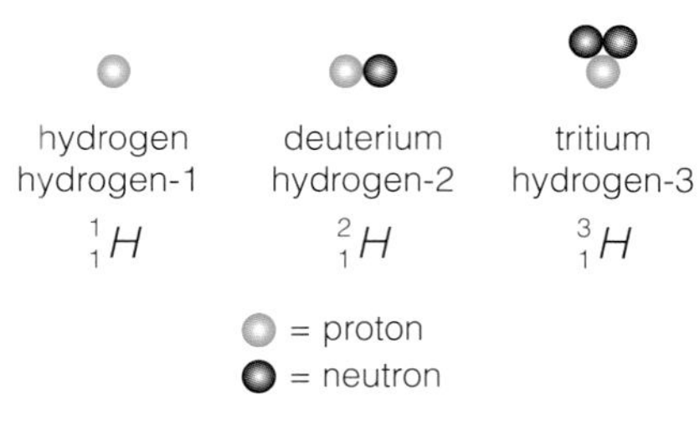

Figure 1.7 The isotopes of hydrogen.

Naturally occurring hydrogen has three isotopes – 'normal' hydrogen, deuterium and tritium – 99.98% of all hydrogen is normal hydrogen-1; less than 0.02% is deuterium, hydrogen-2; and there are only trace amounts of tritium, hydrogen-3.

TEST YOURSELF

Data for test yourself questions is generally given in the questions, but you can find extra information on specific nuclides using an online database such as Kaye and Laby from the National Physical laboratory (NPL), kayelaby.npl.co.uk.

1 Explain what is meant by the **specific charge** of an electron.

2 What are isotopes?

3 Using the data on p.3 obtained by Rutherford, Geiger and Marsden, calculate the (atomic radius : nuclear radius) ratio for gold.

4 Rutherford realised that the nuclear radius that he had calculated from Geiger and Marsden's data from positively charged alpha particle scattering would be a *maximum* value. (This value is now called the charge radius.) More modern methods of determining the nuclear radius of gold, using electron scattering, put the radius at about 5 fm (~ 5×10^{-15} m). Why is there such a difference between the two values?

5 This question is about the element silicon-28, $^{28}_{14}Si$, which is commonly used as a substrate for the manufacture of integrated circuits.

a) How many protons, neutrons and electrons are there in an atom of $^{28}_{14}Si$?

b) The $^{28}_{14}Si$ atom loses two electrons and forms an ion.

i) Calculate the charge of the ion.

ii) State the number of nucleons in the ion.

iii) Calculate the specific charge of the silicon ion.

6 **a)** Germanium atoms are frequently laid onto the top of silicon substrates. About 37% of naturally occurring germanium is germanium-74. Determine the charge of a germanium-74 nucleus.

b) A positive ion with a germanium-74 nucleus has a charge of 4.80×10^{-19} C. Calculate the number of electrons in this ion.

7 Calculate the specific charges of the following:

a) a deuterium nucleus (heavy hydrogen – one proton + one neutron)

b) a carbon-12 nucleus

c) an (oxygen-16)$^{2-}$ ion.

8 Radon-222 is a colourless, naturally occurring radioactive gas.

a) How many protons are there in a nucleus of radon-222?

b) How many neutron are there in a nucleus of radon-222?

c) Write the $^{A}_{Z}X$ notation for radon-222.

9 There are two naturally occurring isotopes of copper: 69% is copper-63, 31% is copper-65.

a) Write the $^{A}_{Z}X$ notations for each of the two isotopes.

b) Calculate the number of neutrons in each of the two isotopes.

c) Calculate the average atomic mass of these two isotopes.

Stable and unstable nuclei

Why are most nuclei stable? Protons are positively charged, so when they are confined close together in a nucleus they should all repel each other and break up the nucleus, so why don't they? If nuclei are stable then there must be another attractive, short-range force that is stronger than the force of electromagnetic repulsion between the protons. In addition, this stronger force must also act between neutrons as well as protons, otherwise it would be extremely easy to remove neutrons from nuclei; at extremely short-ranges the force must be repulsive, otherwise the nucleons inside the nucleus would implode.

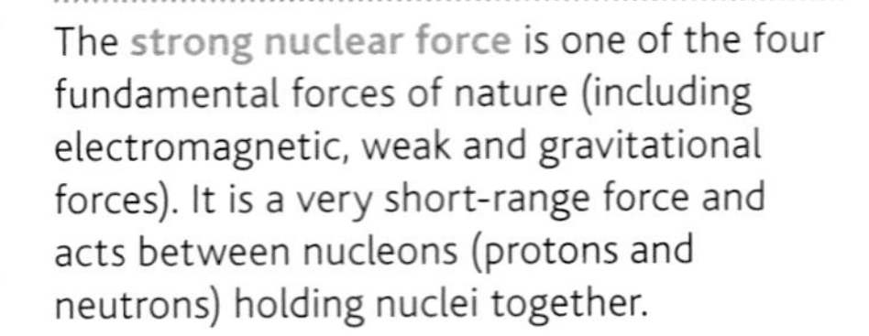

The **strong nuclear force** is one of the four fundamental forces of nature (including electromagnetic, weak and gravitational forces). It is a very short-range force and acts between nucleons (protons and neutrons) holding nuclei together.

The strong nuclear force is one of the four fundamental forces (the other three are electromagnetic, gravitational and weak nuclear) and it acts between nucleons. It is the force that holds nuclei together and keeps them stable. The strong force is attractive up to distances of about 3 fm (3×10^{-15} m) and repulsive below very short-range distances of about 0.4 fm. The graph of force against nucleon separation is shown in Figure 1.8.

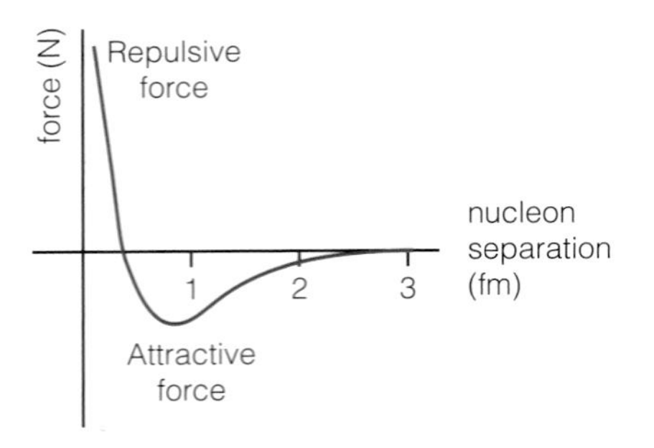

Figure 1.8 The strong nuclear force.

You can see from Figure 1.8, that at separations of about 0.4 fm, the magnitude of the strong nuclear force is zero. The force is neither attractive nor repulsive. This is the equilibrium position for the nucleons (protons or neutrons), where the resultant force on each nucleon is zero. In a stable nucleus, this is the separation of each nucleon. The graph also shows how short range the force is: at a separation of just 3 fm, the strong force has dropped to zero.

Alpha and beta radioactive decay

Not all nuclei are stable. In fact the vast majority of known isotopes are unstable. Unstable nuclei can become more stable by the process of radioactive nuclear decay. Although there are many different decay mechanisms, three types are far more common than all the others. These are alpha, beta and gamma decay. Alpha decay involves the emission of two protons and two neutrons joined together (identical to a helium nucleus). Beta decay involves the decay of a neutron into a proton, and an electron and an antineutrino, which are subsequently emitted from the nucleus. Gamma emission involves the protons and neutrons inside the nucleus losing energy (in a similar way to electrons changing energy level) and emitting a gamma ray photon as part of the process.

The neutrino, ν, is a neutral, almost mass-less fundamental sub-atomic particle that rarely interacts with matter. **Antineutrinos** are the antiparticles of the neutrino. There are three forms of neutrino: the electron-neutrino, ν_e; the muon-neutrino, ν_μ; and the tau-neutrino, ν_τ.

TIP

The antineutrino emitted during beta decay was first suggested to be emitted by the Austrian physicist Wolfgang Pauli in 1930. Beta particles are emitted from a parent nucleus with a range of different energies, up to a maximum value. Pauli suggested that when beta particles are emitted, another particle is emitted at the same time, taking up the balance of the energy. Pauli called these particles 'neutrinos', symbol υ, meaning 'little neutral ones'. He proposed that neutrinos were electrically neutral and almost mass-less (their mass is so small that it has never been determined accurately); they also only feel the weak nuclear and gravitational forces. The anti-neutrino is the antiparticle of the neutrino. It has nuclear properties opposite to that of its particle 'sister'. Antiparticles have the same nuclear symbol as their particle equivalents, but they have a bar over them, hence the symbol for the antineutrino is $\bar{\nu}$.

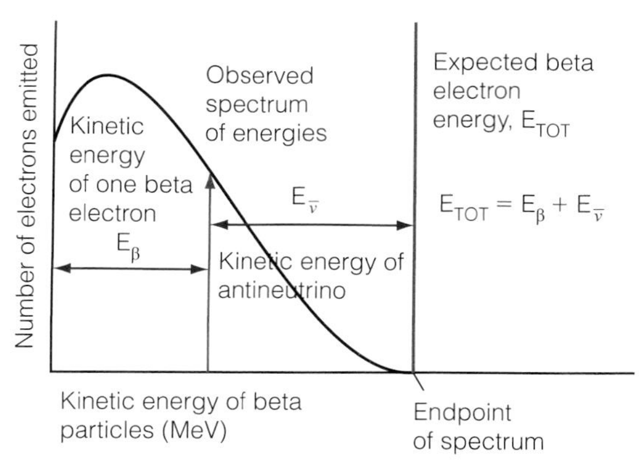

Figure 1.9 A typical beta particle emission spectrum. The total emission energy E_{TOT} is made up of the kinetic energy of the beta particle plus the kinetic energy of the antineutrino.

The observed beta particle spectrum graph (Figure 1.9) shows that beta particles are emitted with a range of energies, E_β, from just over zero up to a maximum value E_{TOT}. E_{TOT} is the expected energy determined by analysing the energies of the parent and daughter nuclides. E_{TOT} is constant for any one radioactive beta-emitting nuclide, so another particle, the antineutrino, must be emitted with an energy $E_{\bar{\nu}}$, which can be any value from just above zero up to a maximum value E_{TOT}. The combined energy of the beta particle and the antineutrino must equal E_{TOT}, so:

$$E_{TOT} = E_\beta + E_{\bar{\nu}}.$$

Detecting nuclear radiation

The cloud chamber

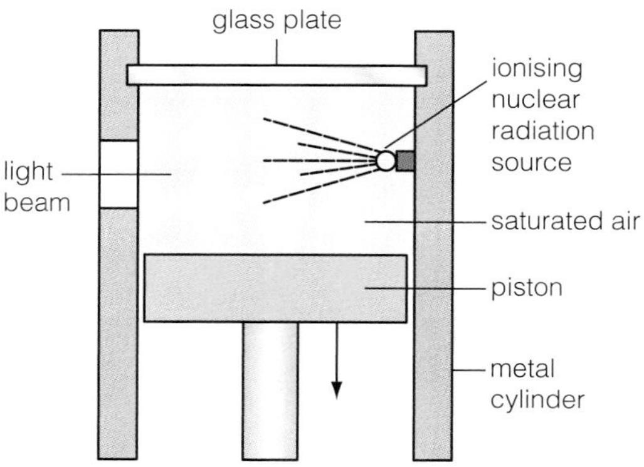

Figure 1.10 The Wilson cloud chamber. (If you are conducting this experiment yourself, follow the CLEAPSS L93 risk assessment.)

One of the earliest ways of detecting ionising nuclear radiation was invented by Charles Wilson in 1911. He came up with the idea of creating clouds artificially in the laboratory, by rapidly expanding air saturated with water vapour inside a sealed chamber. Wilson suspected that clouds form on charged particles in the atmosphere, and he experimented by passing X-rays through his chamber. He discovered that the X-rays left wide cloudy tracks inside the chamber. Wilson then passed alpha, beta and gamma radiation through the chamber and found that the highly ionising alpha particles left broad, straight, definite length tracks; the ionising beta particles left thin, straight or curved tracks (depending on how high their energy was); but the weakly ionising gamma rays left no tracks at all.

The spark counter

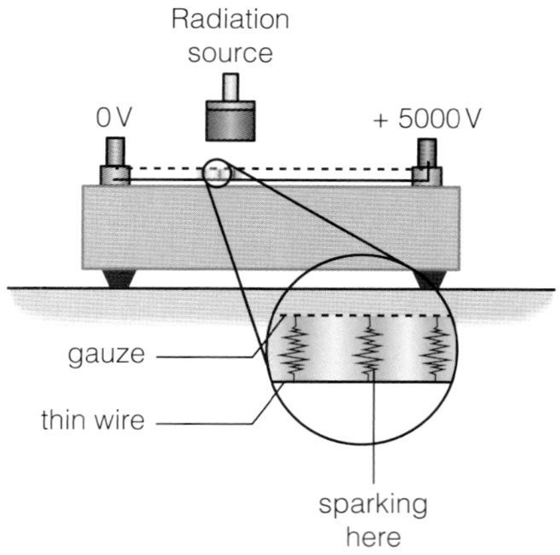

Figure 1.11 A spark counter. (If you are conducting this experiment yourself use a current-limited (5 mA max) EHT and follow the CLEAPSS L93 risk assessment.)

Spark counters only detect highly ionising alpha particles. Beta and gamma radiation do not ionise enough of the air between the metal gauze and the thin wire underneath. When the air particles are ionised by the alpha particles the charged particles produced cause a spark to be formed. That spark jumps across the 5000 V gap between the gauze and the wire. The spark can be seen, heard and counted by an observer or with a microphone. This method of detecting alpha particles is particularly useful for showing that they have a very short range in air.

The Geiger-Muller counter

Figure 1.12 shows a modern Geiger-Muller counter, which is designed to detect and count radioactive emisions automatically, without having to count sparks or scintillations one at a time.

Ionising radiation enters the GM tube through the thin mica window (alpha or beta radiation), or through the window or sides of the counter

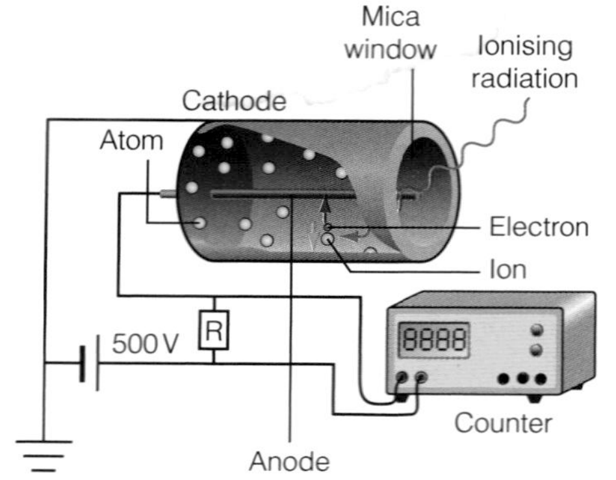

Figure 1.12 A Geiger-Muller (GM) counter.

(gamma radiation). The low-pressure inert gas inside the detector (helium, neon or argon) is ionised producing a cascade of charged particles that are attracted to oppositely charged electrodes. The small pulse of current produced by the moving charge is detected by the electronic counter, which registers a 'count'. Geiger counters have a distinct advantage over spark counters and cloud chambers in that they can detect all three types of radiation.

TEST YOURSELF

10 The pictures to the right show the paths of alpha particles, beta particles and X-rays inside a cloud chamber:

a) Which cloud chamber picture(s) show the following?

i) alpha particle tracks

ii) an X-ray track

iii) a beta particle track

b) Explain why cloud chamber picture C show two types of particles?

11 Explain why a spark counter can easily detect alpha particles but usually cannot detect beta particles or gamma rays.

12 A Geiger counter is used to detect the intensity of alpha particles being emitted from a radium source containing two alpha-emitting isotopes, radium-223 and radium-226. A graph of intensity (counts per minute) against range in air (cm) for this source is shown in Figure 1.17.

Radium-223 emits alpha particles with higher energy than radium-226. Use the graph to determine the range of alpha particles in air emitted by each isotope. Explain your reasoning.

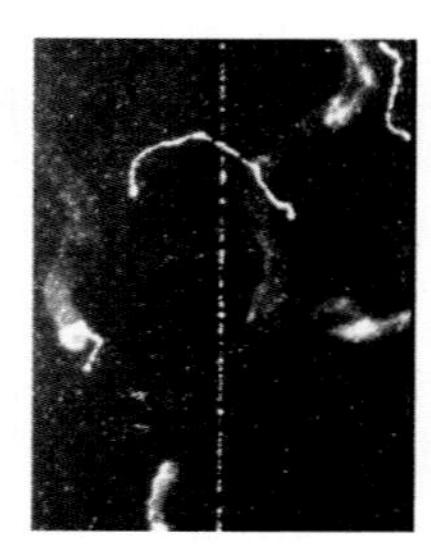

Figure 1.13 Cloud chamber Picture A.

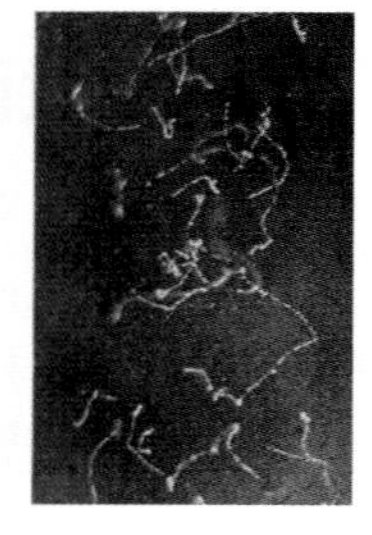

Figure 1.14 Cloud chamber Picture B.

Figure 1.15 Cloud chamber Picture C.

Figure 1.16 Cloud chamber Picture D.

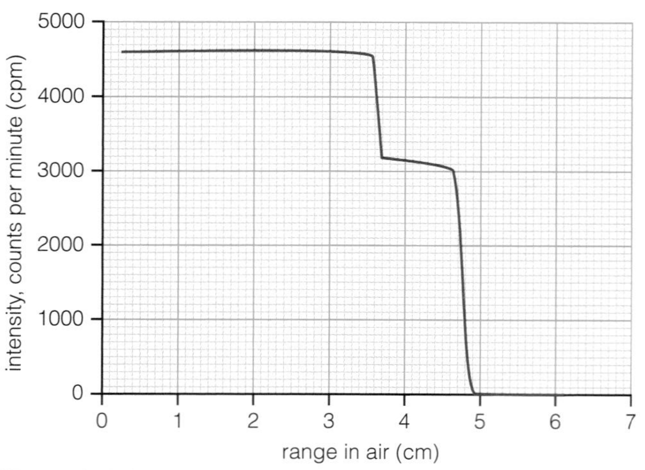

Figure 1.17 Graph of intensity versus range in air.

ACTIVITY

The beta particle emission spectrum of bismuth-210

A student carried out an experiment measuring the kinetic energy of beta particles emitted by the radioisotope bismuth-210. Her measurements are shown below.

Table 1.2

Kinetic energy of beta particles, E ($\times 10^{-15}$ J)	0	16	32	48	64	80	96	112	128	144	160	176	186
Intensity of beta particles (arbitrary units)	10.0	11.5	12.0	11.5	10.0	8.0	6.0	4.4	2.9	1.5	0.6	0.2	0.0

1 Plot a graph of beta particle intensity (y-axis) against beta particle energy (x-axis).
2 A second student measures beta particles being emitted with a second bismuth-210 source with an intensity of 11 arbitrary units. Use the graph to determine the possible kinetic energies of the emitted beta particles from the second source.
3 The maximum kinetic energy of beta particles emitted from bismuth-210 is 186 fJ (186×10^{-15} J). Calculate the energies of the antineutrinos emitted at this intensity.

$^{4}_{2}He$ – *alpha particle* (α)
$^{0}_{-1}e$ – *beta particle* (β^-)
γ – *gamma ray (strictly* $^{0}_{0}\gamma$)
ν – *neutrino (strictly* $^{0}_{0}\nu$)
$\bar{\nu}$ – *antineutrino (strictly* $^{0}_{0}\bar{\nu}$)

Figure 1.18 The $^{A}_{Z}X$ notation for ionising radiation (Note: values of $^{0}_{0}X$ are usually written without the zeros).

TIP

Some nuclei that are unstable and decay via gamma ray emission have the letter 'm' added after their nucleon number, meaning 'metastable'. A good example of this is technetium-99m, a common gamma ray emitter used in nuclear medical imaging. This is covered in next year's book.

Nuclear equations

The $^{A}_{Z}X$ notation can be used to describe the nuclear reactions that take place during radioactive decay.

During alpha, beta and gamma radioactive decay, two simple conservation rules come into play:

- nucleon number, A, is always conserved
- proton number, Z, is always conserved.

In practice this means that the total (addition) value for A before the decay is equal to the total (addition) value of A after the decay (and, similarly, the totals for Z remain the same before and after the decay). Look at the following two examples:

1 A common alpha decay is the decay of radium-226 in granite rocks, producing an alpha particle and the radioactive gas, radon-222. This nuclear decay can be written as:

$$^{226}_{88}Ra \rightarrow {}^{222}_{86}Rn + {}^{4}_{2}He$$

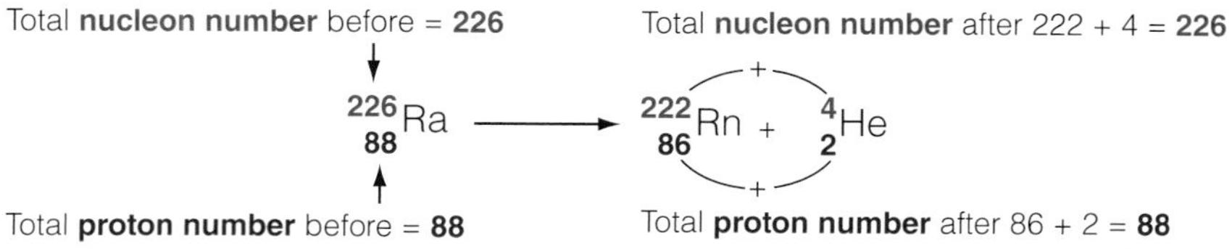

Total **nucleon number** before = **226**

Total **nucleon number** after 222 + 4 = **226**

$$^{226}_{88}Ra \longrightarrow {}^{222}_{86}Rn + {}^{4}_{2}He$$

Total **proton number** before = **88**

Total **proton number** after 86 + 2 = **88**

Figure 1.19 Decay of radium.

2 The same rules apply to beta decay. A good example is the decay of carbon-14, a naturally occurring radio-nuclide used in carbon dating techniques. Carbon-14 decays via beta emission to nitrogen-14, emitting an antineutrino in the process. This decay can be summarised by:

> **TIP**
> It is handy to have access to a database of the different nuclides. The database should give you the values of A and Z, as well as other key data, such as the half-life, abundance and its decay mechanisms. You will find such databases on the web.

Total **nucleon number** before = **14**   Total **nucleon number** after 14 + 0 = **14**

$$^{14}_{6}\text{C} \longrightarrow {}^{14}_{7}\text{N} + {}^{0}_{-1}\beta + \bar{\nu}$$

Total **proton number** before = **6**   Total **proton number** after 7 + (−1) = **6**

Figure 1.20 Decay of carbon-14.

In both these examples, both the nucleon number, A, and the proton number, Z, are conserved during the decay.

TEST YOURSELF

13 Plutonium was the first man-made element. The element is made by bombarding uranium-238 with deuterons (nuclei of deuterium, $^{2}_{1}H$), producing neptunium-238 and two neutrons, ($^{1}_{0}n$); the neptunium-238 then decays via beta decay (where the nucleus emits an electron, $^{0}_{-1}e$) to plutonium-238.

Write nuclear equations summarising the formation of plutonium-238. Uranium has a proton number of 92.

Use a nuclide database, or a periodic table, to write nuclear equations to summarise the following radioactive decays.

14 Americium-241 is an alpha emitter commonly used in smoke detectors; its daughter nuclide is neptunium-237.

15 Strontium-90 is a beta emitter used as a fuel source in radioisotope thermoelectric generators (RTGs) on space probes and lighthouses; it decays to yttrium-90, emitting an antineutrino in the process.

16 Phosphorous-32 is a beta emitter used as a tracer in DNA research.

17 Plutonium-238 is an alpha emitter also used as the fuel source for RTGs.

18 Tritium (or hydrogen-3) is a beta emitter used in some 'glow-in-the dark' paints.

The photon model of electromagnetic radiation

Around the turn of the twentieth century, physicists such as Albert Einstein and Max Planck developed a particle model of electromagnetic radiation to account for two experimental observations that were impossible to explain using classical physics.

Planck developed the concept of a fundamental unit of energy, which became known as a quantum. He proposed that atoms absorb and emit radiation in multiples of discrete amounts that are given by the Planck equation:

$$E = hf$$

where:

f is the frequency of the radiation absorbed or emitted, and

h is a constant, now called the Planck constant.

It was Planck who called these discrete units of energy 'quanta' and the small 'packets' of electromagnetic radiation making up these quanta became known as photons. Measurements of the energy of photons has produced a value for the Planck constant of 6.63×10^{-34} J s. This theory is discussed in more detail in Chapters 3 and 4.

TEST YOURSELF

19 A light-emitting diode (LED) emits light photons of energy 3.6×10^{-19} J.

a) Calculate the frequency of this electromagnetic radiation.

b) Calculate the energy emitted per second (called the output power) of the LED, when it emits 0.9×10^{17} photons each second.

20 LEDs are now used extensively as the light source in torches.

The LEDs used in the torch shown in the diagram above emit blue photons of light. The wavelength of the blue light is 420 nm.

a) Show that the energy of a photon from this LED is about 5×10^{-19} J.

Figure 1.21 LED torch.

LEDs need to be connected to a certain minimum voltage, called the activation voltage, before they emit light. A blue LED needs a higher voltage than a longer wavelength red LED.

b) Explain why a red LED needs a lower voltage than a blue LED.

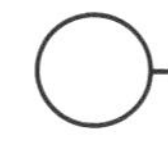

Antiparticles

In 1928, Paul Dirac proposed the existence of the positron, the antiparticle of the electron. He suggested that the positron had the same mass as an electron, but that most of its other physical properties (such as its charge) were the opposite to that of the electron. Dirac thought that positrons were positively charged electrons.

All particles have a corresponding antiparticle. Each particle – antiparticle pair has the same mass and therefore the same rest mass-energy (the amount of energy released by converting all of the mass into energy), but they have opposite properties such as their charge. The positron is one of the few antiparticles that has its own different name; most other anti-particles are described by the word 'anti' in front of the particle name. The other more common antiparticles are the antiproton, the antineutron and the antineutrino. Antiparticles generally have the same symbol as their particle but have a bar drawn over the top of the symbol; for example, the proton has the symbol p and the antiproton has the symbol $\overline{p}$. An exception to this is the positron. It was the first antiparticle to be discovered and is given the symbol e^+.

Antiparticles are not common. The Universe appears to be overwhelmingly dominated by matter. Antimatter only becomes apparent in high-energy particle interactions, for example in the interaction of high-energy cosmic rays with the atmosphere, or in particle accelerator experiments such as the Large Hadron Collider (LHC) at CERN.

Rest mass-energy is the amount of energy released by converting all of the mass of a particle at rest into energy using Einstein's famous mass-energy equation,

$$E = mc^2$$

where m is the rest mass of the particle and c is the speed of light.

Mega electron-volts The energy of nuclear particles is usually given in MeV, mega electron-volts. One electron-volt is a very small amount of energy, equivalent to 1.6×10^{-19} J. This is the same numerical value as the charge on an electron and is defined as the amount of energy needed to accelerate an electron of charge 'e', (1.6×10^{-19} C) through a potential difference of 1 volt. One MeV is a million electron-volts, equivalent to 1.6×10^{-13} J. This unit comes from the definition of the volt. A volt is the amount of energy per unit charge or:

$$\text{volts} = \frac{\text{energy}}{\text{charge}}$$

so

$$\text{energy} = \text{charge} \times \text{volts}$$

Particle–antiparticle interactions

When a particle meets its corresponding antiparticle they will annihilate each other. This means that the total mass of the particle pair is converted into energy, in the form of two gamma ray **photons**.

TIP

Other particle – antiparticle pairs with 'non-bar' symbols are: muon$^+$ (μ^+) and muon$^-$ (μ^-); pion$^+$ (π^+) and pion$^-$ (π^-); kaon$^+$ (K^+) and kaon$^-$ (K^-). The photon and the pion^0 (π^0) are their own antiparticles. You will meet these particles again in Chapter 2.

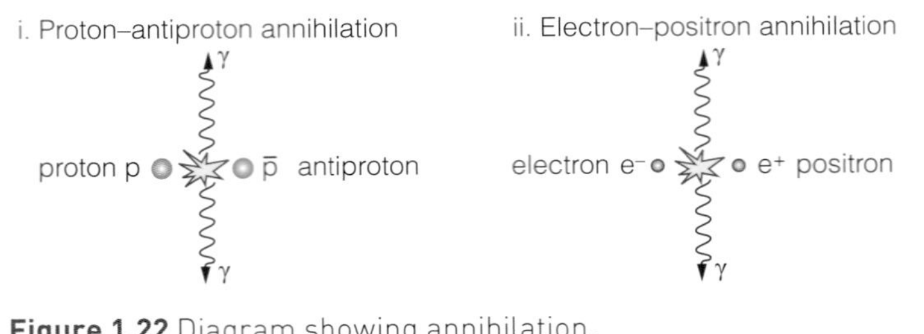

Figure 1.22 Diagram showing annihilation.

When a particle meets its antiparticle and their total mass is converted to energy in the form of two gamma ray photons, the particles annihilate each other.

Two gamma photons are always produced during particle – antiparticle annihilation in order to conserve **momentum**. The total energy of the gamma photons is equal to the total rest energies of the particle – antiparticle pair, in order to conserve energy.

EXAMPLE

The rest energy of an electron is the same as that of a positron and is 0.51 MeV. Calculate the wavelength (in picometres, pm) of the two gamma ray photons produced when an electron and a positron annihilate each other at rest.

Answer

As the electron and the positron are **initially at rest**, their total momentum is zero. This means that the total momentum of the photons produced when the electron/positron annihilates must also be zero. Both the photons will have the same energy and will be moving at the same speed (the speed of light), so they must be moving in opposite directions for their momenta to cancel out to zero.

The total energy of the annihilation will be 1.02 MeV. The two photons that are produced during the annihilation must therefore have an energy equal to 0.51 MeV.

$$0.51\,\text{MeV} = 0.51 \times 1.6 \times 10^{-13}\,\text{J} = 8.2 \times 10^{-14}\,\text{J}$$

To calculate the wavelength of the photons:

$$E = \frac{hc}{\lambda}$$

$$\lambda = \frac{hc}{E}$$

$$\lambda = \frac{6.6 \times 10^{-34}\,\text{Js} \times 3 \times 10^{8}\,\text{ms}^{-1}}{8.2 \times 10^{-14}\,\text{J}}$$

$$= 2.4 \times 10^{-12}\,\text{m}$$

$$\lambda = 2.4\,\text{pm}$$

Pair production

The opposite process to annihilation is called **pair production**. In this process, a photon with enough energy can interact with a large nucleus and be converted directly into a particle – antiparticle pair. The rest energy of an electron (and therefore a positron) is 0.51 MeV, or 8.2×10^{-14} J. In order to create this particle – antiparticle pair, the photon must have enough energy to create both particles, in other words, $(2 \times 0.51) = 1.02$ MeV, or 16.4×10^{-14} J. The wavelength of the photon needed to do this can be calculated by:

$$E = \frac{hc}{\lambda}$$

$$\lambda = \frac{hc}{E}$$

$$\lambda = \frac{6.6 \times 10^{-34}\,\mathrm{J\,s} \times 3 \times 10^{8}\,\mathrm{m\,s^{-1}}}{16.4 \times 10^{-14}\,\mathrm{J}} = 1.2 \times 10^{-12}\,\mathrm{m}$$

$$\lambda = 1.2\,\mathrm{pm}$$

TEST YOURSELF

21 A proton and an antiproton annihilate at rest producing two high-energy photons, each with an energy of 1.5×10^{-10} J. Calculate the frequency and the wavelength of the photons. Compare the wavelength of these photons to the photons produced by electron – positron annihilation. The rest mass-energy of the proton and antiproton is 938 MeV.

22 Every type of particle has its corresponding antiparticle.

a) Write down the name of one particle and its corresponding particle.

b) State one property that the particle and its antiparticle share.

c) State one property that is different for the particle and its antiparticle.

23 Under certain circumstances it is possible for a very high energy photon to be converted into a proton and an antiproton, each with a rest energy of 1.50×10^{-10} J.

a) State the name of this process.

The photon in this process must have a minimum energy in order to create a proton and an antiproton.

b) Calculate the minimum energy of the photon in joules, giving your answer to an appropriate number of significant figures.

c) A photon of even higher energy than that calculated in part (b) is also converted into a proton/antiproton pair. State what happens to the excess energy.

d) Explain why the photon required to produce an electron/positron pair would not be able to produce a proton/antiproton pair.

e) The antiprotons produced during this process have very short lifetimes. Describe what is likely to happen to the antiproton soon after it is formed.

f) Explain why a single photon could not produce a single proton during this process rather than a proton/antiproton pair.

Practice questions

1 Which line represents the correct number of protons, neutrons and electrons in an atom of one of the isotopes of lead $^{208}_{82}Pb$?

	protons	neutrons	electrons
A	82	126	126
B	126	82	126
C	208	82	208
D	82	126	82

2 What is the specific charge of a gold $^{197}_{79}Au$ nucleus? The mass of the gold nucleus is 3.29×10^{-25} kg; the charge on the electron is 1.6×10^{-19} C.

A $3.63 \times 10^{7}\,C\,kg^{-1}$

B $3.84 \times 10^{7}\,C\,kg^{-1}$

C $3.92 \times 10^{7}\,C\,kg^{-1}$

D $9.56 \times 10^{7}\,C\,g^{-1}$

3 Uranium-236 may split into a caesium nucleus, a rubidium nucleus and four neutrons as shown below in the following nuclear equation. What is the value of X for the rubidium nucleus?

$$^{236}_{92}U \rightarrow {}^{137}_{55}Cs + {}^{X}_{37}Rb + 4{}^{1}_{0}n$$

A 92 **B** 95 **C** 98 **D** 99

4 What is the charge, in C, of an atom of $^{15}_{7}N$ from which a single electron has been removed?

A -9.6×10^{-19} C

B -1.6×10^{-19} C

C $+1.6 \times 10^{-19}$ C

D $+9.6 \times 10^{-19}$ C

5 In a radioactive decay a gamma photon of wavelength 8.3×10^{-13} m is emitted. What is the energy of the photon? The speed of light is $3 \times 10^{8}\,m\,s^{-1}$.

A 5.4×10^{-46} J

B 2.4×10^{-13} J

C 57 J

D 2.0×10^{43} J

6 Thorium decays by the emission of an alpha particle as shown in the equation:

$$^{229}_{90}Th \rightarrow {}^{X}_{Y}Ra + \alpha$$

What are the correct values for *X* and *Y*?

	X	Y
A	225	88
B	88	225
C	229	91
D	227	86

7 $^{14}_{6}C$ is a radioactive isotope of carbon. It can form an ion when two electrons are removed from the atom. What is the charge on this ion in coulombs?

A -9.6×10^{-19} C

B -3.2×10^{-19} C

C 3.2×10^{-19} C

D 9.6×10^{-19} C

8 The line spectrum from helium includes a yellow line with a wavelength of 587.6 nm. What is the energy of a photon with this wavelength?

A 3.89×10^{-40} J

B 1.17×10^{-31} J

C 1.13×10^{-27} J

D 3.38×10^{-19} J

9 An alpha particle has a kinetic energy of 9.6×10^{-13} J. The mass of the alpha particle is 6.6×10^{-27} kg. What is the speed of the particle?

A $1.2 \times 10^{7}\,\mathrm{m\,s^{-1}}$

B $1.7 \times 10^{7}\,\mathrm{m\,s^{-1}}$

C $1.5 \times 10^{14}\,\mathrm{m\,s^{-1}}$

D $2.9 \times 10^{14}\,\mathrm{m\,s^{-1}}$

10 ${}^{238}_{92}U$ decays by emitting α and β^- particles in a number of stages to form ${}^{206}_{82}Pb$. How many β^- decays are involved in this decay chain?

A 2 **B** 4 **C** 6 **D** 8

11 a) Name the constituent of an atom which

i) has zero charge (1)

ii) has the largest specific charge (1)

iii) when removed leaves a different isotope of the element. (1)

b) The equation

$${}^{99}_{43}Tc \rightarrow {}^{A}_{Z}Ru + {}^{0}_{-1}\beta + X$$

represents the decay of technetium-99 by the emission of a β^- particle.

i) Identify the particle *X*. (1)

ii) Determine the values of *A* and *Z*. (2)

12 Alpha decay is a process by which an unstable isotope of an element may decay.

a) State what is meant by an **isotope**. (2)

b) Copy and complete this equation for alpha decay: (2)

$${}^{A}_{Z}X \rightarrow {}^{\dots}_{\dots}Y + {}^{4}_{2}He$$

c) Explain why the alpha particle, once outside the nucleus, is unaffected by the strong nuclear force of the parent nucleus. (2)

13 An atom of calcium, ${}^{48}_{20}Ca$, is ionised by removing two electrons.

a) State the number of protons, neutrons and electrons in the ion formed. (1)

b) Calculate the charge of the ion. (1)

c) Calculate the specific charge of the ion. (2)

14 a) Describe how the strong nuclear force between two nucleons varies with the separation of the nucleons, quoting suitable values for the separation. (3)

b) An unstable nucleus can decay by the emission of an alpha particle. State the nature of an alpha particle. *(1)*

c) Copy and complete the equation below to represent the emission of an α particle by a $^{238}_{92}U$ nucleus.

$$^{238}_{92}U \rightarrow {}^{\dots}_{\dots}Th + {}^{\dots}_{\dots}\alpha$$ *(2)*

15 a) Explain what is meant by the specific charge of a nucleus. *(1)*

The incomplete table shows information for two isotopes of uranium.

	Number of protons	Number of neutrons	Specific charge of nucleus...
First isotope	92	143	
Second isotope			3.7×10^7

b) Copy the table and add the unit for specific charge in the heading of the last column of the table. *(1)*

c) Add the number of protons in the second isotope to the second row of the table. *(1)*

d) Calculate the specific charge of the first isotope and write this in the table. *(3)*

e) Calculate the number of neutrons in the second isotope and put this number in the table. *(4)*

Stretch and challenge

16 Figure 1.23 shows an arrangement to measure the Planck constant.

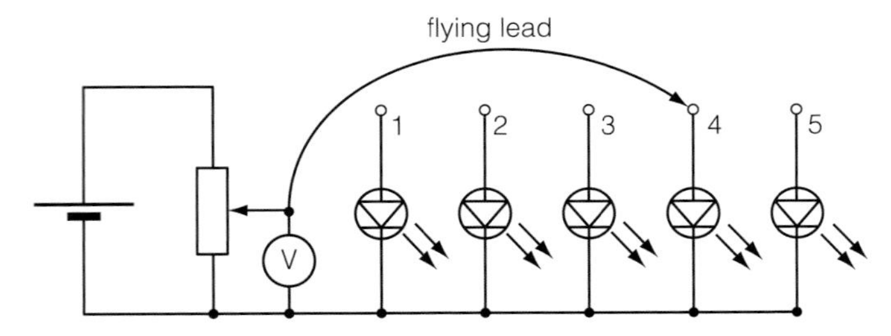

Figure 1.23 Circuit diagram for measuring the Planck constant.

The data obtained from this experiment is shown below.

LED number	Wavelength of emitted photons, λ/nm	Frequency of emitted photons, *f*/Hz	Activation voltage, V_A/V
1 violet	413		3.01
2 blue	470		2.65
3 green	545		2.28
4 yellow	592		2.09
5 red	625		1.98

a) Write a method as a numbered list for this experiment.

b) Copy and complete the table, calculating the frequency of the emitted photons.

c) Plot a graph of activation voltage (y-axis) against photon frequency (x-axis).

d) Draw a line of best fit on the graph and calculate the gradient.

e) Use the equation $eV_A = hf$, together with the gradient of the best fit line, to determine a value of the Planck constant. Show your working.

17 How does the proton number, Z, and the nucleon number, A, of a nucleus change due to:

a) the emission of an alpha particle

b) the emission of a beta particle

c) the fusion with a deuterium nucleus?

18 The potassium isotope $^{42}_{19}K$ disintegrates into $^{42}_{20}Ca$.

a) What are the likely type/s of radiation produced?

b) How many protons, neutrons and electrons are present in an atom of the daughter nucleus $^{42}_{20}Ca$?

19 A muon and an antimuon annihilate each other to produce two γ-rays. Research the data that you need for this question and use it to calculate the minimum energy of the photons.

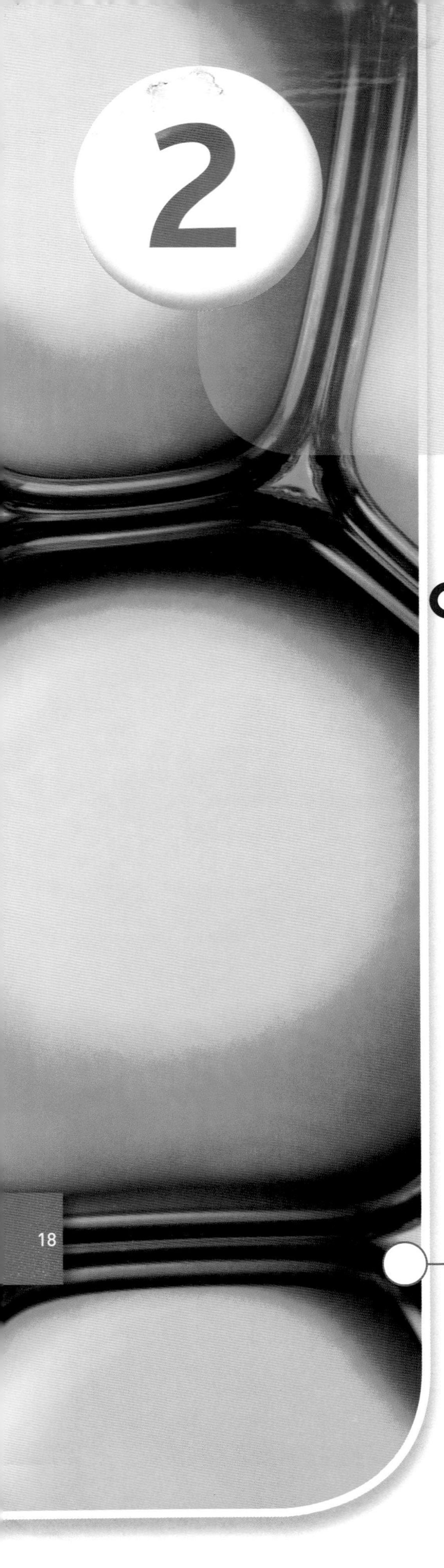

Fundamental particles

PRIOR KNOWLEDGE

- Matter is made of atoms.
- Atoms are made up of a very small central nucleus containing particles called protons and neutrons, surrounded by orbiting electrons.
- Atoms are electrically neutral. Protons carry a positive charge and electrons carry the same magnitude, negative charge. In atoms, the number of protons equals the number of electrons, so their overall charge is zero.
- Protons and neutrons have a mass about 1800 times larger than electrons. Most of the mass of an atom is concentrated in the nucleus due to the protons and neutrons.

TEST YOURSELF ON PRIOR KNOWLEDGE

Data for Test Yourself questions is generally given in the questions but you can find extra information on specific nuclides using an online database such as Kaye and Laby from the National Physical Laboratory (NPL), kayelaby npl.co.uko.

1 a) Calculate the total charge on a sodium-23 nucleus.

b) Calculate the approximate mass of the sodium-23 nucleus.

2 The **atomic** radius of an element can be estimated using the Avogadro Number, $N_A = 6.02 \times 10^{23}$ (the number of particles in 1 mole), the density of the element and the molar mass of the element. The molar mass of carbon-12 is $12.0 \times 10^{-3}\,\text{kg mol}^{-1}$, and carbon-12 (diamond allotrope) has a density of $3500\,\text{kg m}^{-3}$.

a) Calculate the mass of one atom of carbon-12.

b) Calculate the volume of one atom of carbon-12.

c) Calculate the radius of one atom of carbon-12 (volume of a sphere is given by $V = \frac{4}{3}\pi r^3$).

d) The **nuclear** radius of an element is given by the following formula:

$$r = r_0 A^{\frac{1}{3}}$$

where r_0 is a constant equal to $1.25 \times 10^{-15}\,\text{m}$, and A is the nucleon number (number of protons + number of neutrons). Use this information to calculate the radius of a carbon-12 nucleus.

e) Calculate the ratio $\frac{\text{atomic radius}}{\text{nuclear radius}}$ for carbon-12.

The particle garden

The discovery of the three basic sub-atomic particles: the **electron** (in 1897 by JJ Thomson); the **proton** (in 1917 by Ernest Rutherford) and the **neutron** (in 1932 by James Chadwick), seemed to complete the structure of the atom. However, in the 1930s, physicists started to discover a range of new sub-atomic particles, such as the positron and the muon, primarily as the result of experiments on high-energy cosmic radiation from space. The Second World War forced a break in the discovery of new particles as physicists were put to use working on new warfare technology, such as the atomic bomb and radar.

After the end of the Second World War, physicists returned to their research into sub-atomic particles. The huge advance in technology made during the war years led to the discoveries of many new particles and the production of more sophisticated particle accelerators and detectors.

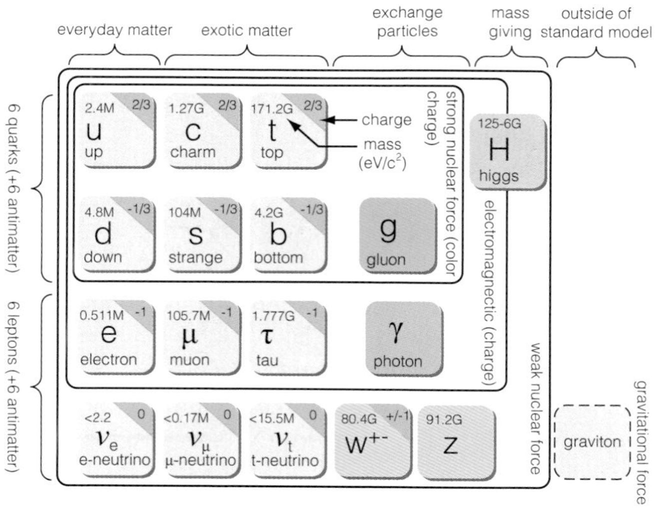

It became obvious that particles were being discovered in 'families', rather like the groups of elements on the periodic table, and new names such as leptons, hadrons and mesons were invented to classify these groups.

In 1967, the Standard Model of particle physics was first described by Steven Weinberg and Abdus Salam, and it is from this that the current picture of the structure and behaviour of matter was constructed. In 2012, the final piece of the Standard Model jigsaw, the Higgs Boson, was 'discovered' by a large team of physicists working on data from the Large Hadron Collider at CERN in Geneva. (CERN stands for *Conseil Européen pour la Recherche Nucléaire or European Council for Nuclear Research.*)

Figure 2.1 The Standard Model of matter showing the 12 fundamental particles and the exchange particles, together with their masses and charges.

The Standard Model

The Standard Model of matter has proved to be very powerful. It successfully describes the huge proliferation of particles discovered during the latter part of the twentieth century. The model is surprisingly simple, consisting of two families of fundamental particles and a set of forces that bind them together.

TIP

The masses of fundamental particles are nearly always given as mass-energy equivalents. Einstein's energy-mass equation, $E = mc^2$, can be rearranged to $m = \frac{E}{c^2}$. As c^2 is a constant value then mass also has the unit eV/c^2. The mass of the electron is 0.51 MeV/c^2, which is equivalent to 9.11×10^{-31} kg. The heaviest fundamental particle, the top quark, has a mass of 171.2 GeV/c^2 or 3.1×10^{-25} kg. Rest mass-energies can therefore be given as a mass (usually in MeV/c^2) or an energy equivalent (usually in MeV).

Fundamental particles

Antiparticles are fundamental particles with the same mass and energy as their particle counterparts, but have opposite properties such as charge. When a particle meets its antiparticle they annihilate (see Chapter 1) converting to gamma ray photons.

The word 'fundamental' in physics has profound meaning. Fundamental particles are particles that appear to have no structure. Fundamental particles cannot be broken down into smaller pieces and they are the basic building blocks of the Universe. When Democritus performed his thought experiment on the nature of matter on a beach in Ancient Greece, his idea was simple – the smallest building block of matter was called an atom, and for about two and a half millennia, this concept dominated science. JJ Thomson's discovery of the electron started the hunt for sub-atomic particles, and the Standard Model appears to finish the hunt with a group of 12 fundamental particles (and their antiparticles). The 12 fundamental particles (and the four exchange particles that hold them together) of the Standard Model can be arranged in many different ways to make up any observed composite particles in the Universe, with the overwhelming proportion of the observed Universe appearing to be made of just three fundamental particles: the electron, the up quark and the down quark. Up and down quarks combine together in threes to make protons and neutrons, and these combine together to form nuclei. The addition of electrons to surround the nuclei forms atoms.

TEST YOURSELF

1. What is the 'Standard Model'?
2. State one similarity and one difference between a 'particle' and its 'antiparticle' (for example, an electron and a positron).
3. What is the conversion factor from MeV into joules?

These questions refer to the Standard Model shown in Figure 2.1.

4. State the six leptons and the six quarks that make up the Standard Model.
5. What is the charge on the following particles?
   a) electron
   b) muon-neutrino
   c) up quark
   d) positron
   e) down quark
   f) strange quark.
6. What are the four most common particles shown in Figure 2.1?
7. What is meant by a fundamental particle?
8. How many quarks are there in a proton?

Quarks

Quarks are fundamental particles that make up particles such as protons and neutrons. They exert the strong nuclear force on one another.

Although you have already met the electron as a fundamental particle, you may not have come across quarks. The concept of quarks as fundamental building blocks of matter was first proposed by Murray Gell-Mann and George Zweig in 1964, as part of the original Standard Model, which initially only contained three 'flavours' of quarks (up, u; down, d; and strange, s). The other three flavours (charm, c; bottom, b; and top, t) were added later. (The bottom quark is also known as the beauty quark.) A basic difference between these quarks is their mass. Figure 2.2 shows the masses of each quark drawn to scale (with the area representing mass). The proton and electron masses are shown for comparison to the left.

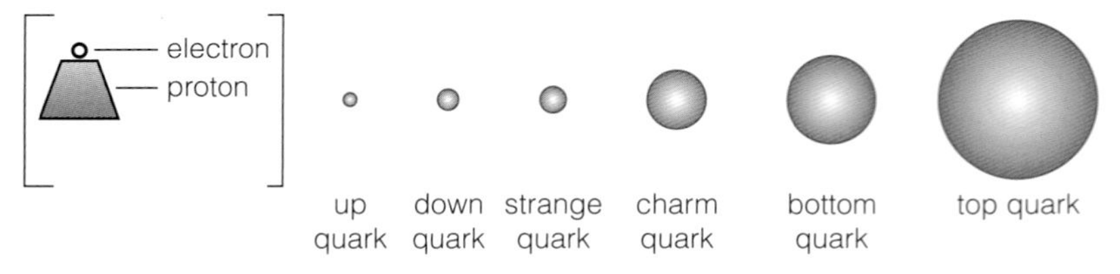

Figure 2.2 The relative masses of the six quarks.

Gluons are one of the four exchange particles of the Standard Model. Gluons act between quarks holding them together. Gluons have an extremely short range of action of about 10^{-15} m.

Most of the mass of a proton is due to the energy of the interactions between the quarks and the gluons that hold the quarks together via the strong nuclear force. These gluons constantly come into and go out of existence as they exchange between the quarks. The energy required to do this is included in the mass of the proton or the neutron. The up and down quarks are much less massive than the other quarks. The strange quark is observed in particles at high altitude due to the interaction of high-energy cosmic rays within the upper atmosphere. The charm, bottom and top quarks are only observed in particle accelerator detectors at extremely high energy.

Deep-inelastic scattering involves firing electrons at protons at very high energies (hence the word 'deep'). Elastic scattering, for example, involves two particles such as two protons colliding with each other and rebounding off each other with the same kinetic energy – rather like two snooker balls hitting each other head-on and rebounding back. The word 'elastic' in this context means that no kinetic energy is lost. Inelastic scattering involves the conversion of kinetic energy into other forms. In this case the electrons penetrate into the proton and interact with the quarks (via exchange of photons); kinetic energy is converted into mass as the proton shatters, producing a shower of other particles. Using the snooker analogy, it would be as if one snooker ball entered the second ball causing it to break into other pieces as the first snooker ball scattered away from it.

The first direct observation of quarks was carried out using **deep-inelastic scattering** of electrons by protons at the Stanford Linear Accelerator (SLAC) facility in 1968. Protons were observed to be made up of two up quarks and a down quark. The strange quark was observed as a result of further experiments at SLAC and the charm quark was observed in 1974. The bottom quark was discovered three years later in 1977, but it took until 1995 for a particle accelerator with enough energy, the Tevatron at Fermilab, to produce interactions involving the top quark.

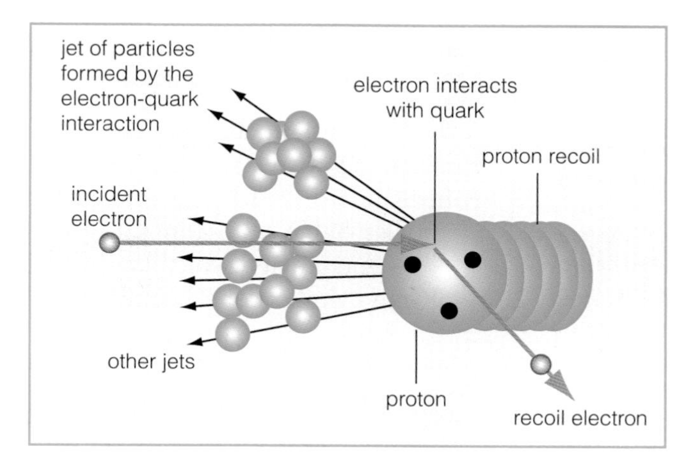

Figure 2.3 Deep-inelastic scattering.

TIP

Make sure you know about the up, u; down, d; and strange, s, quarks (and their antiquarks). You do not need to know about the charm, top or bottom quark, although they could be set as examples in the exam, where all factual knowledge would be provided in the question.

TEST YOURSELF

9 What are the six 'flavours' of quarks and what is the main difference between them?

10 The original deep-inelastic scattering experiments carried out at SLAC in 1968 involved which of the following particle interactions?

A protons and neutrons
B protons and positrons
C protons and electrons
D protons and muons

11 Why are quarks believed to be 'fundamental particles'?

12 Arrange the following particles in order of their masses (heaviest first):
proton, electron, bottom quark, down quark, up quark.

13 Explain what is meant by 'deep-inelastic scattering'.

Fundamental forces

In the same way that the Standard Model describes the 12 fundamental particles, it also describes the four fundamental forces that allow the particles to interact with each other. Indeed the words **force** and **interaction** are interchangeable in this context and mean the same thing. The word interaction is used to describe these forces on a microscopic, quantum scale, because the mechanism used to explain how the forces work is different from the classical 'Newtonian' field theory that is used to describe how forces work in the large-scale, macroscopic world.

In the macroscopic world, charges and masses exert forces on each other due to the electric and gravitational force fields that extend out into space, away from the charge or mass. The patterns of the force field lines show the action of the forces. On the microscopic, quantum scale, classical field theory cannot explain some of the ways that particles interact with each other, and a different 'quantum' interaction approach is used.

Exchange particles are particles involved with the interaction of particles via the four fundamental forces of nature on the quantum scale. Exchange particles are only created, emitted, absorbed and destroyed between the interacting particles.

The Standard Model describes particles interacting by transferring exchange particles. There are three fundamental forces described by the Standard Model, each one with its own distinct exchange particle or particles. Table 2.1 outlines the three fundamental forces of the Standard Model and gravity (which falls outside the Standard Model, but is included in Grand Unification Theories) and their exchange particles.

Table 2.1 Fundamental forces and their exchange particles.

Fundamental force	Nature	Exchange particle(s)
Electromagnetic	Acts between charged particles	Photon
Strong	Acts between quarks	Gluon
Weak	Responsible for radioactive decay (with the exception of alpha decay) and nuclear fusion	W^+, W^-, Z^0
Gravity	Acts between particles with mass	Graviton

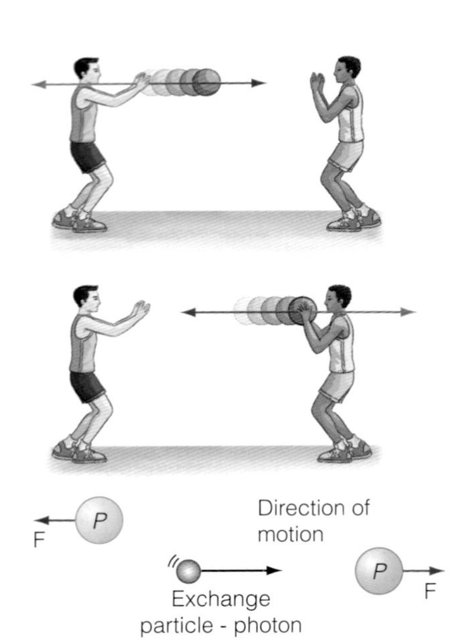

Figure 2.4 Two basketball players 'exchanging' a basketball. This is analogous to two protons exchanging a photon.

On the quantum scale, two protons can interact with each other by exchanging virtual photons via the electromagnetic interaction. The photons are exchanged in both directions but exist for only minute amounts of time – hence the word virtual. A simple model of this interaction could be constructed by two basketball players throwing a basketball back and forward to each other as shown in Figure 2.4. Momentum is exchanged between the players and the ball, and a force is produced.

This model works quite well for illustrating the repulsive force that occurs between two particles with the same charge, but it does not work very well with the attractive forces, such as the strong, weak and gravitational forces, and the electromagnetic attraction between particles with opposite charge. In this case, if you visualise the basketballs attached to elastic cords that stretch tight just before being caught by the catcher, this means that as the basketball is caught it pulls the catcher back towards the thrower as shown in Figure 2.5. To improve the model further, make the elastic cord stiffer as the distance between the catcher and thrower gets smaller, increasing the force of attraction. This makes it more realistic to the quantum world where the attractive forces decrease with increasing separation. Trying to visualise the quantum world is very difficult from our viewpoint.

The electromagnetic force

We are very used to the everyday effects of the electromagnetic force. Not only do we see and feel the effects of static charge, but the electromagnetic

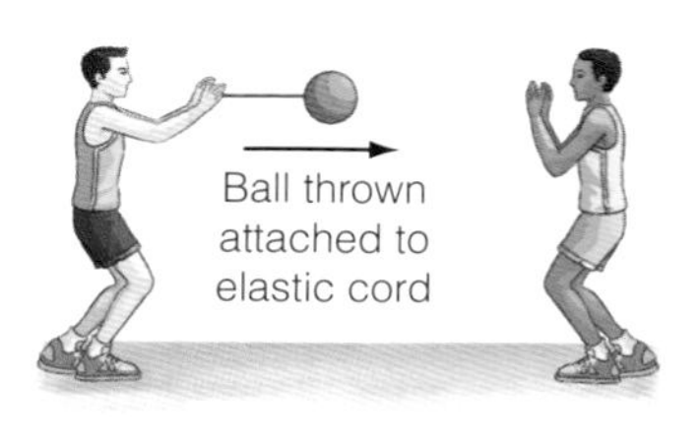

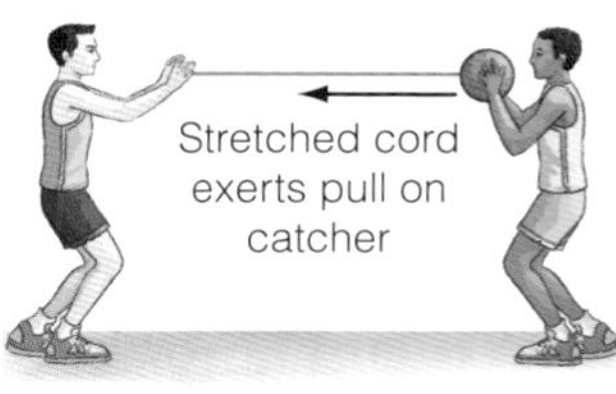

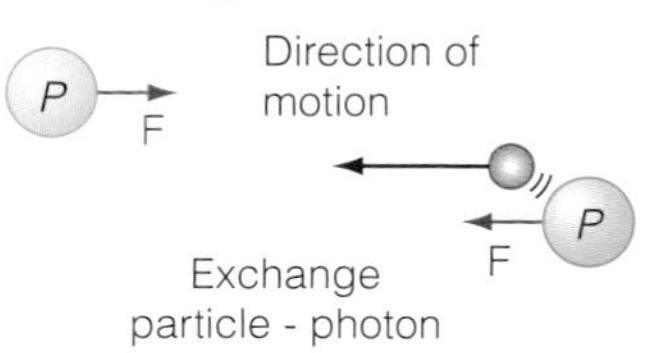

Figure 2.5 A basketball on a cord provides an attractive force.

force is the source of the contact forces between everyday objects. On the quantum scale the electromagnetic interaction occurs between charged particles, most commonly electrons and protons.

Figure 2.6 An electromagnetic interaction.

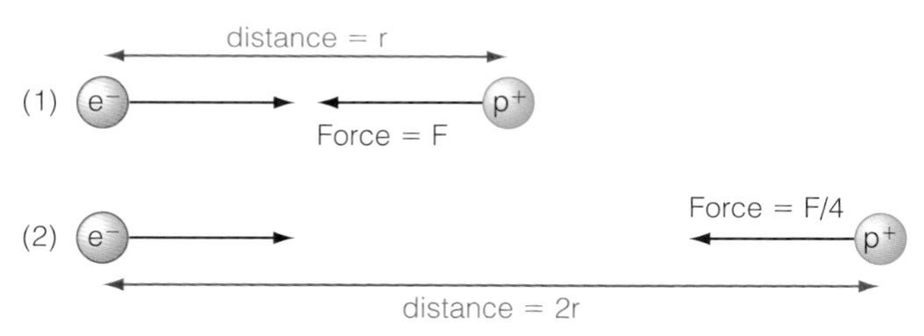

Figure 2.7 As the distance between two oppositely charged particles doubles, then the force reduces by a factor of four.

The electromagnetic interaction exchange particle, the photon, acts over infinite distances. However, the strength of the force decreases with an inverse-square relationship to distance, $\frac{1}{r^2}$. This means that if the distance between the particles is doubled then the force reduces by a factor of four as shown in Figure 2.7.

The photons created during electromagnetic interactions are called **virtual** photons, because they only exist during the time of the interaction. They are created, interact and decay all within the time of the interaction.

The electromagnetic interaction is responsible for most of the behaviour of matter on an atomic and molecular scale. (In fact the other three forces are almost insignificant on this scale.)

The strong nuclear force

The strong interaction is an extremely short-range force (typically acting on the scale of 10^{-14} to 10^{-15} m – the scale of the nucleus) and it acts between quarks. The exchange particle is the gluon.

Although the strong force is very short range, it has a very high magnitude, 137 times larger than the electromagnetic force, hence the name 'strong'. The strong force acts between quarks, so it is the force that holds nucleons such as protons and neutrons together, and it is

Figure 2.8 Murray Gell-Mann, originator of the Standard Model, quarks and gluons.

also the force that holds nuclei together. If the strongest of the fundamental forces, the strong nuclear force, has a magnitude of 1, then the corresponding electromagnetic force would have a magnitude of $\frac{1}{137}$.

The weak force

A version of the weak force was first described in 1933 by Enrico Fermi while he was trying to explain beta decay. The weak force is now known to be responsible for β^- and β^+ radioactive decay, electron capture and electron–proton collisions. There are three exchange particles involved with the weak force, the W^+, W^- and the Z^0, but these were not experimentally verified until 1983. The weak force is about a million times weaker than the strong nuclear force (hence the name), and it acts over an even shorter range, typically 10^{-18} m, which is about 0.1% of the diameter of a proton.

Gravity

The force of gravity is well known to us. We feel its effect at all times. It is the fundamental force that drives the macroscopic behaviour of the Universe as it acts over infinite distances and it acts between masses. On the quantum scale, gravity is the weakest of all the four fundamental forces (typically 6×10^{-39} of the magnitude of the strong force). Although classical, macroscopic, gravitational field theory works very well, the quantum nature of gravity is not very well understood. The proposed name of the exchange particle is the graviton but, on the quantum scale, the graviton would be almost impossible to observe due to the extremely small magnitude of the force of interaction. Huge, planet-scale, detectors would be needed to capture the extremely rare effects of graviton interactions.

Macroscopic means 'large scale', as opposed to the microscopic, quantum scale. Quantum effects operate at distances less than about 100 nm, so anything above this scale is considered macroscopic, and classical (sometimes called Newtonian) physics is applied.

TIP

In the examination, you will not be required to recall information regarding the gluon, Z^0 or the graviton.

TEST YOURSELF

14 What are the four fundamental forces? For each one, state its exchange particle.

15 Why are the fundamental forces called 'interactions' on the quantum scale?

16 Why are exchange particles sometimes called 'virtual particles'?

17 Which of the following particle events **is not** an example of the weak interaction?
beta$^-$ decay
alpha emission
beta$^+$ decay
electron capture

Feynman diagrams

Feynman diagrams are pictorial ways of representing the interactions of quantum particles. They were first introduced by Richard Feynman in 1948. Feynman realised that the interactions of particles on the quantum scale could be represented on paper by a series of arrows and wavy lines, following a set of simple rules. As an example, Figure 2.9 shows the standard Feynman diagram illustrating β^- radioactive decay.

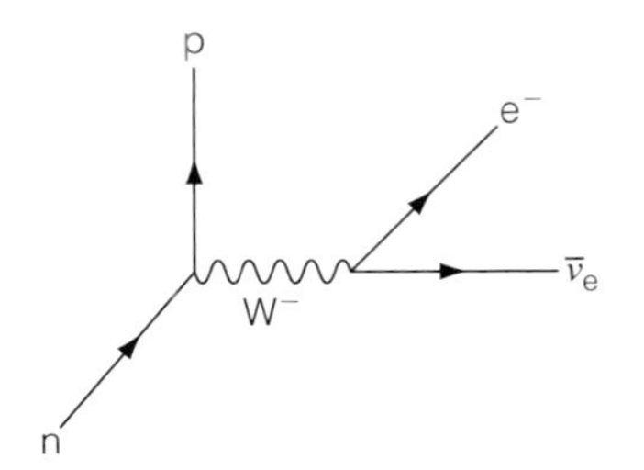

Figure 2.9 Feynman diagram illustrating β^- decay.

Feynman diagrams are generally read from left to right. Figure 2.9 shows a neutron decaying into a proton and a W^- exchange particle, which subsequently decays into an electron and an electron antineutrino. This is an example of the **weak** interaction and can be written as an equation:

$$^1_0n \rightarrow {}^1_1p + {}^{0}_{-1}e + \bar{\nu}_e$$

You can see immediately the advantage of the Feynman diagram over the symbol equation. The Feynman diagram summarises all the parts of the interaction, whereas the equation only tells us what goes into the interaction and what comes out. It tells us nothing about what goes on during the interaction.

Feynman diagram rules

- Particles are represented by straight lines with arrow heads drawn on them.
- Exchange particles are represented by wavy lines.
- Time generally moves on the *x*-axis from left to right (although this is not a hard and fast rule, and many Feynman diagrams have time running vertically).
- Particles are created and annihilated at the vertices between the lines.
- Particles made up of quarks have the quark lines draw parallel and next to each other.
- Exchange particles generally transfer from left to right unless indicated by an arrow above the wavy line.

Feynman diagram examples

Two electrons scattering off each other (Figure 2.10)

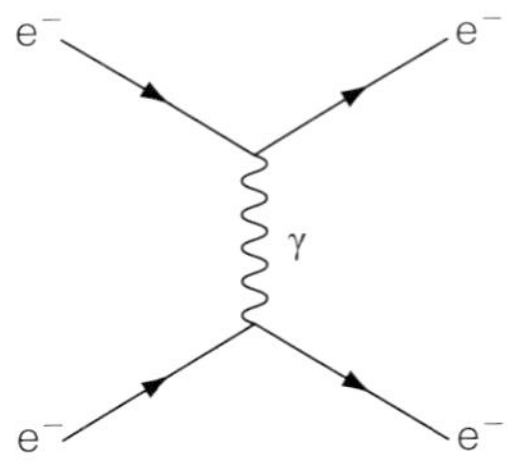

Figure 2.10 Feynman diagram illustrating two electrons meeting.

Two electrons meet, exchange photons and scatter away from each other. The photon symbol γ indicates that this is an example of an **electromagnetic** interaction.

β^+ (positron) radioactive decay (Figure 2.11)

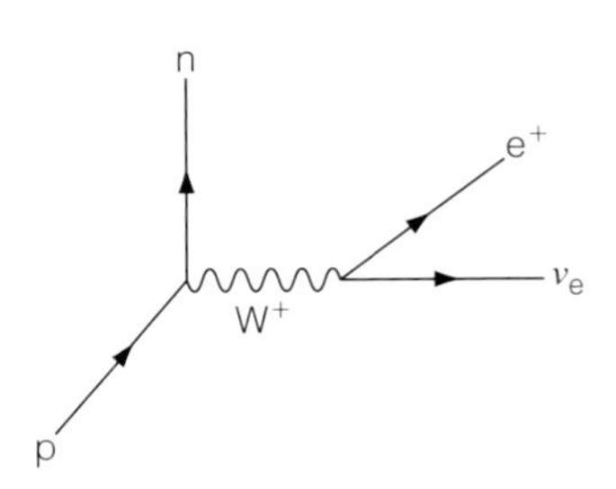

Figure 2.11 Feynman diagram illustrating β^+ (positron) radioactive decay.

In this case, a proton decays into a neutron and a W^+ exchange particle, which subsequently decays into a positron and an electron neutrino. This is another example of the **weak** interaction, (like β^- decay), and is summarised by the equation:

$$^1_1p \rightarrow {}^1_0n + {}^{0}_{+1}e + \nu_e$$

Electron capture (Figure 2.12)

Electron capture is another example of the **weak** interaction. An electron is absorbed by a proton within a nucleus. The proton decays into a neutron and a W^+ exchange particle, which interacts with the electron forming an

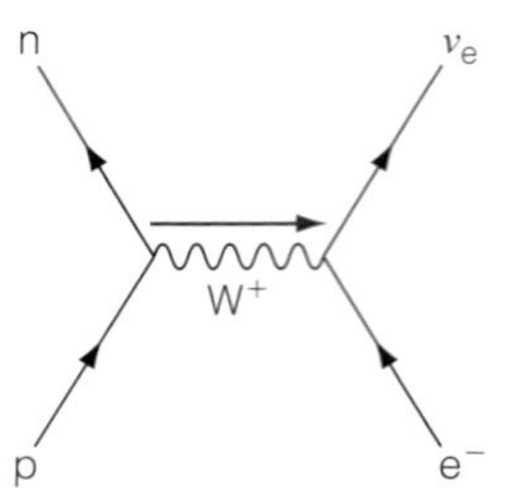

Figure 2.12 Feynman diagram illustrating electron capture.

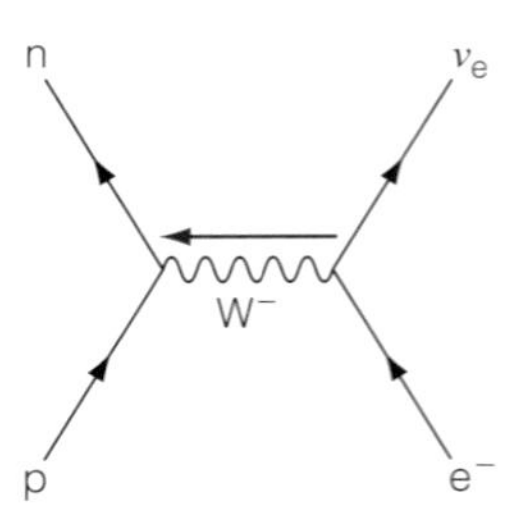

Figure 2.13 Feynman diagram illustrating electron–proton collision.

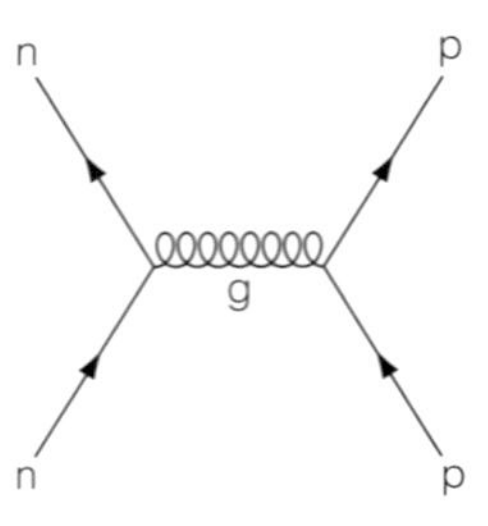

Figure 2.14 Feynman diagram illustrating proton–neutron binding via gluon exchange.

electron neutrino (as the proton acts on the electron the arrow above the exchange particle moves left to right):

$${}^{1}_{1}p + {}^{0}_{-1}e \rightarrow {}^{1}_{0}n + \nu_e$$

Electron–proton collision (Figure 2.13)

An electron and a proton collide transferring a W^- exchange particle, indicating the **weak** interaction; the proton decays into a neutron and the electron decays into an electron neutrino (as the electron collides with the proton the arrow above the exchange particle moves right to left):

$${}^{1}_{1}p + {}^{0}_{-1}e \rightarrow {}^{1}_{0}n + \nu_e$$

TIP

β^- decay is negative so involves e^-, an antineutrino and W^-, whereas β^+ decay is positive so involves e^+, a neutrino and W^+.

Proton–neutron bound by a gluon (Figure 2.14)

A **gluon** is exchanged between a neutron and a proton binding the two particles together (the process repeats over and over again). Notice that the Feynman diagram symbol for a gluon is a different wavy line from that of the photon or the $W^{\pm}/Z$ exchange particles. This is an example of the **strong** interaction.

ACTIVITY

Drawing Feynman diagrams

Write decay equations and draw Feynman diagrams for the following decays.

1. A muon plus (μ^+) decays into a W^+ exchange particle and a muon-antineutrino ($\bar{\nu}_\mu$). The W^+ then decays into a positron (e^+) and an electron neutrino (ν_e).
2. A kaon zero (K^0) decays into a pion minus (π^-) and a W^+ exchange particle, which subsequently decays into a pion plus (π^+).
3. An electron neutrino and an electron antineutrino annihilate into a gamma ray photon.
4. An electron neutrino decays into an electron and a W^+ exchange particle, which subsequently collides with a neutron producing a proton.
5. An electron and a positron annihilate to produce a gamma ray photons, one of which then pair produces an electron and a positron pair.

Classification of particles

Although the Standard Model is used to describe the nature of the matter in the observed Universe, many of the fundamental particles that are part of the model are rarely observed on their own and, even when they are, it is only at extremely high energy. Most of the fundamental particles are seen in combination with others, forming particles that can exist on their own at lower energies. Most of these composite particles were 'discovered' before their constituent fundamental particles and, in the years after the Second World War, names were given to these composite particles and the groups that they seemed to belong to. The particles were arranged into three groups: hadrons, leptons and exchange particles.

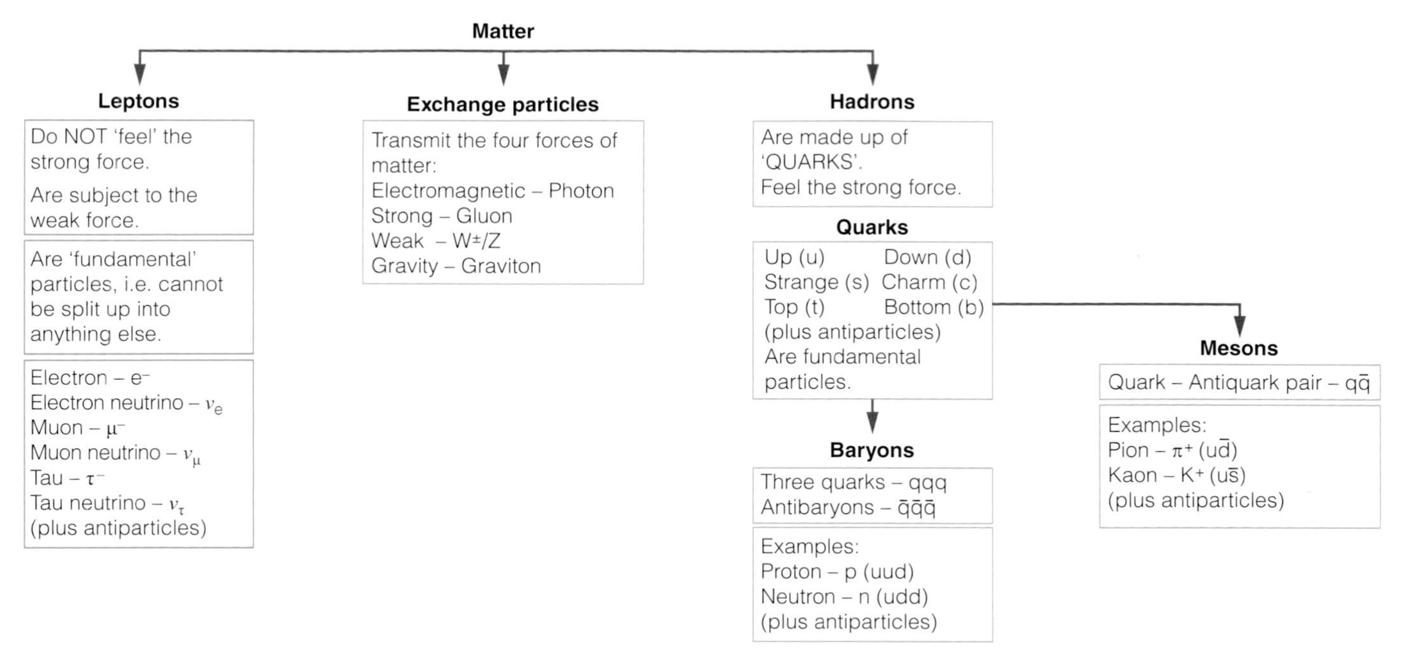

Figure 2.15 The particle garden.

TIP

All the leptons, exchange particles and quarks make up the particles of the Standard Model. However, the tau and tau-neutrino leptons and the charm, top and bottom quarks are not part of the examination specification. Any questions set on the examination papers involving these particles would give you all the information that you need to answer the question.

TEST YOURSELF

18 Use the following list to answer parts (a) to (e).

proton pion muon photon neutron

a) Which particle is a lepton?
b) Which particles are hadrons?
c) Which particles are fundamental particles?
d) Which particle is a meson?
e) Which particle is an exchange particle?

19 State the difference between a baryon and a meson.

20 Which particles feel the weak interaction?

21 Use the following list of particle groups to answer parts (a) to (d).

lepton hadron meson baryon exchange particle.

Which group(s) do the following particles belong to?

a) neutron
b) kaon minus
c) W^+
d) electron antineutrino

> **TIP**
> All leptons have a lepton number L = +1, but a charge $Q = -1(e)$ or 0. All antileptons have a lepton number L = −1 and a charge $Q = +1(e)$ or 0.

> **TIP**
> In the examination, you will only be set questions involving the conservation of lepton number in terms of electron leptons (e^-, e^+, ν_e, $\bar{\nu}_e$) and muon leptons (μ^-, μ^+, ν_μ, $\bar{\nu}_\mu$). **You do not need to know about tau leptons.**

Leptons

Exchange particles aside, matter is arranged into two broad groups with very different properties: leptons and hadrons. **Leptons** are fundamental particles (and are described as part of the Standard Model). The lepton group is made up of the electron, the muon and the tau particles, their respective neutrinos and all their antiparticles (12 particles in total). Leptons **do not** feel the strong force, but they are subject to the **weak** force. All leptons are assigned a quantum number, called a lepton number, L, which distinguishes them as leptons. All the leptons (like the electron) have a lepton number L = +1, all the antileptons (like the positron) have a lepton number L = −1, and all other (non-leptonic) particles have a lepton number L = 0 (zero). In **any** particle interaction, the **law of conservation of lepton number** holds. The total lepton number before an interaction must be equal to the total lepton number after the interaction.

For example, during β^- radioactive decay, lepton number, L, is conserved.

$$^{1}_{1}n \longrightarrow {}^{1}_{1}p + {}^{0}_{-1}e + \bar{\nu}_e$$

$$L: 0 \longrightarrow 0 + (+1) + (-1)$$

$$L: 0 \longrightarrow 0 + 1 - 1 \quad ✓ \quad \text{conserved}$$

Remember – protons and neutrons are not leptons.

Particle **lifetime** is the average time that a particle exists from its creation to its decay. Table 2.2 shows some examples.

Muon decay

Muons are unstable particles with a mass of about 200 times the mass of an electron. Muons have unusually long lifetimes, of the order of 2.2 μs and only the neutron, proton and atomic nuclei have higher lifetimes. All muons decay via the weak interaction into three particles, one of which has to be an electron (or a positron) and the other two particles are neutrinos. The decay equations for the muon and the antimuon are:

$$\mu^- \longrightarrow e^- + \bar{\nu}_e + \nu_\mu$$

$$\mu^+ \longrightarrow e^+ + \nu_e + \bar{\nu}_\mu$$

Table 2.2 Mean lifetimes for some particles.

Particle	Mean lifetime
proton	$>1 \times 10^{29}$ years
electron	$>4.6 \times 10^{26}$ years
(free) neutron	885.7 seconds
muon	2.2×10^{-6} seconds
π^0	8.4×10^{-17} seconds
π^+	2.6×10^{-8} seconds
W^+	1×10^{-25} seconds

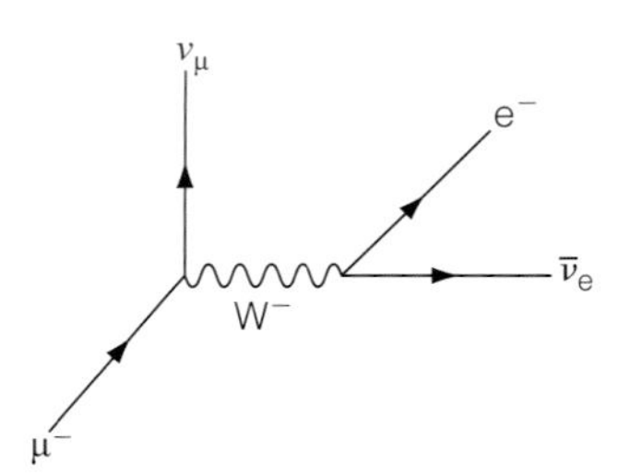

Figure 2.16 Feynman diagram showing muon decay.

The Feynman diagram for the decay of the muon is given in Figure 2.16.

ACTIVITY

The Pierre Auger Observatory and Project AIRES Cosmic Ray Shower Simulations

The following information is taken by kind permission from the COSMUS website and Sergio Sciutto from the AIRES software package:

http://astro.uchicago.edu/cosmus/projects/auger/

and

http://astro.uchicago.edu/cosmus/projects/aires/

'The Pierre Auger Observatory in Malargue, Argentina, is a multinational collaboration of physicists trying to detect powerful cosmic rays from outer space. The energy of the particles here is above 10^{19} eV, or over a million times more powerful than the most energetic particles in any human-made accelerator. This value is about 1 J and, as such, would warm up one cubic centimetre of water by 0.2 °C. No one knows where these rays come from.

Such cosmic rays are very rare, hitting an area the size of a football pitch once every 10 000 years. This means you need an enormous "net" to catch these mysterious ultra high energy particles. The Auger project will have, when completed, about 1600 detectors.

Each detector is a tank filled with 11 000 litres of pure water and sitting about 1.5 km away from the next tank. This array on the Argentinian Pampas will cover an area of about 3000 km^2, which is about the size of the state of Rhode Island or ten times the size of Paris. A second detection system sits on hills overlooking the Pampas and, on dark nights, captures a faint light or fluorescence caused by the shower particles colliding with the atmosphere'.

The cosmic rays that hit the atmosphere create huge showers of particles, particularly leptons; the paths of these particles through the atmosphere has been modelled by Sergio Sciutto's AIRES software and you can investigate these particle showers using the software and the download movie files. A good place to start is the "Five showers" animation on the AIRES page. Investigate the interactive software and the movie files to find out more about how cosmic rays produce particle interactions in the upper atmosphere.

TEST YOURSELF

22 What is the law of conservation of lepton number?

23 The rest-mass of an electron is 9.11×10^{-31} kg. Use this value to estimate the rest-mass of a muon.

24 Use the law of conservation of lepton number to explain why an interaction involving an electron and a positron annihilating and producing two muons **is not** possible.

25 Explain why an electron antineutrino is always produced during β^- emission.

TIP

Remember – you need to know about the following quarks:

up (u) charge, $+\frac{2}{3}e$

down (d) charge, $-\frac{1}{3}e$

strange (s) charge, $-\frac{1}{3}e$

(and their antiquarks, with opposite signs of charge).

Hadrons

Hadrons are particles that are made up of quarks and are therefore subject to the **strong** nuclear interaction. There are two sub-classes of hadrons.

Baryons, such as the proton and the neutron (and their antiparticles), are made up of three quarks (or three antiquarks). The proton comprises uud, with a total charge of +1e and the antiproton comprises $\overline{u}\overline{u}\overline{d}$, with a total charge of −1e.

Mesons, such as the pion and the kaon (and their antiparticles), are made up of a quark–antiquark pair.

> **TIP**
> Make sure you know the quark structure of the following particles:
> - Proton, p, uud (and antiproton, $\overline{u}\overline{u}\overline{d}$)
> - Neutron, n, udd (and antineutron, $\overline{u}\overline{d}\overline{d}$)
> - Pion$^+$, π^+, $u\overline{d}$
> - Pion$^-$, π^-, $d\overline{u}$
> - Pion^0, π^0, $u\overline{u}$ or $d\overline{d}$
>
> (The π^+ and π^- are antiparticles of each other, π^0 is its own antiparticle.)
> - Kaon$^+$, K$^+$, $u\overline{s}$
> - Kaon$^-$, K$^-$, $s\overline{u}$
> - Kaon^0, K^0, $d\overline{s}$ (and antiparticle, $\overline{K^0}$, $s\overline{d}$)
>
> (The K$^+$ and K$^-$ are antiparticles of each other.)

Baryons

As baryons are made up of three quarks, and there are six flavours of quark, there are many different possible baryons. Two of these baryons, the proton and the neutron, are well known and make up most of the mass of the Universe. The other baryons, containing the more massive quarks, are more exotic and are only observed at high energy inside particle detectors such as the LHC, or high up in the atmosphere as the result of interactions of cosmic rays with particles in the upper atmosphere. Each baryon has its own antibaryon – which is made up of the corresponding antiquarks. For example, the sigma baryon, Σ^+, is made up of two up quarks and a strange quark, uus, and the anti-sigma baryon, $\overline{\Sigma^+}$, is made up of two anti-up quarks and an anti-strange quark, $\overline{u}\overline{u}\overline{s}$.

The proton is the most stable and abundant baryon. Spontaneous free proton decay has never been observed and, although some non-Standard Model theories predict that it can happen, the predicted lifetime of the proton is of the order of 10^{34} years. The current measurement of the age of the Universe is only 13.8 billion years (13.8×10^9 years). As the Standard Model has proved to be remarkably robust, then it seems that free protons are stable, and all other baryons will eventually decay into protons. Neutrons also appear to be stable within most nuclei (unless they are β^- radioactive decay emitters), but when they are isolated on their own (free) they have a mean lifetime of 882 seconds (about 15 minutes). The vast majority of all the other baryons have vanishingly short lifetimes, between 10^{-10} and 10^{-24} seconds.

> **TIP**
> Remember charge is also a quantum number like baryon number and is also conserved in particle interactions.
>
> For example, during electron capture a proton inside a nucleus can interact with one of the electrons surrounding the nucleus and it can decay into a neutron and W$^+$ exchange particle, which then interacts with the electron producing an electron neutrino:
>
> $^1_1p + {}^0_{-1}e \rightarrow {}^1_0n + \nu_e$
>
> B: + 1 + 0 → + 1 + 0 ✓ conserved
>
> Q: + 1 + (−1) → + 0 + 0 ✓ conserved

All baryons are assigned a **baryon** quantum **number**, B. All baryons have baryon number B = +1, all anti-baryons have a baryon number B = −1 and all non-baryons have B = 0. Like lepton number, baryon number is also conserved in particle interactions. The total baryon number of all particles before an interaction must equal the total baryon number after the interaction.

As baryons have integer values of baryon number, quarks must have a baryon number of $+\frac{1}{3}$ and antiquarks have a baryon number of $-\frac{1}{3}$. Protons are baryons with a quark structure of uud, so they must have a baryon number of $+\frac{1}{3}+\frac{1}{3}+\frac{1}{3} = +1$, and antiprotons with a quark structure of $\overline{u}\overline{u}\overline{d}$ must have a baryon number of $-\frac{1}{3}-\frac{1}{3}-\frac{1}{3} = -1$.

Feynman diagrams can be drawn involving composite particles such as the proton and the neutron (containing quarks), and they also show how the quarks change during an interaction. The quarks making up the composite particle are shown by arrowed lines drawn parallel and next to each other. The Feynman diagram for β^- decay then becomes:

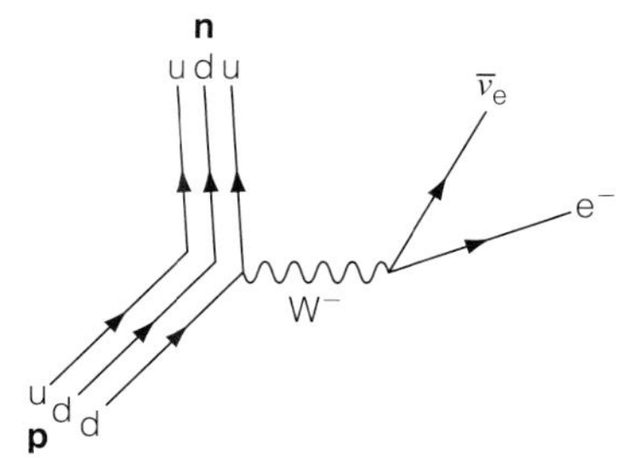

Figure 2.17 Feynman diagram for β^- decay.

In this example, the down quark decays into the W^- exchange particle and an up quark.

The Feynman diagrams for positron emission by protons in terms of quarks is also shown below:

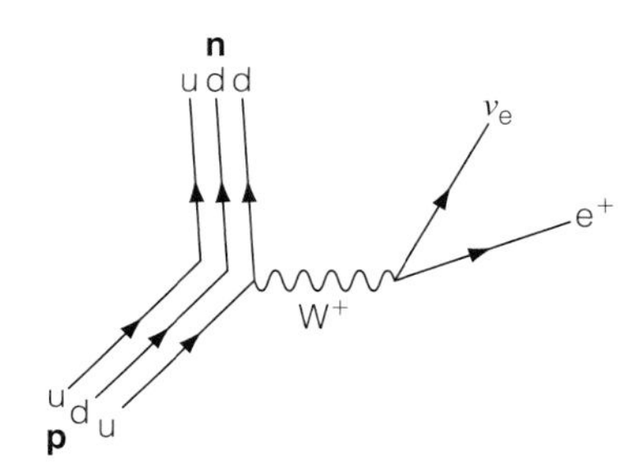

Figure 2.18 Feynman diagram illustrating β^+ (positron) radioactive decay involving quarks.

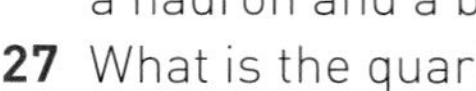

TEST YOURSELF

26 Explain why a proton can be a hadron and a baryon.
27 What is the quark structure of a neutron?
28 During positron emission, the quark structure of a proton changes. Describe this change.
29 Write a nuclear reaction equation for positron emission and use the equation to show that baryon number is conserved.

Mesons

Meson particles were first proposed by the Japanese physicist Hideki Yukawa in 1934 as a way of explaining the strong force holding protons and neutrons together to make nuclei. Yukawa suggested that the strong force was due to the proton and the neutron exchanging mesons. We now know that pions (pi-mesons) are exchanged between protons and neutrons, but that the strong interaction is actually due to the interaction between quarks that make up the protons and neutrons. Pions, being made up of quarks, also feel the strong force and are able to exist outside nucleons and so the strong interaction between the proton and the neutron is due to the pion exchange with the pions acting as a 'force carrier'.

Mesons are hadron particles made up of a quark–antiquark pair.

Mesons are made up of quark–antiquark pairs, $q\overline{q}$, and as with baryons, because there are six quarks and six antiquarks, there are many different possible mesons. Most mesons are high-energy particles and are only seen to exist inside the detectors of particle accelerators, but the pion and the kaon are produced when high-energy cosmic rays interact with the upper atmosphere and can be observed by high-altitude particle detectors. Mesons have a lepton number, L = 0 (they are not leptons) and a baryon number, B = 0 (they are not baryons).

TIP
Although mesons are hadrons, they are not baryons and hence their baryon number is 0.

Pions are combinations of the up, u, and down, d, quarks and their antiparticles. There are three types of pion:

$$\pi^+ = u\bar{d}$$

$$\pi^- = d\bar{u}$$

$$\pi^0 = u\bar{u} \text{ or } d\bar{d}$$

The π^+/π^- mesons are antiparticles of each other and the π^0 is its own antiparticle. All the pions are unstable and decay with lifetimes of the order of 10^{-8} s for the π^+/π^- mesons and 10^{-16} s for the π^0 meson. The π^+/π^- mesons decay into muons and electron neutrinos via the weak interaction:

$$\pi^+ \longrightarrow \mu^+ + \nu_\mu$$

$$\pi^- \longrightarrow \mu^- + \bar{\nu}_\mu$$

The Feynman diagram for π^- decay is shown in Figure 2.19.

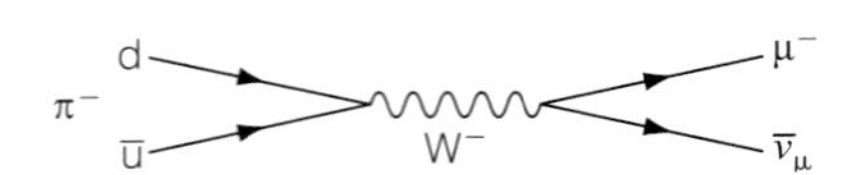

Figure 2.19 Feynman diagram showing pion minus decay.

The π^0 meson decays into two gamma rays.

The other common meson produced by cosmic ray interactions is the kaon. Kaons contain the strange quark (or antiquark), s, which has a charge of $-\frac{1}{3}$e. The quantum number **strangeness**, symbol **S**, is a property possessed by particles containing the strange quark and was coined by Murray Gell-Mann to describe the 'strange' behaviour of particles that are always produced in pairs by the strong interaction but decay via the weak interaction. To explain this, Gell-Mann suggested that strangeness was conserved in the production of strange particles, but not in their decay.

There are four different kaons, (with their strangeness values, S):

$$K^+ = u\bar{s} \ (S = +1)$$

$$K^- = s\bar{u} \ (S = -1)$$

$$K^0 = d\bar{s} \ (S = +1) \text{ and } \bar{K}^0 = s\bar{d} \ (S = -1)$$

TIP

The K^+ particle has a charge of +1 because of the addition of the charges on the u and $\bar{s}$ quarks, $+\frac{2}{3}e + \left|+\frac{1}{3}e\right| = +1e$. The same is true for the charge on the K^- particle, $-\frac{2}{3}e + \left|-\frac{1}{3}e\right| = -1e$.

Kaon pair production occurs via the strong interaction (where strangeness is conserved).

For example, during the high-energy collision of two protons, a K^+/K^- pair is produced:

$$p + p \longrightarrow p + p + K^+ + K^-$$

Strangeness check:

$$0 + 0 = 0 + 0 + (+1) + (-1) = 0 \quad \checkmark \quad \text{conserved}$$

TIP

The strange quark has strangeness −1. The anti-strange quark has strangeness +1.

Kaons are unstable, decaying via the weak interaction with lifetimes of about 10^{-8} s to 10^{-10} s. There are several processes by which the charged kaons can decay:

$$K^+ \longrightarrow \mu^+ + \nu_\mu$$

$$K^+ \longrightarrow \pi^+ + \pi^0$$

$$K^+ \longrightarrow \pi^+ + \pi^+ + \pi^-$$

$$K^- \longrightarrow \mu^- + \bar{\nu}_\mu$$

$$K^- \longrightarrow \pi^- + \pi^0$$

$$K^- \longrightarrow \pi^0 + \mu^- + \bar{\nu}_\mu$$

None of the decay products of these decays contain a strange quark, so strangeness is not conserved.

TEST YOURSELF

30 What is the charge and baryon number of the following particles:

a) proton
b) antiproton
c) neutron
d) electron
e) antineutron
f) positron
g) uds baryon
h) $\bar{u}\bar{u}\bar{s}$ baryon
i) dss baryon

31 The Standard Model and the particle garden arranges matter into the following groups:

A leptons
B quarks
C exchange particles
D baryons
E mesons

Which groups do the following particles (represented by their symbols only) belong to?

a) p
b) n
c) e^-
d) γ
e) μ^+
f) π^-
g) *u*
h) *uds*

32 Which of the following particles are **possible** baryons?

a) $\bar{d}uu$
b) duu
c) sss
d) dud
e) $\bar{d}\bar{u}\bar{s}$
f) sdu
g) ddd
h) $\bar{u}\bar{s}d$

33 Which of the following are **impossible** mesons?

a) ud
b) $d\bar{s}$
c) $u\bar{u}$
d) sd
e) dd
f) $d\bar{d}$

34 What is the baryon and lepton structure of the following atoms?

a) $^{2}_{1}H$
b) $^{4}_{2}He$
c) $^{14}_{6}C$
d) $^{23}_{11}Na$
e) $^{238}_{92}U$
f) $^{294}_{118}Uuo$

ACTIVITY

The Lancaster Particle Physics Package (LPPP)

The LPPP is an online resource for studying the interactions of particles. The package consists of a series of guided computer simulations that take you through the way that particle physicists study particle interactions experimentally. The computer simulations are backed up with basic physics explanations of what is going on inside each simulation. The package contains some material studied in other chapters of this book and at A level. The package can be accessed by using the following link:

www.lppp.lancs.ac.uk

Work your way through the package; you can dip in and out of it throughout the whole of the A level course.

TIP

Remember – Einstein's energy-mass equation, $E = mc^2$, shows that, on the quantum scale, mass and energy are interchangeable – mass can convert to energy and energy can convert back to mass. It is better to talk about mass-energy being conserved on a quantum scale rather than the conservation of mass and the conservation of energy.

Conservation laws

Throughout this chapter you have met several different quantum number conservation laws – properties or physical quantities that are the same after an interaction as they are before the interaction. Three further quantities are also always conserved in any interaction, these are:

- charge, Q
- momentum, p
- mass-energy, $E = mc^2$.

To these we add the quantum number conservation laws:

- lepton number, L
- baryon number, B
- strangeness, S.

With the exception of strangeness, all the other quantities are **always** conserved in **any** interaction. Strangeness is conserved in strong interactions but not in weak interactions (as the strange quark changes flavour).

Some examination questions ask you to decide if a given particle interaction, usually given to you in equation form, can happen or not. Momentum and mass-energy will always be conserved, so all you have to do is to determine if charge, Q; lepton number, L; baryon number, B; and strangeness, S, are conserved. (If strangeness is not conserved this could indicate that the interaction is a weak interaction). Consider the examples shown below.

EXAMPLE

Is this particle interaction possible?

$$p + \overline{\nu}_e \rightarrow e^+ + n$$

Answer

A great way to do this is to construct a table similar to the one below:

Table 2.3

Conservation quantity	Before interaction			After interaction			Quantity conserved?
	p	$\overline{\nu}_e$	Total	e+	n	Total	
Q	+1	0	+1	+1	0	+1	✓
B	+1	0	+1	0	+1	+1	✓
L	0	−1	−1	−1	0	−1	✓
S	0	0	0	0	0	0	✓

In this example all the quantities are conserved, so the interaction is possible.

EXAMPLE

Is this interaction possible?

$$p + e^+ \rightarrow e^- + \Sigma^0 + K^+$$

Answer

The Σ^0 baryon has the following properties: Q = 0; B = +1; L = 0 and S = −1. Using the same table as in the previous example, but adding an extra product column:

Table 2.4

Conservation quantity	Before interaction			After interaction				Quantity conserved?
	p	e^+	Total	e^-	Σ^0	K^+	Total	
Q	+1	+1	+2	−1	0	+1	0	✗
B	+1	0	+1	0	+1	0	+1	✓
L	0	−1	−1	+1	0	0	+1	✗
S	0	0	0	0	−1	+1	0	✓

In this case charge, Q, and lepton number, L, are not conserved, so this interaction is not possible.

TIP

The format of the table can be modified depending on the number of particles involved – you can add extra columns or take them away.

ACTIVITY

Use the conservation laws to decide if the following particle interactions can occur. A table of properties of particles is also shown.

Table 2.5

Particle	Charge, Q	Baryon number, B	Lepton number, L	Strangeness, S
p	+1	+1	0	0
n	0	+1	0	0
e^-	−1	0	+1	0
e^+	+1	0	−1	0
ν_e	0	0	+1	0
$\bar{\nu}_e$	0	0	−1	0
Σ^0	0	+1	0	−1
Σ^-	−1	+1	0	−1
K^+	+1	0	0	+1

1 $p + \pi^- \rightarrow \Sigma^- + K^+$
2 $p + \bar{\nu}_e \rightarrow e^+ + \Sigma^0$
3 $n \rightarrow p + e^+ + \nu_e$
4 $p + e^+ \rightarrow e^- + \Sigma^0 + K^+$
5 $n \rightarrow p + e^- + \bar{\nu}_e$
6 $\pi^- + p \rightarrow n + \pi^0 + \bar{\nu}_e$

What you need to know

- For every type of particle there is a corresponding antiparticle.
- Particles and antiparticles have: rest-mass (in MeV/c^2); charge (in C) and rest-energy (in MeV).
- The positron, the antiproton, the antineutron and the electron antineutrino are the antiparticles of the electron, the proton, the neutron and the electron neutrino respectively.
- The four fundamental interactions are: gravity, electromagnetic, weak and strong. (The strong nuclear force is also known as the strong interaction.)
- Exchange particles are used to explain forces between elementary particles on the quantum scale.
- The virtual photon is the exchange particle for the electromagnetic interaction.
- Examples of the weak interaction are β^- decay, β^+ decay, electron capture and electron–proton collisions.
- The W^+ and W^- are the exchange particles of the weak interaction.
- Feynman diagrams are used to represent reactions or interactions in terms of particles going in and out, and exchange particles.
- Hadrons are particles that are subject to the strong interaction.
- There are two classes of hadrons:
  - baryons (proton, neutron) and antibaryons (antiproton and antineutron)
  - mesons (pion, kaon)
- Baryon number, B, is a quantum number that describes baryons. Baryons have B = +1; antibaryons, B = −1; non-baryons, B = 0.
- Baryon number is always conserved in particle interactions.
- The proton is the only stable baryon and all other baryons will eventually decay into protons.

- Free neutrons are unstable and decay via the weak interaction forming a proton, β^- particle and an electron antineutrino.
- Pions and kaons are examples of mesons. The pion is the exchange particle of the strong nuclear force between baryons. The kaon is a particle that can decay into pions.
- Leptons are particles that are subject to the weak interaction.
- Leptons include: electron, muon, neutrino (electron and muon types) and their antiparticles.
- Lepton number, L, is a quantum number used to describe leptons; leptons have L = +1; antileptons, L = −1; non-leptons, L = 0.
- Lepton number is always conserved in particle interactions.
- Muons are particles that decay into electrons.
- Strange particles are particles that are produced through the strong interaction and decay through the weak interaction (e.g. kaons).
- Strangeness (symbol S) is a quantum number to describe strange particles. Strange particles are always created in pairs by the strong interaction (to conserve strangeness).
- Strangeness is conserved in strong interactions. In weak interactions the strangeness can change by −1, 0 or +1.
- Quarks have: charge $\left(+\frac{1}{3}, -\frac{1}{3}, +\frac{2}{3}, -\frac{2}{3}\right)$, baryon number, $\left(+\frac{1}{3}, -\frac{1}{3}\right)$ and strangeness (+1, 0 or −1).
- Hadrons have the following quark structures:
  - baryons (proton, uud; neutron, udd), antiproton, $\bar{u}\bar{u}\bar{d}$, and antineutron, $\bar{u}\bar{d}\bar{d}$)
  - mesons:
    - pions - Pion$^+$, π^+, $u\bar{d}$; Pion$^-$, π^-, $d\bar{u}$; Pion^0, π^0, $u\bar{u}$ or $d\bar{d}$
    - kaons - Kaon$^+$, K$^+$, $u\bar{s}$; Kaon$^-$, K$^-$, $s\bar{u}$; Kaon^0, K^0, $d\bar{s}$ or $s\bar{d}$.
- During β^- decay a d quark changes into a u quark, and during β^+ decay a u quark changes into a d quark.
- Conservation laws for charge, baryon number, lepton number and strangeness can be applied to particle interactions.

TEST YOURSELF

35 Match the pions to their correct quark structures:

π^0	π^+	π^-
$d\bar{u}$	$u\bar{u}$	ud, $u\bar{d}$

36 Which of the following pion decays is not possible?

a) $\pi^+ \rightarrow e^+ + \nu_e$

b) $\pi^- \rightarrow \mu^- + \bar{\nu}_\mu$

c) $\pi^0 \rightarrow \gamma + \gamma$

d) $\pi^- \rightarrow \mu^- + \mu^+$

e) $\pi^+ \rightarrow \mu^+ + \nu_\mu$

37 Identify the quark structure of the following strange particles from the quark list below.

P uus

Q uds

R $u\bar{s}$

S $s\bar{u}$

T $d\bar{s}$

a) K$^+$

b) K$^-$

c) Λ^0 (lamda^0) baryon

38 Strange quarks, responsible for the long half-lives of strange particles during the weak interaction, were first proposed by Murray Gell-Mann and George Zweig in 1964. As baryon number had already been defined as +1 for protons and neutrons the subsequent discovery of quarks required them to have $+\frac{1}{3}$ and $-\frac{1}{3}$ baryon numbers. Table 2.6 compares some of properties of strange and down quarks and their antiquarks:

Table 2.6

Quark	Baryon number	Charge/e	Strangeness
s	$\frac{1}{3}$	$-\frac{1}{3}$	−1
$\bar{s}$	$-\frac{1}{3}$	$+\frac{1}{3}$	+1
d	$\frac{1}{3}$	$-\frac{1}{3}$	0
$\bar{d}$	$-\frac{1}{3}$	$+\frac{1}{3}$	0

a) Quarks and antiquarks can combine to form four possible mesons. Copy and complete the table, calculating the baryon number, charge and strangeness of the four mesons.

Table 2.7

Quark pair	$s\bar{s}$	$s\bar{d}$	$d\bar{s}$	$d\bar{d}$
Name	phi	kaon^0 (anti-symmetric)	kaon^0 (symmetric)	rho^0
Baryon number				
Charge/e				
Strangeness				

b) The phi and rho^0 mesons have the same properties in the table. Suggest a way that these two mesons could be distinguished from each other.

39 The last quark to be identified experimentally was the top, t, quark in 1995, 18 years after it was first added as a concept to the Standard Model.

a) Suggest why physicists were able to predict the existence of the top quark even though it took 18 years to observe it experimentally.

b) The top quark was observed experimentally by the Tevatron particle accelerator at Fermilab in the USA. The tevatron collided protons and antiprotons with an energy of 1.8 TeV (1.8×10^{12} eV). Although the protons and anti-protons have a high momentum, the sum of the momentums of a colliding pair of particles – a proton and an antiproton – is zero, because they are travelling in opposite directions with the same speed. The resulting collisions produced top quark–antitop quark pairs that were then observed in the particle detectors. Explain why top quark–antitop quark pairs move in opposite directions after they are produced by the collision.

c) The top quark (and antiquark) has an extremely short lifetime (5×10^{-25} s) and decays into a bottom quark and a W$^{\pm}$ exchange particle. This is shown in Figure 2.20.

Identify particles X and Y.

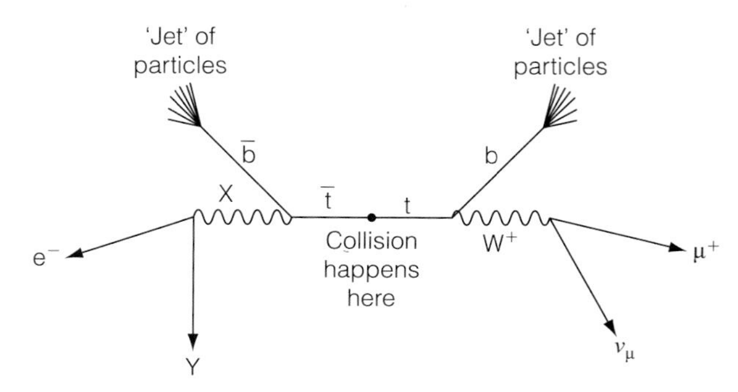

Figure 2.20 Feynman diagram of top–antitop quark pair production.

40 The charm quark was observed experimentally by two particle physics teams at almost the same time in 1974. One team was based at the Brookhaven National Laboratory (BNL) and the other at the Stanford Linear Accelerator (SLAC). Both teams observed a $c\bar{c}$ meson. The BNL team called it the psi (ψ) meson and the SLAC team named it the J meson – since then it has been known as the J/ψ meson. The J/ψ meson has the following properties:

Table 2.8

Charge	Baryon number	Lepton number
0	0	0

The J/ψ meson can decay in many ways; two of these are shown below:

$J/\psi \rightarrow e^+ + e^-$

$J/\psi \rightarrow \mu^+ + \mu^-$

Explain, using conservation laws, how both of these decay equations are correct. You need to consider:

- energy
- momentum
- charge
- baryon number
- lepton number.

Practice questions

1 The group of particles known as hadrons are composed of quarks. There are two sub-groups of hadrons: mesons and baryons.

a) What is the name of the property that defines a hadron? *(1)*

b) State the quark structure of a meson. *(1)*

c) State the quark structure of a baryon. *(1)*

d) The antiproton is the antiparticle of the proton. State **one** way that the proton and the antiproton are similar, and **one** way that they are different. *(2)*

e) What are the following values of an antiproton:

i) its charge *(1)*

ii) its baryon number *(1)*

iii) its quark structure? *(1)*

2 In the Standard Model of Matter, particles can be leptons, hadrons or exchange particles. Here is a list of particles:

electron proton neutron muon pion

a) From the list state the name of **one** lepton and **one** hadron. *(2)*

b) State one difference between leptons and hadrons. *(1)*

c) State the difference in structure between baryons and mesons. *(1)*

d) From the list state the name of **one** baryon and **one** meson. *(2)*

3 The table below shows some basic information about three hadrons.

Table 2.9

Sub-atomic particle	Quark structure	Baryon OR Meson	Relative charge	Baryon number	Strangeness
	$u\bar{d}$	meson		0	
	udd				0
Σ^+, sigma$^+$			+1		-1

a) Copy and complete the table. *(3)*

b) All of the sub-atomic particles shown in the table have a corresponding antiparticle. State one example of a baryon and its antibaryon **not** shown in the table and state their quark structures. *(4)*

c) The electron and the positron are an example of a lepton particle–antiparticle pair. State **one** property of positron that is the same as an electron, and **one** property that is different. *(2)*

4 Figure 2.21 shows two particles interacting. An offset line has been drawn to represent the exchange particle.

a) State the name of the interaction shown and give the name of the exchange particle drawn on the diagram. *(2)*

b) In this interaction, momentum and energy are conserved. Give the name of another quantity that is conserved. *(1)*

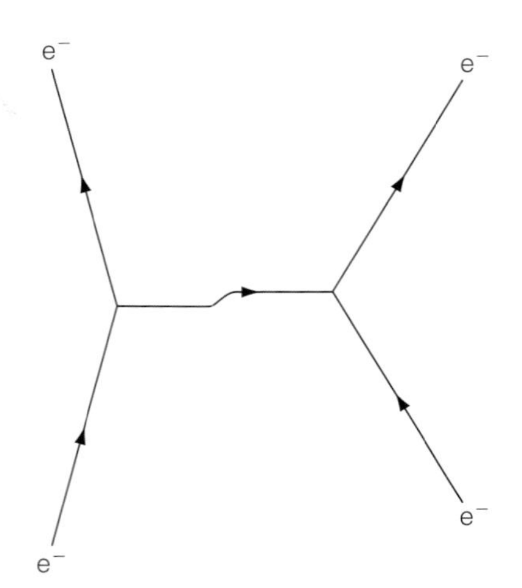

Figure 2.21 Feynman diagram showing two particles interacting.

Figure 2.22 shows another type of interaction. Another offset line has been drawn to represent the exchange particle.

c) Name the particles labelled A, B and C. *(3)*

d) State the name of this type of particle interaction. *(1)*

e) Momentum and energy are also conserved in this particle interaction. Name two other quantities that must be conserved for this interaction to occur and show how they are conserved. *(4)*

f) The Standard Model predicted the existence of the exchange particle involved in this interaction before it had been observed experimentally. Explain why it is important to test the prediction of a scientific model experimentally. *(3)*

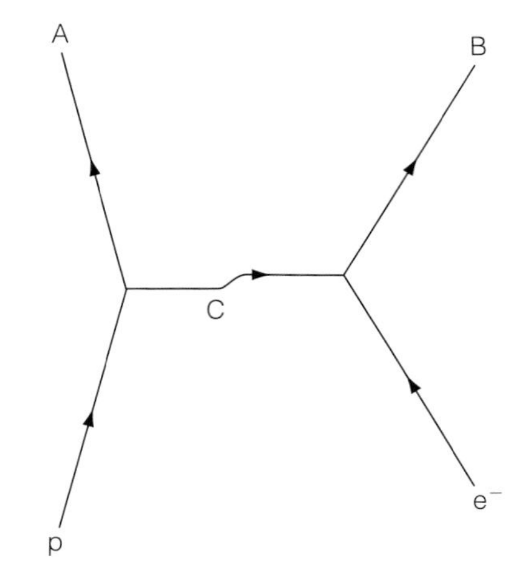

Figure 2.22 Another type of interaction.

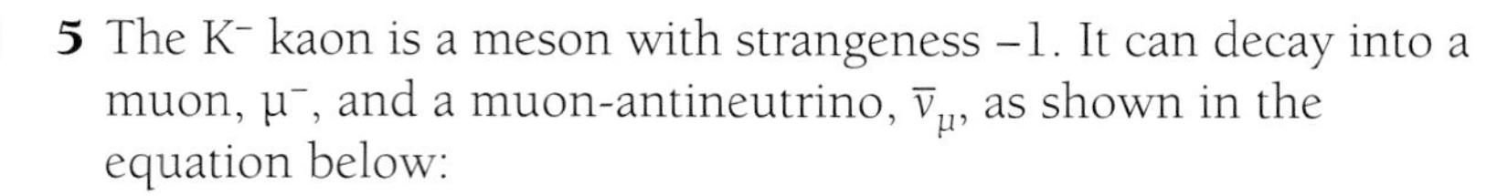

5 The K^- kaon is a meson with strangeness –1. It can decay into a muon, μ^-, and a muon-antineutrino, $\bar{\nu}_\mu$, as shown in the equation below:

$$K^- \rightarrow \mu^- + \bar{\nu}_\mu$$

a) State and explain which particle interaction is responsible for this decay. *(1)*

b) Energy and momentum are conserved in this decay, as are **two** other quantities. State the name of these two quantities. *(2)*

c) State one property of a strange particle, such as the K^- kaon, that makes it different to a non-strange particle, such as a π^- pion. *(1)*

The K^+ kaon is also strange. It **could** decay via either of the following decay equations, however one of these decay equations is not possible:

$$K^+ \rightarrow \mu^+ + \bar{\nu}_\mu$$

$$K^+ \rightarrow \pi^0 + \pi^+$$

d) State and explain which of these decay equations is not possible. *(2)*

The K^- kaon can also decay in the following way, producing a muon, a muon-neutrino and a third particle, X:

$$K^- \rightarrow X + \mu^- + \bar{\nu}_\mu$$

e) State which interaction is responsible for this decay. *(1)*

f) Particle X is identified as a pion. Explain why X must be a meson. *(2)*

g) State the charge on this particle. *(1)*

6 During β^- decay a neutron decays into a proton and an electron antineutrino.

a) Write an equation describing this decay. *(1)*

b) Using conservation laws, explain why an anti-electron-neutrino is produced rather than an electron-neutrino. *(2)*

c) Draw a Feynman diagram for this decay. *(3)*

β^+ decay involves the emission of positrons. ${}^{23}_{12}Mg$ is a positron emitter, decaying into a ${}^{23}_{11}Na$ nucleus, a positron and another particle, X.

d) State the name of the other particle, X. *(1)*

e) State whether each decay product is a baryon or a lepton. *(3)*

During β^+ decay, an up quark decays to a down quark and an exchange particle, which subsequently decays into the positron and the other particle, X.

f) State the quark structure of a neutron and a proton. (2)

g) Draw a Feynman diagram for this decay. (3)

7 Figure 2.23 illustrates the particle interaction known as electron capture.

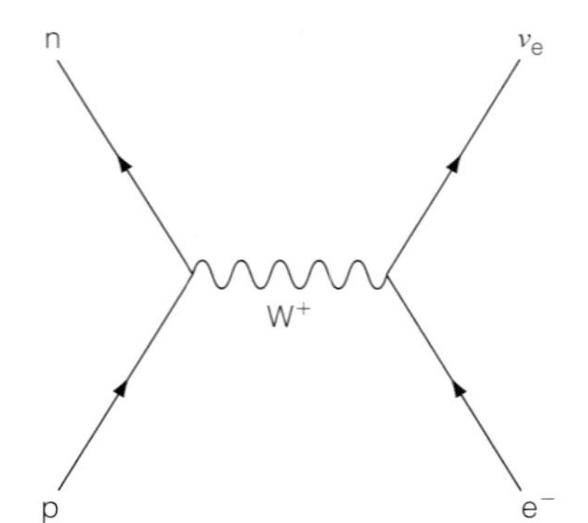

Figure 2.23 Feynman diagram showing electron capture.

a) During electron capture, charge, baryon number and lepton number are all conserved. Show how these three quantities are conserved. (3)

b) The isotope potassium-40, ${}^{40}_{19}K$, is an extremely unusual nuclide because it can decay by all three types of beta decay. It can decay via electron capture **or** positron emission **or** β^- emission. Write an equation summarising **each** decay. *(6)*

8 The proton is an example of a hadron. The positron is an example of lepton.

a) State one similarity and one difference between protons and positrons. (2)

b) There are four forces that act between particles. Exchange particles can be used to explain these forces. Match the correct exchange particle to its force. (3)

Table 2.10

Force	Gravity	Strong	Weak	Electromagnetic
Exchange particle	Graviton	Photon	Gluon	$W^{\pm}$, Z

c) Describe how the force between a proton and a neutron varies with the separation distance between the two particles and quote suitable values for separation distance. (3)

d) Positrons and protons can interact via three of the above forces. Identify the force that cannot exist between the proton and the positron, and explain why they cannot interact in this way. (2)

9 Pions and muons are produced when high-energy cosmic rays interact with gases high up in the Earth's atmosphere. Write an account of how the Standard Model is used to classify pions and muons into particle groups. Your account should include the following:

- the names of the groups of particles that pions and muons belong to
- other examples of particles in these groups
- details of any properties that the particles have in common
- description of the ways that each particle can interact with other particles. *(6)*

Stretch and challenge

10 The Σ^+ baryon has a strangeness of –1.

a) State the quark composition of this particle.

The Σ^+ is unstable and can decay into a meson and another baryon as shown in the equation below:

$$\Sigma^+ \rightarrow \pi^+ + n$$

b) State the names **and** the quark structures of the two decay particles.

c) Apart from momentum and energy, which two other quantities are conserved in this decay?

d) Name a quantity that is **not** conserved in this decay.

e) State the name of the particle interaction shown by the equation.

f) Both of the decay product particles of the Σ^+ baryon are unstable. Write the decay equation for the decay of the n particle.

11 Figure 2.24 shows the decay of a strange quark, s.

a) State which interaction is responsible for the decay of the strange quark.

b) Give names for particles 1, 2 and 3.

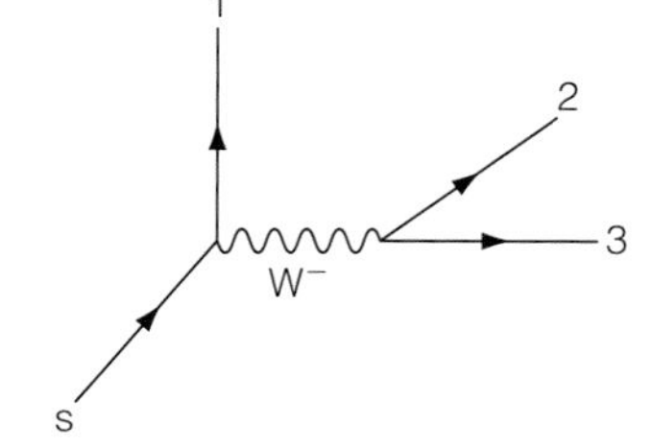

Figure 2.24 Feynman diagram showing the decay of a strange quark.

12 A π^+ meson has a rest mass of 139.6 MeV/c^2 and a mean lifetime of 2.6×10^{-8}s. This meson decays into an antimuon and a muon neutrino.

a) What sort of particle interaction is involved with this decay?

b) Which exchange particle is involved in this decay?

c) Draw a Feynman diagram for this decay.

A particle of rest mass M decays at rest into two high-speed particles, m_1 and m_2, as shown in Figure 2.25.

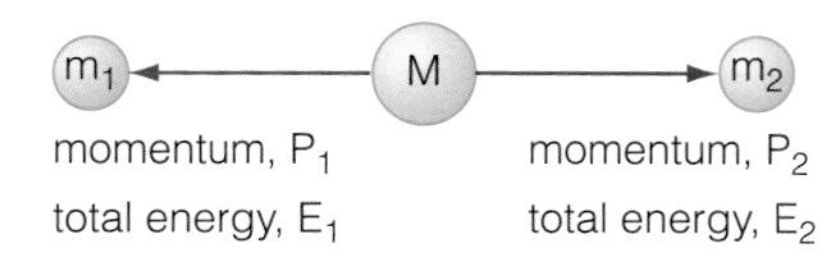

Figure 2.25 Particle decay diagram.

The relativistic version of Einstein's energy-mass equation is given by:

$$E^2 = c^2p^2 + m_0{}^2\, c^4$$

d) Use this relationship to show:

$$E_1 = \frac{(M^2 + m_1{}^2 - m_2{}^2)\, c^2}{2M} \qquad \text{and} \qquad cp_1 = \sqrt{(E_1{}^2 - m_1{}^2c^4)}$$

e) Experiments have shown that the rest mass of an antimuon is 105.7 MeV/c^2, and the Standard Model requires a muon neutrino to be massless. Using this data, and the data for the pion, show that the kinetic energy of the antimuon emitted during the decay is 4.1 MeV.

Electrons and energy levels

PRIOR KNOWLEDGE

- Energy is measured in joules, J. A potential difference of 1 volt transfers 1 joule per coulomb, C, of charge: $1\,V = 1\,JC^{-1}$.
- In the nuclear model of the atom, the atom has a central, massive nucleus, which is positively charged. The nucleus contains protons and neutrons and is surrounded by a cloud of orbiting electrons, which are negatively charged.
- A neutral atom has equal numbers of protons and electrons. When an atom loses or gains an electron, it is ionised. The atom is left with an overall positive charge if it loses electrons, or a negative charge if it gains electrons.
- Light is one form of electromagnetic radiation. Higher frequency radiation, for example gamma rays, has a shorter wavelength compared with lower frequency radiation, for example microwaves.
- In order of decreasing wavelength, the members of the electromagnetic spectrum are: radio waves, microwaves, infrared, visible light, ultraviolet, X-rays, gamma rays.

TEST YOURSELF ON PRIOR KNOWLEDGE

1 How much energy does a potential difference of 6 V transfer to 2 C of charge?

2 A sodium atom has 11 protons. When a particular sodium atom is ionised, it loses electrons.

 a) How many electrons does a sodium atom have?

 b) Is this sodium ion positively or negatively charged?

3 Explain one difference between a sodium atom and a sodium ion.

4 Put these members of the electromagnetic spectrum in order of increasing frequency: gamma rays, infrared, visible light, microwaves.

5 Which colour of visible light has the longest wavelength?

Electromagnetic radiation carries information throughout the Universe to Earth at the speed of light. Under the right conditions, all elements formed since the Universe began, wherever they are in the Universe, can give out the same unique patterns of light, called a spectrum. The spectrum from a star can be analysed in detail to reveal which elements are present in the star. Light from very distant galaxies takes billions of years to reach Earth. When it arrives, the spectrum from each galaxy has been red shifted. When the red shift of the light is measured, the speed and distance of each galaxy from Earth can be calculated, helping us estimate of the age of the Universe.

There are many other uses of spectra. Using line spectra from samples collected at the scene, forensic scientists can identify samples of illegal drugs, or match paint chips from vehicles thought to be involved in

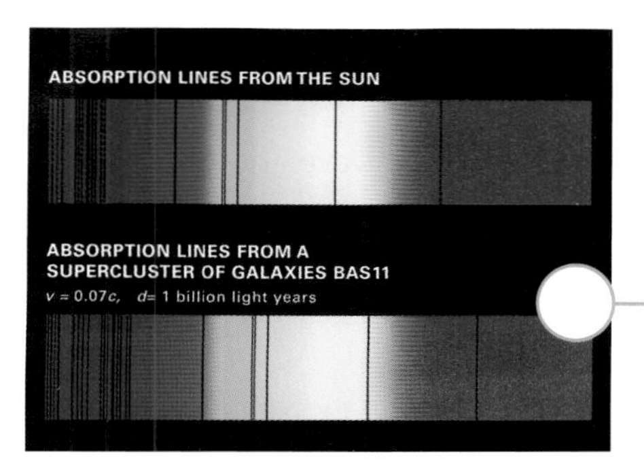

Figure 3.1 The spectrum of our Sun compared with the spectrum from BAS11, a galaxy about 1 billion light years away.

hit-and-run road accidents. One lucky lottery winner was able to collect her winnings after spectral analysis proved her ticket was genuine; a computer error had originally suggested it was fake. In this topic you will learn how line spectra are caused, and the role of photons and electrons in atoms.

Photons

We often see electromagnetic radiation behaving as a wave. For example, waves change wavelength and speed at the boundary between different materials (refraction) and spread through gaps, or around obstacles (diffraction). When electromagnetic waves in the same region overlap, their amplitudes add (superposition). Waves of similar amplitude and wavelength, in the same region, produce interference patterns, which can be detected (see Chapter 6).

Other observations can only be explained by thinking of electromagnetic radiation as a stream of packets (or quanta) of energy called photons. A photon has no mass or charge and is described by its energy, wavelength or frequency. For example, an X-ray photon can have a wavelength of 1 nm and frequency of 3×10^{17} Hz, and a photon of yellow light has a wavelength of 600 nm and frequency of 5×10^{14} Hz. The energy carried by a photon is discussed below.

A **quantum of energy** is a small packet of energy. The word quantum means discrete or separate.

A **photon** is the name given to a discrete packet (quantum) of electromagnetic energy.

Energy of photons

The energy of a photon is proportional to its frequency. A photon of light carries less energy than an X-ray photon because the photon of light has a lower frequency (this is covered further in Chapter 4).

The energy carried by each photon is calculated using:

$$E = hf$$

where E is the energy in joules

h is Planck's constant

f is the frequency of the photon in hertz.

Planck's constant is 6.63×10^{-34} J s.

EXAMPLE

Calculate the energy carried by a photon that has a frequency of 5.60×10^{13} Hz.

Answer

$E = hf$

$= 6.63 \times 10^{-34} \text{ J s} \times 5.60 \times 10^{13} \text{ s}^{-1}$

$= 3.71 \times 10^{-20} \text{ J}$

Since frequency × wavelength equals wave speed, you can rewrite the equation above as $E = \frac{hc}{\lambda}$.

TIP
When working with very large and very small numbers, use the brackets on your calculator wisely. Use the powers of ten to work out the order of magnitude of the answer you should expect.

EXAMPLE

Calculate the energy carried by a photon that has a wavelength of 630 nm.

Answer

$$E = \frac{hc}{\lambda}$$
$$= \frac{6.63 \times 10^{-34}\,\text{J s} \times 3.0 \times 10^{8}\,\text{m s}^{-1}}{630 \times 10^{-9}\,\text{m}}$$
$$= 3.16 \times 10^{-19}\,\text{J}$$

Intensity

The intensity of electromagnetic radiation is the energy transferred per unit time per unit area. The intensity depends on the energy carried by photons, the number of photons transferred each second and the area on which they are incident. The intensity increases:

- when the light source is made more powerful, so more photons are transferred per second; for example, the intensity of light from a 100 W bulb is greater than the intensity from a 10 W bulb if all other factors remain the same
- when each photon transfers more energy; for example, a beam of ultraviolet photons is more powerful than a beam transferring the same number of infrared photons per second
- when the light is incident on a smaller area; for example, when you move closer to a light source, more light energy enters your eye each second, and you sense a greater intensity of light. Intensity follows an inverse square law, so the intensity of light measured from a light source quadruples if you half the distance away from the same light source.

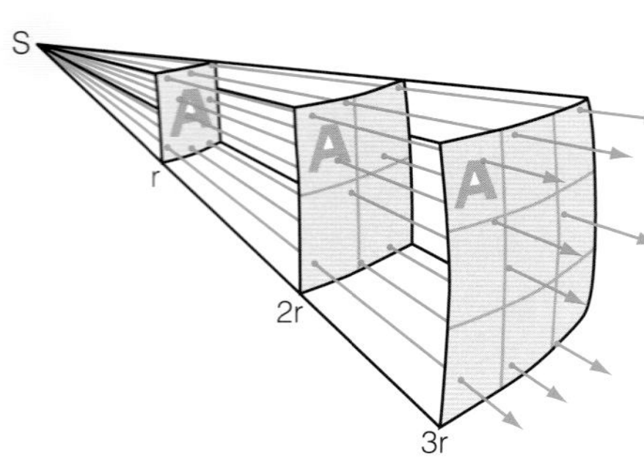

Figure 3.2 Doubling the distance from the light source spreads the energy over four times the area.

Inverse square law applies when a quantity, such as intensity, is inversely proportional to the square of the distance from the source.

EXAMPLE

An 11 W light bulb emits light of average wavelength 550 nm. The bulb is 20% efficient.

Calculate the energy carried by one photon, and hence the number of visible photons emitted each second.

Answer

$$\text{Energy per photon} = \frac{hc}{\lambda}$$
$$= \frac{6.63 \times 10^{-34}\,\text{J s} \times 3 \times 10^{8}\,\text{m s}^{-1}}{550 \times 10^{-9}\,\text{m}}$$
$$= 3.6 \times 10^{-19}\,\text{J}$$

The bulb is 20% efficient.

Power output = power input × efficiency = 11 W × 0.2 = 2.2 W

Power output = number of photons emitted per second × energy per photon

$$\text{Number of photons per second} = \frac{2.2\,\text{W}}{3.6 \times 10^{-19}\,\text{J}}$$
$$= 6 \times 10^{18} \text{ photons per second.}$$

The electron volt

Electrons inside atoms can absorb photons and gain energy. They may gain enough energy to move further from the nucleus into a higher energy level or leave the atom altogether.

An **electron volt (eV)** is a unit of energy equal to 1.6×10^{-19} J. It is the energy gained by an electron when it is accelerated through a potential difference of 1 volt.

The energy carried by a photon and the energy gained by electrons are so small that we use a different unit of energy for these changes. The unit of energy is called the **electron volt**, eV. An electron volt is the energy needed to move an electron through a potential difference of one volt. When charge flows around an electrical circuit, 1 joule of work is done moving 1 coulomb of charge through a potential difference of 1 volt. In the same way, work is done moving electrons in an electric field.

The work done in electron volts is calculated using:

$$W = VQ$$

where

W is the energy transferred in electron volts

V is the potential difference in volts

Q is the electron charge, 1.6×10^{-19} C.

Since the charge on an electron is 1.6×10^{-19} C, the energy transferred per electron volt is 1.6×10^{-19} J.

- To convert joules to electron volts, divide the energy in joules by 1.6×10^{-19} J/eV.
- To convert electron volts to joules, multiply the energy in electron volts by 1.6×10^{-19} J/eV.

EXAMPLE

Calculate the energy of an electron when it moves through a potential difference of 6 V in electron volts and in joules.

Answer

$W = VQ$

$= 6 \times e$

$= 6$ eV

One electron volt equals 1.6×10^{-19} J so the energy in joules is:

$= 6 \times 1.6 \times 10^{-19}$

$= 9.6 \times 10^{-19}$ J

TIP

Check the maths chapter (Chapter 15) if you are not confident working with very small or very large numbers.

TEST YOURSELF

1 State the formula that links the energy of a photon with its frequency. Calculate the energy carried by these photons in joules. Planck's constant is 6.63×10^{-34} J s.

a) an infrared photon with a frequency of 2×10^{13} Hz

b) a visible light photon with a frequency 6×10^{14} Hz

c) an ultraviolet photon with a frequency 9×10^{15} Hz

2 State how to convert joules into electron volts. Convert your answers for question 1 into electron volts.
3 Calculate the frequency of a photon with the energies below. Remember to convert electron volts to joules.
   a) 5.0×10^{-19} J
   b) 2.5 eV
   c) 1975 eV
4 State the formula which relates the energy of a photon to its wavelength. Calculate the wavelength of a photon with these energies.
   a) 8.0×10^{-19} J
   b) 16 eV
   c) 254 eV
5 **a)** Explain what is meant by an electron volt, and how to convert between electron volts and joules.
   b) Calculate the energy needed to move these charged particles through a potential difference of 600 V:
      i) an electron
      ii) a helium nucleus (this contains two protons and two neutrons).

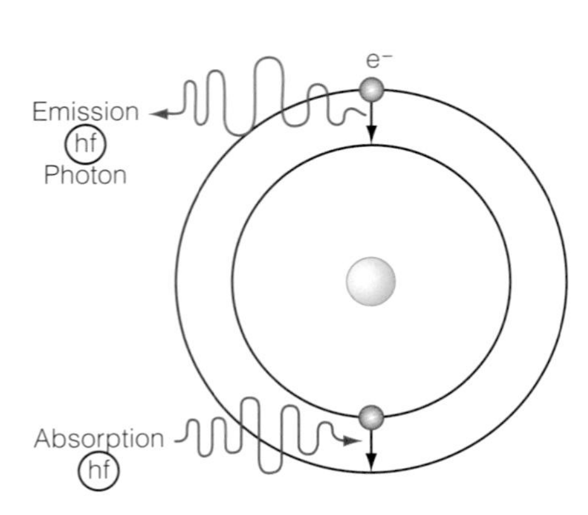

Figure 3.3 Atoms emit photons when electrons move from a higher energy level to a lower energy level. They move from a lower energy level to a higher energy level when they absorb photons.

Absorbing and emitting photons

An electron in an atom gains and loses energy as it moves within the atom. The electron has a combination of kinetic energy and electrostatic potential energy. Because the electron has a negative charge and the nucleus has a positive charge, the electron is attracted to the nucleus and work must be done to move the electron away from the nucleus. This is why the electron has less energy close to the nucleus.

The electron moves further from the nucleus if it gains the right amount of energy by absorbing a photon. We say the electron moves to a higher **energy level**.

If the electron drops from a higher energy level to a lower energy level, it loses its surplus energy by emitting a photon and moves closer to the nucleus.

Electrons can only occupy a small number of separate **energy levels** within an atom. Electrons in an atom move between different energy states or levels when they absorb or emit a photon.

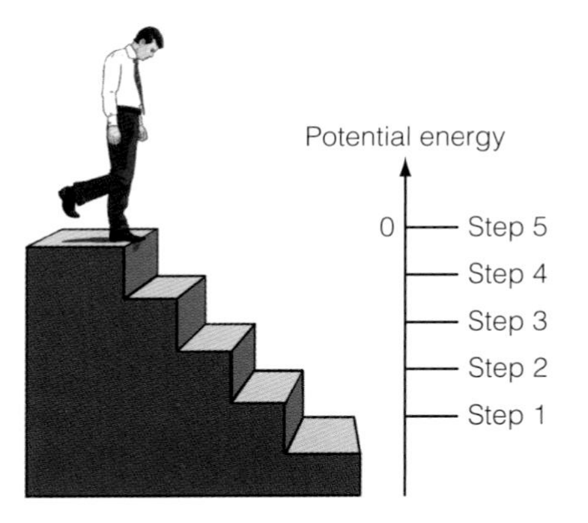

Figure 3.4 There are fixed values of allowed energy when you climb stairs, just as electrons in an atom have fixed values of allowed energy.

Quantised energy levels

An electron can only absorb a specific amount of energy because the possible, or allowed, energies for electrons in an atom are not continuous. Only certain, fixed separate energy levels are allowed. These energy levels are called quantised energy levels because they have fixed energy values.

You come across this idea when you use stairs. You cannot climb part of a step, but must gain or lose gravitational potential energy in precise amounts that match the energy difference between each stair. You can gain or lose energy in larger amounts to move between stairs further apart, but you are still only allowed fixed values of energy. However, while stairs all have the same spacing, the arrangements of energy levels in atoms are more complicated.

Excitation and ionisation

When electrons in an atom are in their lowest energy state, they are in the **ground state**.

When an electron in an atom moves to a higher energy level (above the ground state) after it has absorbed energy, it is in an excited state. **Excitation** has occurred.

If all electrons in an atom are in the lowest energy state, we say the atom is in its ground state.

If an electron, or electrons, in an atom have absorbed energy and moved to higher energy states, we say the atom is excited. Excitation occurs when electrons absorb exactly the right amount of energy to move to higher energy levels. This occurs either by:

- absorbing a photon with the exactly the right amount of energy to move between two levels. A photon is not absorbed if its energy is different from the amount needed for the electron to move between two levels
- absorbing exactly the right amount of energy to move between two levels after colliding with a free electron that has energy equal to or greater than the energy required. The energy gained by the electron in the atom equals the energy lost by the colliding electron. The free electron's kinetic energy after the collision is equal to its kinetic energy before the collision minus the energy transferred to the excited electron in the atom.

Ionisation

Ionisation occurs when an atom gains or loses an electron and becomes charged. It has been ionised.

Ionisation energy is the energy required to remove an electron from its ground state to infinity, i.e. to become detached from the atom.

If an electron in an atom absorbs enough energy to escape the atom completely, we say the atom is ionised. Ionisation occurs when an atom gains or loses an electron and becomes a charged particle called an ion. The ionisation energy is the minimum energy needed to remove the electron from the atom completely.

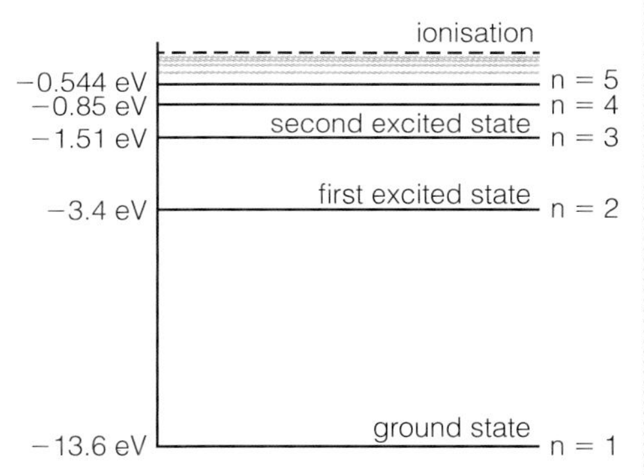

Figure 3.5 When an electron absorbs a photon, it moves into higher excited states or escapes completely (ionisation). This diagram shows a number of possible excited states for a hydrogen atom.

TEST YOURSELF

6 Describe the processes by which electrons gain and lose energy in the atom.

7 Explain the difference between the ground state and an excited state in an atom.

8 Explain the difference between excitation and ionisation.

9 This question refers to the energy level diagram in Figure 3.5. A free electron collides with an electron in a hydrogen atom, which is originally in the ground state, $n = 1$. The ground state electron is moved to the exited state, $n = 2$.

a) How much energy has been transferred to the excited electron? Give your answer in electron volts, and in joules.

The electron now falls to the ground state by emitting a photon.

b) Calculate the wavelength of this photon.

ACTIVITY

Calculating the Planck constant using light-emitting diodes

A student has been asked to measure Planck's constant by investigating the equation:

$$E = \frac{hc}{\lambda}.$$

She sets up the series circuit shown in Figure 3.6 using a variable power supply, light-emitting diode (LED) and a 300 Ω resistor. The potential difference across the LED is measured using a voltmeter.

She increases the potential difference across the LED until the LED is just lit; this potential difference

is measured by the voltmeter. She changes the LED and repeats the experiment using different colours of LEDs; her results are shown in Table 3.1, which compares the wavelength of the light emitted, λ, with the operating p.d., V.

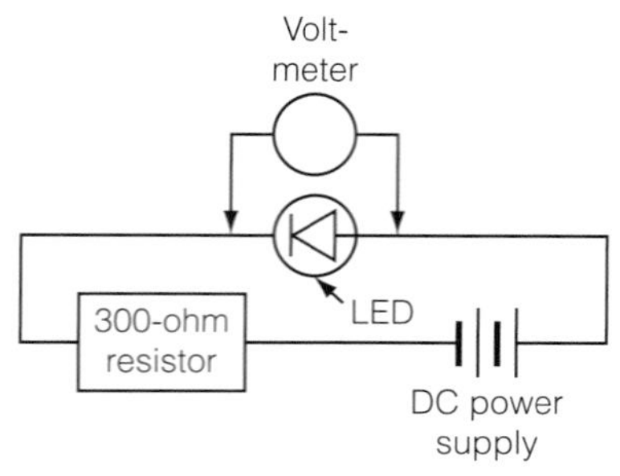

Figure 3.6 Circuit to measure Planck's constant.

The principle behind the operation of an LED is illustrated in Figure 3.7.

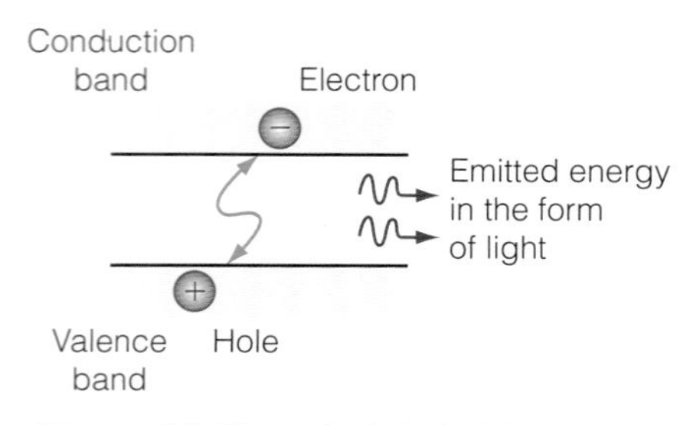

Figure 3.7 The principle behind the operation of an LED.

A semiconductor allows electrons to occupy different energy levels. Energy is needed for electrons to move from the valence band to the conduction band. This energy is supplied by applying a potential difference to the LED. When an electron moves to the conduction band, it leaves a positively charged 'hole' in the valence band. When an electron in the conduction band drops down to the valence band and recombines with a hole, the electron releases surplus energy as light. The energy of these photons equals the difference in energy between the valence band and the conduction band.

Table 3.1

Colour of LED	Wavelength/nm	p.d. when LED is just lit /V
Red	624	1.99
Orange	602	2.06
Yellow	590	2.11
Green	525	2.36
Blue	470	2.64

1 Explain why the blue LED needs a higher p.d. to light than the red LED.

2 Using the data in Table 3.1, plot a graph of the operating p.d., V, across the diodes against $\frac{1}{\lambda}$. Use the gradient of the graph to calculate the Planck constant.

Line spectra and continuous spectra

A **diffraction grating** is a piece of transparent material ruled with very closely spaced lines, used to see the diffraction of light.

A **continuous spectrum** is a spectrum where all frequencies of radiation or colours of light are possible.

A **line spectrum** is a spectrum of discrete coloured lines of light.

Light passing through a fine gap is diffracted, which means that it spreads out after passing through the gap. A diffraction grating is made from a transparent material with many gaps between very closely ruled lines, or ridges. Light passing between each gap in the grating spreads and interferes with light spreading through neighbouring lines. This process splits the light into a spectrum of the colours it contains. The pattern of the coloured light carries information about the light source. You will meet a more detailed explanation of the diffraction grating in Chapter 6.

Sunlight and light from a filament bulb (or incandescent bulb) give a continuous spread of colours that merge into each other. This is a continuous spectrum. A bulb has a solid filament that is heated. Energy levels in a solid overlap, so all energy changes for the electrons are allowed. This means the electrons can emit photons with any energy, producing a continuous spectrum when a solid is heated.

Light from a fluorescent tube produces a spectrum of coloured lines on a dark background. This pattern of lines is called a line spectrum. The pattern of lines is characteristic for certain elements. All fluorescent bulbs have a similar line spectrum because electrons in the mercury vapour inside them have the same excited states regardless of the shape of the bulb.

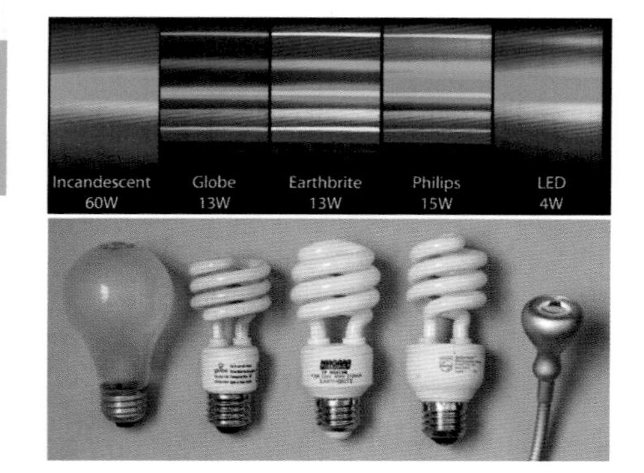

Figure 3.8 Comparing the emission spectra of different light sources.

Emission spectra

An **emission spectrum** is a bright spectrum seen when photons are emitted by atoms.

An emission spectrum is seen when electrons in atoms fall from higher energy levels to lower energy levels, releasing photons. In a gas, mercury vapour for example, electrons can only occupy a small number of discrete energy levels. When a gas is heated, electrons absorb energy and reach higher energy levels. However, electrons stay in these higher states (or excited states) for a short time, before falling back to a lower energy level. When the electron falls to its lower state, it emits a photon whose energy is equal to the difference in energy between the two energy levels. Since only a small number of possible energy levels exist, there is only a small number of possible transitions between them. The photons that are emitted have specific energies, and therefore specific wavelengths and colours. It is these specific colours that we see as lines when the light is diffracted through a diffraction grating.

When you heat salts in a Bunsen flame, sometimes you see different colours. For example, compounds with copper in them emit green light, and sodium compounds a bright yellow light. These colours are determined by electrons falling from one energy level to another, emitting the particular colour of light specific to that element.

Absorption spectra

An **absorption spectrum** is a spectrum of dark lines seen on a coloured background produced when a gas absorbs photons.

An absorption spectrum can be seen when light shines through a gas, and electrons in the atoms absorb photons corresponding to the possible energy transitions. All other photons pass through, as they cannot be absorbed. The dark lines of the spectrum correspond to the wavelengths of the possible energy transitions for the electrons of the gas atoms. The electrons become excited, moving from lower energy levels to higher energy levels. In fact, the electrons then fall back to their original energy state, releasing photons as they do so, but the spectrum still appears to have black lines as these photons are emitted in all directions.

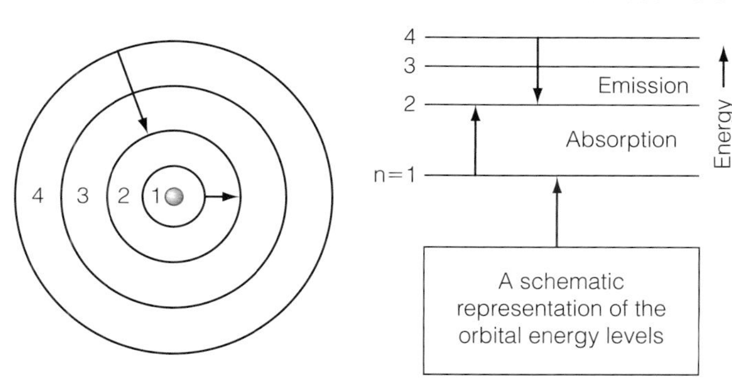

Figure 3.9 An energy level diagram.

Fraunhofer lines: an absorption spectrum

Joseph von Fraunhofer realised that the continuous spectrum from the Sun was overlaid with about 570 dark absorption lines, now known as Fraunhofer lines.

Processes within the Sun create a continuous spectrum of radiation that passes through cooler gases in the Sun's outer layers. Electrons, which are bound to atoms in the cooler gases, absorb photons at specific frequencies and then re-emit the photons in all directions when returning to their ground state. This is why we see a dark line surrounded by the rest of the continuous spectrum. The continuous spectrum is caused by radiation emitted by free electrons.

TEST YOURSELF

10 A beam of sunlight passes through the Sun's atmosphere. Explain why dark lines are seen in the continuous spectrum from the sunlight. (These lines are called Fraunhofer lines.)

11 Describe the difference between an emission spectrum and an absorption spectrum.

12 Explain whether an emission spectrum is caused by atoms in their ground state or by atoms in excited states.

The energy of spectral lines

When an electron moves to a lower energy level, it emits a photon. The energy lost by the electron equals the energy of the photon. This energy is calculated as:

$$E_2 - E_1 = hf$$

where E_1 and E_2 are the energies of energy levels 1 and 2 (in J), h is the Planck constant (in J s) and f is the frequency of radiation (in Hz).

Many spectral line diagrams show the energy in electron volts. One electron volt is 1.6×10^{-19} J.

Figure 3.10 shows energy levels in a hydrogen atom.

- The ground state is labelled as $n = 1$, and the first excited energy level is $n = 2$ and so on.
- The ionisation energy is 13.6 eV (moving the electron from its ground state up to $n = \infty$).
- The arrows indicate the electron moves to a lower energy level, emitting photons. An emission spectrum would be seen.
- The energy is always negative as we define the zero point of energy is when the electron is at infinity. This energy gets lower as the electron gets closer to the nucleus.
- The energy of the green line is the difference in energy between $n = 4$ and $n = 2$, that is –0.85 eV – (–3.41 eV), or 2.56 eV.
- 2.56 eV is equivalent to 4.1×10^{-19} J.

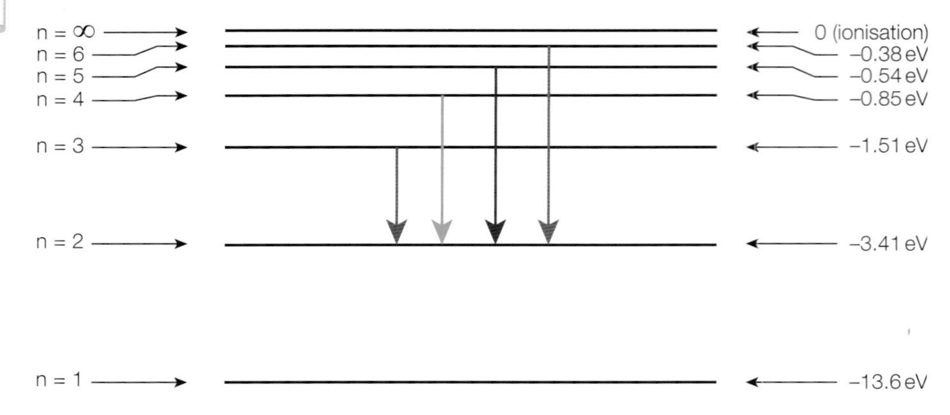

Figure 3.10 Spectral lines in a hydrogen atom.

EXAMPLE

Use Figure 3.10 to calculate the frequency of a photon released when an electron moves from energy level $n = 6$ to $n = 2$ in a hydrogen atom.

Answer

The energy of the photon is the difference in energy between level 6 and level 2, which is –0.38 eV – (–3.41 eV) = 3.03 eV.

To convert energy in electron volts to energy in joules, multiply by the electron charge, e, 1.6×10^{-19} C. This gives an energy of 4.85×10^{-19} J.

To find f, rearrange the equation $E = hf$

$$f = \frac{E}{h}$$

$$= \frac{4.85 \times 10^{-19}\ \text{J}}{6.63 \times 10^{-34}\ \text{J s}}$$

$$= 7.3 \times 10^{14}\ \text{Hz}$$

TEST YOURSELF

Use Figure 3.10 to answer these questions.

13 What is the ionisation energy of hydrogen in electron volts?

14 How would the diagram be different if photons were absorbed, not emitted?

15 Which energy levels does an electron move between if it emits a photon with each of these energies?
a) 0.31 eV **b)** 1.9 eV **c)** 1.13 eV

16 Calculate the frequency of a photon emitted when an electrons falls from:
a) energy level $n = 5$ to $n = 2$
b) energy level $n = 6$ to $n = 5$.

17 Explain which of these photon energies can be absorbed by the electron: 10.1 eV, 0.16 eV, 1.20 eV and 0.47 eV.

The fluorescent tube

Fluorescence is when a substance absorbs short wavelength electromagnetic radiation and emits it as longer wavelength radiation.

A **fluorescent tube** is a type of light bulb that gives out light when its inner coating fluoresces.

Thermionic emission happens when free electrons are released from a heated filament.

If you are at school or college, the room you are in probably uses fluorescent lights. Fluorescence in these lights occurs when electrons absorb photons of ultraviolet radiation, and move to a higher energy level. When the excited electrons fall back to the lower energy level, energy is released as visible light.

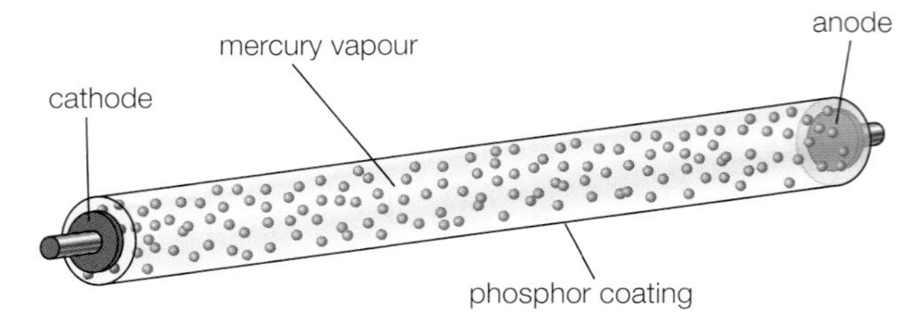

Figure 3.11 The main components of a fluorescent tube.

A fluorescent tube is a glass tube filled with mercury vapour and coated inside with fluorescent materials called phosphors. When the light is switched on, the cathode is heated causing thermionic emission. Thermionic emission occurs when a heated cathode releases free electrons from its surface. The free electrons have a range of energies. A potential difference of 500 V, applied across ends of the glass tube, accelerates the electrons from the cathode to the anode through the mercury vapour. If the free electrons collide with mercury atoms inelastically, some energy may be transferred from the free electrons to the mercury atoms. These atoms may be ionised or excited, provided the free electrons transfer enough kinetic energy. High-energy electrons cause ionisation, and lower energy electrons cause excitation. As the mercury atoms in the vapour become ionised (lose electrons), a mixture of ions and free electrons is created; this is called a plasma. When the electrons in the excited mercury atoms return to their ground state, they release photons of ultraviolet radiation. These photons strike the phospors in the coating and are absorbed. The energy is re-emitted as visible light, and some energy is transferred as heat.

Plasma is a mixture of ions and electrons in a gas.

TEST YOURSELF

18 Describe the purpose of these parts of a fluorescent tube: electrode, mercury vapour, phospor coating.

19 Explain why a person viewing a fluorescent bulb though a diffraction grating observes specific peaks in the spectrum.

20 Explain why the spectrum from a fluorescent lamp and an incandescent bulb are different.

21 The diagram shows some allowed energy levels for mercury. Use the diagram to calculate:

a) the energy of photons emitted for each of the three electron transitions from excited states to the ground state

b) the wavelength of the emitted photons.

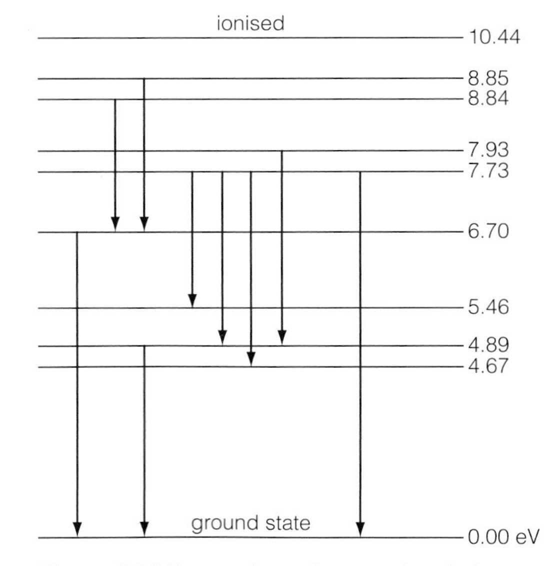

Figure 3.12 Some allowed energy levels for mercury.

Practice questions

1 Figure 3.13 shows some energy levels, in eV, of an atom.

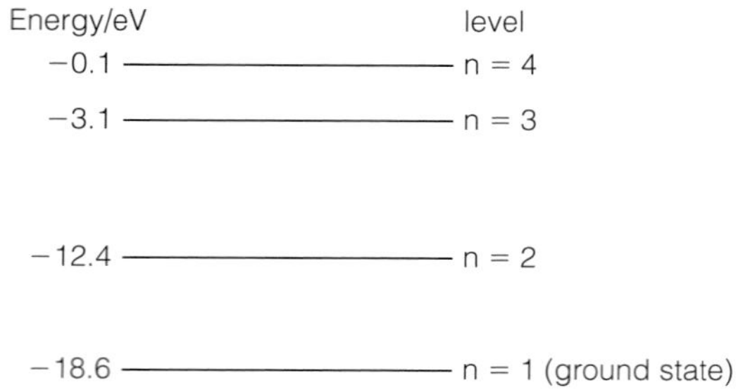

Energy/eV	level
−0.1	n = 4
−3.1	n = 3
−12.4	n = 2
−18.6	n = 1 (ground state)

Figure 3.13

An emitted photon has an energy of 9.92×10^{-19} J. Which transition was responsible for this photon?

A n = 4 to n = 1

B n = 3 to n = 1

C n = 2 to n = 1

D n = 3 to n = 2

2 An atom emits light of wavelength 2.0×10^{-7} m. What is the energy, in J, of a photon with this wavelength?

A 5.0×10^{-19} J

B 9.9×10^{-19} J

C 1.2×10^{-18} J

D 9.9×10^{-18} J

3 Figure 3.14 shows part of the energy level diagram for a hydrogen atom.

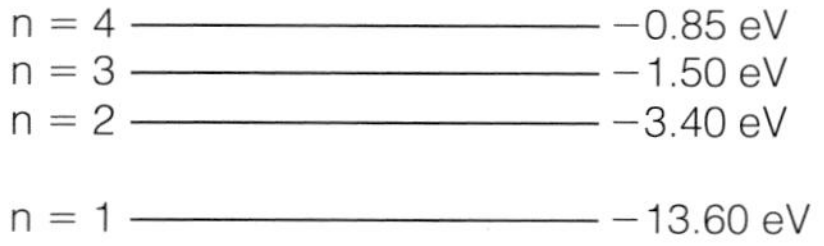

n = 4	−0.85 eV
n = 3	−1.50 eV
n = 2	−3.40 eV
n = 1	−13.60 eV

Figure 3.14

What is the ionisation energy of the atom in J?

A 3.04×10^{-19} J

B 4.08×10^{-19} J

C 2.04×10^{-18} J

D 2.18×10^{-18} J

4 The electron in a hydrogen atom (as shown above) is excited to the n = 4 level. How many different frequencies of photons would this produce in the hydrogen emission spectrum?

A 3

B 4

C 6

D 7

5 Read the following statements.

1 The frequency and wavelength of light are inversely proportional.

2 As the energy of electromagnetic radiation increases, its frequency decreases.

3 An atom can be excited by emitting light energy.

4 An excited atom can return to a lower energy level by emitting light energy.

Which of the statements are true?

A 2 and 3

B 1 and 4

C 1 and 3

D 3 and 4

6 A student is using LEDs in an electronics project. The wavelength of the LEDs are given in the table. What is the energy of the lowest energy photon emitted by the LEDs in the project?

Wavelength / $\times 10^{-9}$ m
530
600
680

A 1.5×10^{-19} J

B 2.2×10^{-19} J

C 2.9×10^{-19} J

D 3.7×10^{-19} J

7 An electron is accelerated through a large potential difference and gains a kinetic energy of 36 keV. What is this energy expressed in joules?

A 2.88×10^{-18} J

B 5.76×10^{-18} J

C 2.88×10^{-15} J

D 5.76×10^{-15} J

8 A photon has an energy of 4.14 keV. What is the frequency of the photon?

A 5.86×10^{13} Hz

B 1.00×10^{15} Hz

C 1.06×10^{18} Hz

D 6.27×10^{33} Hz

9 Which of the following graphs best describes how photon energy varies with the wavelength of the photon?

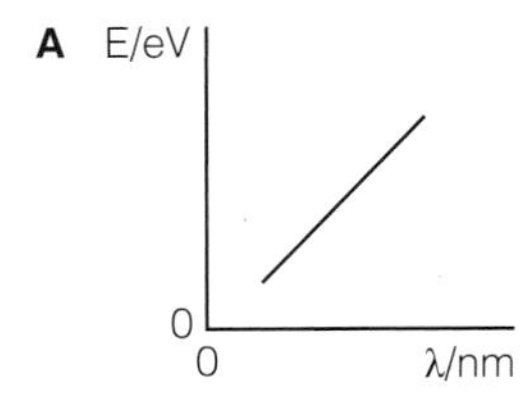

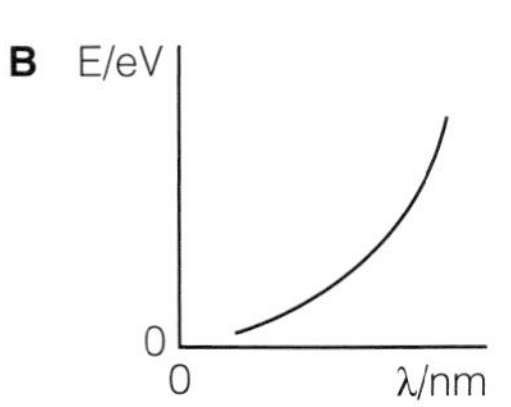

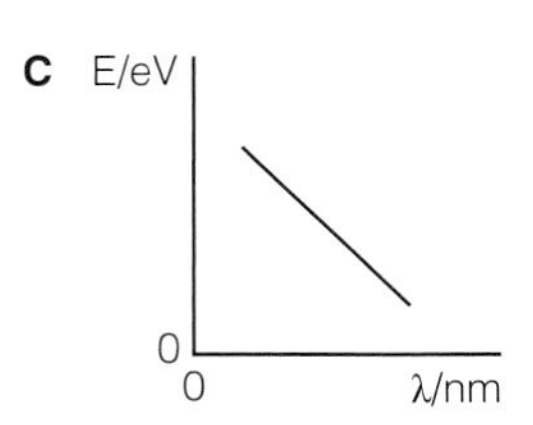

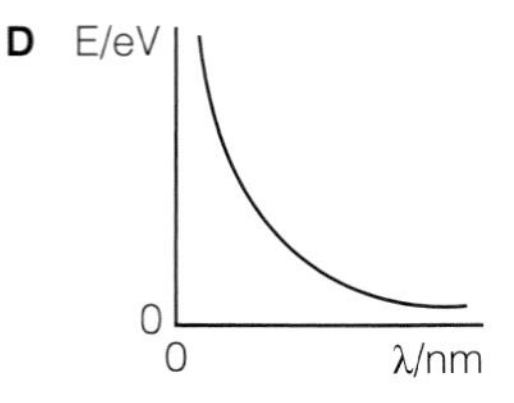

Figure 3.15

10 Figure 3.16 represents the three lowest energy levels of an atom.

Which of the diagrams in Figure 3.17 best represents the emission spectrum that could arise from electron transitions between these energy levels?

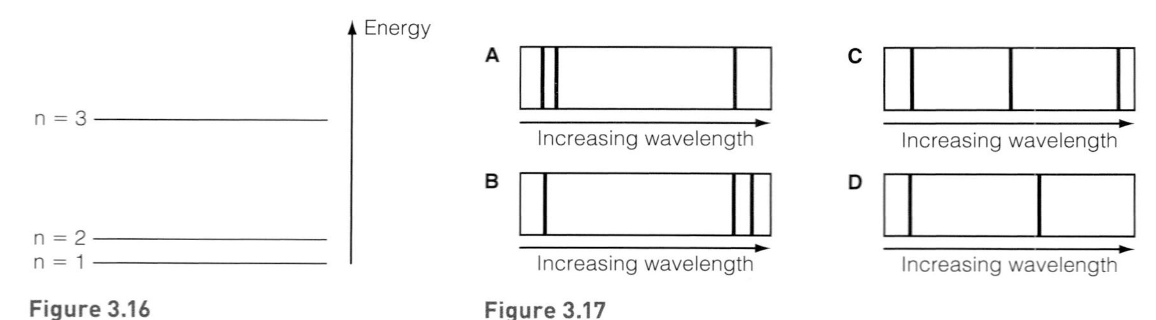

Figure 3.16 **Figure 3.17**

11 Figure 3.18 shows the lowest four energy levels of a hydrogen atom.

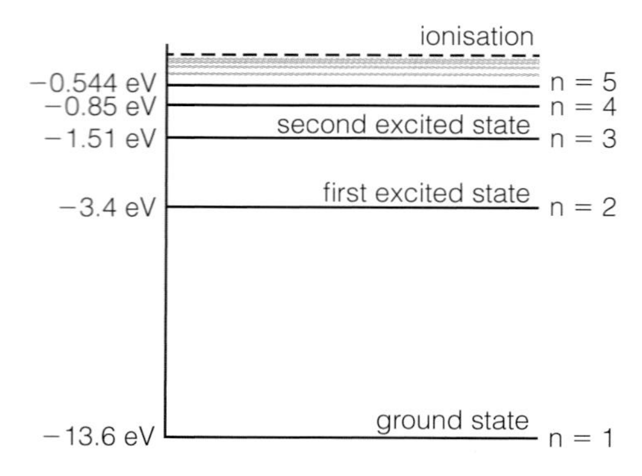

Figure 3.18 Energy levels of a hydrogen atom.

An electron in a hydrogen atom is excited from the ground state to the $n = 2$ energy level. The atom then emits a photon when the electron returns to the ground state.

a) Explain what must happen to the electron for it to move from the ground state to its excited state. *(2)*

b) Calculate the frequency of the photon and state which part of the electromagnetic spectrum it belongs to. *(3)*

12 A fluorescent tube contains mercury vapour at low pressure. When the mercury atoms are excited then fall back to their ground state the atoms emit electromagnetic radiation.

a) What is meant by the ground state of an atom? *(1)*

b) Explain how mercury atoms can become excited in a fluorescent tube. *(3)*

c) What is the purpose of the cathode in a fluorescent tube? *(2)*

d) Explain why only specific peaks in the spectrum are observed. *(2)*

13 Some of the energy levels of a tungsten atom are shown in Figure 3.19.

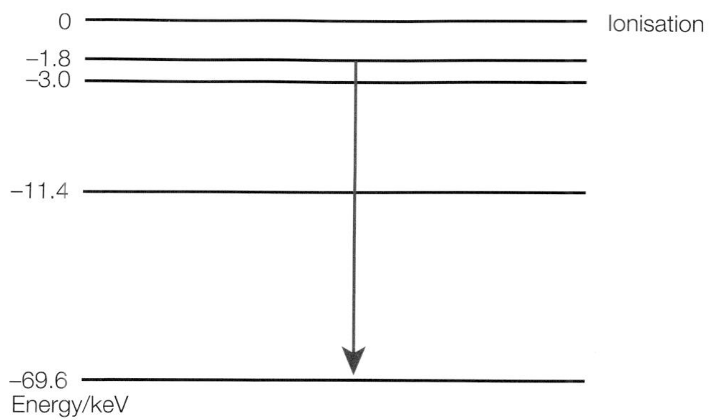

Figure 3.19 Some of the energy levels of a tungsten atom.

a) Calculate the wavelength of a photon emitted as an electron moves from the energy level −1.8 keV to energy level −11.4 keV. *(3)*

b) Calculate the frequency of the photons required to ionise the tungsten atoms. *(3)*

c) State one transition which emits a photon of a shorter wavelength than that emitted in the transition from level $n = 4$ to level $n = 3$. *(2)*

d) i) An electron collided with an atom, exciting the atom from the ground state to the level −11.4 keV. The initial kinetic energy of the incident electron is 3.2×10^{-15} J. Calculate its kinetic energy after the collision. *(2)*

ii) Show whether the incident electron can ionise the atom from its ground state. *(2)*

e) State the ionisation energy of tungsten in joules. Give your answer to an appropriate number of significant figures. *(3)*

14 Photons of more than one frequency may be released when electrons in an atom are excited above the $n = 2$ energy level.

a) Sketch a diagram showing the lowest three energy levels in an atom to explain why photons of more than one frequency can be released. You do not need to include specific energy values. *(3)*

b) Use your diagram to explain how many different frequencies can be produced. *(2)*

15 Fluorescent tubes are filled with low-pressure mercury vapour. Describe the purpose of the main features of a fluorescent tube and use this to explain how visible light is generated.

The quality of your written communication will be assessed in this question. *(6)*

16 Figure 3.20 shows the energy levels in a sodium atom. The arrows show some of the transitions that occur.

a) What is ionisation energy of the sodium atom?

b) The most common transitions are 589.6 nm and 589 nm, giving the strongest spectral lines. What colour do you expect a sodium discharge tube to be?

c) Use the diagram to determine the the wavelength of light emitted for these transitions: 4s to 3p; 3d to 3p; 3p to 3s.

d) Which transitions gives rise to a spectral line of wavelength 620 nm?

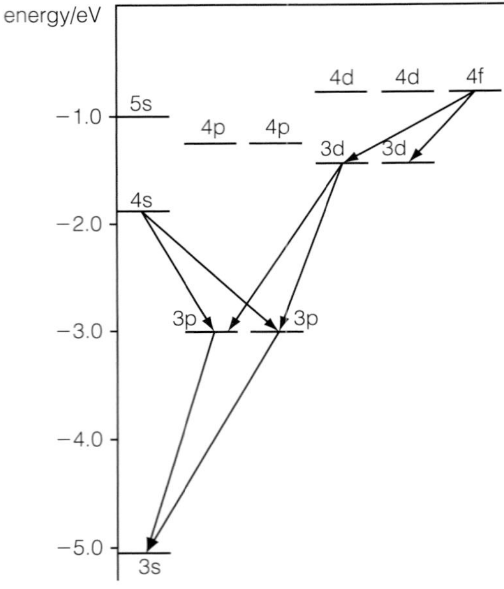

Figure 3.20 The energy levels in a sodium atom.

Stretch and challenge

17 X-ray machines produce photons with wavelengths between 0.01 nm and 10 nm by colliding electrons with a tungsten metal target. X-rays with energies above 5 keV are categorised as hard x-rays. These x-rays pass through most tissues. Soft x- rays usually have energies in the range 100 eV to 5 keV, and are totally absorbed in the body. Mammograms are routine x-rays images used to detect tumours in the breast before symptoms are observed.

a) Explain whether hard x-rays or soft x-rays are most useful for medical diagnosis, giving two reasons

18 A successful mammogram uses the lowest x-ray dose that can pass through tissues, while still producing clear contrast between tumours and surrounding tissue. Low energy x-rays increase contrast but are less penetrating; higher energy x-rays reduce contrast but pass well through tissues. Many mammograms uses energies of about 24 keV

a) Calculate the wavelength of a typical mammogram x-ray

b) Explain whether bone x-rays should have a higher or lower wavelength than for mammograms.

4

Particles of light

PRIOR KNOWLEDGE

- Electromagnetic waves are transverse waves with a range of wavelengths. In order of increasing frequency, members of the electromagnetic spectrum are radio waves, microwaves, infrared, visible, ultraviolet, X-rays, gamma rays.
- When atoms are ionised, they lose or gain electrons. Electromagnetic waves can ionise atoms when an electron absorbs a photon. Electromagnetic waves with a higher frequency and shorter wavelength are more ionising.
- Diffraction occurs when waves spread through a gap or around an obstacle. The wavefront is more curved when the wavelength is about the same size as the gap, and diffraction is greatest.
- Electrons are negatively charged particles. The electron's mass is about 2000 times less than that of a proton or neutron.
- Electrical current is a flow of charged particles.
- In classical mechanics, momentum is calculated using: momentum = mass × velocity.
- Velocity is a vector, so the direction of motion must be specified.
- Kinetic energy = $\frac{1}{2}$mass × velocity^2.
- Energy is conserved. It cannot be created or destroyed, but must be accounted for at all stages of a process.
- 1 electron volt = 1.6×10^{-19} J
- The energy of a photon is given by: $E = hf$, where h is Planck's constant and f the frequency of the electromagnetic radiation.

TEST YOURSELF ON PRIOR KNOWLEDGE

1 List the members of the electromagnetic spectrum in order of increasing frequency, giving one use for each member.

2 Sketch the shape of wavefronts diffracting through a gap when the gap is:
   a) much larger than the wavelength
   b) approximately equal in size to one wavelength.

3 a) Compare the kinetic energy and momentum of a proton and an electron travelling at the same velocity.
   b) Compare the kinetic energy of a proton and an electron with the same momentum.

4 Describe the change in momentum of an atom when it bounces off a surface and travels in the opposite direction at the same speed.

5 Describe the energy transfers that occur in a torch circuit when it is turned on.

Have you ever wondered how photovoltaic (solar) cells work? Light-sensitive material in photovoltaic cells absorbs photons from sunlight. When electrons in the material absorb this energy, the electrons are released and can generate electrical power. Other devices, such as CD players and bar code scanners, work in a similar way using light sensitive cells.

Charge coupled devices (CCD) are digital picture sensors that also use photodiodes. They are found in digital cameras, astronomical telescopes and digital X-ray machines. A CCD device is a grid of thousands of photodiodes. Each photodiode absorbs photons from the light shining on it and emits electrons. More electrons are released if brighter light shines on the photodiode. The signals from all the photodiodes combine to produce signals that change in real-time.

All of these applications make use of quantum effects, which you will learn about in this chapter.

The photoelectric effect

The photoelectric effect was first observed in the 1880s. This was a highly significant discovery because it caused scientists to think about light in an entirely new way. At that time light was thought to be a wave, and wave theory proved to be sufficient to explain many properties of light, such as reflection, refraction and diffraction, which you will meet in Chapter 5.

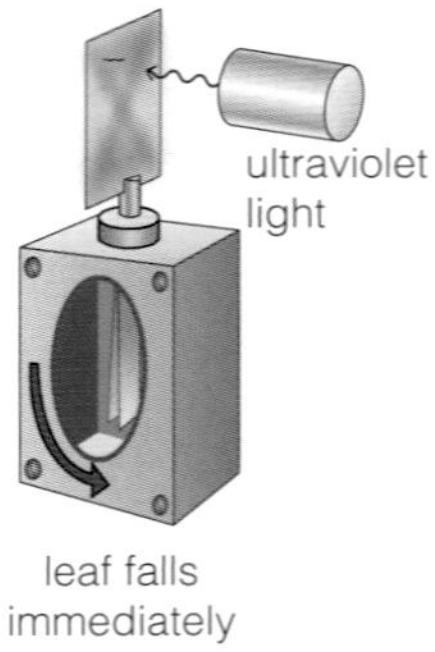

ultraviolet light

leaf falls immediately

Figure 4.1 Demonstrating the photoelectric effect of ultraviolet light on a negatively charged gold leaf electroscope.

The photoelectric effect can be demonstrated using the apparatus shown in Figure 4.1. A freshly cleaned sheet of zinc is placed on top of a negatively charged gold leaf electroscope. When a weak source of ultraviolet light is directed towards the zinc sheet, the electroscope discharges immediately. However, when the gold leaf electroscope is charged positively, as shown in Figure 4.2, the electroscope remains charged. This is evidence that the ultraviolet light knocks electrons out of the metal surface. A zinc plate with a negative charge can release electrons but this is not possible when the plate is positively charged. Electrons which are emitted from the surface of a metal are called **photoelectrons**.

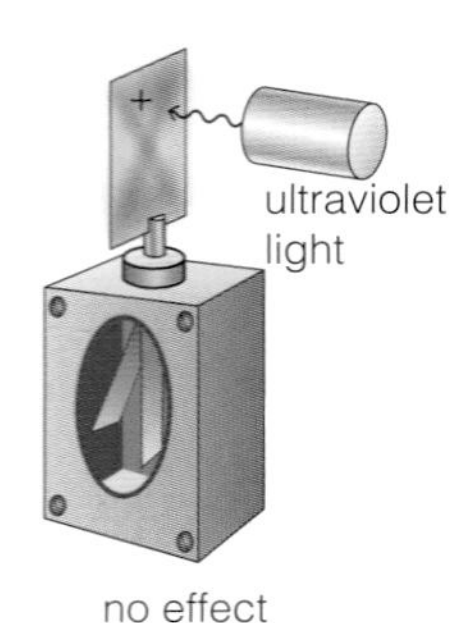

Figure 4.2 Demonstrating the photoelectric effect of ultraviolet light on a positively charged gold leaf electroscope.

Figure 4.3 shows that a very bright white light shining on the zinc sheet does not discharge the electroscope, nor does a very intense red laser. This result surprised the physicists of the 1880s. They expected the electroscope to discharge when a strong light source was incident on it. At that time, physicists were familiar with the thermionic emission of electrons from a heated metal wire. The electrons in such a wire gain enough energy for them to escape from the attractive forces of the atoms in the wire. This is a similar process to evaporation. When water molecules have enough kinetic energy, they are able to escape from the body of the fluid.

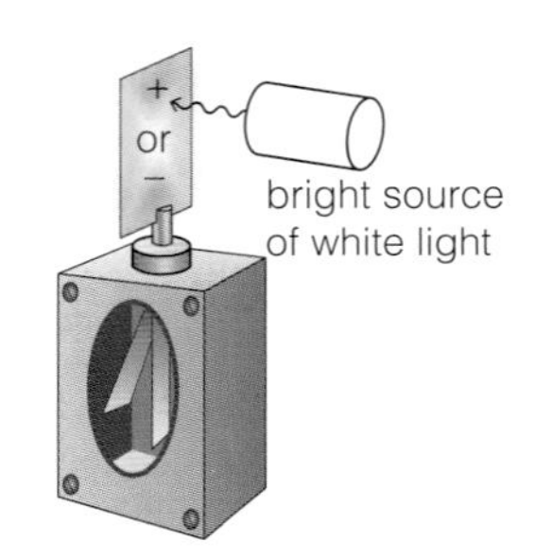

Figure 4.3 Demonstrating the photoelectric effect–bright source of white light.

Using wave theory, physicists had argued that shining a bright light on the zinc sheet would be rather like heating it. After a small delay, the light falling on the zinc would heat it, and some electrons would be emitted from the surface.

The problem was that the bright white light did not discharge the electroscope, while the weak ultraviolet light began to discharge the electroscope immediately.

(Note: if you are conducting an experiment like this yourself this yourself, you will need a 280 nm (UVC) lamp. Do not shine UV light on skin or eyes.)

A **quantum** of light is a small indivisible package of energy. This is the smallest unit of energy for a particular frequency of light. The total energy emitted from a source of light is a multiple of this quantum of energy.

An explanation with quantum theory

In 1900 Max Planck suggested a solution that would solve the problem with his quantum theory. His idea was that light and other types of electromagnetic radiation are emitted in indivisible small packets of energy, which he called quanta (a single packet of energy is called a quantum). You met this idea in Chapter 3, and you will recall that the quanta are also known as photons. Planck argued that the energy of a quantum of light is proportional to its frequency, and is given by the equation:

$$E = hf$$

where E is the energy of the photon in joules, f is the frequency of the light in Hz, and h is the Planck constant, which has a value of 6.6×10^{-34} J s.

This theory provides a solution to the problem of the electroscope. For the electron to be removed from the surface of the metal, it must absorb a certain minimum amount of energy. Planck suggested that when light falls on to the surface of the metal, one photon interacts with one electron, and that the photon's energy is absorbed by the electron. The ultraviolet photons have enough energy to knock an electron out of the metal; so, as soon as the ultraviolet photons hit the zinc sheet in Figure 4.1, electrons are removed from the metal and the electroscope immediately begins to discharge. However, the photons of visible light have a lower frequency and, when they are absorbed by an electron, the electron does not have sufficient energy to escape from the metal surface, so the electroscope does not discharge, as shown in Figure 4.3.

Further photoelectric experiments

The previous section described a simple demonstration of the photoelectric effect using ultraviolet light and an electroscope. There are many more experiments that can be done using different metals and a wide range of frequencies of incident light. It is also possible to measure the kinetic energies of the photoelectrons after they have been emitted from the surface of the metal. Here is a summary of these experimental observations.

Threshold frequency is the minimum frequency of light required to cause the photoelectric effect.

Intensity is rate of energy transfer per unit area.

A **photoelectron** is an electron emitted during the photoelectric effect.

- No photoelectrons are emitted if the frequency of light is below a certain value. This is called the **threshold frequency**
- The threshold frequency varies for different materials. Visible light removes photoelectrons from alkali metals, and calcium and barium. Other metals require ultraviolet radiation, which has a higher frequency than visible light.
- More photoelectrons are emitted as the **intensity** of light increases, but only if the frequency of the light used is above the threshold value.
- **Photoelectrons**, emitted from a particular metal, have a range of energies. Their maximum kinetic energy depends on the frequency of the incident light so long as the frequency of this light is above the threshold frequency.

TEST YOURSELF

1 a) Describe the main features of the photoelectric effect.

b) State two experimental observations of the photoelectric effect that were evidence that light does not behave as a wave.

2 A 2.5 mW laser pointer is powered by one 1.5 V cell delivering a current of 6.0 mA.

a) Calculate the efficiency of the laser pointer.

b) If the wavelength of light produced is 532 nm, calculate the number of photons released per second.

3 Explain why this statement is incorrect: 'The photoelectric effect is only seen using ultraviolet light.'

4 The threshold frequency of copper is 1.1×10^{15} Hz. Describe what is observed if a sheet of copper, which is placed on the top of a negatively charged electroscope, is exposed to light of frequency:

a) 1.0×10^{15} Hz

b) 1.2×10^{15} Hz

5 a) Why was the photoelectric effect significant when it was first observed?

b) The photoelectric effect is only seen with zinc using ultraviolet light, but can be seen when visible light shines on sodium. Compare the energy needed to release photoelectrons from the different metals.

c) Explain how the kinetic energy of photoelectrons released from sodium changes when ultraviolet light is used instead of visible light.

6 Explain whether a photon of ultraviolet radiation carries more or less energy than a photon of gamma radiation.

7 A laser pen emits red light of frequency 4.3×10^{14} Hz. In terms of photons, explain the changes if:

a) a laser pen emitting more intense light is used

b) a laser pen emitting green light (frequency 5.5×10^{14} Hz) with the same intensity is used.

8 Give two examples of quantities that are quantised.

A model to explain the photoelectric effect

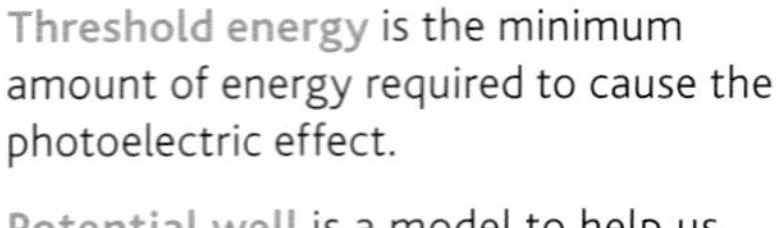

Threshold energy is the minimum amount of energy required to cause the photoelectric effect.

Potential well is a model to help us understand why electrons require energy to remove them from a metal. An electron has its least potential energy in the potential well.

- To explain threshold energy, we can describe electrons in a metal as being trapped in a potential well. This type of well is a well of electrostatic attraction. The 'depth' of the well represents the least amount of energy an electron requires to escape from the metal. For photoelectrons, this amount is the threshold energy. It is a bit like a person trying to jump out of a hole. The person can only jump out if they make one jump that is high enough. They cannot escape from the hole by making several small jumps. In the same way, an electron also needs enough energy to escape from the metal, which is gained when the electron absorbs a photon. If the energy from the photon matches or exceeds the threshold energy then the electron escapes from the potential well and is released as a photoelectron.
- The energy needed to escape from the potential well is different for different metals, so the threshold energy varies for these metals. For example, in alkali metals the potential well is small, so the threshold energy is small and photoelectrons are released using visible light. The potential well also varies within the metal. Electrons at the surface require less energy to escape compared to electrons deeper within the metal, which may be bound more strongly to atoms. The threshold energy is the minimum energy required to remove an electron from the metal.
- Planck's formula for the energy of a photon, $E = hf$, explains the threshold frequency. Below the threshold frequency, photons do not carry enough energy to release a photoelectron so the photoelectric effect is not observed. Electrons cannot be released by absorbing two photons, just as the person cannot escape from a hole with two smaller jumps. Each photon must carry at least the threshold energy in order to release an electron.
- As light intensity increases, the number of photons increases but the energy per photon stays the same. Because there are more photons, more electrons can be released, so long as the frequency of the radiation is above the threshold frequency.
- The photon's energy does work to release the photoelectron and give it kinetic energy. The maximum kinetic energy of a photoelectron is the difference between the photon's energy and the threshold energy. This is why the maximum kinetic energy of photoelectrons increases with frequency.

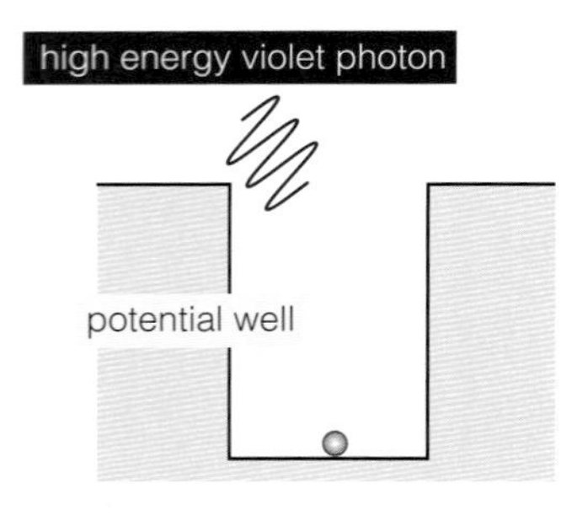

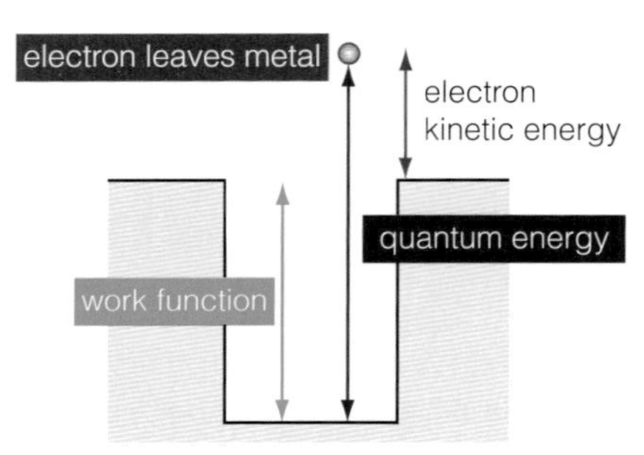

Figure 4.4 An electron can leave a potential energy well if it absorbs enough energy by absorbing a photon.

TEST YOURSELF

9 Describe how threshold energy can be explained using the potential well model for a metal.

10 Use Table 4.1 to calculate the energy carried by a photon of each colour of light in joules and in eV.

Table 4.1 Frequencies and wavelengths for photons of different colours.

Colour	Frequency (10^{14} Hz)	Wavelength (nm)
Yellow	5.187	578
Green	5.490	546
Blue	6.879	436
Violet	7.409	405
Ultraviolet	8.203	365

11 The threshold energy for sodium is 2.29 eV and for selenium is 5.11 eV. Calculate the threshold energy of each metal in joules.

12 Explain why the maximum kinetic energy of photoelectrons is not proportional to the frequency of radiation.

Einstein's photoelectric equation

Energy from a photon is used to release a photoelectron and give the photoelectron kinetic energy. Conservation of energy means that energy from the photon equals the threshold energy plus kinetic energy of the photoelectron.

Work function is the least energy needed to release a photoelectron from a material. This equals the threshold energy.

The work function of a material is the least energy needed to release a photoelectron from a material. The work function has the symbol ϕ.

Applying the conservation of energy when energy, hf, is transferred from the photon to a photoelectron:

$$hf = \phi + \frac{1}{2}mv_{max}^2$$

where h is Planck's constant

f is the frequency of the radiation in hertz

ϕ is the work function of a material in joules

$\frac{1}{2}mv_{max}^2$ is the maximum kinetic energy of photoelectrons, with mass m and maximum velocity v_{max}.

This equation is called Einstein's photoelectric equation. Electrons emitted from the material's surface have the maximum kinetic energy because energy is not used moving to the surface. Electrons from deeper in the material have less kinetic energy as some energy is used moving to the surface.

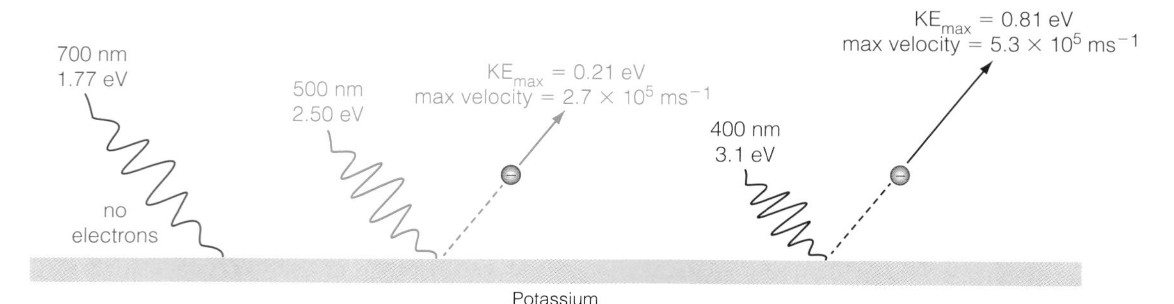

Figure 4.5 Potassium's work function is 2.29 eV. Surplus energy (above the work function) becomes the kinetic energy of emitted electrons.

TIP

Energy (joules) = energy (eV) x charge (C)

For a photon:

- E = hf, or
- E = hc/ λ

A photon of energy 1 eV has a frequency of about 2×10^{14} Hz and a wavelength of about 10^{-6} m.

EXAMPLE

a) Calculate the minimum frequency of radiation that causes the photoelectric effect for aluminium. The work function for aluminium is 4.08 electron volts.

b) Calculate the maximum kinetic energy (in electron volts) of a photoelectron if light of wavelength 240 nm shines on the surface.

Answer

a) Convert this energy in electron volts to joules:

$$4.08\,\text{eV} \times 1.6 \times 10^{-19}\,\text{C} = 6.53 \times 10^{-19}\,\text{J}$$

Use $E = hf$ to calculate the frequency that matches this energy:

$$f = \frac{E}{h} = \frac{6.53 \times 10^{-19}\,\text{J}}{6.63 \times 10^{-34}\,\text{J s}}$$

$$f = 9.8 \times 10^{14}\,\text{Hz}$$

b) Calculate the energy of a photon if its wavelength is 240 nm:

$$E = hf = \frac{hc}{\lambda}$$

$$= \frac{6.63 \times 10^{-34}\,\text{J s} \times 3 \times 10^{8}\,\text{m s}^{-1}}{240 \times 10^{-9}\,\text{m}}$$

$$= 8.28 \times 10^{-19}\,\text{J}$$

Convert joules to electronvolts:

$$E = \frac{8.28 \times 10^{-19}\,\text{J}}{1.6 \times 10^{-19}\,\text{J/eV}} = 5.18\,\text{eV}$$

Using Einstein's photoelectric equation, $KE_{max} = E - \phi$

Subtract the work function from energy carried by the photon to find KE_{max}

$$KE_{max} = 5.18\,\text{eV} - 4.08\,\text{eV}$$

$$= 1.10\,\text{eV}$$

Using the work function

When a material's work function is less than 3.1 eV, visible light can release electrons. For example, the work function for calcium is 2.87 eV, potassium is 2.29 eV, and caesium is 1.95 eV.

When a material's work function is less than 1.77 eV, infrared light can release electrons. **Semiconductors** are materials that have been treated so they have a low work function and respond to visible light and infrared radiation. Semiconductors are found in many applications, for example in CCD devices in digital cameras, and in photovoltaic cells.

A **semiconductor** is a material with conductivity between a metal and an insulator.

TEST YOURSELF

13 Using your answers to questions 10 and 11:

a) For each of the metals, sodium and selenium, explain which colours of light release photoelectrons corresponding to the threshold energy for each metal.

b) For each colour of light, calculate the maximum kinetic energy in eV of photoelectrons. Refer to the table in question 10, p. 61.

14 Visible light releases electrons from the surface of caesium but not from iron. Ultraviolet light releases electrons from both metals. Explain which metal has the larger work function.

15 The work function of silver is 4.26 eV:

a) Calculate the threshold frequency for silver.

b) Ultraviolet light of frequency 1.50×10^{15} Hz shines on a silver surface. Explain whether photoelectrons are emitted from this metal's surface.

c) Calculate the maximum kinetic energy of the photoelectrons. (Hint: calculate the energy from incident photons first.)

16 The work function of calcium is 2.87 eV.

a) Calculate the threshold wavelength of light that matches the work function of calcium.

b) Light of wavelength 700 nm falls on the surface of a piece of calcium. Discuss whether electrons are emitted from the surface.

c) Light of wavelength 400 nm now shines on the surface of the calcium. Calculate the maximum kinetic energy of electrons emitted from the surface.

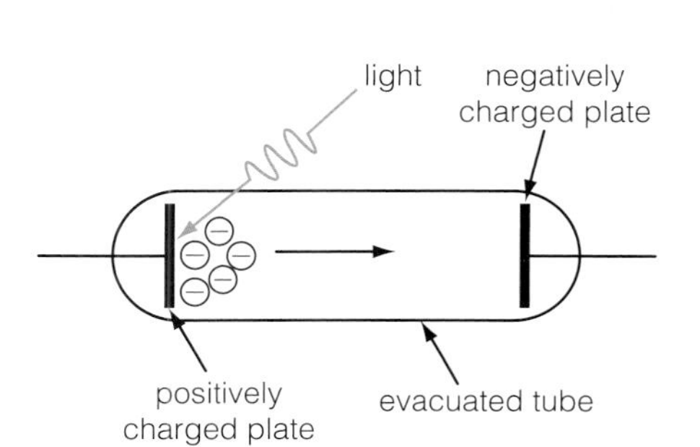

Figure 4.6 Photoelectric cell.

Stopping potential

Figure 4.6 shows a photoelectric cell. When light strikes the metal surface on the left photoelectrons are emitted for some wavelengths of light. The photoelectrons that are emitted can be detected by a sensitive ammeter when they reach the electrode on the right of the tube.

By making the potential of the right-hand electrode negative with respect to the left-hand electrode, the photoelectrons can be turned back, so that they do not reach the right-hand electrode. At this point, the current is zero, and we say that we have applied a stopping potential to the electrons. The stopping potential gives a measure of the electrons' kinetic energy, because the work done by the electric field, eV, is equal to the photoelectron's kinetic energy.

So the electron's kinetic energy may be calculated from the equation:

$$\frac{1}{2}mv^2 = eV_{stop}$$

EXAMPLE

Use Figure 4.7 to answer these questions:

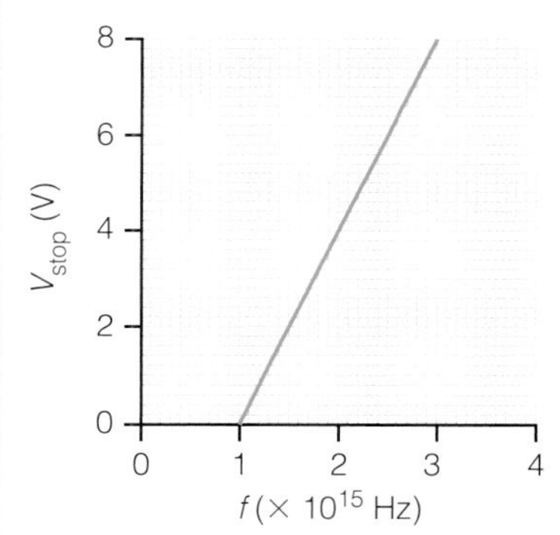

Figure 4.7 Graph to show how stopping potential varies with frequency.

a) Explain why the stopping potential changes with frequency.

b) Calculate the work function for this material.

c) Calculate the maximum kinetic energy of photoelectrons when the incident radiation is of frequency 2.5×10^{15} Hz.

Answer

a) Higher frequency photons transfer more energy to each photoelectron. The surplus energy above the work function is changed to kinetic energy of the photoelectron

b) The work function equals hf when the line intercepts the x-axis. hf $= 6.63 \times 10^{-34}$ J s^{-1} $\times 1 \times 10^{15}$ Hz
$= 6.63 \times 10^{-19}$ J

c) Using the graph, the stopping potential is 6V when the incident radiation is of frequency 2.5×10^{15} Hz. Using $KE_{max} = eV_{stop}$ gives $KE_{max} = e \times 6V = 6$ eV or 9.6×10^{-19} J

Electron diffraction

Since waves behave as particles, scientists wondered if particles could behave as waves. Diffraction can be seen when waves spread round an obstacle or through a gap and is a defining wave property. If scientists could observe diffraction patterns using particles such as electrons, this would prove that particles also behave as waves.

The wavelength of particles

If particles behave as waves, they must have a wavelength. In 1923, Louis de Broglie suggested how to calculate the wavelength of particles using the particle's momentum. The momentum of a photon,

$$p = mc.$$

Substituting $p = mc$ in the equation $E = mc^2$ gives

$$E = pc \text{ or } p = \frac{E}{c}$$

Since $E = hf$ then

$$p = \frac{hf}{c} = \frac{h}{\lambda}$$

The **de Broglie wavelength** for a particle is the Planck constant divided by a particle's momentum; it represents the wavelength of a moving particle, or $\lambda = \frac{h}{p}$.

De Broglie proposed that this relationship would hold for an electron, or any particle. He calculated that the de Broglie wavelength of electrons travelling about 0.1 to 1% of the speed of light is similar to the wavelength of X-rays.

Diffraction using crystals

Diffraction is greatest when the wavelength of the wave is roughly equal to the size of the gap it passes through. The wavelength of X-rays is of the same order of magnitude as the spacing between ions in many crystals. Diffraction images were first produced using X-rays travelling through crystals in 1912. Since the wavelength of electrons is similar to the wavelength of X-rays, the same crystals should diffract X-rays and electrons. In 1927, electron diffraction was seen in experiments by Davisson and Germer, then repeated in other laboratories around the world.

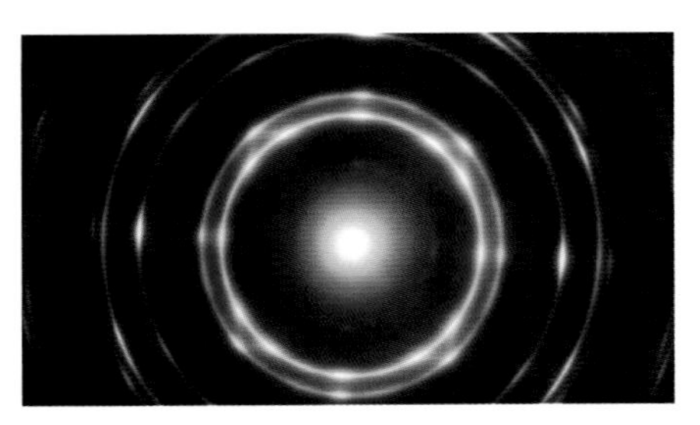

Figure 4.8 X-ray diffraction.

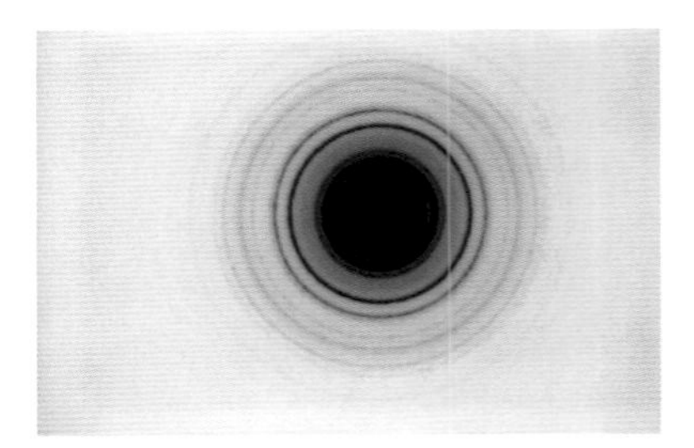

Figure 4.9 Electron diffraction.

ACTIVITY

Diffraction rings

In the diffraction rings experiment, a beam of electrons is fired from an electron gun. The electrons pass through a thin graphite screen in which the carbon atoms are arranged in a lattice, acting as a diffraction grating. As the electrons pass through the lattice, they diffract and interfere. Many individual diffraction patterns are produced, which combine to form a single pattern of diffraction rings. A voltage across the electron gun accelerates the electrons.

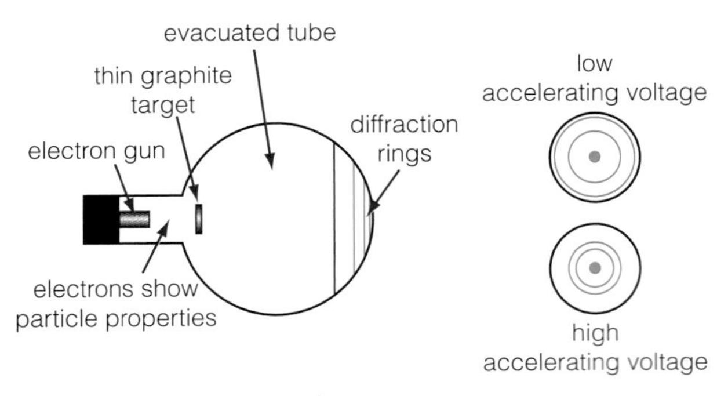

Figure 4.10 Diffraction rings apparatus.

The screen glows where the electrons strike it and the diameter of the rings can be measured. When the rings are closer together, electrons have deviated less from their original path so less diffraction has occurred.

1. Explain why a diffraction pattern is produced.
2. What does the diffraction pattern prove about the spacing in the graphite lattice and the wavelength of the electrons?
3. Calculate the kinetic energy of the electron when the accelerating voltage is:
   a) 2000 V
   b) 4000 V
4. Calculate the speed of the electron for each voltage.
5. Calculate the wavelength of the electron for each voltage.
6. Use your answer to question 5 to explain why there is less diffraction when the accelerating voltage is higher.

Calculating the de Broglie wavelength

The de Broglie wavelength is the wavelength of a moving particle. The wavelength is calculated as Planck's constant divided by the momentum of the particle using this equation:

$$\lambda = \frac{h}{mv}$$

where λ is the de Broglie wavelength in metres
h is Planck's constant
m is the mass of the particle in kg, v is the velocity of the particle in $m\,s^{-1}$.

This equation shows that faster moving particles have a shorter wavelength.

This equation can also be written as:

$$\lambda = \frac{h}{p}$$

where p is the momentum of the particle, mass × velocity.

EXAMPLE

Calculate the wavelength of an electron moving at 0.1% of the speed of light. The mass of the electron is 9.1×10^{-31} kg.

Answer

$$\lambda = \frac{h}{mv}$$

$$= \frac{6.63 \times 10^{-34}\,J\,s}{(9.1 \times 10^{-31}\,kg \times 3 \times 10^{5}\,m\,s^{-1})}$$

$$= 2.4 \times 10^{-9}\,m$$

Wave–particle duality

Wave–particle duality is the idea that matter and radiation can be described best by sometimes using a wave model and sometimes using a particle model.

Wave–particle duality is the idea that particles such as electrons and electromagnetic radiation can behave as particles as well as waves. Electron diffraction proves that electrons do diffract and interfere, just as waves do. The photoelectric effect can only be explained using a particle model of electromagnetic radiation. It was no longer possible to use the same model to explain the behaviour of electrons in an electric field and the behaviour of electrons when they diffract. The idea of wave–particle duality and the quantum theory were radically different to conventional thinking of the time. The quantum theory completely changed our understanding of physics and led us to a better understanding of the microscopic world.

TEST YOURSELF

17 A crystal of rock salt diffracts X-rays but not visible light.

a) State the conditions for diffraction.

b) X-rays have a wavelength of about 1 nm and visible light has a wavelength of about 400 nm to 700 nm. Suggest the approximate spacing in the crystal lattice.

18 a) Compare the de Broglie wavelength for a neutron with the de Broglie wavelength of an electron moving at the same speed.

b) Use your answer to (a) to suggest why neutrons do not produce diffraction patterns from the same crystals that electron diffraction patterns are clearly seen with, if the neutrons and electrons are travelling at the same speed.

19 Calculate the de Broglie wavelength of:

a) a 70 kg person running at $5\,m\,s^{-1}$

b) a proton travelling at $1 \times 10^{6}\,m\,s^{-1}$. The proton mass is 1.67×10^{-27} kg

c) Comment on your answer to (a).

d) Suppose Planck's constant suddenly became 100 J s. Explain what would happen to a class of students as they left a classroom to go for lunch.

20 Explain which of these are evidence for wave–particle duality:

a) the diffraction of neutrons, and the knowledge that a neutron can have kinetic energy and momentum

b) polarisation of X-rays

c) diffraction of microwaves.

Practice questions

1 Which of the following is not true about the photoelectric effect?

A For most metals UV light is needed for the photoelectric effect to occur.

B A bright light causes more electrons to be emitted than a faint light provided the frequency is above the threshold frequency.

C Higher frequency light emits electrons with higher kinetic energies.

D A faint light contains very little energy so it takes a few minutes before electrons are emitted from the metal it is shining on.

2 Monochromatic light of wavelength 4.80×10^{-7} m is shone on to a metal surface. The work function of the metal surface is 1.20×10^{-19} J. What is the maximum kinetic energy, in J, of an electron emitted from the surface?

A 2.94×10^{-19} J

B 4.14×10^{-19} J

C 5.34×10^{-19} J

D 6.25×10^{14} J

3 Electrons travelling at a speed of 5.00×10^{5} m s^{-1} exhibit wave properties. What is the wavelength of these electrons?

A 7.94×10^{-13} m

B 2.42×10^{-12} m

C 1.46×10^{-9} m

D 6.85×10^{8} m

4 Radiation from a mercury lamp has an energy of 2.845 eV. What is the wavelength of the radiation?

A 355.0 nm

B 435.9 nm

C 453.5 nm

D 554.2 nm

5 Ultraviolet light shines on a metal surface and photoelectrons are released. The effect of doubling the intensity of the ultraviolet light while keeping the frequency constant is to release what?

A Twice as many photoelectrons, and double their maximum kinetic energy.

B The same number of photoelectrons with double the maximum kinetic energy.

C Twice as many photoelectrons with the same maximum kinetic energy.

D The same number of photoelectrons with the same maximum kinetic energy.

6 Some electrons have a de Broglie wavelength of 1.80×10^{-10} m. At what speed are they travelling?

A 2.20×10^{3} m s^{-1}

B 4.04×10^{6} m s^{-1}

C 8.04×10^{6} m s^{-1}

D 1.02×10^{8} m s^{-1}

7 The work function of caesium is 1.95 eV. Electrons are produced by the photoelectric effect when a caesium plate is irradiated with electromagnetic radiation of wavelength 434 nm. What is the kinetic energy of the emitted photons?

A 1.46×10^{-19} J

B 2.35×10^{-19} J

C 2.92×10^{-19} J

D 4.16×10^{-19} J

8 What is the wavelength of an electron that has been accelerated through a potential difference of 9.0 kV?

A 1.2×10^{-26} m **C** 5.2×10^{-21} m

B 1.7×10^{-22} m **D** 1.3×10^{-11} m

9 Monochromatic ultraviolet radiation with a photon energy of 3.2×10^{-19} J falls on the surface of a metal. Photoelectrons are emitted from the metal with a maximum kinetic energy of 2.1×10^{-19} J.

What is the work function of the metal?

A 0.7×10^{-19} J **C** 1.5×10^{-19} J

B 1.1×10^{-19} J **D** 5.3×10^{-19} J

10 A proton and an electron have the same velocity. The de Broglie wavelength of the electron is 3.2×10^{-8} m. What is the de Broglie wavelength of the proton?

A 1.7×10^{-11} m **C** 6.4×10^{-8} m

B 5.5×10^{-9} m **D** 2.3×10^{4} m

11 When light shines on a clean copper surface, photoelectrons with a range of kinetic energies up to a maximum of 2.4×10^{-20} J are released. The work function of copper is 4.70 eV.

a) State what is meant by a photoelectron. *(2)*

b) Explain why the kinetic energy of the photoelectrons has a maximum value. *(1)*

c) Calculate the wavelength of the light. Give your answer to an appropriate number of significant figures. *(4)*

12 As a result of work on the photoelectric effect and electron diffraction, a model of wave–particle duality was suggested.

a) State one effect that can only be explained if electrons have wave properties. Details of apparatus used are not required. *(1)*

b) The theory of wave–particle duality modified existing ideas of the behaviour of particles and waves. Explain, with a relevant example, the stages involved when an existing scientific theory can be modified or replaced with a new theory. The quality of your written communication will be assessed in this question. *(6)*

13 When a clean metal surface in a vacuum is irradiated with ultraviolet radiation of a high frequency, electrons are emitted from the metal.

a) Explain what is meant by the threshold frequency. *(2)*

b) Explain why electrons may be emitted from the surface of one metal but not from the surface of a different metal when light of the same frequency shines on each surface. *(2)*

c) Explain what is meant by the term *work function*, and state its unit. *(1)*

14 Electrons can exhibit wave properties.

a) Explain what is meant by wave–particle duality. (2)

b) Calculate the de Broglie wavelength of electrons travelling at a speed of $3.2 \times 10^4\,m\,s^{-1}$. The mass of an electron is $9.11 \times 10^{-31}\,kg$. (3)

c) The proton has a mass equal to 1836 times the mass of an electron. Calculate the speed of protons with the same de Broglie wavelength as the electrons in part (b). (2)

15 Light energy of frequency $1.4 \times 10^{15}\,Hz$ is incident on each square centimetre of an aluminium surface at a rate of $3.0 \times 10^{-7}\,J\,s^{-1}$.

a) Calculate the energy of an incident photon. (2)

b) Calculate the number of photons incident per second per square centimetre of the aluminium surface. (2)

16 Describe the main features of the photoelectric effect and explain how these provided evidence for the particle nature of light. The quality of your written communication will be assessed in this question. (6)

17 A freshly cleaned zinc plate has a work function of 3.6 eV.

a) i) Explain what is meant by the term work function. (2)

ii) Ultraviolet light of wavelength 280 nm is now shone on to the metal surface. Show that photoelectron will be emitted from the zinc surface. (3)

b) The zinc plate is now charged to a potential of +2 V relative to the surroundings. What is the minimum wavelength of light which will cause photoelectrons to be emitted now? (3)

Stretch and challenge

18 A teacher demonstrates an experiment to show electron diffraction (see Figure 4.11). A metal filament is heated, releasing electrons. A high voltage accelerates electrons through a vacuum in an evacuated tube. As the electrons pass through the carbon disc, the carbon lattice forming the disc diffracts them. The diffraction rings seen on the phosphor layer are caused by different layers of atoms.

a) Explain why the electrons are diffracted as they pass through the carbon target.

b) The electrons are accelerated through 5000 V. Calculate:

i) their kinetic energy

ii) their velocity

iii) use your answer to calculate the de Broglie wavelength of the electrons.

c) The potential difference is halved. Predict how these quantities change:

i) kinetic energy

ii) momentum of electrons

iii) de Broglie wavelength.

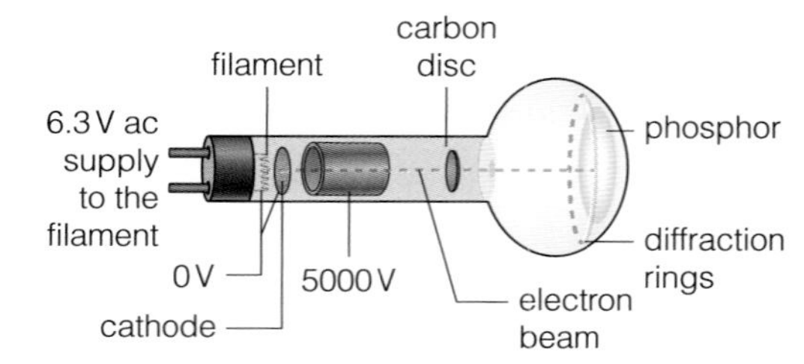

Figure 4.11 Using a diffraction tube to diffract electrons. For safety, the 5000 v current is limited to 5 mA.

d) When the potential difference decreases, the diameter of the rings increases. Suggest why.

e) Suggest why a magnet held near the screen changes the pattern seen.

19 Young's double slit experiment was a famous experiment you will learn about in more detail in Chapter 6. It demonstrated that waves passing through two very narrow slits interfere with each other, causing a pattern of alternating regions of darkness and brightness called fringes. These patterns can only be seen with waves.

a) Compare the images in Figure 4.12 which show the pattern obtained by passing protons, electrons and very low intensity l ight, one photon at a time, through double slits.

b) Suggest why the patterns look similar and how you could use this as evidence for wave–particle duality.

c) Explain whether the slit separation would be the same in all experiments.

(a)

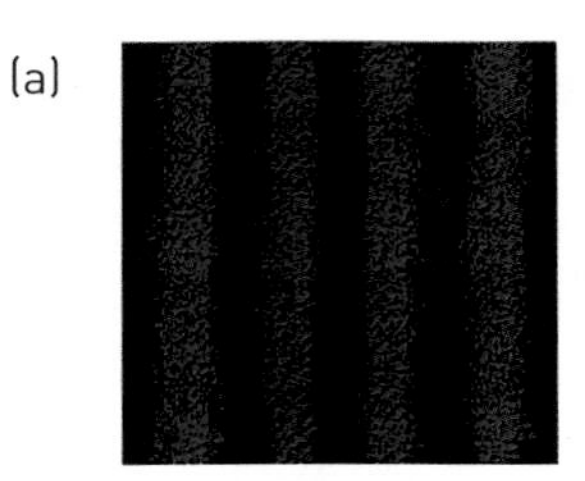

(b)

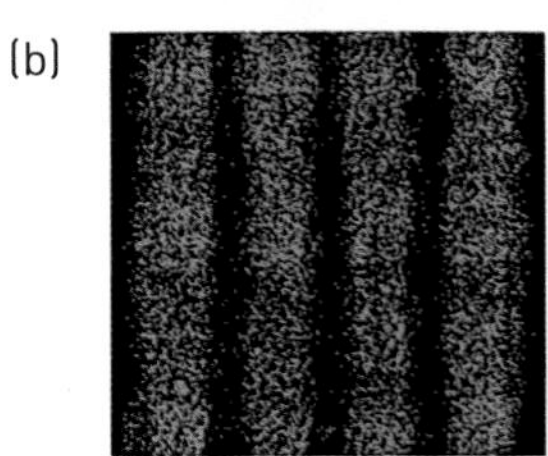

(c)

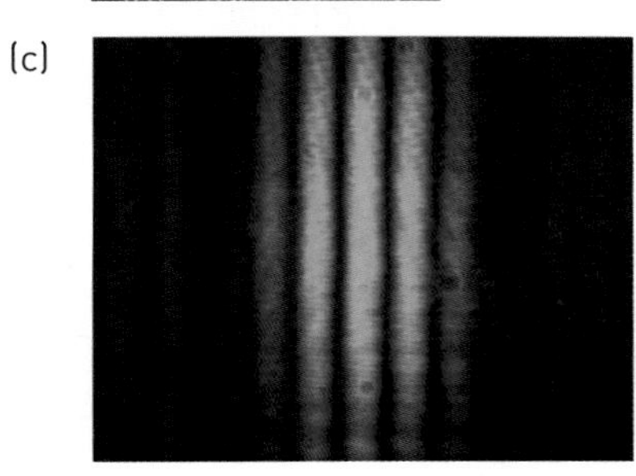

Figure 4.12 Images showing the patterns obtained when a) electrons b) protons and c) single photons pass through double slits.

Waves

PRIOR KNOWLEDGE

- Wave motion transfers energy and information without transferring matter.
- Vibrations or oscillations in transverse waves are perpendicular to the direction of energy transfer, for example electromagnetic waves.
- Vibrations or oscillations in longitudinal waves are in the same direction as the energy transfer; for example sound waves.
- Frequency (in Hz) is the number of cycles per second; wavelength (in m) is the distance between equivalent points in successive cycles. Amplitude is the maximum displacement from the equilibrium position.
- The wave equation states that the speed of a wave (in $m\,s^{-1}$) equals its frequency (in Hz) multiplied by its wavelength (in m).

  $$v = f \times \lambda$$

- Refraction is the change of wave speed and direction at a boundary.
- Reflection occurs when a wave bounces off a surface. The angle of reflection equals the angle of incidence.
- Total internal reflection occurs when all wave energy reflects off the inside surface of a material.
- Diffraction is the spreading of waves around an obstacle or through a gap.

TEST YOURSELF ON PRIOR KNOWLEDGE

1 Two transverse waves travel on a slinky. Both waves have the same frequency but one wave has double the amplitude of the other wave. Make a sketch to show these waves, using the same axes: the y-axis shows the transverse displacement of the waves, and the x-axis the distance along the slinky. Label the amplitude for each wave.

2 Sketch two waves with the same amplitude but one wave has double the frequency of the other wave. Label the wavelength for each wave.

3 Write down one similarity and one difference between reflection and refraction.

4 a) Calculate the speed of a sound wave: its wavelength is 2 m and its frequency is 170 Hz; give your answer in $m\,s^{-1}$.

b) Sound waves from the same source now travel in water at $1500\,m\,s^{-1}$. The frequency does not change. Calculate the new wavelength.

5 Draw a diagram to show how a mirror can be used to see around a corner. Include the incident ray and the reflected ray, and label the angle of incidence and the angle of reflection.

6 Write down two similarities and one difference between transverse waves and longitudinal waves.

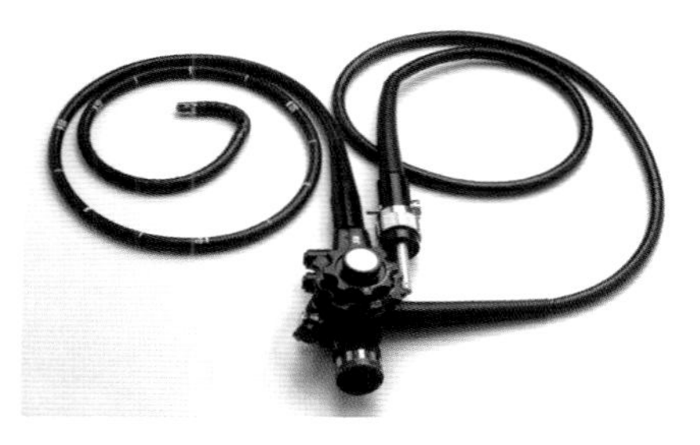

Figure 5.1 An endoscope.

Optical fibres transmit information using an electromagnetic wave, which travels down a fibre by means of repeated reflections off the surface of the fibre. This is called total internal reflection. Cable TV networks and communications networks transmit information through optical fibres by converting electrical signals to digital pulses of light or infrared. These travel through the fibre by total internal reflection. Endoscopes also use optical fibres to examine inside a patient without needing to operate. Optical fibres inside a light cable transmit light into the patient; reflections are transmitted back through the fibre optic cable and sent to a display screen to show a real-time image.

What are progressive waves?

A progressive wave is a wave that travels through a substance or space, transferring energy.

Waves are caused by a vibrating source. A progressive wave is an oscillation or vibration that transfers energy and information. Only energy is transferred, although the substance is disturbed as the waves pass through. The particles oscillate about their fixed positions but do not move to a different place. As you hear your teacher speak, a sound wave travels through the air from your teacher to your ear. The air does not move from your teacher's mouth to your ear, but the air oscillates backwards and forwards.

Figure 5.2 The energy spreads out from the centre.

Waves are represented in two different ways:

- A wave front joins points on a wave that are at the same point of the cycle as their neighbours.
- For waves caused by a stone dropped in water, the ripples are wave fronts, and the rays radiate outwards from where the stone fell into the water. You can see wave fronts spreading out from a point in Figure 5.3.
- A ray is an imaginary line showing the direction the wave travels in. It joins the position of the wave source to the wave fronts.

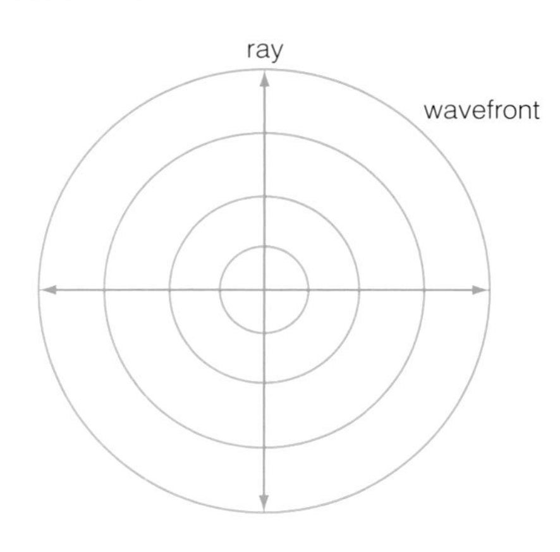

Figure 5.3 Water spreading out from a ripple.

How do particles move in waves?

Mechanical waves on a string cause particles to vibrate perpendicular to the direction energy is transferred. The amplitude, is the maximum displacement from the particle's undisturbed position and the larger the amplitude is, the more energy is transferred. The distance between wave peaks is the wavelength, λ. λ is the distance between two equivalent points in successive cycles.

Amplitude is the maximum displacement from the undisturbed position (for a wave).

The distance between wave peaks is the wavelength.

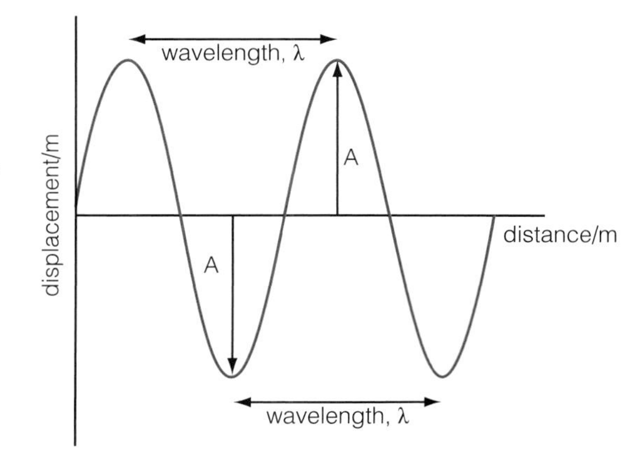

Figure 5.4 Wave terminology.

Frequency is the number of cycles per second, measured in Hz.

Hertz is the unit of frequency, equivalent to s^{-1}.

Period is the time for one complete cycle, measured in seconds.

The frequency of a wave, measured in hertz, is the number of cycles or vibrations per second. The time per cycle is the period of a wave and is measured in seconds. A wave with a frequency of 10 Hz has 10 cycles per second, each one taking 0.1 seconds. Period and frequency are related using:

$$T = \frac{1}{f}$$

where T = period in s

f = frequency in Hz

Since 1 Hz is a very low frequency we also measure hertz in kHz (10^3 Hz), MHz (10^6 Hz) and GHz (10^9 Hz).

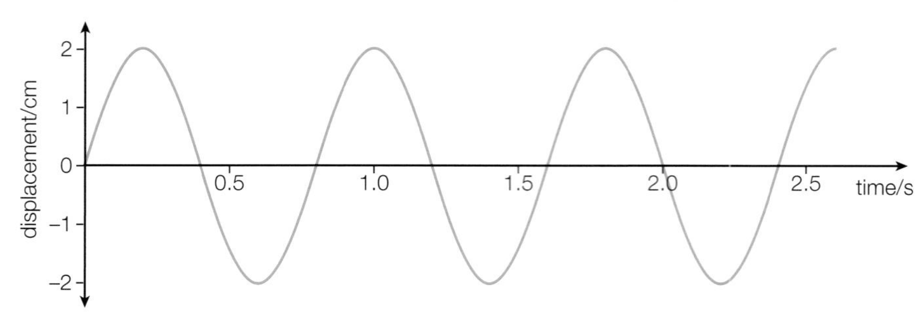

Figure 5.5 The time period for one oscillation is 0.8 seconds.

The wave equation

Different waves travel at very different speeds, but in each case the speed is calculated the same way using the equation:

$$\text{speed} = \frac{\text{distance}}{\text{time}}$$

For a wave, the wavelength, λ, is the distance travelled in one cycle and the time to complete one cycle is the period, T.

This means that:

$$\text{wave speed} = \frac{\text{wavelength}}{\text{period}}$$

The period is the time (in seconds) for one complete cycle, and frequency is the number of cycles per second. This gives frequency, $f = \frac{1}{T}$. Substituting for $\frac{1}{T}$ in the equation above gives:

wave speed = frequency (Hz) × wavelength (m)

$$c\ (\text{m s}^{-1}) = f\ (\text{Hz}) \times \lambda\ (\text{m})$$

This is known as the wave equation.

TIP

Be careful when you are using the wave equation as c can be used to represent wave speed but it is also often used to represent the speed of light, $3 \times 10^8\ \text{m s}^{-1}$.

EXAMPLE

A microwave oven operates at a frequency of 2450 MHz and produces waves of wavelength 12.2 cm. Use this information to calculate the speed of microwaves.

Answer

wave speed = frequency (Hz) × wavelength (m)

$= 2450 \times 10^6\ \text{Hz} \times 0.122\ \text{m}$

$= 2.9 \times 10^8\ \text{m s}^{-1}$

TIP

Remember to convert units to metres, seconds, hertz, m s^{-1}.

Particles along a wave that move **in phase** move in the same direction with the same speed. The particles have the same displacement from their mean position. Particles in phase are separated by a whole number, n, of wavelengths, $n\lambda$.

Particles along a wave that move **out of phase** are at different points in their cycle at a particular time.

Particles along a wave that move in **antiphase** move in opposite directions at the same speed. The particles have opposite displacements from their mean position. Particles moving in antiphase are separated by a distance of a whole number, n, of wavelengths plus an extra half wavelength, $n\lambda + \frac{\lambda}{2}$.

Phase difference

A wave repeats its cycle at regular time intervals. The **phase** of a wave describes the fraction of a cycle completed compared to the start of the cycle. Particles in parts of a wave that are moving at the same speed in the same direction are **in phase**. They are **out of phase** if they are at different points in their cycle at a particular time. Particles in parts of a wave that move in opposite directions and at exactly the same speed are moving in **antiphase**, or completely out of phase.

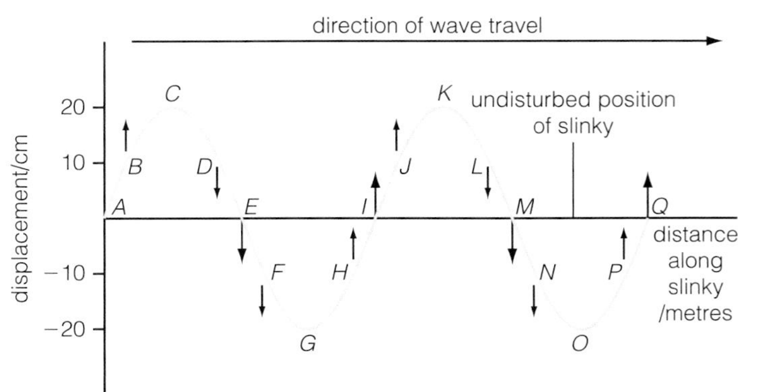

Figure 5.6 The displacement of a slinky along its length. The arrows on the graph show the direction of motion of the slinky, the wave is travelling from right to left.

One wavelength is the shortest distance between two points that are moving in phase. Points that are a whole number of wavelengths apart move in phase; points that are half a wavelength or 1.5 wavelengths or 2.5 wavelengths, etc. move exactly out of phase (or in antiphase) with each other.

Figure 5.6 shows a transverse wave travelling on a slinky. In this diagram,

- points E and M are in phase as they move in the same direction at the same speed and are at the same point in their cycle. These three points are out of phase with all other points
- points E and I are moving in antiphase, because they are moving in opposite directions with the same speed.

Phase difference is measured as a fraction of the wave cycle between two points along a wave, separated by a distance x.

$\Phi = \frac{2\pi x}{\lambda}$ rad

Any **phase difference** can be measured as an angle in degrees or radians. The motion of many waves can be described using a sine function. Since a sine function repeats itself over an angle of 360° or 2π radians, we say the phase change of a wave during one complete cycle is 360° or 2π radians. You can read more about radian and degree measure in Chapter 15.

The maths of phase differences

One complete rotation involves turning through 2π **radians**, so 2π (or 6.28) radians is equivalent to 360°.

One complete rotation of a circle involves turning through an angle of 360°. You can also measure this angle in **radians**. One complete rotation involves turning through 2π radians, so 2π radians is equivalent to 360°.

The motion of the wave in Figure 5.6 is sinusoidal with a time period of T. When time t, equals T, one cycle has been completed, so the value of the angle in the sine function must be 2π radians. This happens when $t = T$ for the angle, $\frac{2\pi t}{T}$.

We can describe the vertical displacement of the particles in the wave using an equation of the form:

$$y = A \sin \left(\frac{2\pi t}{T}\right)$$

Since $f = \frac{1}{T}$, this can also be written as

$$y = A \sin (2\pi f t)$$

where y is the vertical displacement at a time t, A is the amplitude of the wave, and f is the frequency of the oscillations.

At a different point, the oscillations will be out of phase. We describe this phase difference as a fraction of the angle 2π. So points that are half a wavelength apart have a phase difference of π (points B and F, for example, in Figure 5.6), and points three-quarters of a wavelength apart have a phase difference of $\frac{3\pi}{2}$ (points C and I, for example, in Figure 5.6). In general, two points separated by a distance x on a wave have a phase difference of:

$$\Phi = \frac{2\pi x}{\lambda}$$

where Φ is the phase difference, measured in radians.

We also talk about phase differences in terms of a fraction of a cycle. For example, a phase difference of $\frac{1}{4}$ of a cycle is a phase difference of $\frac{\pi}{2}$ radians, and a phase difference of $\frac{1}{2}$ a cycle is a difference of π radians.

EXAMPLE

What is the phase difference between two points along a wave separated by a distance of 3.5λ?

Answer

$$\Phi = \frac{2\pi x}{\lambda}$$

$$= 2\pi \left(\frac{3.5\lambda}{\lambda}\right)$$

$$= 7\pi$$

But this phase difference can be reduce to π (or half a cycle), because a phase difference of 6π is the same as a phase difference of 0. Points that are three wavelengths apart, on a wave train, move in phase.

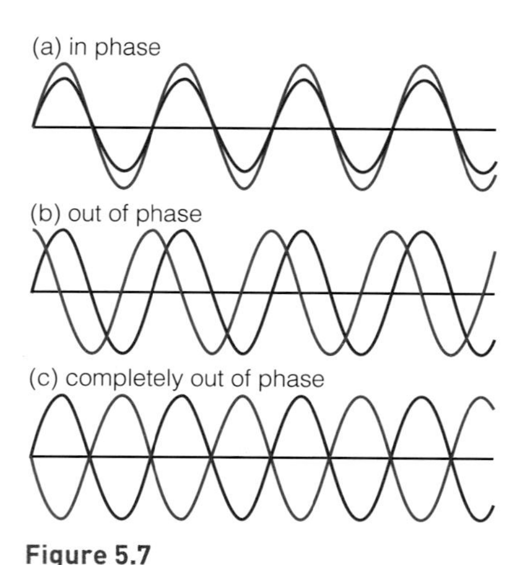

Figure 5.7

Figure 5.7 show two waves travelling in the same direction. They are in phase when, at all places, the particles in each wave move up and down together at the same speed and in the same direction, as in Figure 5.7(a).

The waves are out of phase when the particles at the same distance along the wave train, do not move together in the same direction and with the same speed. This is shown in Figure 537(b).

In Figure 5.7(c) the particles in the each wave, at the same position on the wave train, move exactly out of phase, or in antiphase. Now the particles on each wave move in the opposite direction but with the same speed.

Phase change on reflection

When a wave is reflected off the surface of a denser medium, it undergoes a phase change of 180°. This is illustrated in Figure 5.8.

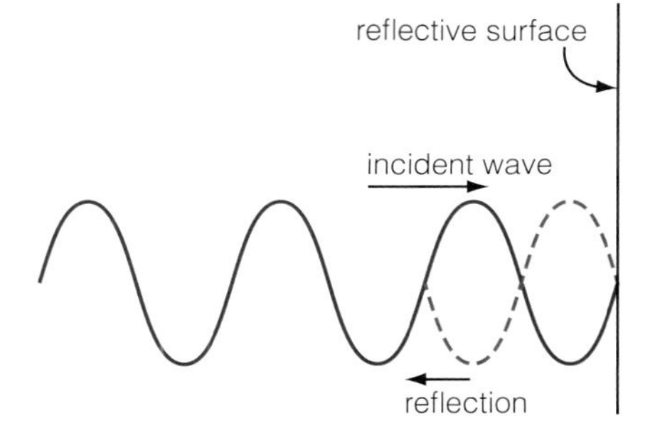

Figure 5.8 When a wave reflects from a denser material, there is a phase change of 180°.

TEST YOURSELF

1 A student uses a rope, fixed at one end, to demonstrate transverse waves. A knot is tied at the midpoint of the rope. One end of the rope is fixed and the student can move the other end of the rope. Describe how the displacement of the knot changes during one complete cycle, starting from its equilibrium position.

2 a) A student reads that some radio navigation systems use waves of frequency 15 kHz. Calculate the period of a radio wave of this frequency.

b) A pulse of these radio waves is emitted for 0.03 s. How many complete waves are emitted during this time?

3 A signal generator, attached to a loudspeaker, generates a sound wave with a wavelength of 3 m. The sound wave travels at 330 m s^{-1} in air.

a) Calculate the frequency of the sound wave.

b) The loudspeaker from the signal generator is now placed underwater. The frequency of the sound wave does not change. What is the wavelength of this sound wave underwater? The speed of sound in water is 1500 m s^{-1}.

4 Calculate the smallest phase difference in degrees and radians

a) for two points along a wave that are $\frac{1}{4}$ of a cycle out of phase

b) for two points along a wave that are $\frac{3}{4}$ of a cycle out of phase.

5 Calculate the smallest distances, in terms of wavelength, between two points along a wave, which are:

a) π out of phase

b) $\frac{\pi}{3}$ out of phase

c) $\frac{2\pi}{3}$ out of phase.

6 Some students are using a slinky spring to create transverse waves. The amplitude (maximum displacement) of the transverse wave at point X on a slinky is 3 cm when the time is 0 s. The period of the wave cycle is 4 s.

a) State the displacement at the point X when the time t = 1 s, 2 s, 3 s and 4 s.

b) Calculate the displacement at the point at a time of 0.5 s.

Longitudinal waves and transverse waves

In **longitudinal waves** particles vibrate parallel to the direction that energy travels in.

Sound waves, primary seismic waves and pulses moving along a slinky spring are all examples of longitudinal waves. **Longitudinal waves** are waves where the wave source vibrates parallel to the direction in which the wave travels, so particles in the medium also vibrate parallel to the direction of energy travel.

In transverse waves vibrations are at right angles to the direction that energy travels in.

Electromagnetic waves are transverse waves all of which travel at the speed $3 \times 10^8 \, m \, s^{-1}$ in a vacuum.

Electromagnetic waves and secondary seismic waves are examples of transverse waves. Transverse waves are created when the wave source vibrates at right angles to the direction the wave travels in, so vibrations in the medium are at right angles to the direction that energy travels in. Electromagnetic waves are transverse waves all of which travel at the speed $3 \times 10^8 \, m \, s^{-1}$ in a vacuum. The electric field and magnetic fields of the electromagnetic wave vibrate at right angles to each other and to the direction of transmission. The wavelength of an electromagnetic wave is the shortest distance between two points where the electric field (or magnetic field) is in phase.

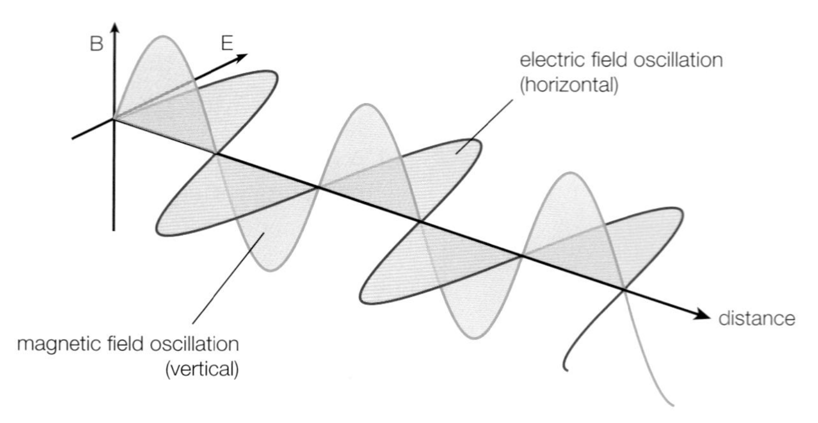

Figure 5.9 The energy in electromagnetic waves is carried by oscillating electric and magnetic fields.

Mechanical waves cannot travel through a vacuum, but need a medium to travel through.

Mechanical waves need a medium to travel through and can travel as transverse or longitudinal waves. Seismic waves, sound waves and waves on a string are examples of mechanical waves.

ACTIVITY

Slinky springs

You can investigate longitudinal and transverse mechanical waves using a slinky spring.

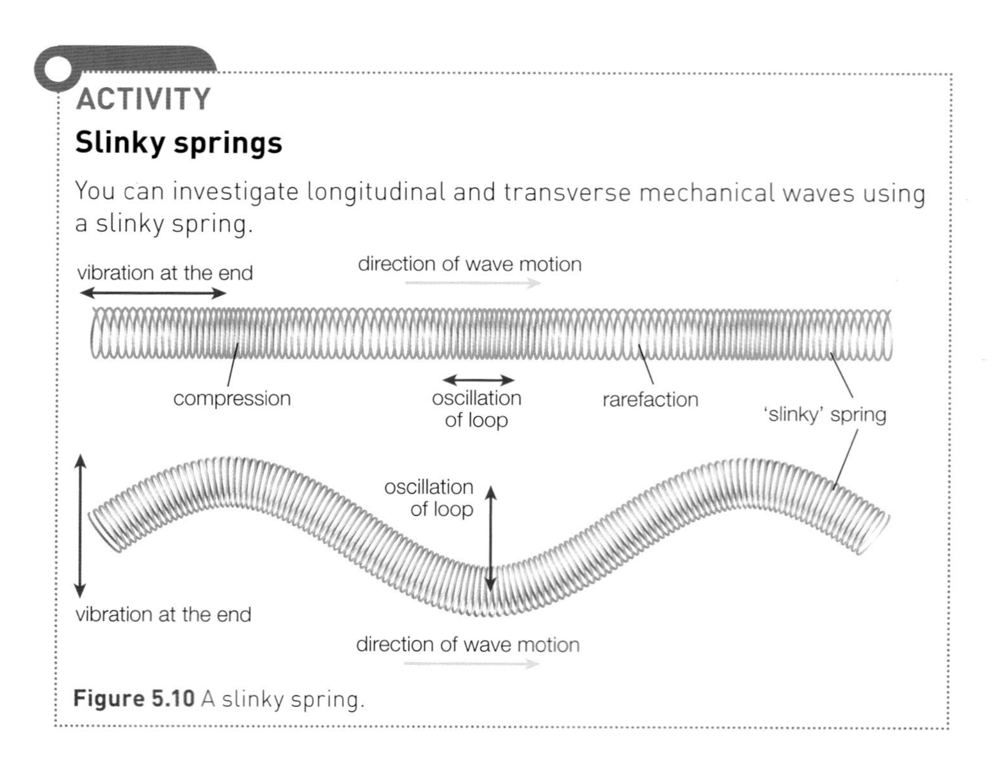

Figure 5.10 A slinky spring.

More about longitudinal waves

Sound waves are longitudinal waves produced by vibrations that move backwards and forwards along the line of wave progression. These vibrations produce regions of high and low pressure. The regions of high pressure are called **compressions** and the regions of lower pressure are called **rarefactions**. The wavelength of a longitudinal wave is the distance between successive compressions, or the distance between successive rarefactions.

> **TIP**
> A longitudinal wave shown on an **oscilloscope** looks similar to a transverse wave although the processes creating it are different. Compressions and rarefactions in the sound wave vibrate a microphone, which generates an electrical signal. The signal on the oscilloscope shows the form of these compressions and rarefactions.

trace shown on an oscilloscope

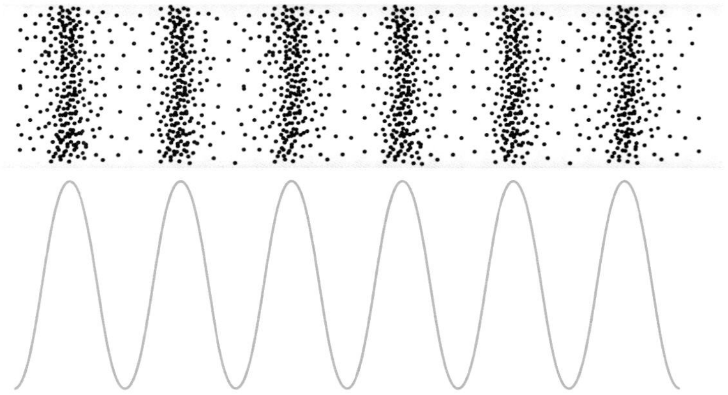

Figure 5.11 This diagram shows the changes in air pressure at a microphone diaphragham, measured over a period of time.

When sound travels through a solid, energy is transferred through inter-molecular or inter-atomic bonds. Sound travels quickly through solids because the bonds are stiff and the atoms are close together. In gases the energy is transferred by molecules colliding, so the speed of the sound depends on the speed of the molecules. Sound waves travel fastest in solids (5100 m s^{-1} in aluminium), less quickly in liquids (1500 m s^{-1} in water) and even slower in gases (340 m s^{-1} in air). Sound waves are mechanical waves, so they cannot travel through a vacuum.

Electromagnetic waves

All electromagnetic waves are similar in that they all travel at the speed of light, and their energy is carried by oscillating electric and magnetic fields. However, the properties of the waves vary considerably with their wavelength so we normally consider the spectrum as seven main groups shown in Figure 5.12.

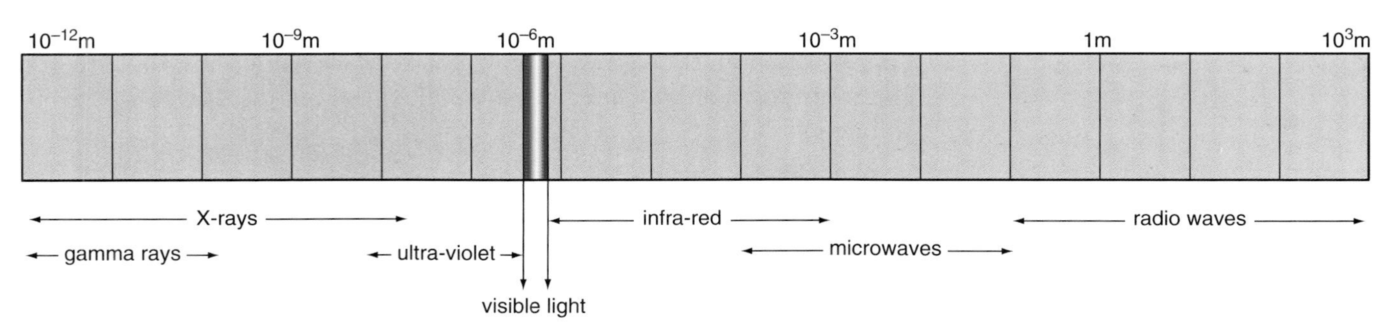

Figure 5.12 The electromagnetic spectrum.

Effect of electromagnetic radiation on living cells

The effect of electromagnetic radiation on living cells depends on the intensity, duration of exposure and frequency of the radiation. Radio waves, microwaves and infrared radiation do not cause ionisation, but may have a heating effect. Ionising radiation includes shorter wavelength electromagnetic radiation: ultraviolet radiation, X-rays or gamma rays. If living cells absorb ionising radiation, DNA molecules in the cells may be damaged, which can lead to mutations or cancer. (The ionising effect of short wavelength radiations can be understood by the particle nature of waves; photons of ultraviolet light, X-rays and gamma rays have sufficient energy to remove electrons from atoms. This is explained in Chapter 3.)

- Radio waves have little effect on living cells.
- In microwave ovens, water and fat molecules in food absorb microwaves effectively. The molecules are forced to vibrate, heating the food. Microwave ovens used in many homes and businesses produce the frequencies most strongly absorbed by water, 2450 MHz.
- Molecules that absorb infrared radiation vibrate more, increasing their internal energy. The heating effect is used for cooking and heating.
- We can see because light-sensitive cells in the retina at the back of our eyes absorb visible light.
- Ultraviolet radiation absorbed by skin causes a tan as a molecule called melanin develops in these cells. Exposure to ultraviolet radiation raises the risk of skin cancer but, if a person is tanned, the melanin absorbs some UV radiation, reducing the amount of UV reaching cells deeper in the tissues. Skin cells exposed to ultraviolet radiation also produce vitamin D.
- Gamma rays and X-rays have similar properties, but gamma rays come from radioactive decay while X-rays are created when electrons strike a metal target. X-rays are used for medical imaging as they can penetrate soft tissues but are absorbed by dense cells such as bones. High doses of radiation kills cells, for example from exposure to high-intensity, high-energy radiation or exposure for a long duration. Gamma rays are used in radiotherapy and to sterilise surgical instruments.

Communication uses of electromagnetic radiation

Electromagnetic waves are used in many different situations to transmit information. Our radios, televisions and phones all depend on electromagnetic waves carrying programmes and messages. In deciding which wavelength to use for a particular form of communication, consideration must be given to how much a wave will be absorbed by the atmosphere or how much a wave will spread out due the effects of diffraction.

- Radio waves transmit TV and radio programmes by superimposing information from the programme on to a carrier wave, changing (modulating) either its amplitude or its frequency. The receiver is tuned to the frequency of the carrier wave and converts the signal into sound and images. Radio telescopes are massive, land-based telescopes that receive radio signals from bodies in space that can penetrate through the atmosphere.

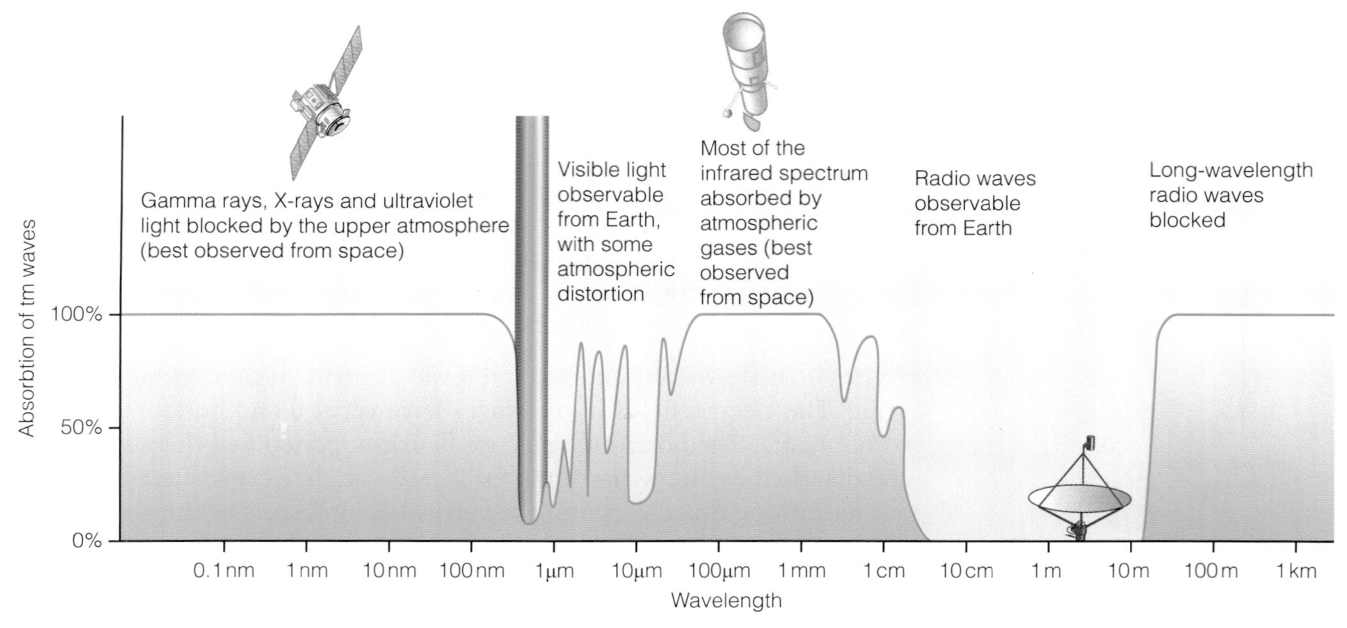

Figure 5.13 Astronomical telescopes can be based on earth or in space.

- Some microwave frequencies penetrate the atmosphere and are used for satellite communication and in mobile phone networks. Microwaves have a shorter wavelength than radio waves and diffract less, so transmitters and receivers need a straight line of sight.
- Infrared and visible radiation can carry data in optical fibres. Remote controls often use infrared radiation as it only travels a short distance in air.
- Gamma rays, X-rays and short-wavelength ultraviolet radiation do not penetrate the atmosphere, so space-based telescopes are needed to investigate these wavelengths in the Universe.

Polarisation

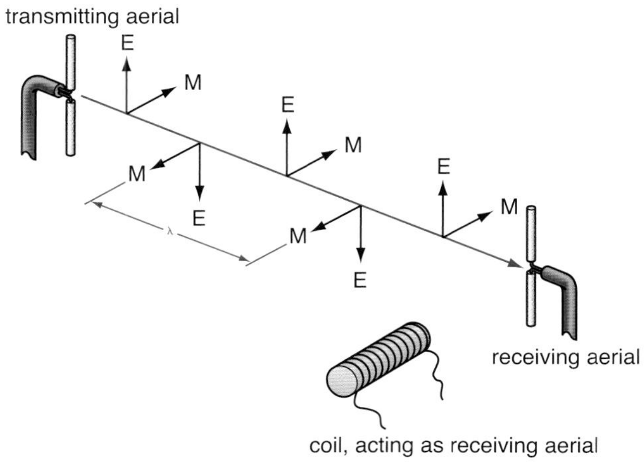

Figure 5.14 shows a simple way to demonstrate polarisation using laboratory equipment. In the diagram you can see a radio aerial transmitting 30 cm radio waves, which are received by another aerial placed a metre away. A high-frequency (1 GHz) signal is applied to the transmitting aerial, which causes electrons to vibrate up and down. Therefore, the electromagnetic waves that are transmitted only have their electric fields oscillating vertically (and their magnetic fields oscillating horizontally). When the fields of an electromagnetic wave only oscillate in one direction, the wave is said to be a polarised wave. In this case we say the wave is vertically polarised, because the electric fields are confined to the vertical plane. (By convention the direction of polarisation is defined as the direction of the electric field oscillation.) We can show the polarisation of the waves by turning the receiving aerial: when the receiving aerial is placed vertically, as in the diagram, radio waves are detected, because they cause electrons in the vertical aerial to oscillate up and down; when the aerial is turned to a horizontal direction, no signal is detected, because the electric field is in the wrong direction to make the electrons move along the aerial.

transmitting aerial
receiving aerial
coil, acting as receiving aerial

Figure 5.14 A simple way to demonstrate polarisation.

When the oscillations of the wave are confined to one plane the wave is a polarised wave. For example, in an electromagnetic wave the electric field might be confined just to the vertical plane. These waves are said to be vertically polarised.

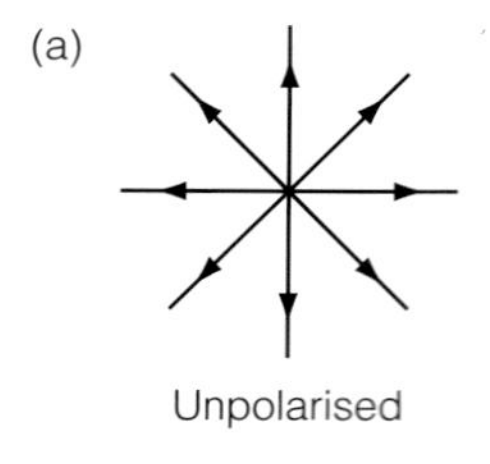

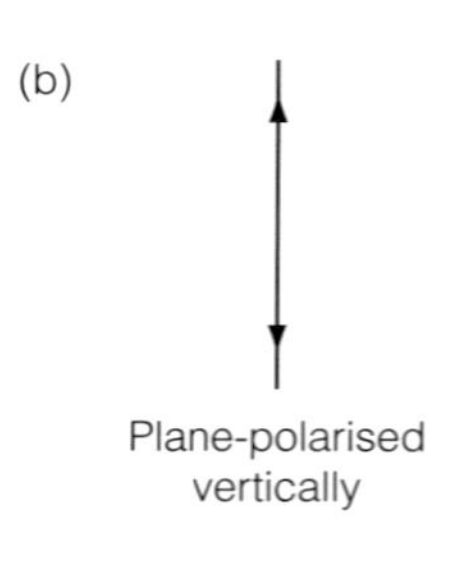

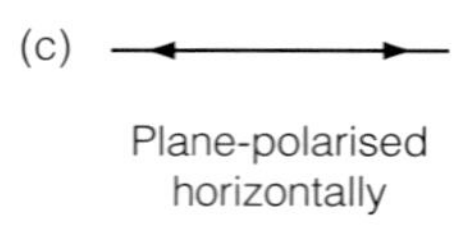

Figure 5.15 Unpolarised and polarised light. These lines show the direction of the oscillating electric fields.

Light is an example of an electromagnetic wave that is usually unpolarised when it is transmitted. Light waves are produced when electrons oscillate in atoms producing electromagnetic waves of frequency about 5×10^{14} Hz. Since electrons in atoms can vibrate in any direction, the electric and magnetic fields of light oscillate in any direction – so such light is unpolarised. Figure 5.15 shows the electric fields of unpolarised and polarised light (the magnetic fields are omitted for clarity): in part (a) light is unpolarised because the electric field oscillating in any direction. In part (b) the light is vertically polarised, and in part (c) the light is horizontally polarised.

Figure 5.16 shows unpolarised microwaves incident on a metal grille. The electric fields of the waves have vertical and horizontal components. The horizontal electric field components of the waves are absorbed by the wires of the grille because they cause electrons to oscillate along this direction. The vertical components of the electric field are not absorbed by the grille, so the waves that emerge from the grille are polarised with their electric field vertical. Polarisation is a property of transverse waves only. Longitudinal waves cannot be polarised as the particles in a longitudinal wave always oscillate along the direction of energy transfer.

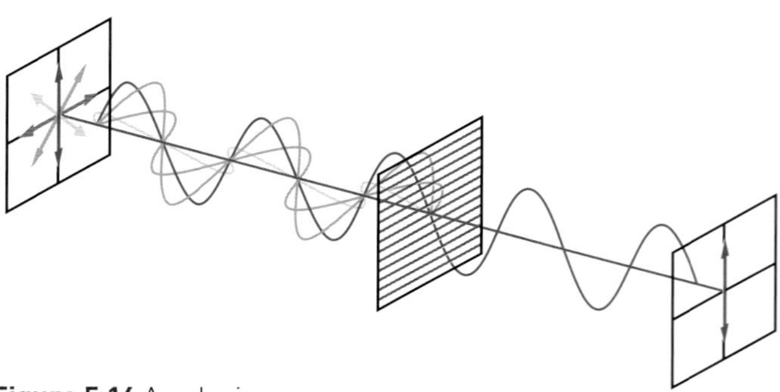

Figure 5.16 A polariser.

Polarisation effects

Light is polarised when it passes through a polarising filter. The filter only allows through electric field oscillations in one plane because the filter absorbs energy from oscillations in all other planes. The metal grille shown in Figure 5.16 is a good model for a polarising light filter, such as that shown in Figure 5.17a. Long molecules of quinine iodosulphate are lined up, and electrons in these molecules affect the light in the same way as the microwaves are affected by the grille. Polarised light is less intense than unpolarised light because only half the energy is transmitted through the filter. If a second polarising filter is held at right angles to the original filter, all oscillations are blocked and no light is transmitted. This is called crossing the polarisers.

If a second polarising filter is held at right angles to the original filter, this is called crossing the polarisers.

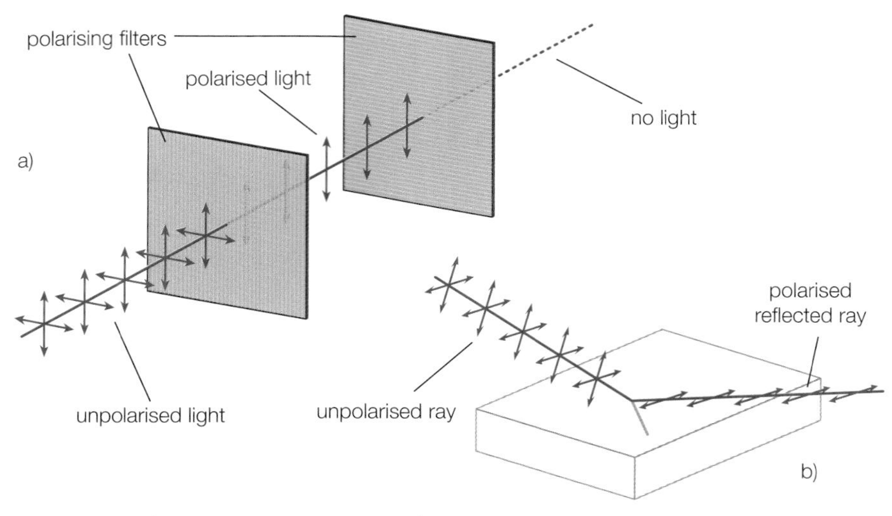

Figure 5.17 a) A polarising filter and **b)** reflected light is sometimes polarized.

Light reflecting from a surface can be polarised by reflection. At some angles of incidence, the only reflected rays are rays whose electric fields oscillate in one direction.

Figure 5.18 A scene without polarising sunglasses (left) and with polarising sunglasses (right). Polarised sunglasses do not transmit reflections from the water surface, making it easier to see into the water.

Polarisation applications

Polarising sunglasses include lenses with polarising filters that are orientated so that they reduce the glare of reflected light. When light has been reflected it is partially polarised. The filters are orientated so that they cut out light reflecting from horizontal surfaces, such as water and snow. This reduces eyestrain for skiers and makes it much safer for drivers. Using polarising filters to reduce the glare of reflected light also makes it far easier to see into water.

Outdoor television and FM radio aerials must be correctly aligned for the best reception. Transmitters generate plane-polarised electromagnetic waves, which are picked up most effectively by a receiver with the same alignment. It is possible to reduce interference between nearby transmitters if one of the two transmitters is aligned vertically, and the other is aligned horizontally as shown in Figure 5.19.

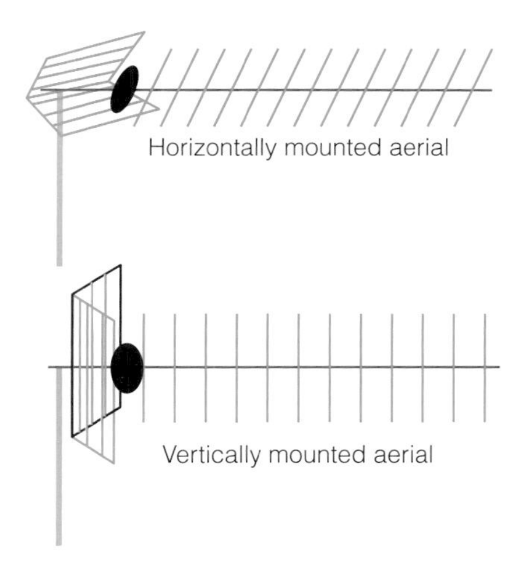

Figure 5.19 A horizontally mounted aerial absorbs horizontally polarised signals efficiently.

TEST YOURSELF

7 a) Describe how you could use a slinky spring to demonstrate the nature of

i) transverse waves

ii) longitudinal waves.

b) Describe how you could show that light is a transverse wave.

8 Explain how the properties of microwaves make them suitable for

a) cooking food

b) satellite communication.

9 Explain why the alignment of a TV aerial affects the strength of the signal received.

10 State one everyday use of a polarising filter, explaining why it is useful.

11 Earthquakes produce transverse and longitudinal waves that travel through the Earth. The diagram shows the displacement of particles of rock at a particular time for different positions of a transverse wave.

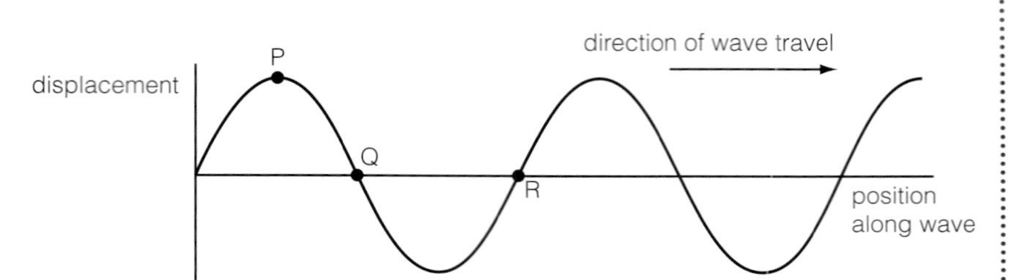

a) i) State the phase difference between points P and R.

ii) Describe the difference in the motion of the rock at points Q and R.

b) Describe the motion of the rock and point P over the next complete cycle.

c) A seismologist detects a wave that is polarised. Explain what the seismologist can deduce from this.

d) The frequency of the seismic wave is 0.65 Hz and its speed is $4.8\,km\,s^{-1}$.

i) Calculate the time period of the wave.

ii) Calculate the wavelength of the wave.

ACTIVITY

Polarisation of glucose solution

A student investigates how glucose solution affects the plane of polarisation of light passing through it. The student is told that glucose solution rotates the plane of polarisation of polarised light. The angle of rotation is measured using two polarising filters, one placed below the analyser and one placed above. Both polarising filters are aligned so transmitted light has maximum intensity when there is no solution. Glucose solution is poured into the analyser and one polarising filter is rotated until the light intensity is a maximum again.

The student uses the same concentration of glucose each time but changes the depth of the solution.

Here is a table of her results.

Depth (cm)	Angle of rotation (1)/°	Angle of rotation (2)/°
2	8	11
4	21	25
6	30	29
8	42	38
10	54	52

Figure 5.20 Polarisation of glucose solution.

1 Find the mean of the student's results

2 The student is told that the equation linking these results is:

measured angle of rotation = specific angle of rotation per decimetre × solution depth in decimetres

Plot a suitable graph to show if her results obey this relationship.

3 Use the graph to calculate the specific angle of rotation.

Refraction

What is refraction?

Refraction is the change in direction at a boundary when a wave travels from one medium to another.

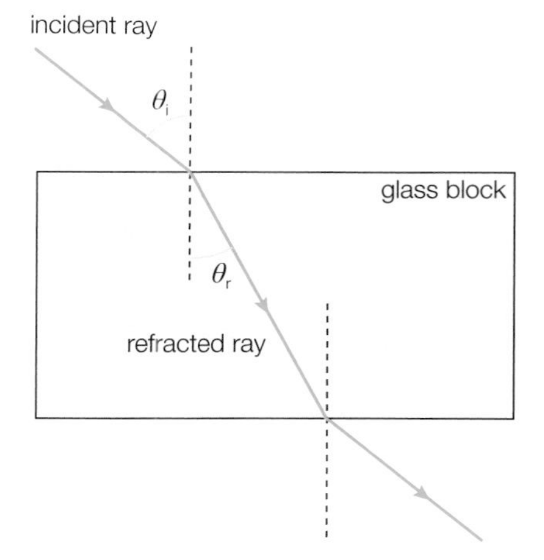

Figure 5.21 Light in air travels faster than light in the glass block. Light is refracted at each boundary.

Waves change speed as they travel from one medium into another. When light travels at any angle other than along the normal, the change in speed causes a change in direction. This is called refraction. All waves refract, including light, sound and seismic waves. The frequency of a wave does not change when it travels from one medium to another but its wavelength does. When light travels from air into a medium such as glass or water, the wave slows down and the wavelength gets shorter.

- Waves arriving along the normal do not change direction because all parts of the wave front reach the boundary simultaneously.
- Waves arriving at any other angle change direction as some parts of the wave front arrive earlier and slow down before the rest of the wave front.
- Waves change direction towards the normal when they slow down on entering a medium. Waves change direction away from the normal when they speed up on entering a medium.

Angle of incidence is the angle between the incident ray and the normal (θ_i in Figure 5.22). The normal is an imaginary line at right angles to the boundary between two materials.

Angle of refraction is the angle between the refracted ray and the normal (θ_r in Figure 5.22).

Figure 5.22 shows light entering three different mediums from air, all at the same angle of incidence (the angle between the incident ray and the normal). The light changes direction least when it enters water, and the most when it enters diamond. This is because the light travels fastest in water, slower in glass and, of the three mediums, the slowest in diamond. Light refracts more when it enters diamond from air than it does when it enters glass from air. The angle of refraction is least in diamond. Diamond is said to have a higher refractive index than glass.

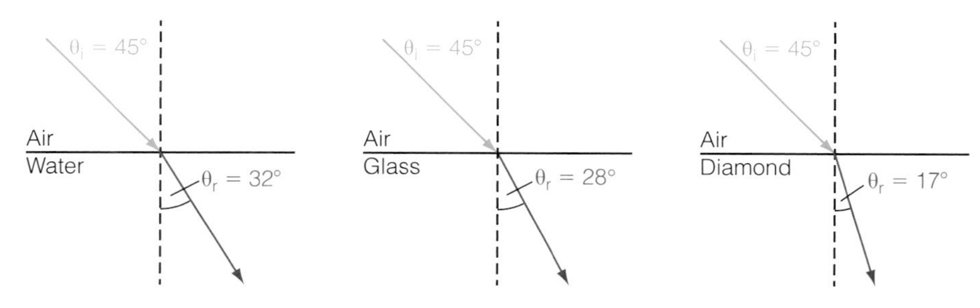

Figure 5.22 Angle of refraction for different materials. In each case the angle of incidence is 45°, but the angle of refraction is different for each material.

Refractive index

Refractive index is the ratio of a wave's speed between two materials. It is normally quoted for light travelling from a vacuum into a material.

The refractive index, n, is the ratio of the wave's speed between two materials. Because refractive index is a ratio it has no units. A refractive index is calculated using:

$$n = \frac{c}{c_s}$$

where n is the refractive index

c is the speed of light in a vacuum

c_s is the speed of light in the material.

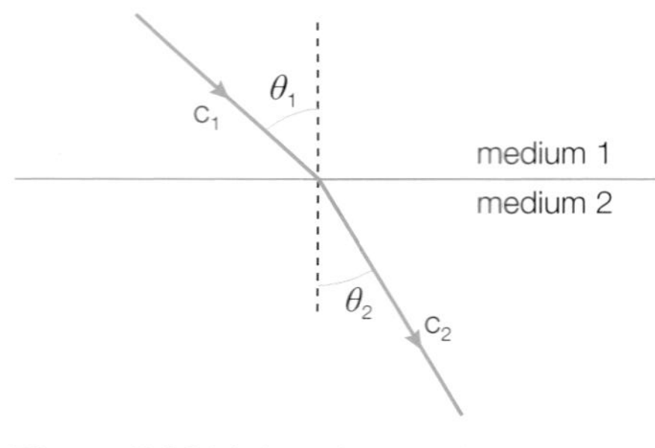

Figure 5.23 Light refracted between two materials.

However, the speed of light in a vacuum is almost exactly the same as the speed of light in air. This leads to two useful and accurate approximations:

- the refractive index of air ≈ 1
- the refractive index of a material

$$n \approx \frac{\text{speed of light in air}}{\text{speed of light in the material}}$$

The refractive index of window glass is about 1.5, which means that the speed of light in air is about 1.5 times faster than it is in glass. The refractive index of water is 1.33, so light travels 1.33 times faster in air than it does in water.

When light travels from one material to another (other than air) we can define the relative refractive index between them as follows:

$${}_1n_2 = \frac{c_1}{c_2}$$

where ${}_1n_2$ = refractive index between materials 1 and 2

c_1 is the speed of light in material 1

c_2 is the speed of light in material 2.

However, if we divide the top and bottom of the equation above by the speed of light in air, c, we get:

$${}_1n_2 = \frac{(c_1/c)}{(c_2/c)}$$

Since $\frac{c_1}{c} = \frac{1}{n_1}$ and $\frac{c_2}{c} = \frac{1}{n_2}$

it follows that:

$${}_1n_2 = \frac{(1/n_1)}{(1/n_2)} = \frac{n_2}{n_1}$$

The advantage of this formula is that we only need to know one refractive index for a material, and we can calculate a new refractive index when light passes between any pairs of materials other than air.

EXAMPLE

Calculate the refractive index when light passes from:

a) water to glass
b) glass to water.

Answer

a) ${}_1n_2 = \frac{n_2}{n_1}$

$= \frac{1.33}{1.5}$

$= 0.89$

b) ${}_1n_2 = \frac{n_1}{n_2}$

$= \frac{1.5}{1.33}$

$= 1.125$

This last example leads to an important result, which shows that the refractive index travelling from water to glass is the reciprocal of the refractive index when light travels from glass to water.

In general:

$${}_1n_2 = \frac{1}{{}_2n_1}$$

Law of refraction

For many hundreds of years scientists have studied refraction and have been able to predict the angles of refraction inside transparent materials. Willebrord Snellius was the first person to realise that the following ratio is always constant for all materials:

$$\frac{\sin \theta_i}{\sin \theta_r} = \text{constant}$$

where θ_i is the angle of incidence and θ_r is the angle of refraction. This is now known as Snell's law.

At a later date it was understood that this constant ratio is the refractive index between the two materials that the light passes between.

More generally, Snell's law of refraction is stated as:

$$_1n_2 = \frac{\sin\theta_1}{\sin\theta_2}$$

where

$_1n_2$ = refractive index between materials 1 and 2

θ_1 = angle of incidence in material 1

θ_2 = angle of refraction in material 2

or, since

$$_1n_2 = \frac{n_2}{n_1}$$

it follows that:

$$\frac{n_2}{n_1} = \frac{\sin\theta_1}{\sin\theta_2}$$

and

$$n_1 \sin\theta_1 = n_2 \sin\theta_2$$

where

n_1 = refractive index of material 1

n_2 = refractive index of material 2

EXAMPLE

The speed of light is $3.00 \times 10^8\,\mathrm{m\,s^{-1}}$ in air, and $2.29 \times 10^8\,\mathrm{m\,s^{-1}}$ in ice.

a) Calculate the refractive index of ice.

b) Calculate the angle of refraction for a ray of light that is incident on the air–ice boundary at 35°.

Answer

a) refractive index $= \frac{c_1}{c_2}$

$$= \frac{3.00 \times 10^8}{2.29 \times 10^8}$$

$$= 1.31$$

b) Snell's law states that:

$$_1n_2 = \frac{\sin\theta_1}{\sin\theta_2}$$

$$1.31 = \frac{\sin 35}{\sin\theta_2}$$

$$\theta_2 = \sin^{-1}\left(\frac{\sin 35}{1.31}\right) = 26°$$

EXAMPLE

Light travelling in water is incident on a glass surface at an angle of 37°. What is the angle of refraction of the light in the glass of refractive index 1.5?

Answer

$$n_1 \sin\theta_1 = n_2 \sin\theta_2$$

$$1.33 \times \sin 37° = 1.5 \sin\theta_2$$

$$\sin\theta_2 = \left(\frac{1.33}{1.5}\right) \times \sin 37°$$

and

$$\sin\theta_2 = 0.533$$

$$\theta_2 = 32°$$

Total internal reflection

Total internal reflection of light is the complete reflection of light at a boundary within a material that has a higher refractive index than its surroundings.

Critical angle is the angle of refraction for which the angle of incidence is 90°.

Total internal reflection is the complete reflection of waves back inside a medium at a boundary with a second material in which the wave travels faster. For example, light can be totally internally reflected off the inside of a glass block in air. Light travels more slowly in glass than air. (Air has a lower refractive index than glass.)

In the example of light passing from glass to air, the angle of refraction is larger than the angle of incidence. When the angle of refraction reaches 90°, refracted light travels along the boundary and the angle is called the critical angle, θ_c. When the angle of refraction reaches 90° all light is refracted internally. At this point the refracted angle is 90° and $\sin\theta = 1$.

When light is in a material with refractive index n_1 and is incident on a boundary with medium which has refractive index n_2, we can use Snell's law to predict the critical angle as follows:

$$n_1 \sin\theta_1 = n_2 \sin\theta_2$$

$$n_1 \sin\theta_c = n_2$$

since $\theta_2 = 90°$

$$\sin\theta_2 = 1$$

so

$$\sin\theta_c = \frac{n_2}{n_1}$$

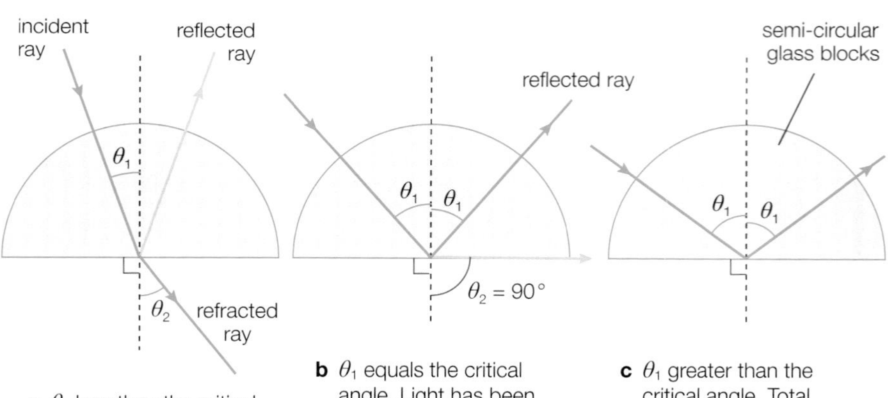

Figure 5.24 Total internal reflection.

EXAMPLE

The refractive index of ice is 1.31. Calculate the critical angle for light striking a boundary between ice and air.

Answer

$$\sin\theta_c = \frac{n_2}{n_1}$$

$$\sin\theta_c = \frac{1}{1.31} = 0.763$$

So

$$\theta_c = 50°$$

EXAMPLE

Light travelling inside flint glass with a refractive index of 1.58 is incident at boundary with a layer of paraffin, which has a refractive index of 1.44. What is the critical angle for the glass–paraffin boundary?

Answer

$$\sin\theta_c = \frac{n_2}{n_1}$$

$$\sin\theta_c = \frac{1.44}{1.58} = 0.91$$

So

$$\theta_c = 66°$$

TEST YOURSELF

12 a) Calculate the angle of refraction for a ray of light travelling from water into air, when the angle of incidence is 24°. The refractive index of water is 1.33.

b) Calculate the critical angle for water.

13 Light travels from water into oil, approaching the boundary at an angle of 30°. The refractive index of the oil is 1.52 and the refractive index of water is 1.33.

a) Calculate the angle the refracted ray makes with the boundary.

b) Calculate the ratio speed of light in water (c_w) : speed of light in oil (c_o).

14 When light passes from air into a crown glass prism, it disperses into a visible spectrum from red to violet light. The refractive index of red light in crown glass is 1.509 and the refractive index of violet light is 1.521. Calculate the difference in refracted angle for each colour if the light originally approaches the boundary at 40°.

15 Diamond has a refractive index of 2.4.

a) Calculate the critical angle of diamond.

b) Explain what a jeweller must do to a diamond to make it sparkle.

c) Calculate the critical angle for rays leaving a diamond when it is submerged in water with a refractive index of 1.33.

16 A light ray leaves a fish in a fish tank and is incident on the glass aquarium wall at an angle of 40°. At what angle of refraction does the ray emerge into the air on the other side of the glass? The refractive index of water is 1.33.

17 Read the following passage then answer the questions below.

Figure 5.25 shows a cross-section of the Earth with seismic waves spreading out from an earthquake at the top of the diagram. Two types of wave are shown: p-waves (primary or pressure waves) are longitudinal waves; s-waves (secondary or shear waves) are transverse waves. The Earth has three major layers: the solid mantle is its outer layer; below that is the outer core, which is molten or liquid rock; and the deepest layer is the solid inner core. Waves are refracted as they pass into the Earth's centre, and there is marked change of direction when the waves pass from the mantle into the outer core.

a) Explain why pressure waves can travel through the outer core but shear waves cannot.

b) i) Explain how the diagram shows that p-waves slow down when they pass from the mantle to the outer core.

ii) Explain how the diagram shows that both p-waves and s-waves travel more quickly as they travel deeper into the mantle.

c) P-waves travel with a velocity of 12.5 $km\,s^{-1}$ as they reach the mantle/outer-core boundary. Use the information in the diagram to calculate the speed of the p-waves as they enter the outer core from the mantle.

d) Calculate the critical angle for p-waves as they pass from the outer core into the mantle. Comment on the angles you see in the diagram.

e) A p-wave with a frequency of 0.2 Hz reaches the mantle/outer-core boundary. Use the information in part (c) to calculate the wavelength of the wave.

f) The diagram shows a region that is called the 'shadow zone'. The theory of refraction predicts that no seismic waves will reach this zone. However, weak seismic waves are detected in this zone. A seismologist suggests waves may reach the shadow zone due to diffraction. Explain how this might happen.

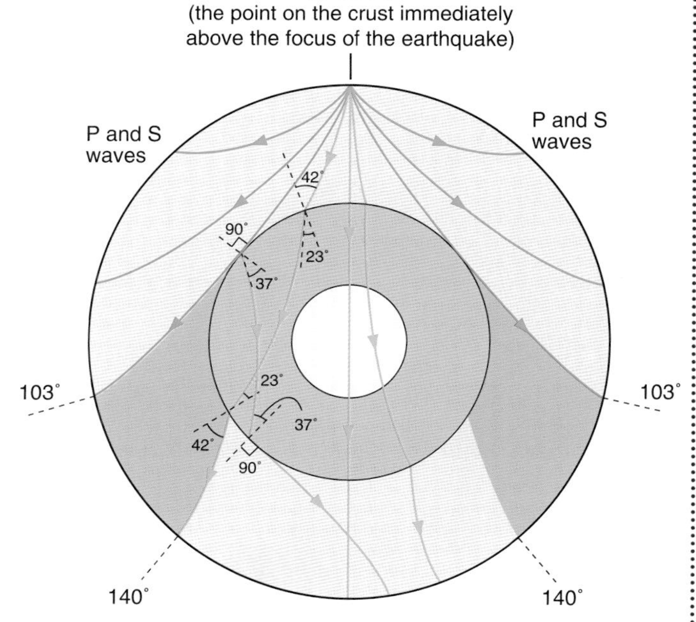

Figure 5.25 Seismic waves.

Optical fibres

Optical fibre is a thin glass (or plastic) fibre that transmits light.

An optical fibre is a thin glass (or plastic) fibre that transmits light or infrared radiation. These waves travel through the glass but are trapped inside by repeated total internal reflection. This is possible if each reflection has an angle of incidence larger than the critical angle. The critical angle depends on the ratio between the refractive index of the optical fibre and its cladding (coating).

Step index optical fibre is an optical fibre with a uniform refractive index in the core and a smaller uniform refractive index for the cladding.

A step index optical fibre has a central core with a uniform refractive index, while the cladding has a different, smaller, refractive index. By choosing a suitable material for the core and cladding, only certain wavelengths of light or infrared radiation can travel through the fibre by total internal reflection.

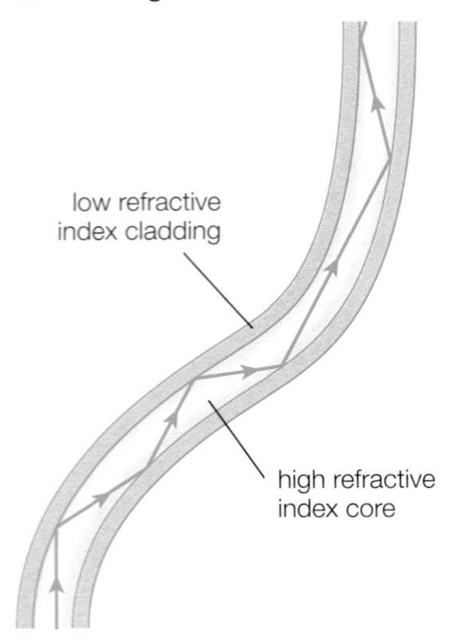

Figure 5.26 Optical fibre.

Material and modal dispersion

Material dispersion is the spreading of a signal caused by the variation of refractive index with wavelength.

Pulse broadening occurs when the duration of a pulse increases as a result of dispersion in an optical fibre.

Modal dispersion is the spreading of a signal caused by rays taking slightly different paths in the fibre.

A spectrum seen using a prism is caused by dispersion. Different colours of light travelling through the glass slow down by different amounts. The refractive index varies with wavelength. A similar effect happens inside optical fibres. As a signal travels within an optical fibre, it disperses. Two types of dispersion occur in step index optical fibres:

- material dispersion occurs because the refractive index of the optical fibre varies with frequency. Different wavelengths of light in the signal travel at different speeds. This causes a sharp pulse to spread into a broader signal. Therefore, the duration of each pulse increases. This is called pulse broadening. Pulse broadening is a problem because it limits the maximum frequency of pulses and therefore the bandwidth available.

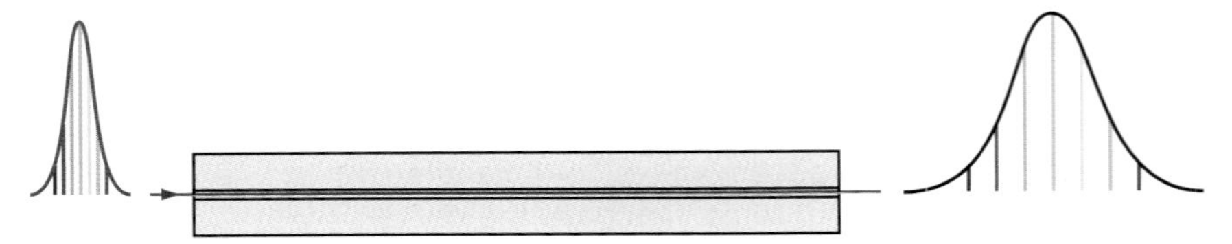

Figure 5.27 Material dispersion: different frequencies have a different refractive index.

- modal dispersion occurs when rays inside an optical fibre take slightly different paths. Rays taking longer paths take longer to travel through the fibre, so the duration of the pulse increases and the pulse broadens. Modal dispersion is significant in multimode fibres, because these fibres are broad enough to allow rays to take different paths. For communications, monomode fibres are used. These have a very narrow core, so that light is very nearly confined to one single path along the axis of the cable.

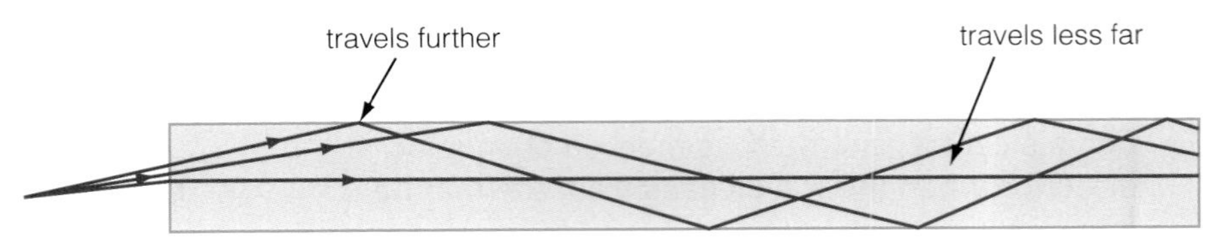

Figure 5.28 Modal dispersion: rays can take more than one path in the fibre.

Absorption

Absorption occurs when energy from a signal is absorbed by the optical fibre in which it travels.

Some wavelengths of light are absorbed strongly in materials that are used to make optical fibres, so the signal strength falls. It is important to make an optical fibre from a material with low absorption at the wavelength used to send signals. The wavelengths commonly used are 650 nm, 850 nm and 1300 nm. It may also be necessary to amplify the signal if it travels long distances through the optical fibre.

TEST YOURSELF

18 Optical fibres have a core surrounded by cladding made from a different material.
 a) Describe the function of the core and of the cladding for an optical fibre.
 b) Suggest at least one suitable property for each material.

19 The refractive index of the cladding of an optical fibre is 1.52, and the core has a refractive index of 1.62.
Calculate the critical angle for light incident on the core–cladding boundary.

20 Describe one similarity and one difference between modal dispersion and material dispersion.

21 a) Describe one way an optical fibre can be designed to minimise multipath dispersion.
 b) Explain why it is important to minimise pulse broadening.

22 Figure 5.29 shows three clear pieces of glass, X, Y and Z, which are all joined together. Each piece of glass has a different refractive index. A ray of light passes through the blocks as shown.
 a) State two conditions necessary for light to undergo total internal reflection at a boundary between two transparent media.
 b) The refractive index of glass X is 1.73. Calculate the speed of light in this glass.
 c) Show that angle θ is about 34°.
 d) The refractive index of glass Z is 1.36. Calculate the critical angle between glass X and glass Z.
 e) Explain what happens to the ray once it strikes glass Z.

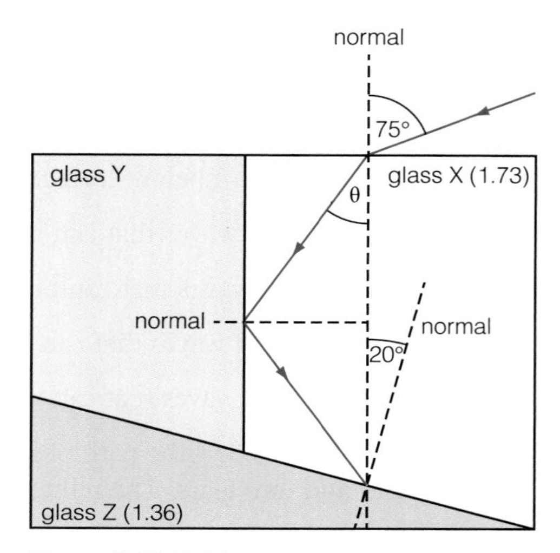

Figure 5.29 Multiple refractions.

Practice questions

1 A prism for which all angles are 60° has a refractive index of 1.5. What angle of incidence at the first face will just allow total internal refraction at the second face?

A 48° **C** 28°

B 42° **D** 60°

2 Two polarising filters are aligned to transmit vertically polarised light. They are held in front of a source of horizontally polarised light. The filter closest to the light source is rotated by 45°. The intensity of light passing through the filters

A does not change **C** increases to maximum intensity

B increases **D** decreases

3 Electromagnetic radiation and sound waves are two types of waves. Which statement below correctly describes both of them?

A progressive waves that can be polarised and reflected

B transverse waves that can be polarised and refracted

C longitudinal waves that can be refracted and reflected

D progressive waves that can be refracted and reflected

4 Figure 5.30 shows the path of light travelling from air, through water and into glass. The refractive index of glass is 1.50.

The refracted angle in glass, θg, is

A 25° **C** 40°

B 30° **D** 66°

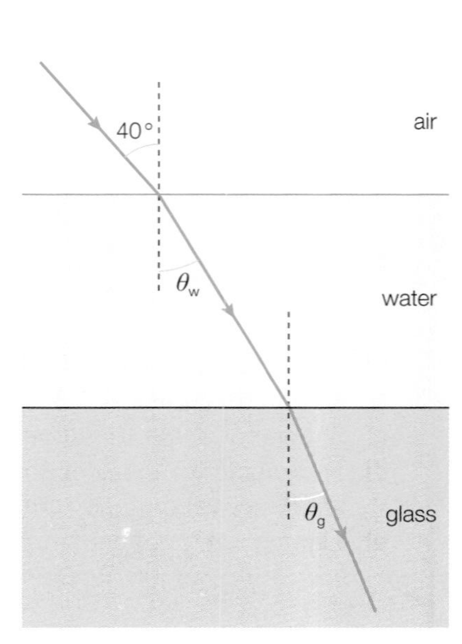

Figure 5.30 Light travelling from air, through water and into glass.

5 A ship is stationary on the ocean. As the ocean waves move past the ship, the front of the ship rises and falls with a period of 10 s. When the crest of a wave is under the front of the ship, an adjacent trough is under the back of the ship. The ship is 60 m long. What is the speed of the ocean waves relative to the sea bed?

A 6 m s^{-1} **C** 70 m s^{-1}

B 12 m s^{-1} **D** 120 m s^{-1}

6 A wave of frequency 2.5 Hz travels along a string with a speed of 20 m s^{-1}. What is the phase difference between the oscillations at points 2 m apart along the string?

A $\frac{\pi}{4}$ rads **C** π rads

B $\frac{\pi}{2}$ rads **D** 2π rads

7 A transverse seismic wave is produced by an earthquake. The wave travels through rock. Figure 5.31 shows the position of particles in the rock at an instant as the wave passes.

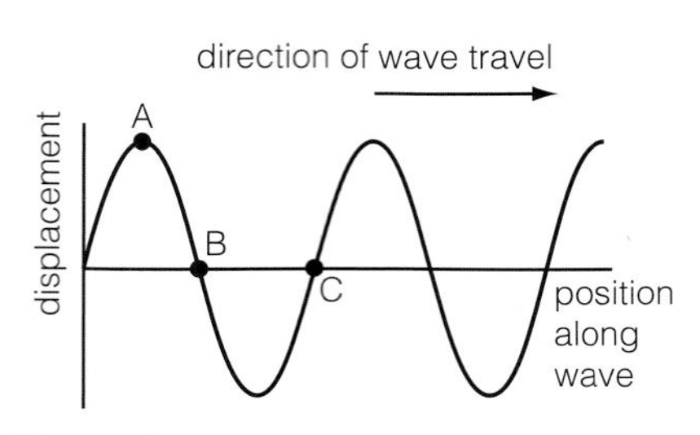

Figure 5.31 The position of particles in the rock at an instant as the wave passes.

Which line shows the correct phase difference between A and B, and B and C?

	Phase difference / rads	
	between A and B	**between B and C**
A	$\frac{\pi}{4}$	$\frac{3\pi}{2}$
B	$\frac{\pi}{2}$	π
C	$\frac{\pi}{2}$	$\frac{3\pi}{2}$
D	$\frac{3\pi}{4}$	π

8 The frequency of the seismic wave shown in question 7 is measured and found to be 6.0 Hz. The wave speed is $4.5 \times 10^3\,\text{m s}^{-1}$. What is the wavelength of the seismic wave?

A 750 m **C** 13 500 m

B 7 500 m **D** 27 000 m

9 Figure 5.32 shows a typical glass step index optical fibre used for communications. What is the refractive index of the core?

A −3.1 **C** 1.47

B 0.04 **D** 1.49

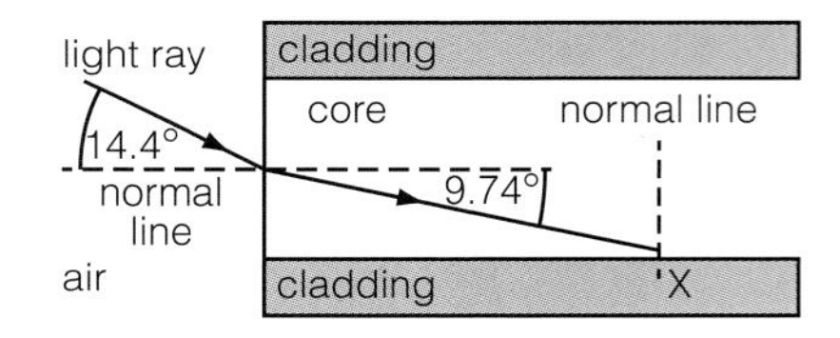

Figure 5.32 An optical fibre.

10 A light ray travels from a slab of air into amber. The refractive index of amber is 1.56. Which of the following pairs gives possible values for the angle of incidence and the angle of refraction?

	Angle of incidence	Angle of refraction
A	20°	16°
B	40°	20°
C	60°	34°
D	80°	40°

11 The diagram shows an a.c. supply. This is a three-phase supply, so three voltages are supplied, running at the same time but out of phase with each other.

a) The supply has a frequency of 50 Hz. Calculate its time period. *(1)*

b) State the phase difference between phase 2 and phase 3 in terms of radians. *(1)*

c) Calculate the time between phase 1 having a maximum value and phase 3 having a maximum value. *(1)*

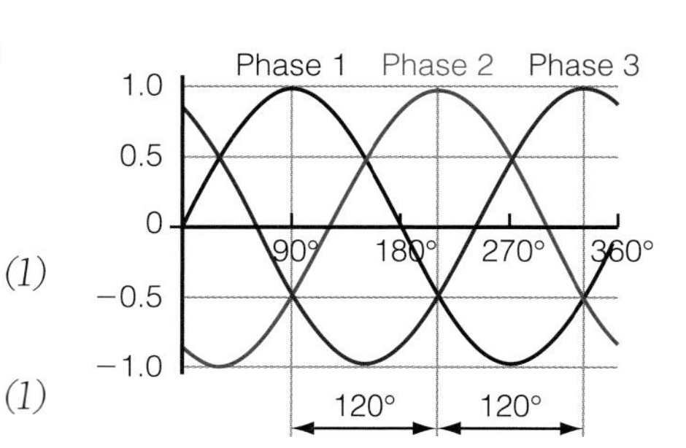

Figure 5.33 An a.c. supply.

12 Submarines communicate underwater using very low-frequency radio waves. These waves travel at approximately $3.33 \times 10^7\,\text{m s}^{-1}$ in seawater. One underwater system uses a frequency of 82 Hz.

a) Calculate the wavelength of radio waves of this frequency, giving the unit. *(2)*

A dolphin near the submarine sends out a series of clicking sounds of frequency 100 kHz. The speed of sound waves in water is $1500\,\text{m s}^{-1}$. The time for the dolphin to detect the sounds reflecting off the submarine is 100 ms.

b) How far away is the submarine? *(3)*

c) Compare the two methods of communicating underwater. *(4)*

13 Figure 5.34 shows a glass sculpture made by Colin Reid. In one picture, a reflection of the internal surface is visible; from a slightly different angle the reflection is not visible.

a) State the effect that makes this possible. *(1)*

b) Explain what has changed to change the appearance of the sculpture. *(4)*

Figure 5.34 Two pictures of a glass sculpture made by Colin Reid.

14 A prism (not drawn to scale) has a refractive index of 1.50, and internal angles of 60° and 120°. Light is incident on the face shown at an angle of 30° to the normal, which is shown as a dotted line. Use calculations to show the path of the light through the prism.

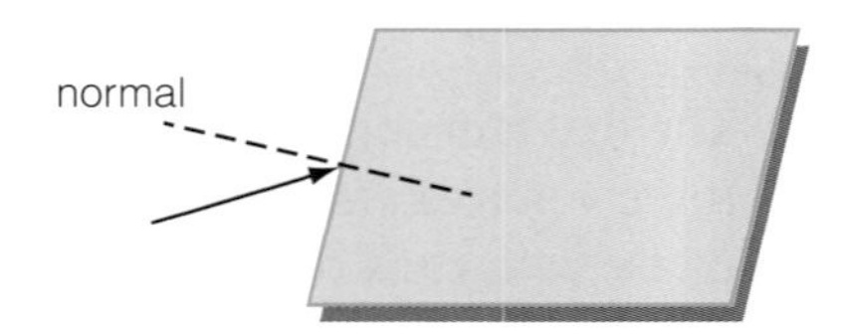

Figure 5.35 A prism.

15 Figure 5.36 shows an optical fibre.

a) Name the process by which light is trapped in the optical fibre. *(1)*

b) The refractive index of the core is 1.60 and the refractive index of the cladding is 1.50. Calculate the critical angle for light travelling in the optical fibre. *(3)*

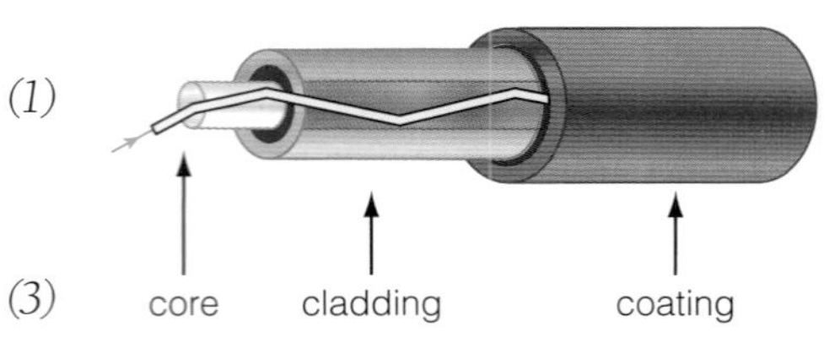

Figure 5.36 An optical fibre.

c) A multimode optical fibre has a wider diameter core than a single mode fibre, and carries more information.

i) Explain what is meant by pulse broadening. (2)

ii) Explain why a multimode optical fibre is more likely to suffer from pulse broadening than a single mode optical fibre. (2)

iii) Explain one implication of pulse broadening. (2)

d) Explain why it is important to select the material for the core very carefully. (5)

16 A student measures the refractive index of glass using Snell's law. The student measured the angle of incidence and angle of refraction for rays travelling through the glass block.

Angle of incidence	Angle of refraction	Angle of refraction (mean)
0	0, 0	0
10	5, 6	5.5
15	10, 10	10
20	12, 14	13
25	15, 18	16.5
30	20, 20	20
35	23, 24	23.5
40	25, 25	25
45	29, 30	29.5
50	32, 32	32
55	34, 33	33.5

a) The student kept the same number of decimal places in the raw data. Explain why this is acceptable. (1)

b) The student has calculated the mean. Explain why the calculated data has not been presented correctly in the table. (1)

c) Calculate the values of $\sin i$ and $\sin r$, and plot these on a graph. (5)

d) The ratio $\sin i : \sin r$ equals the refractive index, n, for the material. Use your graph to calculate the refractive index for the material. (4)

e) Calculate the maximum uncertainty in the readings. (2)

f) Describe two difficulties with carrying out this experiment, and how the student could overcome these. (4)

g) The angles are measured using a protractor to a precision of $\pm 0.5°$. For the angle θ, measured to be 20°, calculate the range of the values of $\sin \theta$ and the percentage uncertainty in $\sin \theta$.

17 The Fermi gamma ray space telescope is a satellite that orbits Earth at a height of 550 km. Communications with the satellite from Earth use electromagnetic radiation.

a) Calculate the minimum time it takes a signal to reach the satellite. (3)

b) Suggest, with reasons, a suitable type of electromagnetic radiation to communicate with the satellite. (2)

c) Explain two advantages of using a space-based telescope orbiting Earth to observe gamma ray bursts from distant galaxies. (2)

The Arecino Observatory includes a telescope with a 305 m diameter dish. It detects wavelengths between 3 cm and 1 m.

d) Describe the two advantages of basing the telescope on Earth. (2)

Stretch and challenge

18 A fibre optic cable has a core of refractive index n_{co}, and is surrounded by cladding of refractive index n_{cl}.

a) Write an expression for the critical angle, θ_c, for a pulse of red light travelling inside the core.

The fibre optic cable is made from glass whose refractive index varies with wavelength. The table gives information about the refractive index of the cladding and the core for light of different wavelengths.

Wavelength	Refractive index (core)	Refractive index (cladding)
400 nm	1.635	1.470
700 nm	1.602	1.456

b) Calculate the difference in critical angle for red light and for green light.

An endoscope is a device that uses optical fibres to look inside the body. One endoscope uses the material described in part a) for a light guide used to illuminate inside a patient's body. The diagram, which is not to scale, shows the core of the optical fibre of diameter 0.20 mm.

A ray of white light travelling along the axis of the fibre reaches the region where the fibre is coiled into an arc of radius R.

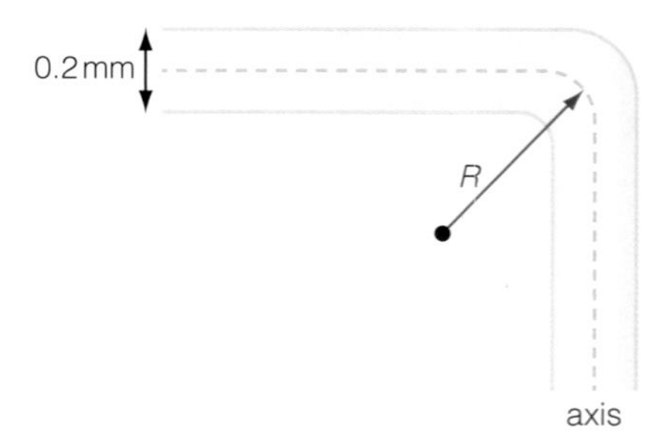

Figure 5.37

c) Write down the condition needed for all light to remain in the fibre.

d) Calculate the smallest value of R, which allows the ray to remain within the fibre

e) Explain why it is important to use a cladding material for optical fibres used in an endoscope

19 The speed of longitudinal waves, c on a spring is given by

$$c = \sqrt{(kl/\mu)}$$

where k is the spring constant, l is the stretched length of the spring and μ is the mass per unit length.

a) By substituting the dimensions of the quantities in this equation, show that the dimensions of the spring constant are MT^{-2}

b) The slinky has a mass of 0.6 kg and is stretched by 3 m using a force of 9 N. Calculate the speed of longitudinal waves in the slinky. Assume the spring had negligible length initially.

c) Show that doubling the length of the stretched slinky does not affect the time for the wave to travel from one end of the slinky to the other.

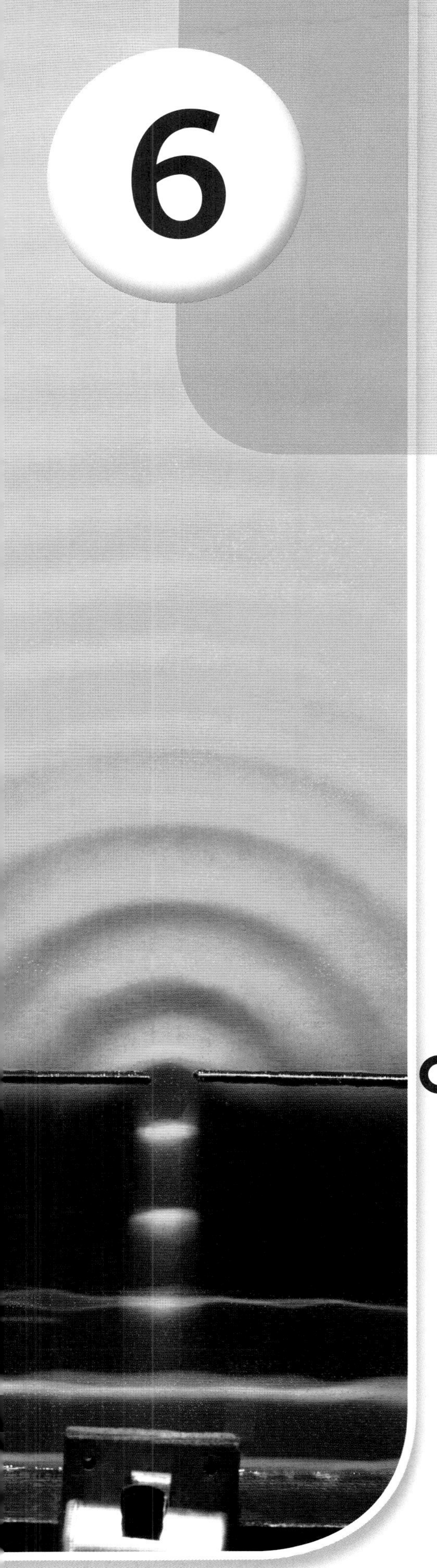

6 Combining waves

PRIOR KNOWLEDGE

- The wave equation states that the speed, v, of a wave (in m s^{-1}) equals its frequency, f (in Hz), multiplied by its wavelength, λ (in m).

  $v = f \times \lambda$
- The frequency of a wave, $f = \frac{1}{T}$, where T is the period of the wave.
- The phase of a wave describes the fraction of a cycle completed compared to the start of the cycle.
- Phase difference for a single wave compares different points along the wave at the same time.
  - i Parts of a wave are in phase when particles in the medium move at the same speed in the same direction.
  - ii Parts of a wave that move in different directions and speeds are out of phase.
  - iii Parts of a wave moving in opposite directions at the same speed are in antiphase.
- In one complete cycle, waves travel a distance of one wavelength.
- The shortest distance between two points moving in phase is one wavelength, λ.
- The phase difference between two points a wavelength apart is 2π or 360°.
- The phase difference, Φ, between two points along a wave, separated by a distance x, is given by:

  $$\Phi = \frac{2\pi x}{\lambda} \text{ rad}$$
- Waves diffract when they go through a small gap. The angle of diffraction is small when the wavelength is small in comparison with the gap width; the angle of diffraction is large when the wavelength is comparable in size to the gap width.

TEST YOURSELF ON PRIOR KNOWLEDGE

1. Calculate the time period of a wave of frequency 50 Hz.
2. Calculate the frequency of a wave if its time period is 2×10^{-3} s.
3. What is the phase difference between two points along a wave separated by a distance of 2.75 wavelengths?
4. When two points of a wave are in phase, what can you say about their motion?
5. Explain why a radio receiver can detect radio waves (wavelength 30 m to 1000 m) emitted by a transmitter on the opposite side of a hill that is between the receiver and transmitter.
6. Explain why we can hear but not see around the corner of buildings (wavelength of sound is 0.3 m to 10 m; wavelength of light is 400 nm to 700 nm).
7. What is the speed of radio waves of wavelength 3 km and frequency 100 kHz?

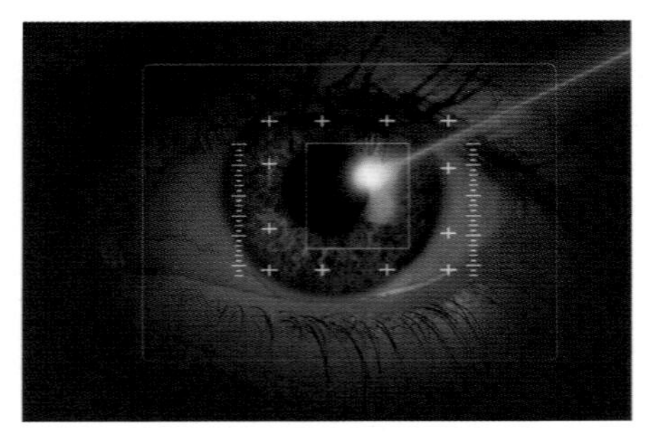

Figure 6.1 Permanent changes to your vision are possible using laser surgery.

Lasers have impacted our lives since they were developed in the 1960s. The word laser is an acronym standing for Light Amplification by Stimulated Emission of Radiation. You learned about emission of photons in Chapter 3. The great intensity of laser beams has several practical uses: laser surgery allows corrections to eyesight by reshaping the cornea on the outer surface of the eye; surgeons can cut through flesh using a laser instead of a scalpel. The accuracy and focus of the beam makes lasers useful in DVD players, scanners and optical disc drives. Much scientific research relies on the precise, polarised, coherent beam of radiation from a laser, which is created using two mirrors in an optical cavity to produce stationary waves. You will learn about stationary waves in this chapter, and other effects caused when two waves interact.

Superposition

When two waves of the same type meet at the same point and overlap, the resultant displacement of the oscillations is the vector sum of the displacements of each wave. This phenomenon is called the superposition of waves. You can see water waves superpose when you watch waves on the seashore: when two peaks of waves meet, they combine to make a larger wave.

Superposition is when two waves of the same type (e.g. sound waves) overlap and interact. The displacement of the medium, where the waves overlap, is the vector sum of the two wave displacements.

Figure 6.2 shows the superposition of two water waves of the same amplitude, A, in phase and in antiphase. The resultant amplitudes are 2A when the waves are in phase, and zero when the waves are in antiphase. When water waves overlap in phase, larger displacements of the water occur; when waves overlap in antiphase, the water can remain still. The idea of wave superposition is used to reduce unwanted noise. Figure 6.3 shows the principle. Unwanted sound, from some nearby source, is received by a microphone and fed into noise-cancellation circuitry. This circuitry inverts the noise (which is the same as changing its phase by 180°), so that the original sound and the newly added waveform cancel each other out. A worker can be protected from the sound by wearing headphones that soften a loud noise, which has the potential to damage their hearing.

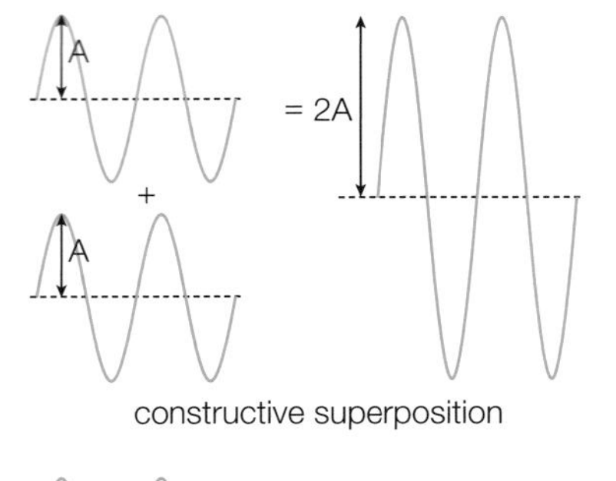

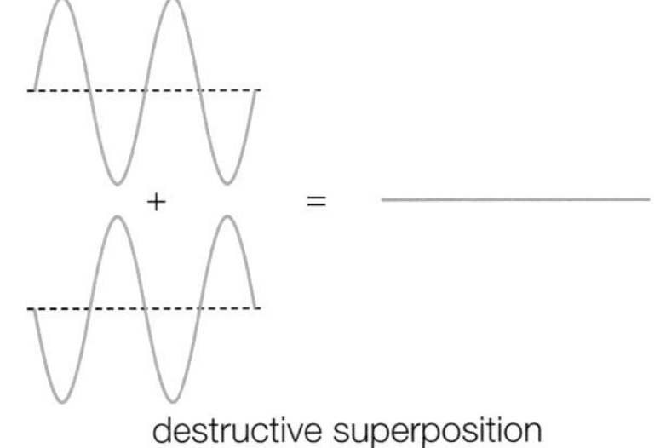

Figure 6.2 The superposition of two water waves.

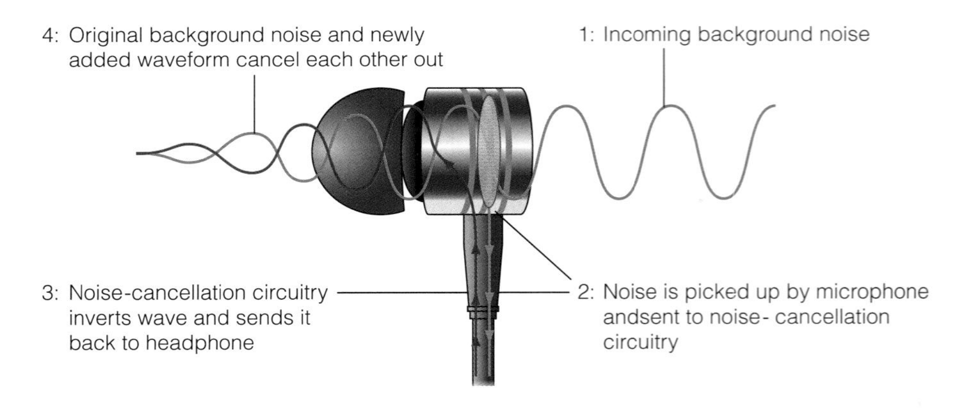

Figure 6.3 How noise cancellation works.

Superposition of waves only occurs between the same types of wave. Two water waves can superpose, but light and sound waves cannot superpose. We can notice the superposition of water waves as the process takes place relatively slowly. Often when two sources of waves overlap, we do not notice the superposition, as the frequency of the waves might be high and any pattern of superposition lasts for a very short time. Under some circumstances, two sources of waves can overlap in such a way

Interference is the name given to the superposition of waves from two coherent sources of waves. Interference is constructive if waves are in phase, or destructive if waves are in antiphase (out of phase by 180°).

Two waves are coherent when they have a fixed phase difference and have the same frequency.

that positions of high and low intensity are produced in fixed positions. Here fixed patterns of wave superposition occur in the same place over long periods of time, so that we can observe them. A fixed pattern of superposition is called interference.

Interference

Two sources of waves can overlap and produce constructive and destructive regions of superposition. However, a stable pattern or superposition (or interference) will only occur if the two sources of waves are coherent. Two sources are coherent when the waves from each source have the same wavelength (and, therefore, the same frequency), and there is also a constant phase relationship between the two sources. For example, light which meets at a point from two coherent sources can produce regions of bright light (or maxima) where light interferes constructively, and regions of darkness (or minima) where light interferes destructively.

Interference patterns from sound waves

A single sustained note from a signal generator played through two loudspeakers creates interference patterns. The loudness of the sound heard by a person walking in front of the speakers varies from loud to quiet on a regular spacing due to the pattern of constructive and destructive interference. When coherent sound waves are in phase, the sound is louder because of constructive interference.

The path difference is the difference in distance travelled by the two waves produced by the loud speakers. Path difference is usually measured in multiples of wavelength.

If the waves are in phase when they leave the speakers, they are in phase at any point where they have travelled the same distance or where their path difference is a whole number of wavelengths, $n\lambda$. These sound waves, initially in phase, are out of phase at any point where the path difference is a whole number of wavelengths plus a half wavelength, $(n + \frac{1}{2})\lambda$. Points along a wave are in phase if they are separated by a whole number of wavelengths, and in antiphase when the distance between them is a half wavelength, one and a half wavelengths, and so on.

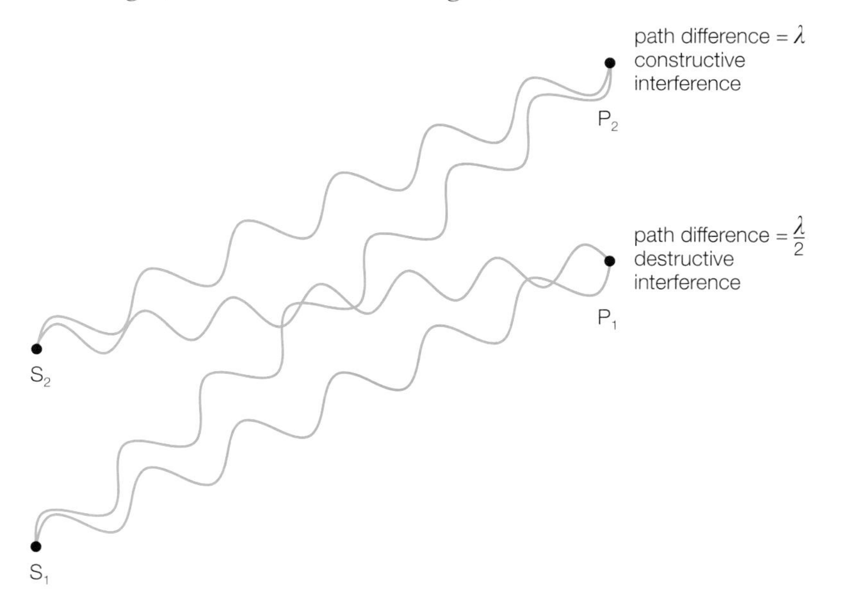

Figure 6.4 Path difference.

EXAMPLE

A student walks between two loudspeakers each playing a note of frequency 300 Hz. The waves are in phase as they leave the loudspeakers. The speed of sound in air is $340\,m\,s^{-1}$.

a) Calculate the smallest path difference for waves from each loud speaker when the student hears
   i) constructive interference
   ii) destructive interference.

b) Now calculate the next smallest path difference for
   i) constructive interference
   ii) destructive interference.

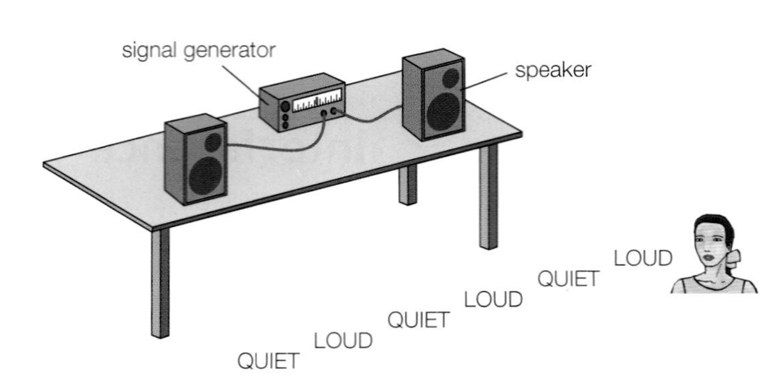

Figure 6.5 An experiment demonstrating interference of sound waves.

Answers

a i) Condition for constructive interference is a path difference of $n\lambda$. The shortest path differences occur when $n = 0$ so the shortest path difference is 0 m.

ii) Condition for destructive interference is a path difference of $(n + \frac{1}{2})\lambda$. For the minimum path difference, $n = 0$.

The path difference is $\frac{\lambda}{2} = \frac{1.13}{2}$

$= 0.57\,m$.

b i) The next smallest path difference for constructive interference, is $n = 1$

So the path difference is

$$\lambda = \frac{c}{f}$$

$$= \frac{(340\,m\,s^{-1})}{300\,Hz}$$

$$= 1.13\,m$$

ii) The next path difference for the first minimum occurs at a distance of $1.5\lambda = 1.70\,m$. $(n = 1)$

EXAMPLE

The student in the previous example stands in a region of destructive interference. Explain what the student would expect to hear

a) if one loudspeaker is covered with a cushion

b) if the connections to one loud speaker are reversed.

Answers

a) The student will hear the sound get louder as there is no destructive interference, but only the sound from one speaker.

b) Changing the connections round makes sound from the two speakers move out of phase with each other. A path difference of $\frac{\lambda}{2}$ now makes the waves in phase again. So there is constructive interference and the sound becomes louder.

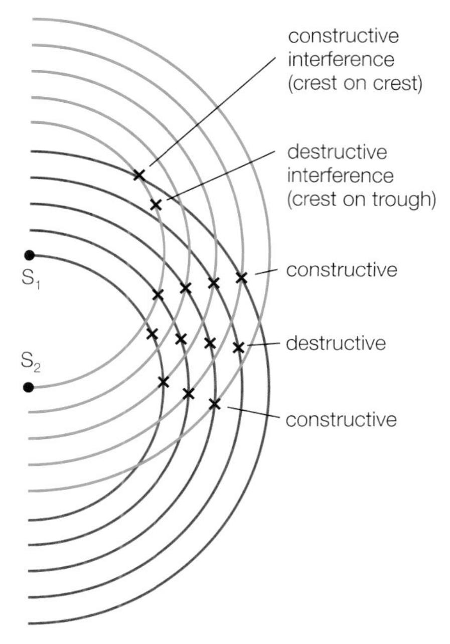

Figure 6.6 Interference in a ripple tank.

Young's double slit experiment–Required practical number 2

Young's double slit experiment demonstrates interference between coherent light sources, thus showing the wave nature of light. The experiment uses two coherent sources of light waves, produced from a single source of light, which then pass through two very narrow, parallel slits. The light diffracts (spreads) through the slits, producing an interference pattern of fringes on a screen. Interference occurs because the waves overlap and superpose in a stable pattern.

Light has such a short wavelength it is difficult to see its interference patterns. This is why Young's double slit experiment works best using a blacked out room and a very bright white light source, or a laser as a source of intense single-wavelength (monochromatic) light. The slits must be very narrow and less than 1 mm apart. An interference pattern can be seen where patches of bright light alternate with regions of darkness. These correspond to areas of constructive and destructive interference. These patterns are called **fringes**. The interference fringes are visible on a screen placed at least 1 m away from the slits. If the screen is further away, the fringe separation increases but their appearance becomes fainter.

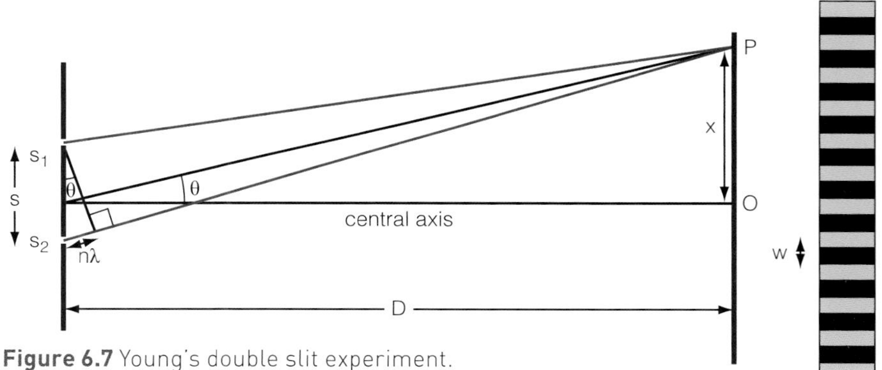

Figure 6.7 Young's double slit experiment.

Light from each slit travels a slightly different route to the screen, creating a path difference as shown in Figure 6.7. Dark fringes occur where there is destructive interference (the path difference between the two slits is $(n + \frac{1}{2})\lambda$). Bright fringes occur where there is constructive interference (the path difference between the two slits is $n\lambda$).

Fringes are ordered. The order of the central bright fringe is $n = 0$. The order of the fringes closest to the central fringe is $n = 1$; the order of the next pair of fringes is $n = 2$, etc.

Referring to Figure 6.7, the condition for constructive interference (bright fringes) is:

$$s \sin \theta = n\lambda$$

where s = spacing between the slits

θ = angle from the beam towards the screen

n = a whole number

λ = wavelength of light.

The extra distance travelled by the waves leaving S_2 is $s \sin \theta$, and for constructive interference this distance (or path difference) must be a whole number of wavelengths, $n\lambda$.

We can also express the sepration of the fringes in terms of the angle θ.

Refering again to Figure 6.7, we can see that:

$$\frac{X}{D} = \tan\theta$$

where X is the distance to the nth fringe from the midpoint, and D is the distance from the slits to the screen on which we see the interference fringes.

However, since $D \gg X$, the angle θ is small.

Then, the small angle approximation gives:

$$\sin\theta \approx \tan\theta$$

since

$$\sin\theta = \frac{n\lambda}{s}$$

and

$$\tan\theta = \frac{X}{D}$$

it follows that (for small angles):

$$\frac{n\lambda}{s} = \frac{X}{D}$$

We can use this formula to predict the fringe spacing, w, between two adjacent bright fringes. When $n = 1$ the distance X becomes the spacing, w, between adjacent bright fringes.

So

$$\frac{\lambda}{s} = \frac{w}{D}$$

and

$$w = \frac{\lambda D}{s}$$

EXAMPLE

A student shines monochromatic light of wavelength 580 nm through two slits 0.25 mm apart. A screen is positioned 2.3 m away. What is the separation of the bright interference fringes seen on the screen?

Answer

$$w = \frac{\lambda D}{s}$$

$$w = \frac{(580 \times 10^{-9}\,\text{m}) \times (2.3\,\text{m})}{(2.5 \times 10^{-4}\,\text{m})}$$

$$= 5.3 \times 10^{-3}\,\text{m or } 5.3\,\text{mm}$$

LASER DANGER

Lasers produce a very intense light. In schools, the most powerful lasers that are allowed are class 2 lasers which must have a power output of less than 1 mw. You should never look directly at a laser or its reflection, or allow it to reflect from shiny surfaces into anyone's eyes.

TEST YOURSELF

1 a) Radiowaves are emitted from two aerials in phase. At a position x, the path difference between the two aerials is 3 wavelengths. Explain whether there will be constructive or destructive interference of the waves at x.
   b) Explain how your answer to part (a) would change if the waves were out of phase as they left the aerial.
   c) Coherent soundwaves are emitted from two loudspeakers in phase. A microphone is placed at a position where the path difference between the two loudspeakers is 5.5 wavelengths. What type of interference is detected?
   d) Explain how your answer to part (d) would change if the waves were out of phase as they left the loudspeaker.
2 Compare the different conditions needed for superposition and for interference.
3 Explain what is meant by two sources of coherent waves, giving an example.
4 Two radio wave transmitters, that are separated by a distance of 6.0 m, emit waves of frequency 1 GHz. The waves leave the transmitters in phase. The waves are directed into a field where they are detected by a man holding an aerial at a distance of 200 m.
   a) Describe what the man detects on his aerial as he walks along a line XY, which is parallel to the line joining the two transmitters.
   b) The man stands at a point of maximum intensity, when the phase of one of the transmitters is advanced by $\frac{1}{4}$ of a cycle. How far does he have to walk along XY to detect a maximum again? Does it matter which way he walks?
5 A Young's slit experiment uses light of wavelength 600 nm.
   a) Explain why laser light is often used for Young's slit experiments.
   b) If the two slits are 0.2 mm apart, and the screen is 3.0 m away, calculate the separation of the fringes.
   c) What changes could be made to increase the separation of the fringes?
6 A microwave transmitter transmits waves with a wavelength of 2.8 cm through slits in a metal plate that are 5.8 cm apart. A probe is placed 30 cm from the slits and detects regions of high and low intensity. Calculate the separation of the regions of high intensity.
7 In a Young's slit experiment, a student measures the separation between six bright fringes to be 3.0 mm. The light used in the experiment has a wavelength of 600 nm, and the slits are placed 36 cm from the screen. Calculate the slit separation.
8 The diagram below shows two waves approaching each other on a string. Describe the displacement of the string at points A and B over the next 5 seconds.

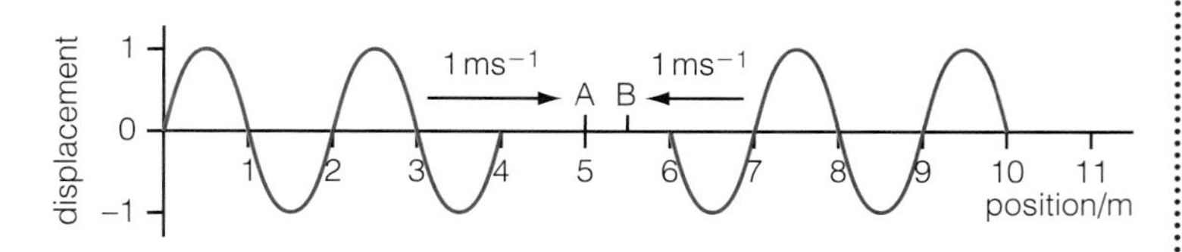

Figure 6.8 Two waves approaching each other on a string.

Stationary waves

A **stationary (or standing) wave** is a wave formed by the superposition of two progressive waves of the same frequency and amplitude travelling in opposite directions.

If you watch a plucked guitar string closely, you will see that it appears to be in two fixed positions at once. In fact, the string is vibrating rapidly between two positions. What you see is called a stationary (or standing) wave. Stationary waves can be set up in many situations including using a guitar string.

Creating stationary waves

Stationary waves are created when two progressive waves of the same frequency and amplitude moving in opposite directions superpose. This often occurs when reflections of a progressive wave superpose with the original wave. When waves reflect off a rigid boundary, the reflected waves are out of phase with the original wave by 180° and travelling in the opposite direction, but have the same amplitude and frequency.

TIP

The maximum displacement of all the particles along a stationary wave is not the same for stationary waves. Particles at nodes do not move at all, but particles at antinodes vibrate at the maximum amplitude.

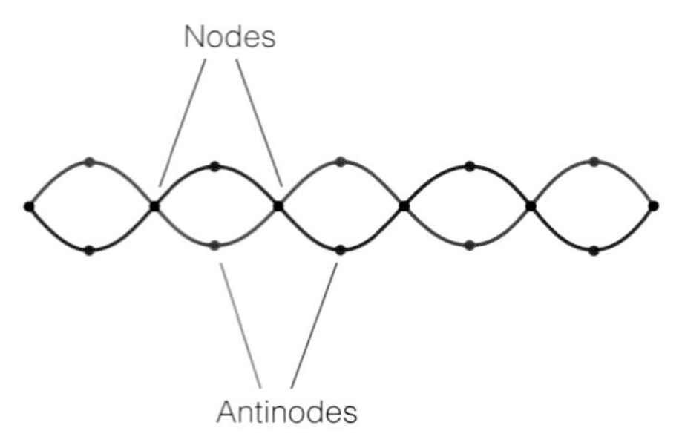

Figure 6.9 A stationary wave.

A **node** is a point of zero amplitude on a stationary wave.

An **antinode** is a point of maximum amplitude on a stationary wave.

Stationary waves only form on a guitar string at specific frequencies. Where the waves are in phase, the displacements add to form a peak or a trough of double the original amplitude. Where the waves are in anti-phase, their displacements cancel out. For stationary waves, the positions of maximum and minimum amplitude remain in the same places, with particles, at antinodes, vibrating rapidly between their positions of maximum displacement. This is why the waves don't appear to progress along the string.

A stationary wave alternates between the positions shown by the red wave and the blue wave in Figure 6.9.

Nodes and antinodes

Nodes and antinodes are at fixed points along the guitar string. At a node, the amplitudes of the two progressive waves moving in opposite directions always cancel out so the particles do not oscillate at all. At an antinode, the amplitudes add together and the guitar string is displaced between the peak and trough during a cycle. The amplitude of the peak or trough is double the amplitude of the two progressive waves. Nodes are always separated by half a wavelength, as are antinodes. The displacements of particles in positions between the nodes and antinodes vary but do not remain zero or reach the amplitude of an antinode during the cycle according to their positions.

How stationary waves form

Figures 6.10 (a) to (d) show how stationary waves form on a guitar string. Wave 1 and wave 2 are progressive waves travelling in opposite directions along the string with the same frequency and amplitude. The amplitude of the wave 3 is the amplitude of the two waves superimposed.

The diagrams show a sequence of snapshots of the wave at different times during one complete cycle.

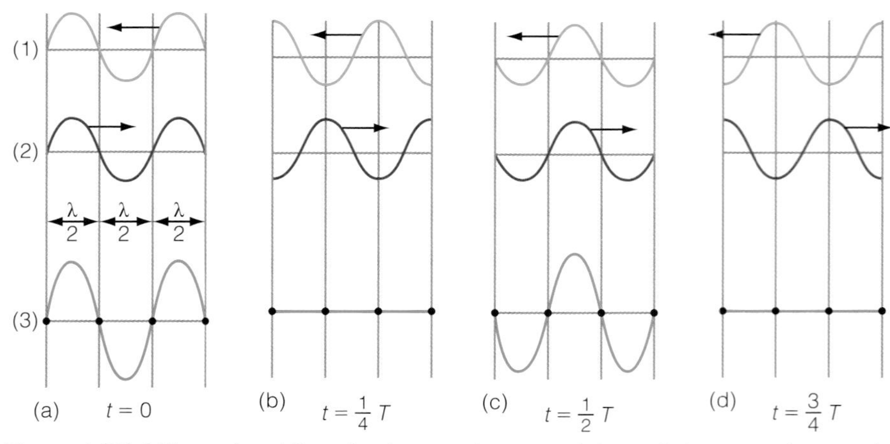

Figure 6.10(a) Waves 1 and 2 are in phase at the start of the cycle but move in opposite directions. There is constructive superposition and their displacements add. The combined displacement of both waves is double their original displacement.

Figure 6.10(b) Waves 1 and 2 are in antiphase after quarter of a cycle. The particles at all points are displaced in opposite directions. Their displacements cancel out as the superposition is destructive. The combined displacement of both waves is zero.

Figure 6.10(c) Waves 1 and 2 are in phase halfway through their cycle. Compared to the start of the cycle, the position of peaks and troughs is reversed. There is constructive superposition and their displacements add. The combined displacement of both waves is double their original displacement.

Figure 6.10(d) Waves 1 and 2 are in antiphase three-quarters of the way through a cycle. Their displacements cancel out as the superposition is destructive. The combined displacement of both waves is zero.

A **harmonic** is a mode of vibration that is a multiple of the first harmonic.

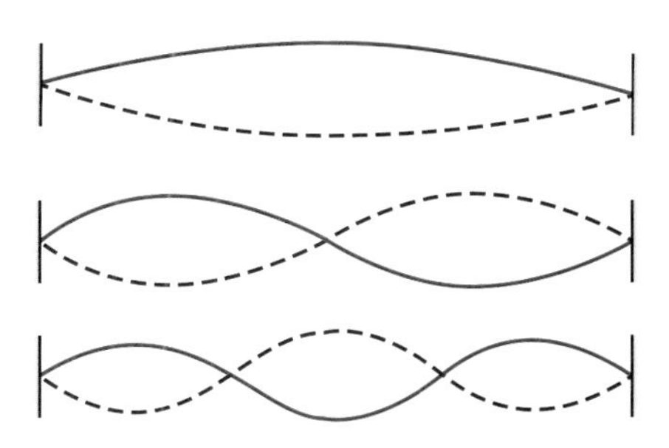

Figure 6.11 The three lowest harmonics for a guitar string.

Harmonics

A guitar string can support different modes of vibration for stationary waves. The different modes of vibration are called harmonics. Harmonics are numbered: the first harmonic is the mode of vibration with the longest wavelength. The second harmonic is the mode of vibration with the next longest wave.

The frequency of the vibration is found using

$$f = \frac{v}{\lambda}$$

where f = frequency of the harmonic in Hz

v = speed of wave in $m\,s^{-1}$

λ = wavelength of harmonic in m

In Figure 6.11,

- the first harmonic on a string includes one antinode and two nodes. Stringed instruments have nodes at each end of the string as these points are fixed. For a guitar string of length l, the wavelength of the lowest harmonic is $2l$; this is because there is one loop only of the stationary wave, which is a half wavelength. Therefore the frequency is:

$$f_1 = \frac{v}{\lambda} = \frac{v}{2l}$$

- the second harmonic has three nodes and two antinodes; the wavelength is l and frequency is:

$$f_2 = \frac{v}{\lambda} = \frac{v}{l}, \text{ or } 2f_1$$

- the third harmonic has four nodes and three antinodes; the wavelength is $2L/3$ and frequency is:

$$f_3 = \frac{v}{\lambda} = \frac{3v}{2l} \text{ or } 3f_1$$

The first harmonic on a string

For waves travelling along a string in tension, the speed of a wave is given by:

$$v = \sqrt{\frac{T}{\mu}}$$

where:

T = tension in the string in N

μ = mass per unit length of the string in $kg\,m^{-1}$

Using Figure 6.11, you can see that the first harmonic for a stationary wave on a string is half a wavelength. The wavelength of the first harmonic is $2l$ for a string of length l.

We can combine the equation above and the wave equation as follows:

$$c = f$$

$$\text{so } \sqrt{\frac{T}{\mu}} = f$$

$$= f \times 2l$$

Therefore:

$$f = \frac{1}{2l}\sqrt{\frac{T}{\mu}}$$

This equation gives the frequency of the first harmonic where:

f = frequency in Hz

T = tension in the string in N

μ = mass per unit length of the string in $kg\,m^{-1}$

l = the length of the string length in m

EXAMPLE

A string is stretched between two points 0.45 m apart.

a) Calculate the frequency of the first harmonic when the tension in the string is 80 N and the mass per unit length is $3.2\,g\,m^{-1}$.

b) Predict the effect on frequency of:

i) doubling the length of the string

ii) doubling the tension in the string.

Answers

a) $f = \frac{1}{2l}\sqrt{\frac{T}{\mu}}$

$= \frac{1}{2 \times 0.45\,m} \times \sqrt{\frac{80\,N}{3.2 \times 10^{-3}\,kg\,m^{-1}}}$

= 175.7 Hz or 176 Hz to 3 sf

b) i) Frequency is proportional to $\frac{1}{l}$, so doubling the length halves the frequency (to 88 Hz).

ii) Frequency is proportional to $\sqrt{T}$, so doubling the tension increases the frequency by a factor of $\sqrt{2}$ (to 248 Hz).

Stationary waves: Sound waves and musical instruments

Longitudinal waves such as sound waves also form stationary waves. Sound waves are pressure waves with particles vibrating in the direction the wave travels. Stationary wave diagrams for sound waves show the amplitude for particles vibrating longitudinally in the air column. The amplitude is greatest at the open end of pipes, where there is an antinode. The particles cannot vibrate at a closed end and so there is a node. If the pipe has two open ends, the stationary wave has at least two antinodes, at either end of the pipe.

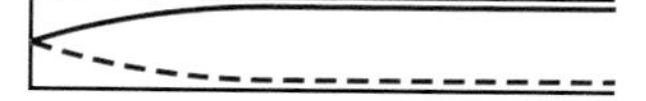

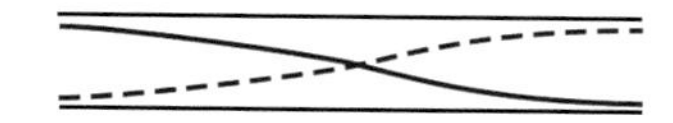

Figure 6.12 Stationary waves for sound waves in open-ended tubes. When there is one open end, there is an antinode at the open end and a node at the closed end. In a pipe that is open at both ends, there is a node in the middle and antinodes at both ends.

Musical instruments include stringed instruments, where stationary waves form on the string when it is plucked or bowed. Stationary waves also form when the air column in organ pipes or wind instruments is forced to vibrate. The harmonic frequencies available to players will vary with different instruments, even if the air column is the same length, depending on whether the instrument has one or both ends open.

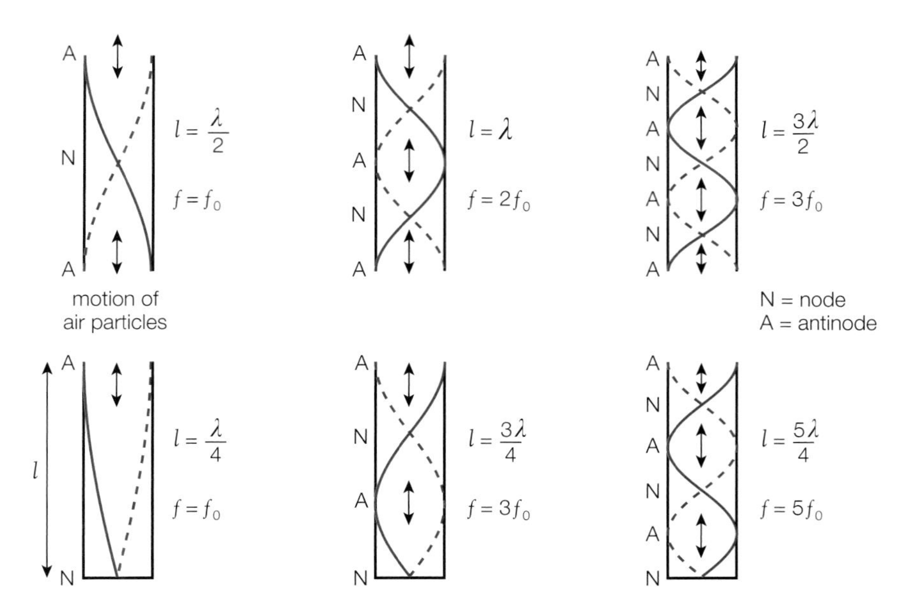

Figure 6.13 Different notes are created from the different harmonics of stationary waves in air columns.

REQUIRED PRACTICAL 1

Investigation into the variation of the frequency of stationary waves on a string with length, tension and mass per unit length of the string

Note: This is just one example of how you might tackle this required practical.

A student used the equipment shown in Figure 6.14 to investigate how the frequency of stationary waves on a wire varied with tension in the wire. The student fixed a length of wire at one end and passed the other end over a pulley. A hanger was attached to the end of the wire. The tension in the wire was changed by adding masses to the hanger.

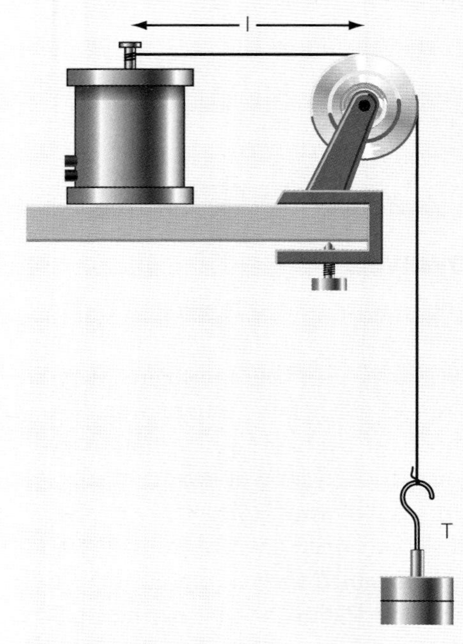

Figure 6.14

The student placed a microphone connected to a frequency analyzer near the wire. When the wire was plucked, the frequency of the sound could be read off the frequency analyzer. The student sometimes observed several frequencies, but only recorded the lowest frequency.

The student was told the mass per unit length of wire was 0.16 g/m.

1. Suggest why the string is passed over a pulley, and not over the edge of the table.

The student obtained these results with a length, l, of 0.4m:

Mass on hanger (kg)	Frequency (Hz) 1st reading	Frequency (Hz) 2nd reading	Frequency (Hz) mean
0.20	140	135	
0.40	195	200	
0.60	240	240	
0.80	280	278	
1.00	310	310	
1.20	340	340	
1.40	365	368	
1.60	392	390	

2) Calculate the average frequency using the data

3) Calculate the frequency using the equation

$$f = \frac{1}{2l}\sqrt{\frac{T}{\mu}}$$

4) Suggest, with reasons, the maximum percentage uncertainty in this experiment. You will need to suggest the resolution of the equipment used in the experiment.

Extension

Measure the change in frequency of standing waves if wires of different length are used or if a wire with a different mass per unit length is used.

ACTIVITY

Measuring the wavelength of sound waves

A student uses a set-up called a Kundt's tube to measure the speed of sound in air. The Kundt's tube is placed horizontally, and a fine powder is sprinkled along its length. One end of the tube is closed; the other end has a loudspeaker connected to a signal generator to produce notes of different frequency inside the tube.

The pitch of the note is changed until the sound from the tube increases to a maximum, and a stationary wave is set up inside the tube. The powder moves with air and settles into piles corresponding to nodes.

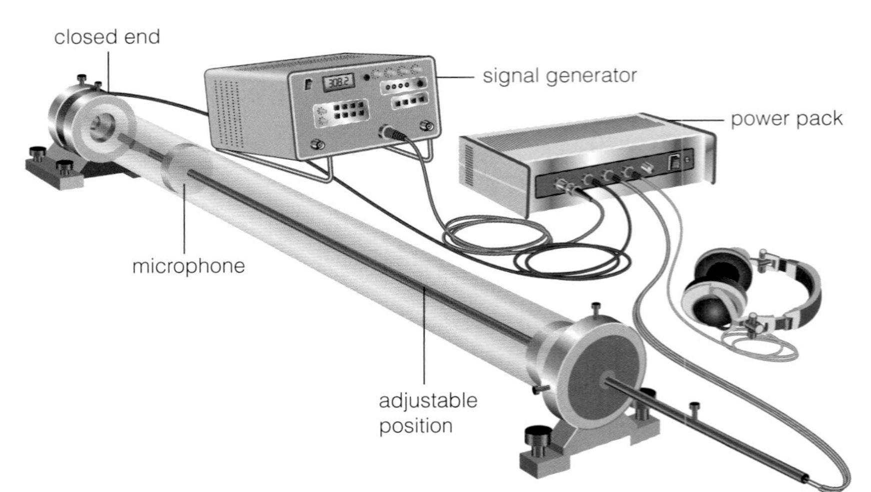

Figure 6.15 Kundt's tube apparatus.

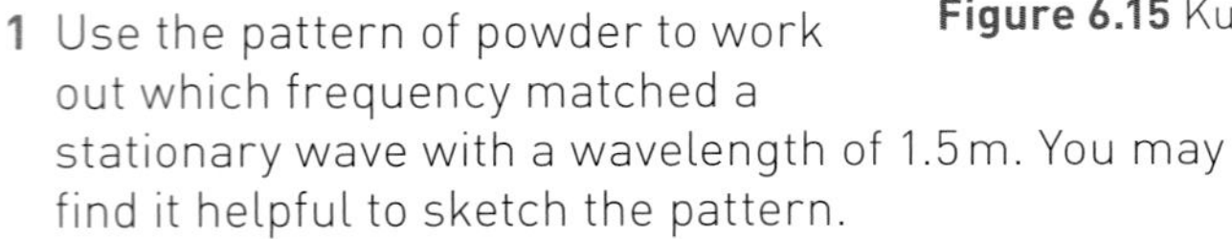

1 Use the pattern of powder to work out which frequency matched a stationary wave with a wavelength of 1.5 m. You may find it helpful to sketch the pattern.
2 Calculate the number of stationary waves in the tube for each frequency.
3 Calculate the wavelength of the stationary wave and speed of sound for each frequency.
4 The speed of sound in air is about 340 m s^{-1}. Comment on your answer and any reason for differences.

Pitch of note/Hz	Length of tube/m	Where powder piled up/cm	Where there was no powder/cm
118	1.5	75	0, 150
235	1.5	38 and 112	0, 75, 150
357	1.5	25, 75, 125	0, 50, 100, 150
179	1.5	0, 100	50, 150

Stationary waves: microwaves

Microwaves and other electromagnetic waves can interact and form stationary waves. Microwaves in microwave ovens are generated using a device called a magnetron. Since microwaves reflect off metals, they are directed from the magnetron and reflect off metallic inner surfaces to ensure they spread evenly throughout the oven. However, stationary waves still tend to develop, resulting in over-cooked food at the antinodes and undercooked food at the nodes. This is why most microwave ovens have a rotating turntable. Grated cheese or chocolate covering a plate placed in a microwave oven with the turntable disabled melt in places about 6 cm apart. These places correspond to antinodes for the first harmonic. This information tells us that the wavelength of the microwaves is about 12 cm.

ACTIVITY

Interference patterns from microwaves

Electromagnetic waves are carried by oscillating electric and magnetic fields. For clarity of explanation we will consider only the electric fields in this chapter. When two microwaves overlap and superpose, interference can take place. Where the electric fields of the two waves are in phase, there is constructive interference and regions of high intensity. The intensity is low where the electric components are out of phase. The changes in intensity are detected using a probe and this information can be used to calculate the wavelength of the microwaves.

A student sets up a microwave transmitter, microwave detector and aluminum plate placed about 40 cm from the transmitter as shown in Figure 6.16. Microwaves that reflect off the metal plate interfere with the transmitted microwaves. The interference pattern has regions of low intensity where the waves are out of phase, and regions of high intensity where the waves are in phase. She uses a receiving aerial, or probe, to detect nodes.

The student discovers that the distance between successive regions of destructive interference is 1.4 cm.

1 By considering the path difference between the wave reaching the receiver directly from the transmitter, and the wave reaching the receiver after it has been reflected from the metal plate, deduce the wavelength of the microwaves.
2 The student takes measurements to check the wavelength. Describe how she could reduce uncertainties in her measurements.
3 Explain the student's could explain her observations, using ideas about stationary waves.

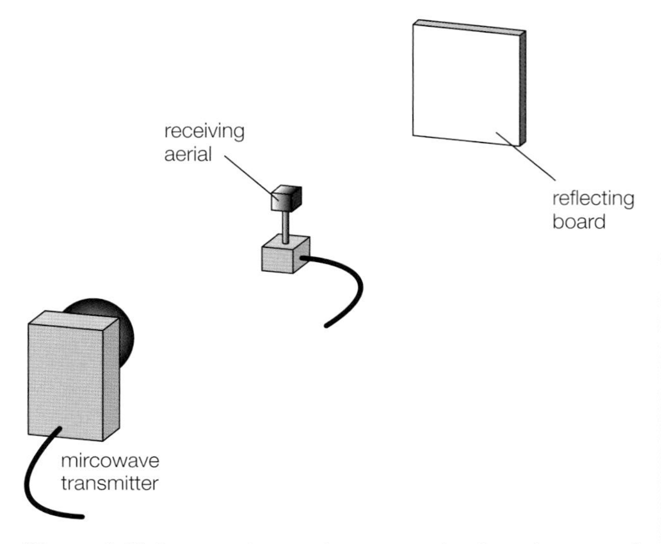

Figure 6.16 An experiment demonstrating interference of microwaves.

TEST YOURSELF

9 State and explain one difference between:
  a) a stationary wave and a progressive wave
  b) a node and an antinode.

10 State three conditions that are needed for a stationary wave to be et up.

11 Two progressive waves travelling in opposite directions originally have the same amplitude. Explain why the amplitude of an antinode in a stationary wave is double the amplitude of the progressive waves.

12 Describe how the displacement of a particle at an antinode changes during one complete wave cycle.

13 A stationary wave is set up on a rope of length 3 m. Write down the wavelength of:
  a) the first harmonic
  b) the third harmonic.

14 a) A stationary sound wave is set up in a hollow pipe with both ends open of length 1.4 m. Write down the wavelength of:
    i) the first harmonic
    ii) the second harmonic.
  b) One end of the pipe is now closed. Write down the wavelengths of:
    i) the first harmonic
    ii) the second harmonic.

15 A student blows across the top of a bottle, setting up a stationary wave, and creating a note. The bottle is 20 cm high.

a) What is the wavelength of the first harmonic?
b) The speed in air is $340\,m\,s^{-1}$. What is the frequency of the note?
c) The student fills half the bottle with water. Describe the note that the student hears now when he blows across the top of the bottle.

16 a) A string has a mass per unit length of $5\,g\,m^{-1}$. It has a length of 0.9 m and has a first harmonic frequency of 170 Hz. Calculate the tension in the string.
b) The tension is increased by a factor of 4. Calculate the frequency of the first and second harmonics now.

17 It is possible to calculate the diameter of a metal wire by using information about its harmonic frequencies. Work through this question to find out how.
A steel wire is fixed at one end and the other end is stretched over a pulley, with a mass of 4 kg hung on the end. The wire is set vibrating at a frequency of 83 Hz, and the length of wire between the fixed end and the pulley is 63 cm. At this frequency a stationary wave is set up, which has two nodes between the pulley and the fixed end.
a) Calculate the wavelength of the waves.
b) Calculate the speed of the waves.
c) Calculate the tension in the wire.
d) Use your answers to parts (b) and (c) to calculate the mass per unit length of the wire.
e) Given that the density of the steel wire is $7800\,kg\,m^{-3}$, calculate the diameter of the wire.

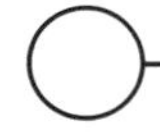

Diffraction

When waves pass through a gap or move past an obstacle, they spread out. This is called **diffraction**.

A single source of light can also create interference patterns. This effect occurs because of **diffraction**. Diffraction occurs when waves spread around an obstacle or through a gap. It affects all waves, and is why we can hear conversations around corners even if we cannot see the people speaking. Diffraction can be shown easily using a ripple tank, and is most pronounced if the wavelength is of a similar order of magnitude as the obstacle or gap. The wavelength of diffracted waves does not change, but the waves become curved and spread into shadow regions around the obstacle or gap.

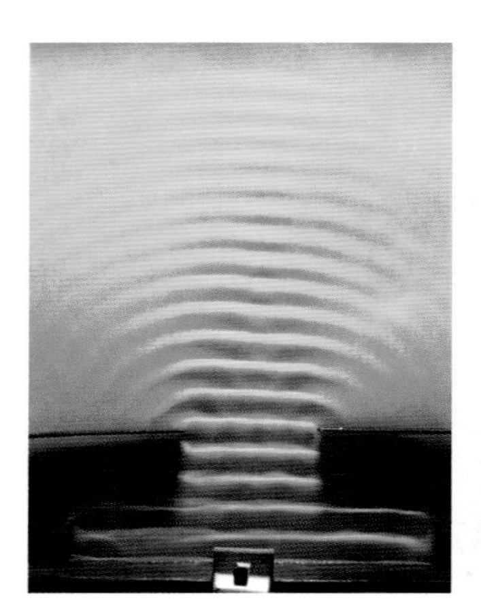

Figure 6.17 Diffraction of water through a gap.

Diffraction of light

Visible light has a very short wavelength (400×10^{-9} m to 700×10^{-9} m) so diffraction is only significant if the slit is very narrow. When light diffracts through a narrow slit, a fringe pattern is seen as a bright central fringe surrounded by other less-bright fringes either side. These fringes, called maxima, are due to constructive interference of light. The dark regions in between (minima) are due to destructive interference of light. The interference pattern is due to light from one part of the slit diffracting, overlapping and then interfering with light that has diffracted from other parts of the slit. When the path difference between the top half of the slit and the bottom half of the slit is half a wavelength, destructive interference occurs and the light intensity is zero.

The width of the central diffraction maximum depends on the wavelength of light and the slit width. At the edges of the central maxima, light leaving the top half of the slit interferes destructively with light leaving the bottom half, so:

$$\frac{\lambda}{2} = \frac{a}{2}\sin\theta$$

or the first minimum occurs at an angle given by:

$$\sin\theta = \frac{\lambda}{a}$$

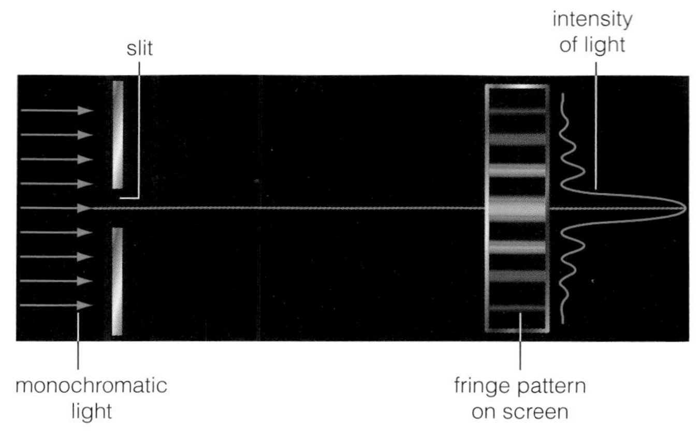

Figure 6.18 Diffraction fringes from a single slit.

where λ is the wavelength of light, a is the slit width and θ is the angle of diffraction.

As the wavelength of light increases, the central maximum becomes wider because, as λ increases, $\sin\theta = \frac{\lambda}{a}$ increases.

As the slit width increases, the central maximum becomes narrower because, as a increases, $\sin\theta = \frac{\lambda}{a}$ decreases.

If a source of white light is used, the fringes are coloured because white light is a mixture of different colours of light. Since the central maximum depends on the ratio $\frac{\lambda}{a}$, the central maximum is broader for red light than it is for blue light. Red light has a longer wavelength than blue light.

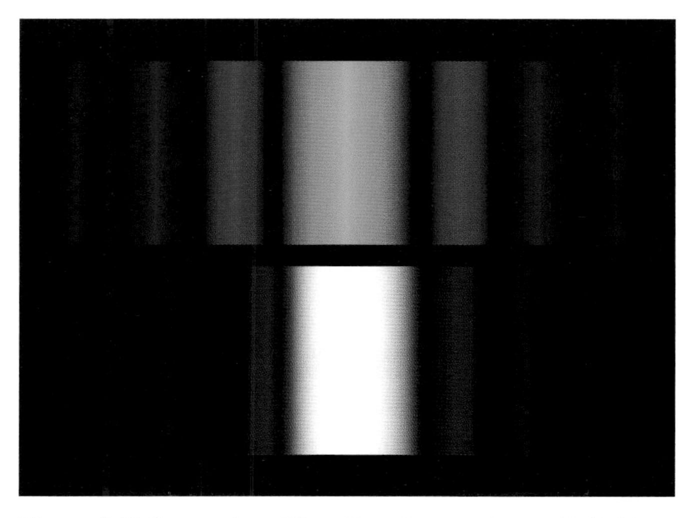

Figure 6.19 Comparing diffraction of monochromatic light and white light.

Diffraction gratings–Required practical number 2

A **diffraction grating** has thousands of slits spaced very closely together.

We can see interference patterns using a diffraction grating. A diffraction grating has thousands of slits spaced very closely together. The slits are very narrow so that the light diffracts through a wide angle. The pattern is a result of light overlapping (or superposing) and interfering from a great number of slits.

When light passes straight through the slits, the path difference between each slit is zero, so the light from each slit is in phase and the overlapping

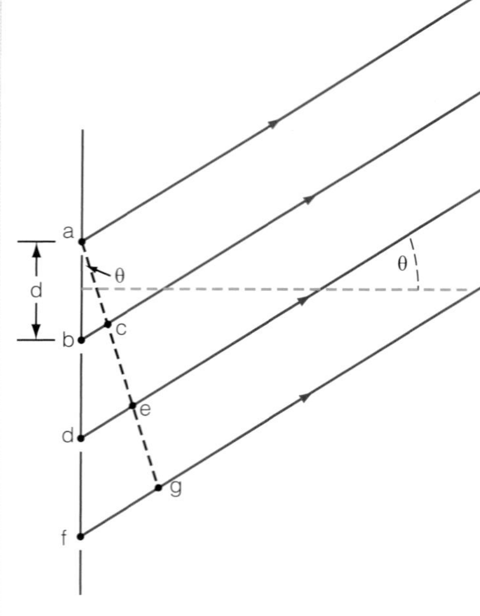

Figure 6.20 Finding the path difference for a diffraction grating.

waves combine to form a maximum intensity. This maximum is called the central maximum. It can also be called the zeroth order maximum because there is 0 path difference between each slit. Such a maximum is easily seen on a screen when a laser is shone through a grating. A diffraction grating also produces other maxima on either side of the central maximum. These occur when the waves leaving the slits arrive in phase on the viewing screen. The condition for constructive interference is that the path difference for waves leaving adjacent slits must be a whole number of wavelengths. The first order maximum is seen when the path difference for waves leaving adjacent slits is one wavelength; the second order maximum is seen when the path difference for waves leaving adjacent slits is two wavelengths, and so on.

For the first order maxima:

- interference is constructive
- the path difference of light from adjacent slits is λ
- light travels at angle θ_1 to the direction of incident light.

Applying this information to the triangle *abc*,

$$\sin\theta_1 = \frac{bc}{ab} = \frac{\lambda}{d}$$

This means that

$$\sin\theta_1 = \frac{\lambda}{d}$$

where λ is the wavelength of light in m

θ_1 is the angle between the 0th and 1st order maxima

d is the spacing between slits in m.

For second order fringes, the path difference is 2λ and light travels at angle θ_2 to the direction of incident light. The equation becomes:

$$\sin\theta_2 = \frac{2\lambda}{d}$$

In general, the condition for maxima to occur is given by:

$$\sin\theta_n = \frac{n\lambda}{d}$$

where n is a whole number, and the order of the maximum.

EXAMPLE

Calculate the angle of a third order maximum for light of wavelength 500 nm passing through a diffraction grating with 600 lines per mm.

Answer

The spacing of the grating, d, is $\frac{1 \times 10^{-3}\,\text{m}}{600} = 1.67 \times 10^{-6}\,\text{m}$

The order, n, is 3; the wavelength, $\lambda = 500 \times 10^{-9}\,\text{m}$

$$\sin\theta_n = \frac{n\lambda}{d}$$

$$= \frac{3 \times 500 \times 10^{-9}\,\text{m}}{1.67 \times 10^{-6}\,\text{m}}$$

$$= 0.898$$

$\sin^{-1} 0.898 = \theta$

So $\theta = 64°$

Applications of diffraction gratings

Diffraction gratings are used to separate light of different wavelengths in great detail (high resolution). This is useful when the diffraction grating is part of a spectrometer and used for investigating atomic spectra in laboratory measurements. Diffraction gratings are also used in telescopes to analyse light from galaxies. CDs and DVDs use diffraction gratings but they reflect light from a grating rather than transmitting it. The light source is a laser installed inside the CD or DVD drive.

TEST YOURSELF

18 Why is diffraction of light best seen using very narrow slit sizes?

19 Refer to the photograph in Figure 6.19, which shows a diffraction pattern for green light when it passes through a narrow slit. Describe how the fringe pattern changes as the angle increases away from the central maximum.

20 A student is looking at a diffraction pattern from a single slit. He makes the slit narrower. How does this affect the fringe pattern he sees?

21 Light of wavelength 500 nm passes through a diffraction grating with 500 lines per nm. Calculate the angle for the second order maximum seen for the light.

22 Calculate the spacing of a diffraction grating when the angle of a second order maximum for light (λ = 400 nm) is 20°.

23 A diffraction grating has 400 lines per mm. White light is shone through the grating.

a) Describe the diffraction pattern produced by this grating.

There is a second order maximum for red light of wavelength 690 nm at the same angle as a third order maximum for blue light of wavelength 460 nm.

b) Calculate this angle.

c) How many orders of the blue light and how many orders of the red light does this grating produce?

Practice questions

1 When two waves are initially in phase, their path difference at any point where the waves are in antiphase is:

A $n\lambda$

B $n\lambda + \frac{\lambda}{2}$

C $n\lambda + \frac{\lambda}{4}$

D 0

2 The separation between a node and the adjacent antinode in a stationary transverse wave on a rope is:

A $\frac{\lambda}{4}$

B $\frac{\lambda}{2}$

C λ

D 2λ

3 Which statement is not always true about coherent waves?

A Coherent waves have the same frequency.

B Coherent waves have the same wavelength.

C Coherent waves have a constant phase difference.

D Coherent waves have the same amplitude.

4 In an experiment to investigate nodes and antinodes produced by stationary microwaves, a microwave probe detects antinodes 3.0 cm apart. What is the wavelength of the microwaves?

A 3.0 cm

B 1.5 cm

C 6.0 cm

D 12.0 cm

5 A pipe, open at both ends, has a first harmonic of 30 Hz. The frequency of the second harmonic is:

A 60 Hz

B 15 Hz

C 120 Hz

D 40 Hz

6 When a wave reflects off a denser medium, its phase change is:

A 90°

B 180°

C 360°

D 45°

7 Which of these statements is incorrect about interference patterns for light?

A Dark fringes are caused by the interference of waves with a phase difference of 180°.

B Dark fringes are the result of constructive interference.

C Light fringes are caused by the interference of waves with a path difference of a whole number of wavelengths.

D Dark fringes are the result of destructive interference.

8 Light of wavelength 500×10^{-9} m shines through a pair of narrow slits 0.5 mm apart. The separation of fringes on a screen 2 m away is:

A 2×10^{-4} m

B 1.25×10^{-3} m

C 2×10^{-3} m

D 0.125 m

9 A diffraction grating has 500 lines per mm.

a) Calculate the spacing of the lines on the grating. *(1)*

b) Calculate the angle of the 3rd order diffraction lines when light of wavelength 633 nm shines through the grating. *(3)*

c) Explain if it is possible to see the 4th order diffraction lines. *(2)*

10 Young's double slit experiment is set up using slits set 0.3 mm apart and a screen 2 m from the slits. A student shines light of wavelength 557 nm from a krypton source through the slits and measures the fringe pattern.

a) Calculate the width of ten fringe spacings. *(3)*

b) Describe the changes the student could make to increase the separation of the fringes. *(3)*

11 A few people can see a spectrum caused by diffraction through their eyelashes.

a) Explain the conditions needed for a diffraction pattern like this to be seen. *(2)*

b) The wavelength of visible light is 400 nm to 700 nm. Your eyelashes have a separation of about 0.1 mm. Use this information to estimate the angles at which the 4th order spectrum will be seen. *(4)*

c) Comment on your answer. *(2)*

12 a) Explain what is meant by *a stationary wave*. *(2)*

b) i) The frequency of the microwaves generated in a microwave oven is 2.4 GHz. Calculate the wavelength of the waves. *(2)*

ii) Using your answer to part (i), calculate the distance between antinodes inside the oven. *(1)*

c) A person placed a plate of grated cheese inside the microwave oven, switched off the turntable and turned the oven on for about 30 seconds. State and explain what would happen, and how a person could use this to verify the frequency of the microwaves. *(2)*

13 Discuss the formation of stationary waves in an open pipe. You should:

- include a diagram of the first harmonic, explaining where nodes and antinodes are found
- describe how nodes and antinodes form
- explain what conditions are needed for the formation of stationary waves.

The quality of written communication will be assessed in your answer. *(8)*

14 Coherent light of wavelength 460 nm shines on a pair of parallel slits; a pattern of dark and bright fringes is seen on a screen some distance away.

a) State one reason why the experiment was important when it was first performed 200 years ago. *(1)*

b) Explain why coherent waves are necessary for this experiment. *(2)*

c) Calculate the ratio of slit width to screen distance if the fringes are 1.8 mm apart. *(3)*

15 **a)** A laser emits *monochromatic light*. Explain what is meant by monochromatic light. *(1)*

Blue light from a laser travels through a single slit. The intensity graph for the diffracted blue light is shown.

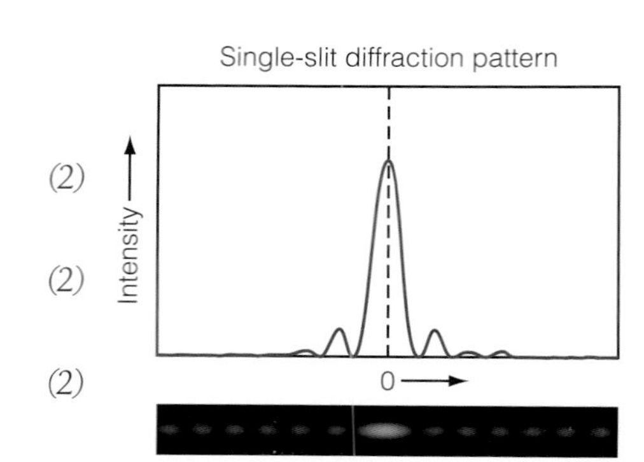

Figure 6.21

b) Explain how the intensity graph changes if a different laser that emits red light is used. *(2)*

c) State and explain one risk of working with lasers, describing how to reduce this risk. *(2)*

d) The red laser light is replaced by a non-laser source emitting white light. Describe the new appearance of the pattern. *(2)*

16 Monochromatic light of wavelength λ shines on a diffraction grating, which has lines separated by a distance *d*.

a) Derive the formula: $d \sin \theta_n = n\lambda$ for light of wavelength λ shining through a diffraction grating of spacing *d* where θ_n is the angle at which the *n*th order maximum occurs. *(4)*

b) Describe and explain two ways that the pattern of maximum intensities changes as the wavelength increases. *(2)*

c) Calculate the wavelength of light for 3rd order fringes. The spacing of the grating is 1×10^{-6} m and one sharp maximum is seen at a diffraction angle of 70°. *(2)*

17 A grating with slit separation of 2×10^{-6} m is used analyse a helium spectrum. A blue line and a red line are observed very close to each other at 42°. What order are these lines? (Range of red light is 645 nm to 700 nm; range of blue light is 440 nm to 490 nm.) *(5)*

18 A grating has 300 lines per mm and is used with a laser that emits light of wavelength 7×10^{-7} m.

a) What is the slit spacing of the grating? *(1)*

b) If the light is incident at right angles, how many orders of diffraction are visible and in what directions do they occur? *(4)*

Stretch and challenge

19 Compare the three ways of observing interference patterns: Young's double slits, diffraction gratings and diffraction through a single slit, including a discussion of why the patterns form.

20 White light travels through a diffraction grating. Use calculations to work out the lowest order spectrum when colours overlap with the next order spectrum. The wavelength of red light is 700 nm and violet light is 400 nm.

21 A stellar interferometer is located on a cliff top, 100 m high. It is used to study 1.5 m waves from the sun.

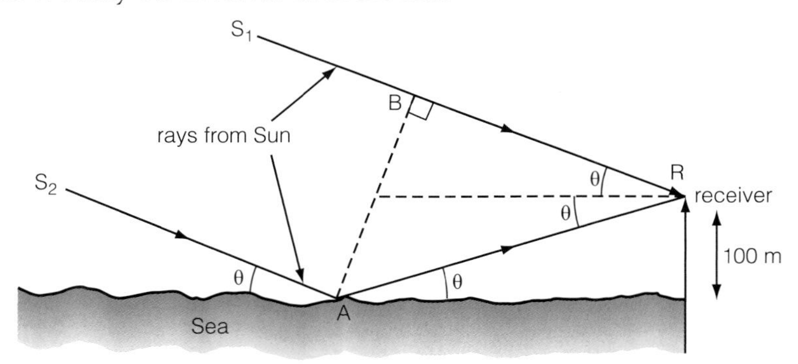

Figure 6.22

a) Using the letters on the diagram, name the path difference that may give rise to interference.

b) The path difference is $200 \sin \theta$. Write down the condition for the receiver to detect an intensity maximum. Remember that there is a phase change for waves reflecting off the sea. (Mathematicians might like to prove this result for the path difference.)

c) Explain why the intensity goes through a series of maxima and minima as the sun sets.

d) Will the changes in maxima and minima be most rapid when the sun is directly overhead or when it is nearly set? Explain your answer.

22 Explain why a diffraction grating with 100 lines per mm produces a few sharply defined maxima when laser light shines through it, whereas the same laser shining through just two slits, produces equal areas of light and dark. (Hint: think of part of the grating – perhaps 100 lines. If the path difference between the first and second slit is $\frac{\lambda}{100}$, what is the path difference between the first and 51st slit?)

23 Two microscope slides, length L, are placed flat with one slide on top of the other.

One end of the top slide is raised through a small distance D, enclosing a wedge of air between the two slides.

Monochromatic light shines on to the slides and an interference pattern of parallel, equally spaced fringes can be seen.

a) Light reflects in two places: from the bottom face of the upper slide and from the upper face of the lower slide. Which reflected ray undergoes a phase change of 180°?

b) Write down the general condition for dark fringes to be seen in this interference pattern.

c) The reflection takes place where the thickness of the air wedge is a distance, d. Write down an expression for the path difference when dark fringes can be seen in terms of distance d, and wavelength λ.

d) Use your answer to part (c) to predict how the fringe pattern would change if:

i) the light source does not change but the air gap is filled with a transparent liquid

ii) the only change is to increase the wavelength of light.

7 Introduction to mechanics

PRIOR KNOWLEDGE

- A scalar quantity has only size, for example a mass of 2 kg.
- A vector quantity has both size and direction, for example a 5N force acting downwards.
- A force is a push or a pull. Forces are measured in newtons, N.
- The size of a turning effect is called a turning moment or torque.
- Turning moment = force applied × perpendicular distance from the pivot.
- Newton's first law of motion states that when no force acts, or balanced forces act on an object, that object will remain at rest or continue to move in a straight line at a constant speed.

Figure 7.1 Formula 1 racing cars moving at speed around a corner.

Figure 7.2 A modern bridge with central pillar chords.

Engineers who build bridges, or who design Formula 1 cars, need a profound understanding of mechanics. Engineers use the laws of physics, and their knowledge of the strength and flexibility of materials, to calculate the size and shape of material required at each stage of design and construction. In this chapter you will meet ideas about the vector nature of forces, and also the concept of turning moments. Both these ideas are used by design and construction engineers.

TEST YOURSELF ON PRIOR KNOWLEDGE

1. Explain why speed is a scalar quantity and velocity is a vector quantity.
2. Make an estimate of the size of the following forces:
   a) The weight of an apple.
   b) The push on a pedal when you cycle.
   c) The drag on a car moving at 20 $m\,s^{-1}$.
3. A parachutist falls at a constant speed of 5 $m\,s^{-1}$. Which of the following statements is true?
   a) The pull of gravity on the parachutist is greater than the drag.
   b) The pull of gravity on the parachutist is less than the drag.
   c) The pull of gravity on the parachutist is the same size as the drag.
4. **a)** Explain why it is easier to undo a nut using a spanner with a long handle.
   b) Calculate the turning moment when a force of 30 N is applied at right angles to the handle of a spanner with a length of 0.25 m.

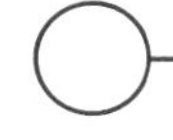

Scalar and vector quantities

A **scalar quantity** is one which has size only.

A **vector quantity** is one which has size and direction.

A **resultant vector** is the vector that results from adding two or more vectors in a vector sum (e.g. resultant force or resultant displacement).

A scalar quantity is one that only has size. Scalar quantities include: mass, temperature, energy, distance, speed and time. We say that 'room temperature is 19°C'; it makes no sense to say '19°C westwards'.

A vector quantity is one that has magnitude (size) and direction. Vector quantities include: force, acceleration, velocity and displacement. Velocity and speed seem like similar quantities, as they both can be measured in $m\,s^{-1}$. But, if you write that a car is travelling at 30 $m\,s^{-1}$, you are talking about a speed; when you write that a car is travelling at 30 $m\,s^{-1}$ due east, you have used a vector quantity, which is called velocity. In a similar way, the word displacement is used to describe a distance travelled in a particular direction.

You have learnt to add or subtract simple vectors. In Figure 7.3 (a) two forces acting in the same direction add up to make a larger resultant force of 200 N, and in Figure 7.3 (b) two forces acting in opposite directions make a smaller resultant force of 300 N.

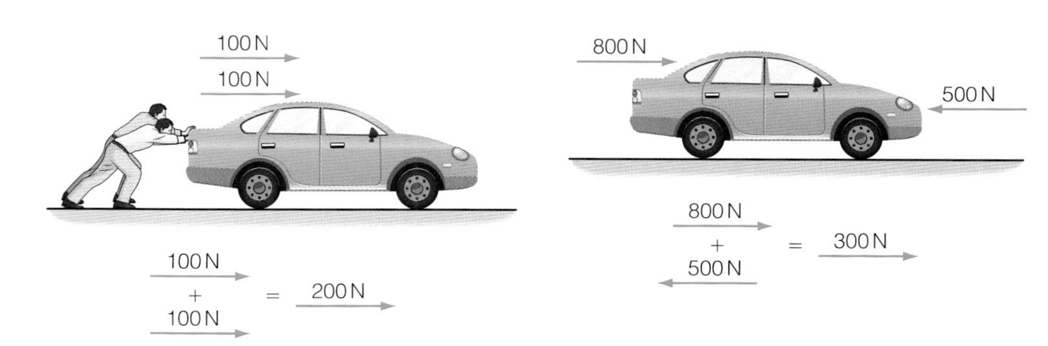

Figure 7.3 (a) two forces acting in the same direction and (b) two forces acting in opposite directions.

TEST YOURSELF

1 Name three scalar and three vector quantities, including at least one of each that do not appear on this page or on page 123.
2 Which of the following are vector quantities and which are scalars: density, volume, momentum (mass × velocity), weight, gravitational field strength, electrical resistance, potential difference?
3 A motorist drives from Abinton to Barmouth via Chine Roundabout (see Figure 7.4). The journey takes 0.8 hours. Calculate:
   a) the distance travelled and the displacement of the car.
   b) the average speed of the car during the journey and the average velocity of the car during the journey.

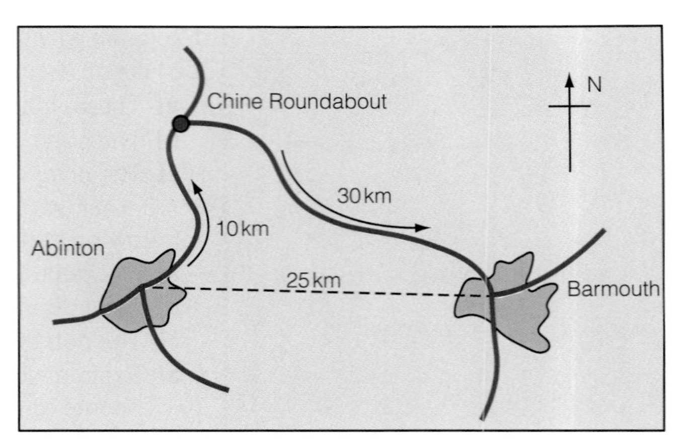

Figure 7.4 Route map of journey from Abinton to Barmouth via Chine Roundabout.

The addition of vectors

The method of adding vectors is illustrated by the following example.

EXAMPLE

A rambler walks a distance of 8 km travelling due east, before walking 6 km due north. Calculate their displacement.

Answer

Displacement is a vector quantity, so we must calculate its magnitude and direction. This calculation, for vectors at right angles to each other, can be done using Pythagoras' theorem and the laws of trigonometry.

$$(AC)^2 = (8\text{ km})^2 + (6\text{ km})^2$$

$$(AC)^2 = (64 + 36)\text{ km}^2$$

$$\Rightarrow AC = 10\text{ km}$$

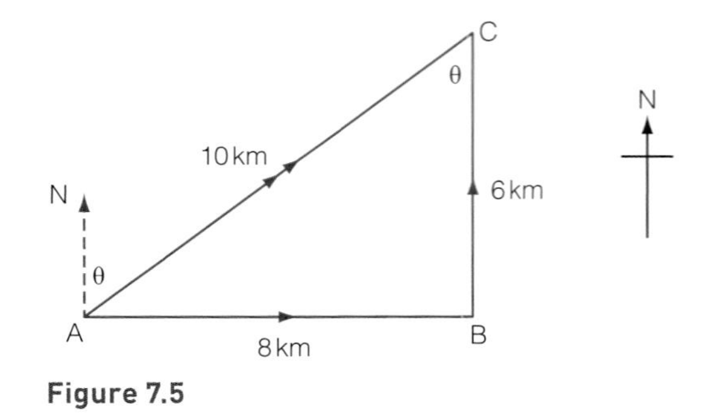

Figure 7.5

Directions can be calculated on a bearing from due north; this is the angle θ, in Figure 7.5, which is also the angle θ, in the triangle.

$$\tan\theta = \frac{8\text{ km}}{6\text{ km}}$$

$$\Rightarrow \theta = 53°$$

The displacement is 10 km on a bearing of 53°.

You are expected to be able to calculate vector magnitudes when two vectors are at right angles, but the calculations are harder when two vectors are separated by a different angle. Now the magnitude can be calculated using a scale drawing.

EXAMPLE

In Figure 7.6a, a liner is being pulled by two tugs. What is the magnitude and direction of the resultant force?

Answer

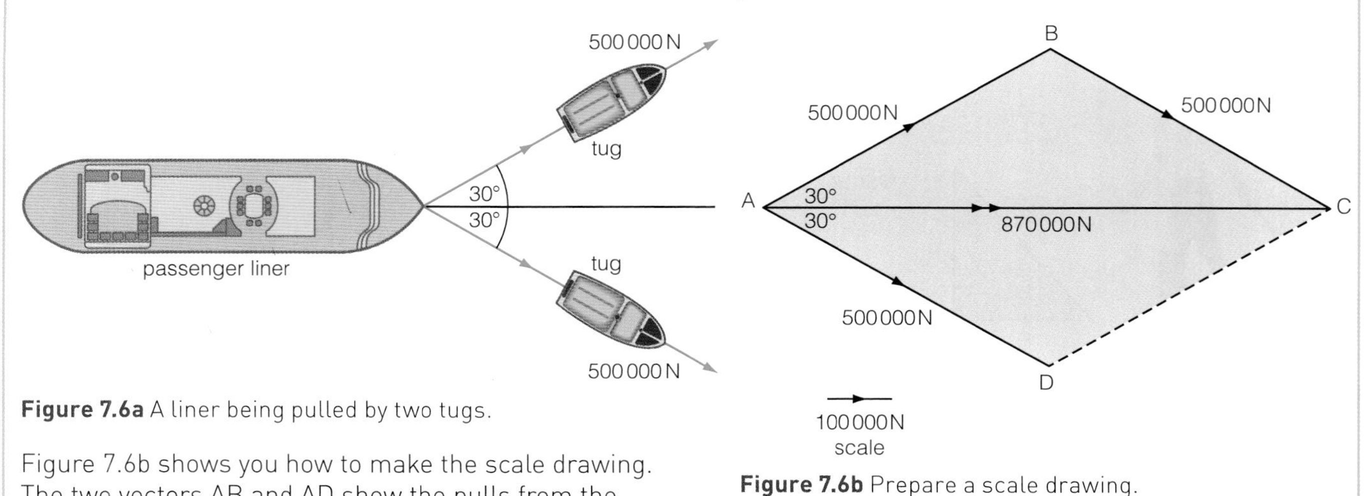

Figure 7.6a A liner being pulled by two tugs.

Figure 7.6b shows you how to make the scale drawing. The two vectors AB and AD show the pulls from the two tug boats; these can be added by simply adding the vector BC (which is parallel to line AD) to vector AB. The resultant force is represented by the vector AC, which can be measured to be 870 000 N.

Figure 7.6b Prepare a scale drawing.

TEST YOURSELF

4 A plane flies due north with a speed of 420 km h^{-1}; a cross wind blows in a westerly direction with a speed of 90 km h^{-1}.

a) Calculate the plane's velocity relative to the ground.

b) Calculate the time taken for a 600 km flight. On another day, the plane flies due west with a velocity of 400 km h^{-1}, and the wind blows from a bearing of 45° with a speed of 100 km h^{-1}.

c) Make a scale drawing to calculate the plane's velocity.

d) Calculate the time taken for a 600 km flight.

5 Calculate the resultant force on the plane shown in Figure 7.7.

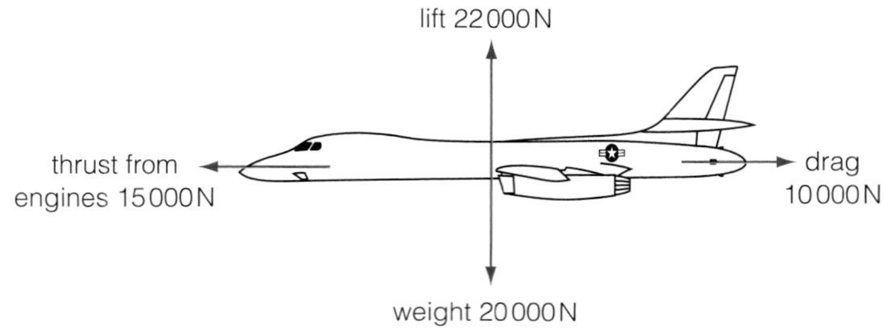

Figure 7.7

6 A light aeroplane, whose air speed is 60 m s^{-1}, is flying in a westerly gale of 40 m s^{-1}. (Westerly means that the wind is coming from the west.)

Draw vector diagrams to calculate the plane's ground speed, when:

a) The pilot keeps the plane heading east.

b) The pilot keeps the plane heading north (the plane's nose is pointing north, but it does not travel north because the wind blows it sideways).

c) The pilot now keeps the plane flying north. In which direction is the plane headed in this case?

The resolution of vectors

Resolving a vector Split it into two mutually perpendicular components that add up to the original vector.

Figure 7.8a shows a passenger pulling his wheelie bag at the airport. He pulls the bag with a force, F, part of which helps to pull the bag forward and part of which pulls the bag upwards, which is useful when the bag hits a step.

The force, F, can be resolved into two components: a horizontal component F_h and a vertical component F_v. A vector can be resolved into any two components that are perpendicular, but resolving a vector into vertical and horizontal components is often useful, due to the action of gravity.

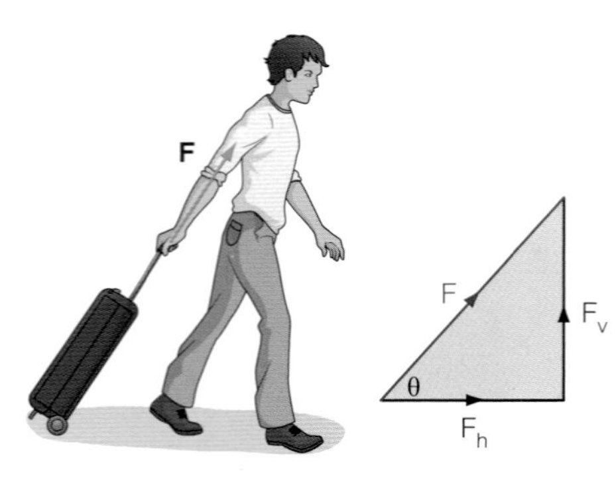

Figure 7.8a and **Figure 7.8b**

Using the laws of trigonometry:

$$\sin\theta = \frac{F_v}{F}$$

and

$$\cos\theta = \frac{F_h}{F}$$

The vertical component of the force is: $F_v = F\sin\theta$

The horizontal component of the force is: $F_h = F\cos\theta$

TIP

You must make sure you understand your trigonometrical formulas to resolve vectors.

EXAMPLE

In Figure 7.9, a ball is travelling at a velocity of $25\,\mathrm{m\,s^{-1}}$ at an angle of 30° to the horizontal. Calculate the horizontal and vertical components of the ball's velocity.

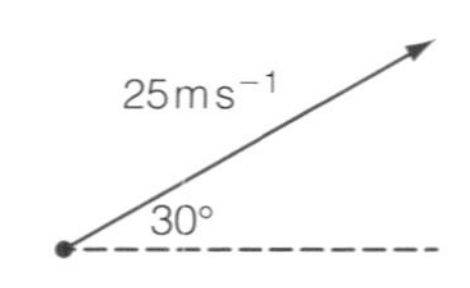

Figure 7.9

Answer

$$V_h = (25\,\mathrm{m\,s^{-1}}) \times \cos 30°$$
$$= 21.7\,\mathrm{m\,s^{-1}}$$
$$V_v = (25\,\mathrm{m\,s^{-1}}) \times \sin 30°$$
$$= 12.5\,\mathrm{m\,s^{-1}}$$

An inclined plane

Figure 7.10 shows a car at rest on a sloping road. The weight of the car acts vertically downwards, but here it is useful to resolve the weight in directions parallel (||) and perpendicular (⊥) to the road. The component of the weight parallel to the road provides a force to accelerate the car downhill. There are other forces acting on the car, which are dealt with in Figure 7.12.

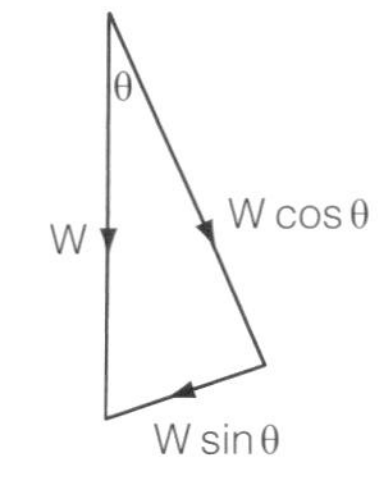

The component of the weight acting along the slope is:

$$W_{||} = W\sin\theta$$

The component of the weight acting perpendicular to the slope is:

$$W_{\perp} = W\cos\theta$$

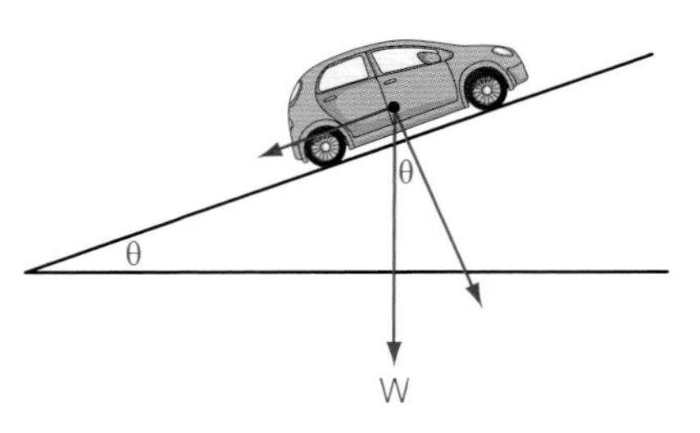

Figure 7.10

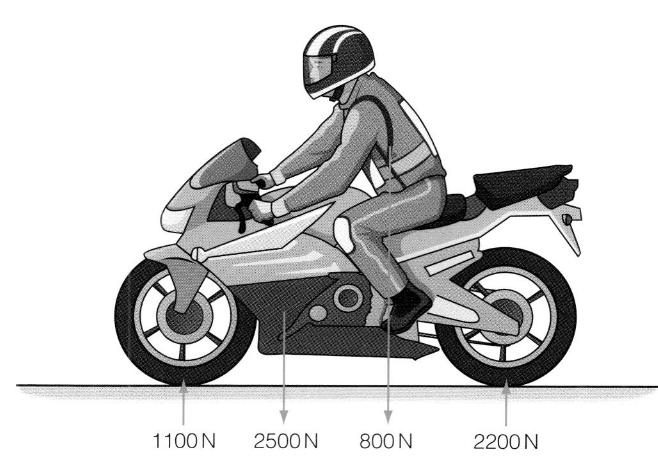

Figure 7.11

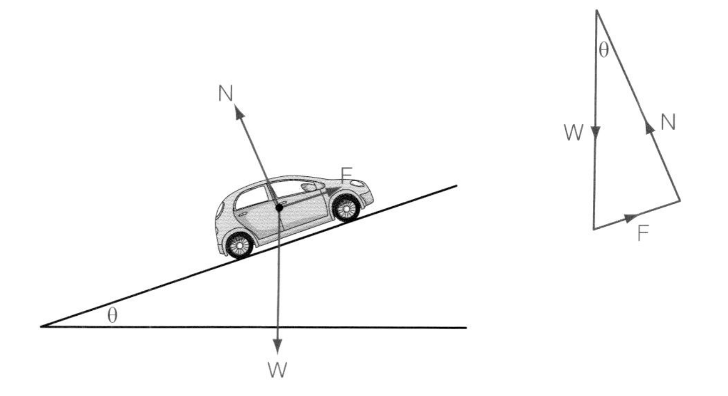

Figure 7.12

Forces in equilibrium

You met the idea of balanced forces in the prior knowledge section. In Figure 7.11 the motorcyclist has a weight of 800 N and the bike has a weight of 2500 N – a total of 3300 N. Those forces are balanced by the two contact forces exerted by the road, so the bike remains at rest. This is an example of the application of Newton's first law of motion.

The example in Figure 7.11 is easy to solve as all of the forces act in one direction. Figure 7.12 shows all the forces acting on the stationary car first shown in Figure 7.10. Newton's first law of motion applies to this situation too, although the forces do not all act in a straight line. However, the vector sum of the forces must still be zero. Three forces act on the car: its weight, a normal reaction perpendicular to the road, and a frictional force parallel to the road.

The triangle of forces shows that the three forces – W, N and F – add up to zero, so there is no resultant force to accelerate the car. We can also explain why the car has no resultant force acting on it by resolving the forces along and perpendicular to the plane.

Along the plane: $F = W \sin\theta$

At right angles to the plane: $N = W \cos\theta$

So the two components of the weight are balanced by friction and the normal reaction.

ACTIVITY

The apparatus shown in Figure 7.13 provides a practical way for you to test the theory discussed in the last section. Three weights are set up in equilibrium. A central weight, W_3, is suspended by two light strings that pass over two frictionless pulleys. These strings are attached to the weights W_1 and W_2. Since point O is stationary, the three forces acting on that point, T_1, T_2 and W_3, must balance. The tensions T_1 and T_2 are equal to W_1 and W_2 respectively.

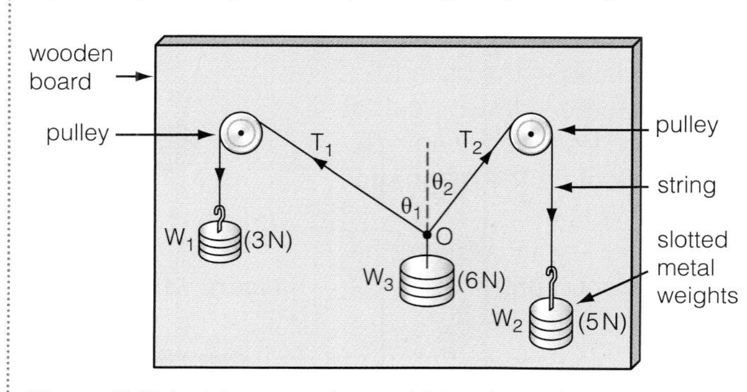

Figure 7.13 In this example $\theta_1 = 60°$ and $\theta_2 = 30°$.

We can demonstrate the balance of forces by resolving the forces acting on point O vertically and horizontally. The forces must balance in each direction.

So, resolving horizontally: $T_1 \sin\theta_1 = T_2 \sin\theta_2$

And resolving vertically: $T_1 \cos\theta_1 + T_2 \cos\theta_2 = W_3$

We can check using the example numbers in the diagram:

$$T_1 \sin\theta = 3\,\text{N} \sin 60 = 2.6\,\text{N}$$

$$T_2 \sin\theta = 5\,\text{N} \sin 30 = 2.5\,\text{N}$$

So the horizontal components balance to within a reasonable experimental error.

Resolving vertically:

$$3\,\text{N} \cos 60 + 5\,\text{N} \cos 30 = 5.8\,\text{N}$$

And, to within a reasonable experimental error, the vertical components balance too; this is close to the weight of 6 N shown in the diagram.

a) Discuss how you would measure the angles θ_1 and θ_2 accurately.

b) Discuss what other sources of error might contribute to the size of errors seen in these checks.

c) A student now carries out some experiments and in each case tries to calculate the unknown weights or angles shown in Table 7.1. Copy and complete the table, filling in the gaps. Suggest what weight the student actually used; in each case she used weights that were a whole number of newtons.

Table 7.1

W_1/N	W_2/N	W_3/N	θ_1	θ_2
6	6		60°	60°
8	8		40°	40°
7		14	42°	28°
6	8	10	53°	

TEST YOURSELF

7 An oil tanker is pulled into harbour by two tugs. Each tug applies a force of 180 000 N at an angle of 37° to the forwards direction of the tug. Calculate the resultant force on the tug. You may calculate this or use a scale drawing.

8 Three men are trying to move a piano. One pushes southwards with a force of 100 N; the second pushes westward with a force of 173 N; the third pushes in a direction 60° east of north with a force of 200 N. Calculate the resultant force on the piano.

9 A helicopter rotor provides a lift of 180 000 N when the blades are tilted 10° from the horizontal.

a) Resolve the lift into horizontal and vertical components.

b) The helicopter is flying level. Calculate its mass.

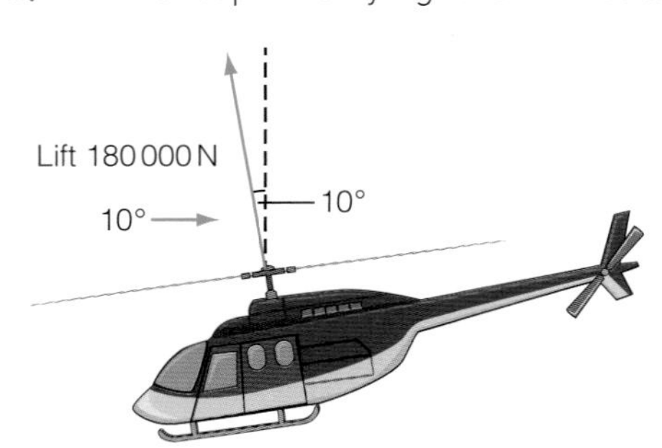

Figure 7.14

10 A cyclist climbs a 1 in 5 slope as shown in Figure 7.15. The combined weights of the cyclist and the bike are 950 N. The frictional forces acting on the cyclist are 40 N.

Figure 7.15

a) By resolving the weights of the cyclist and the bike, calculate the force acting down the slope.

b) Calculate the force he must exert on the road to climb at a constant speed.

11 A massive ball is suspended on a string 1 m long, and pulled sideways by a force of 1 N through a distance of 30 cm.

a) Calculate the angle θ shown in Figure 7.16.

b) Make a scale drawing (in the form of a vector triangle)of the three forces that act on the ball: its weight W, the tension in the string T, and the sideways pull of 1 N. Use the triangle to calculate the magnitude of W and T.

c) Check your answers by calculation.

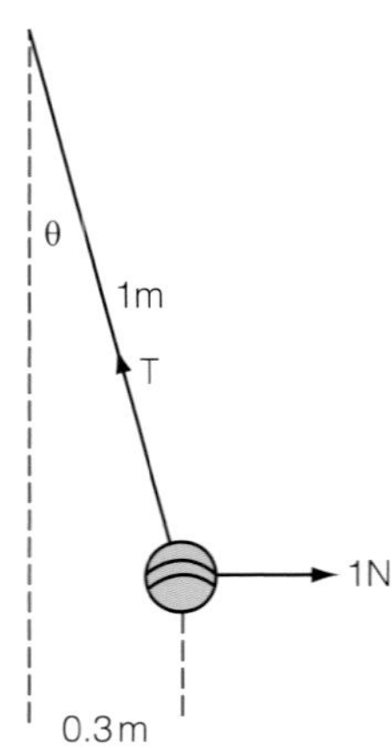

Figure 7.16

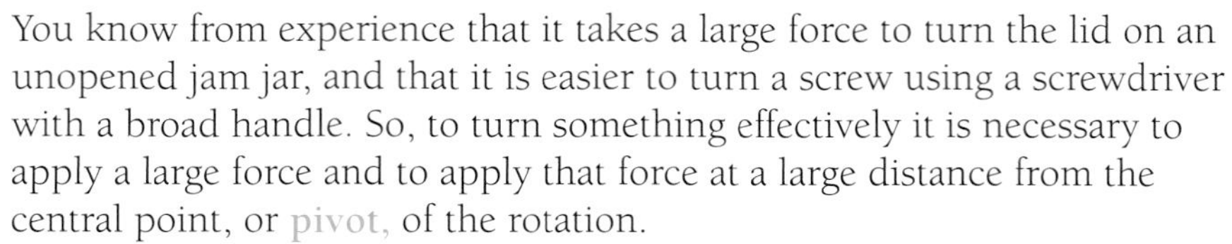

Turning moments

Introducing moments

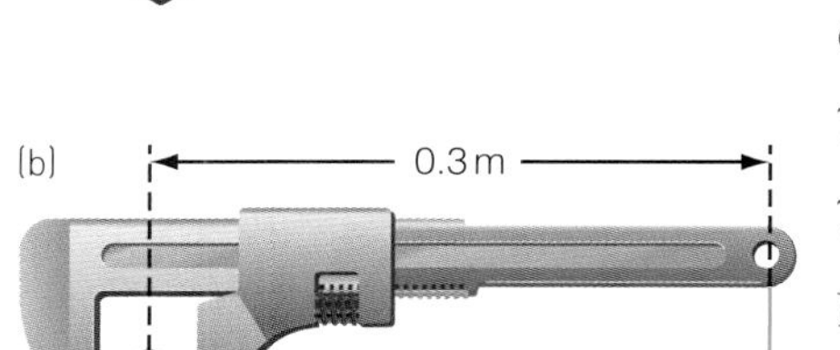

(b) 0.3 m 100 N

Figure 7.17

You know from experience that it takes a large force to turn the lid on an unopened jam jar, and that it is easier to turn a screw using a screwdriver with a broad handle. So, to turn something effectively it is necessary to apply a large force and to apply that force at a large distance from the central point, or pivot, of the rotation.

The size of the turning effect is called a turning moment or torque.

Turning moment = force applied × perpendicular distance from the pivot

Figure 7.17a and Figure 7.17b illustrate the importance of the perpendicular distance. In figure 7.17a there is no turning effect when the force acts through the pivot. However, in Figure 7.17b:

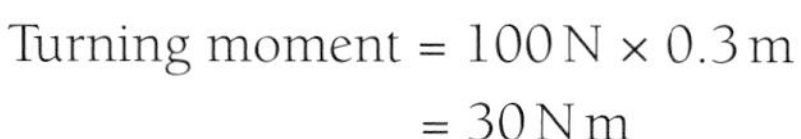

$$\text{Turning moment} = 100\,\text{N} \times 0.3\,\text{m}$$
$$= 30\,\text{N m}$$

Moment The force × perpendicular distance from the pivot. The unit of a moment is N m or N cm.

Pivot The point about which an object rotates. In Figure 7.18 the support of the see-saw is the pivot.

You know from Newton's first law of motion that an object will remain at rest if the forces on it balance. However, if the body is to remain at rest without translational movement, or rotation, then the sum of the forces on it must balance, and the sum of the moments on the object must also balance.

Another way of expressing Newton's first law is to say that when the vector sum of the forces adds to zero, a body will remain at rest or move at a constant velocity.

Equilibrium An object is in equilibrium, when the sum of the forces = 0 and the sum of the turning moments = 0.

Figure 7.18 shows two children on a see-saw. The see-saw has a weight of 200 N, which can be taken to act through its support. We can show that the see-saw is in equilibrium as follows.

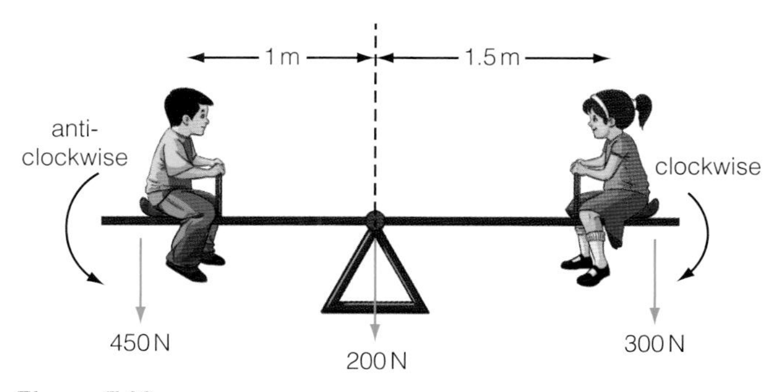

Figure 7.18

The anticlockwise turning moment of the boy = 450 N × 1.0 m = 450 N m

The clockwise turning moment of the girl = 300 N × 1.5 m = 450 N m

Since the clockwise and anticlockwise moments balance (or add up to zero), one condition for equilibrium is met.

The total weight of the see-saw and the children, which acts on its support is: 450 N + 300 N + 200 N = 950 N. The support provides an upwards force of 950 N on the see-saw, so the resultant force acting on it is zero and the see-saw is in equilibrium.

TEST YOURSELF

12 State the unit of turning moment.

13 Figure 7.19 shows three children on a see-saw.

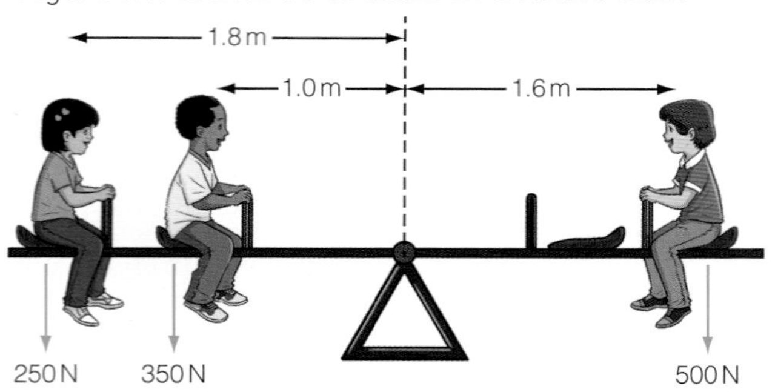

Figure 7.19

a) Use the principle of moments to show that the see-saw is in equilibrium.

b) State the size of the upward force exerted by the pivot on the see-saw. Assume the see-saw itself has negligible weight.

14 Figure 7.20 shows a pair of wire cutters.

a) Explain why the cutters have long handles.

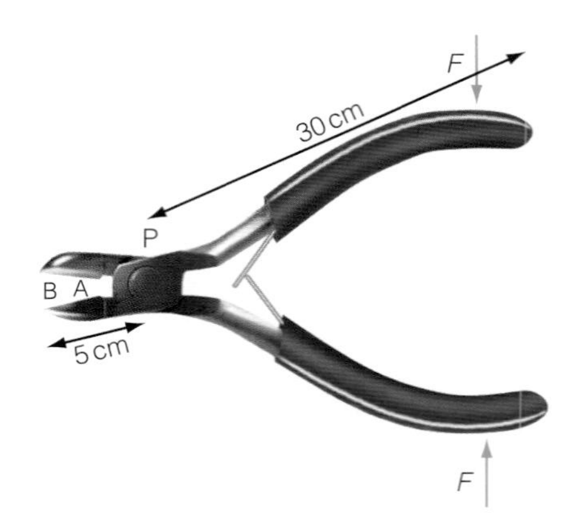

Figure 7.20

b) You want to cut some thick wire. Use your knowledge of moments to establish whether it is better to cut the wire at point A or B.

c) A wire, which needs a force of 210 N to cut it, is placed at B, 5 cm from the pivot. How big a force will you need to exert at the ends of the handle to cut the wire?

Centre of mass

Figure 7.21a shows a rod in equilibrium on top of a pivot. Gravity acts equally on both sides of the rod, so that the clockwise and anticlockwise turning moments balance. This rod is equivalent (mathematically) to another rod that has all of the mass concentrated into the midpoint; this is known as the centre of mass. In Figure 7.21b the weight acts down through the pivot, and the turning moment is zero.

The **centre of mass** is the point in a body around which the resultant torque due to the pull of gravity is zero.

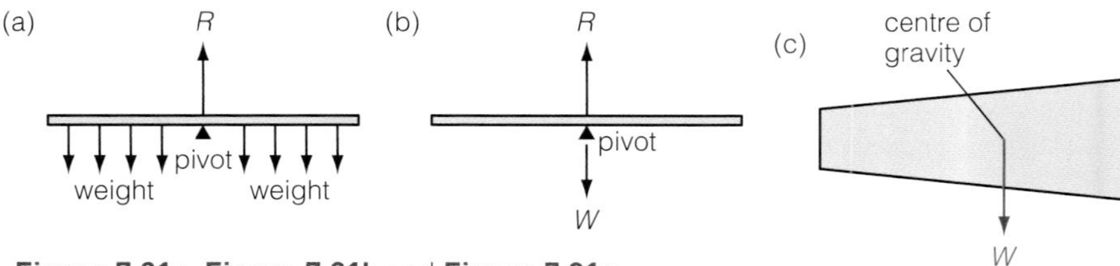

Figure 7.21a, Figure 7.21b and **Figure 7.21c**

The centre of mass is the point in a body around which the resultant torque due to the pull of gravity is zero.

This means that you can always balance a body by supporting it under its centre of mass. In figure 7.21c the tapered block of wood lies nearer to the thicker end.

Figure 7.22 The rocking toy is stable because its centre of gravity lies below the pivot.

Moments in action

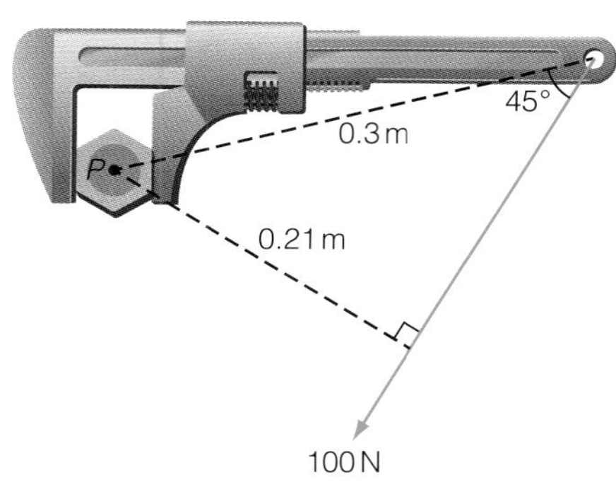

Figure 7.23

When we look at problems in real situations they are not always as straightforward as the examples above. Figure 7.23 shows the action of a force to turn a spanner, but the line of action lies at an angle of 45° to the spanner. How do we determine the turning moment now?

This can be done in two ways.

First, a scale drawing shows that the perpendicular distance between the line of the force and the pivot is 0.21 m, so the turning moment is: 100 N × 0.21 m = 21 N m.

Secondly, the turning moment can be calculated using trigonometry, because the perpendicular distance is (the length of the spanner) × sin θ.

$$\Rightarrow \text{Moment} = F \times l \times \sin\theta$$
$$= 100\,\text{N} \times 0.3\,\text{m} \times \sin 45$$
$$= 21\,\text{N m}$$

EXAMPLE

Figure 7.24 shows a man standing on a small bridge; he has a weight of 800 N and the bridge has a weight of 1200 N, which acts through the midpoint (centre of mass) of the bridge. How can we use the principle of moments to calculate the reaction forces, R_1 and R_2, which support the bridge?

Answer

If we calculate the moments about A, then the clockwise moments of the two weights are balanced by the anticlockwise moment of R_2.

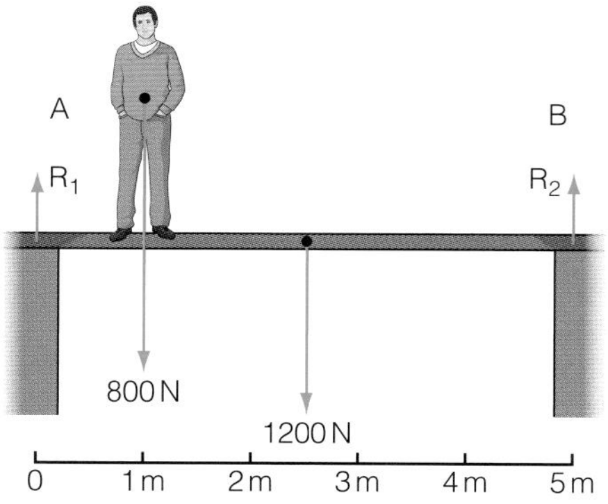

It follows that:

$$R_2 \times 5\,\text{m} = 800\,\text{N} \times 1\,\text{m} + 1200\,\text{N} \times 2.5\,\text{m}$$

$$R_2 \times 5\,\text{m} = 3800\,\text{N m}$$

$$R_2 = 760\,\text{N}$$

Figure 7.24

But the sum of the forces, in the vertical direction, must also balance:

$$R_1 + R_2 = 800\,\text{N} + 1200\,\text{N} = 2000\,\text{N}$$

$$R_1 = 2000\,\text{N} - 760\,\text{N} = 1240\,\text{N}$$

R_1 is larger than R_2 because the man is standing closer to A than to B.

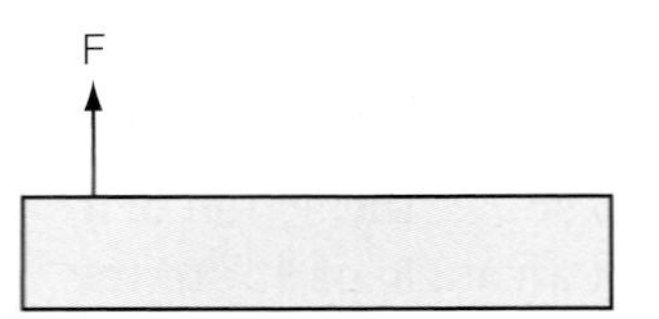

Figure 7.25

Couple A pair of forces that provide a turning effect but no translational movement. They act in opposite directions, are parallel, but do not act along the same line.

Couples

In Figure 7.25 a force is applied to an object. This will have two effects. The first is to set the object off moving in the direction of the force (translation); the second is to start the object rotating about its centre of gravity. If the intention of applying the force is just to rotate the object, applying one force is inefficient because it sets the object moving along too. To avoid this problem we often apply two forces to the object that are parallel but in opposite directions; this is called a couple.

A couple is shown in Figure 7.26a. You apply a couple when you turn a steering wheel; the two forces turn the wheel but exert no translational force. If you take one hand off the wheel (which you should do only to change gear), you can still exert a couple, but the second force is applied by the reaction from the steering wheel (see Figure 7.26b).

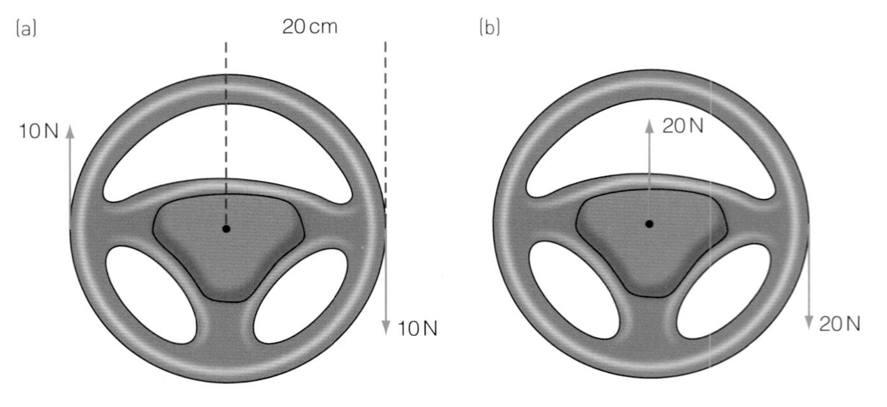

Figure 7.26a and **Figure 7.26b**

TEST YOURSELF

15 Figure 7.27 shows a window cleaner carrying a ladder and bucket.

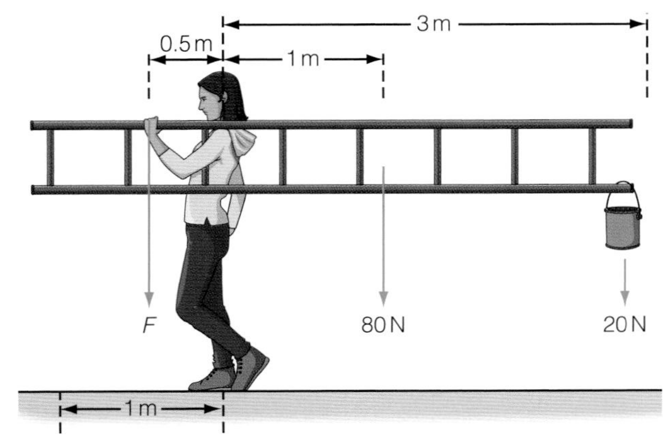

Figure 7.27

a) Calculate the force, F, she must exert to balance the ladder.

b) Now calculate the reaction from her shoulder to support the ladder.

The bucket is now moved to the other end of the ladder, 1 m away from her shoulder.

c) Calculate the reaction force now.

d) Explain how the woman can arrange the bucket and ladder so that the reaction force is only 100 N.

16 The man shown in Figure 7.24 now stands 2 m away from A. By taking moments about B, calculate the reaction forces R_1 and R_2.

17 Calculate the two couples in Figures 7.26a and 7.26b; comment on the answers.

18 The unit of a turning moment is N m. The joule, which is the unit of work, can also be written as N m. Explain why a turning moment is never described as a joule. Hint: how do you calculate the work done when a turning moment moves an object?

19 Figure 7.28 shows a windsurfer in action. The sail experiences a lift, L, which can be represented by a force of 500 N as shown; this acts at an angle of 30° and at a height of 1.5 m

above the board. The surfer has a weight of 650 N. How far out does he have to lean to balance the board?

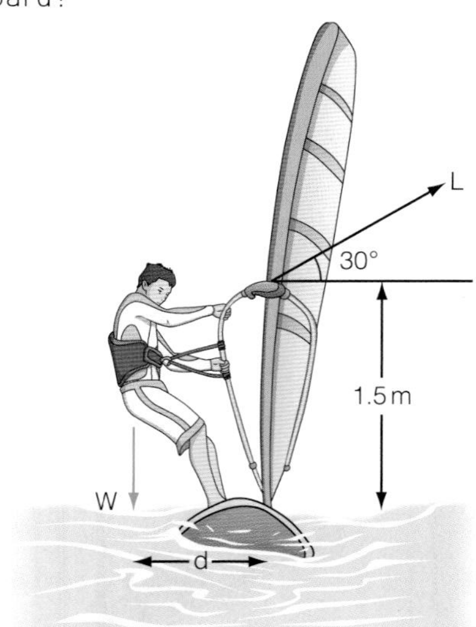

Figure 7.28

20 A uniform beam AC is 6 m long. It is supported at two points, B and C, where AB = 1 m, BC = 5 m. The weight of the beam is 250 N.

a) Calculate the reaction forces from the supports on the beam at points B and C. To start solving this problem, take moments about point B.

A man of mass 850 N stands on the beam at a point D. The reaction forces from the supports at B and C are now equal.

b) State the size of the two reaction forces now.

c) Calculate the distance AD.

21 The Royal Engineers build a small bridge to cross a river. The bridge is 18 m long and is supported at each end by two pillars of concrete. The bridge is of uniform cross section along its length. The weight of the bridge span is 32 000 N. A jeep with its occupants, with a weight of 12 000 N, is 6 m from one end. Calculate the forces on the pillars.

Practice questions

1 Which of the following quantities is a scalar?

A acceleration **C** kinetic energy

B momentum **D** force

2 The direction and size of two forces acting on a body are shown in Figure 7.29. The sum of these forces acts along the line AC. Which of the following pairs gives the correct magnitude of the sum of the forces, and the angle θ?

A 10.0 N, 34° **C** 7.2 N, 48°

B 7.2 N, 34° **D** 8.4 N, 42°

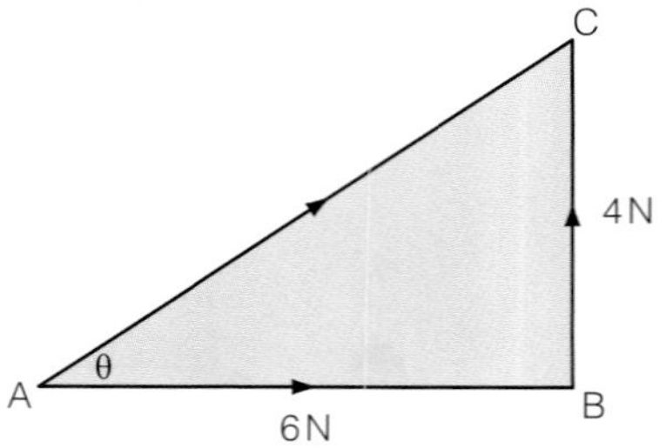

Figure 7.29

3 In Figure 7.30, a car accelerating down a slope under the action of gravity; its engine is turned off. The mass of the car is 800 kg, and the frictional force acting up the slope against the car is 720 N. Which of the answers below gives the correct acceleration of the car down the slope?

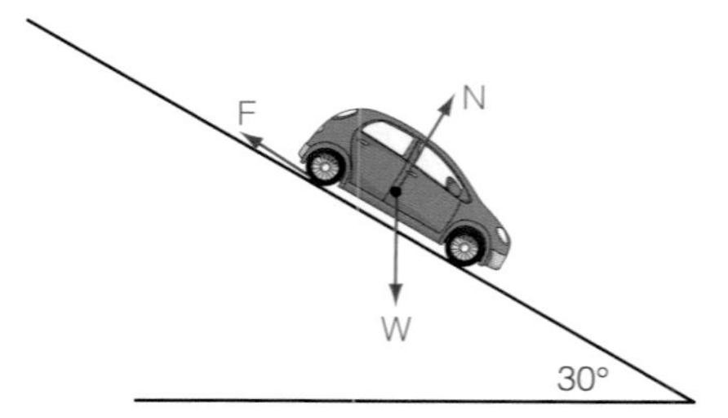

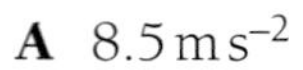

A $8.5\,m\,s^{-2}$ **C** $4.9\,m\,s^{-2}$

B $7.6\,m\,s^{-2}$ **D** $4.0\,m\,s^{-2}$

Figure 7.30

4 Which of the following values gives the value for the normal force N, that the road exerts on the car in the previous question?

A 7800 N **C** 3900 N

B 6800 N **D** 3400 N

5 Figure 7.31 shows a metal ball supported by a string and held at an angle of 20° to the vertical by a horizontal force of 1.5 N. Which of the following possible answers is the correct value for the weight of the ball?

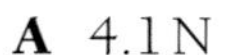

A 4.1 N **C** 1.4 N

B 0.6 N **D** 4.4 N

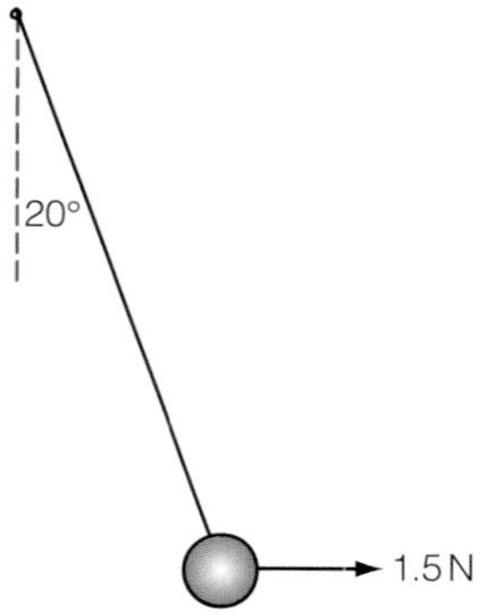

Figure 7.31

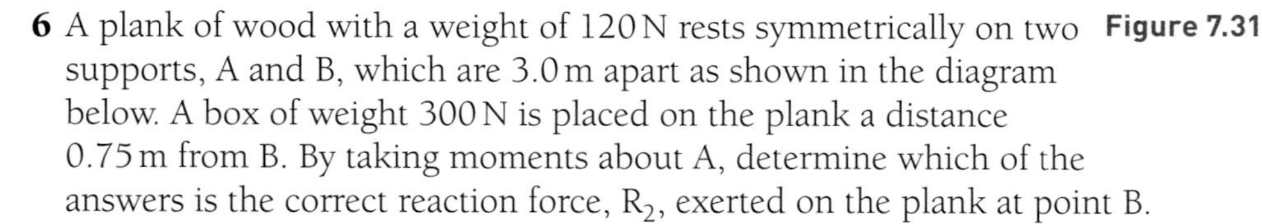

6 A plank of wood with a weight of 120 N rests symmetrically on two supports, A and B, which are 3.0 m apart as shown in the diagram below. A box of weight 300 N is placed on the plank a distance 0.75 m from B. By taking moments about A, determine which of the answers is the correct reaction force, R_2, exerted on the plank at point B.

A 285 N **C** 195 N

B 225 N **D** 180 N

7 The weight is now moved to the right of B, to a point where the plank begins to tip up about B. Which of the answers below gives the correct distance between the box and point B?

A 0.50 m **C** 0.75 m

B 0.60 m **D** 1.20 m

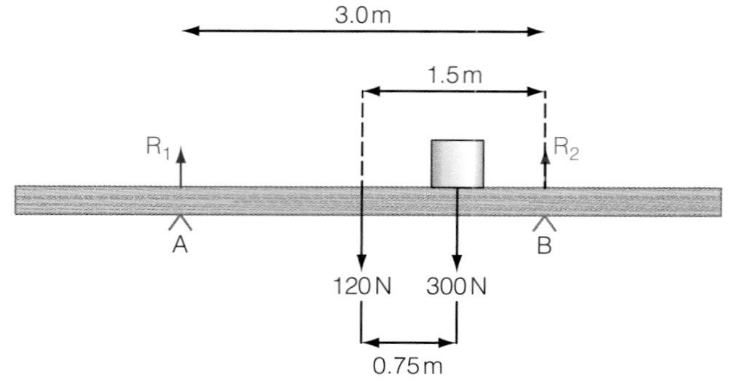

Figure 7.32

8 A piece of wood of weight 2.4 N is balanced on a support as shown in Figure 7.33. Which of the answers gives the correct position of the wood's centre of gravity?

A 0.8 m to the right of the support

B 0.5 m to the right of the support

C 0.4 m to the right of the support

D 0.1 m to the left of the support

Figure 7.33

9 Figure 7.34 shows a uniform ruler which is supported at its midpoint. Two forces act on either side of the ruler as shown: a force, F, acting at an angle of 37° and a weight of 3 N on the other side. The ruler is in equilibrium. Which of the following answers is the closest to the value of the force, F?

A 12 N **C** 6 N

B 9 N **D** 4 N

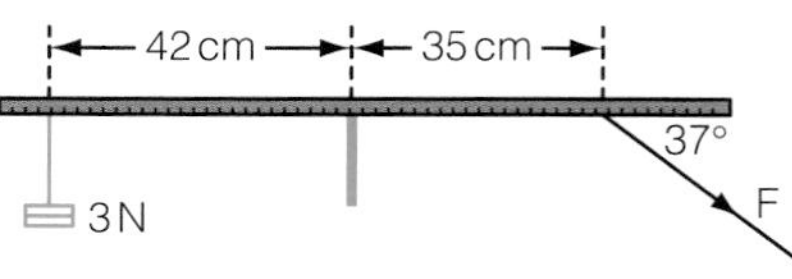

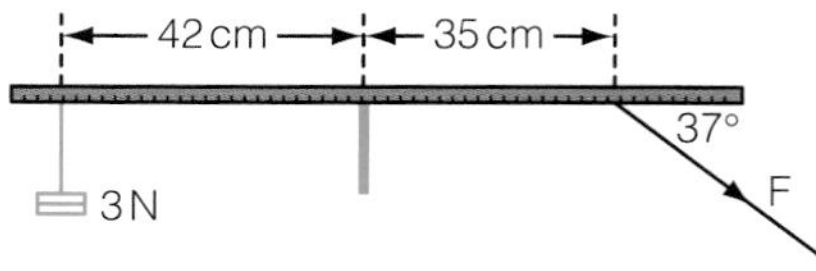

Figure 7.34

10 The ruler has a weight of 3 N. Which of the following answers is the closest to the vertical cimponent of the reaction force exerted on the ruler by the pivot?

A 12 N **C** 9 N

B 10 N **D** 6 N

11 Copy and complete the table below to show the fundamental SI units of the listed quantities, and to show whether each quantity is a vector or a scalar quantity.

Table 2.10

Quantity	Fundamental SI units	Type of quantity
Force	$kg\,m\,s^{-2}$	vector
Kinetic energy		
Acceleration		
Displacement		
Power		

12 A sprinter is set for the start of a race. His weight, W, is 780 N and the reaction from the track on his hands, R_1, is 320 N.

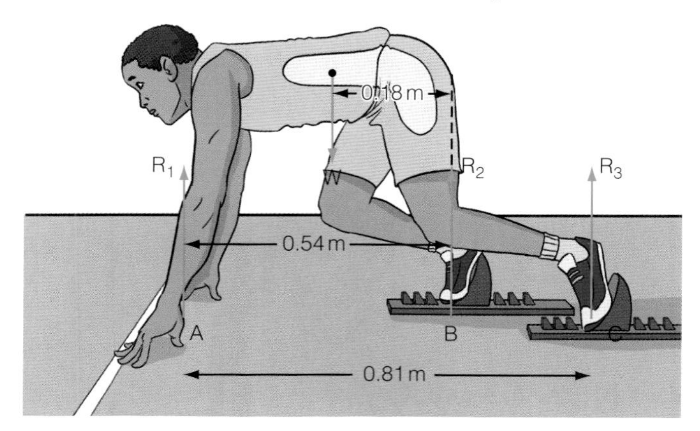

Figure 7.35

By taking moments about B, calculate:

a) the reaction R_3 *(3)*

b) the reaction R_2. *(1)*

13 A truck has a weight of 16 000 N and a load of 8000 N.

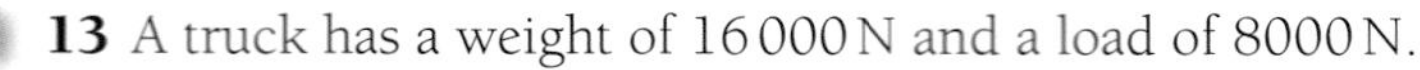

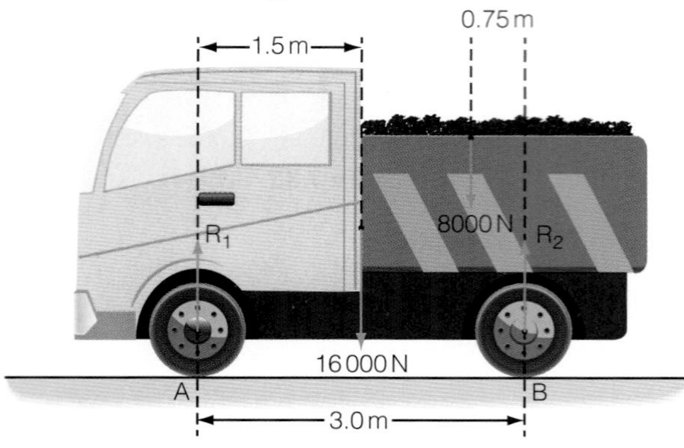

Figure 7.36

a) Calculate the combined moments of the load and the weight of the truck about point B. *(2)*

b) State the principle of moments. *(1)*

c) Use the principle of moments to calculate the reaction force R_1, and then calculate R_2. *(3)*

d) State the values of R_1 and R_2 when the truck carries no load. *(1)*

14 A golfer decides to determine the position of the centre of gravity of one of his clubs. He balances it as shown in the diagram. The mass of the club is 350 g.

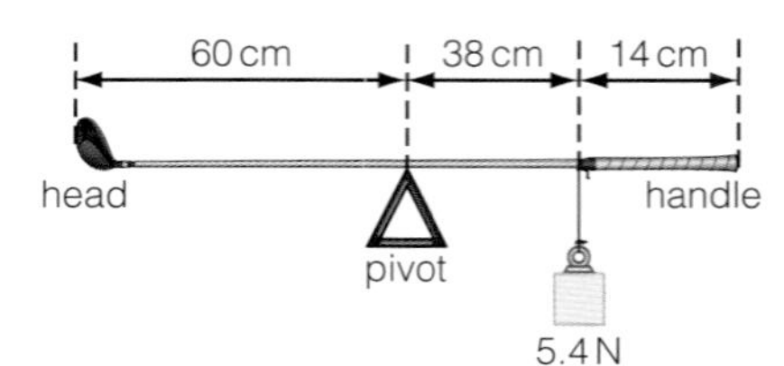

Figure 7.37

a) Explain why this balancing arrangement produces a stable position of equilibrium. *(1)*

b) Calculate the weight of the club. *(1)*

c) Use the information to determine the position of the centre of gravity from the head end of the club. *(2)*

The golfer now hits the ball with a speed of $45\,\mathrm{m\,s^{-1}}$ at an angle of 40° to the horizontal.

d) Calculate:

i) the vertical component of its velocity *(2)*

ii) the horizontal component of its velocity. *(2)*

The ball is in flight for 5.9 s.

e) Determine the distance the ball travels before it lands on the ground. *(2)*

15 Figure 7.38 shows the forces acting on a kite that is in equilibrium.

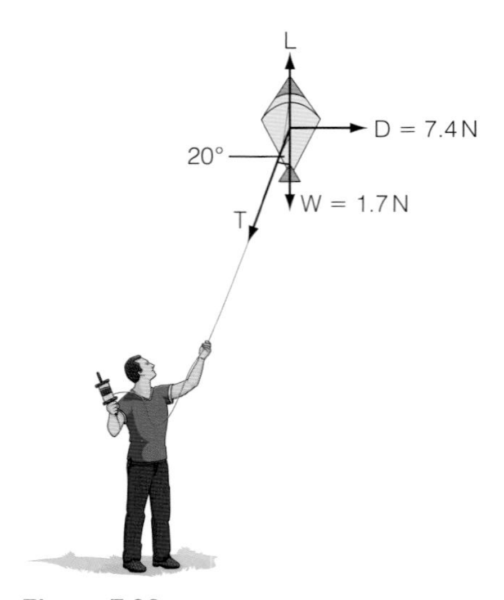

Figure 7.38

a) Use the information in the diagram to calculate the magnitude of the tension, T, in the string. *(3)*

b) Now calculate the magnitude of the lift on the kite, L. *(3)*

16 Here is a cable car which is stationary on a wire.

a) Draw a diagram for the point A (where the car is suspended from the cable) to show how the three forces are in equilibrium. *(3)*

b) Either by scale drawing, or by calculation, determine the tension, T, in the cable. *(4)*

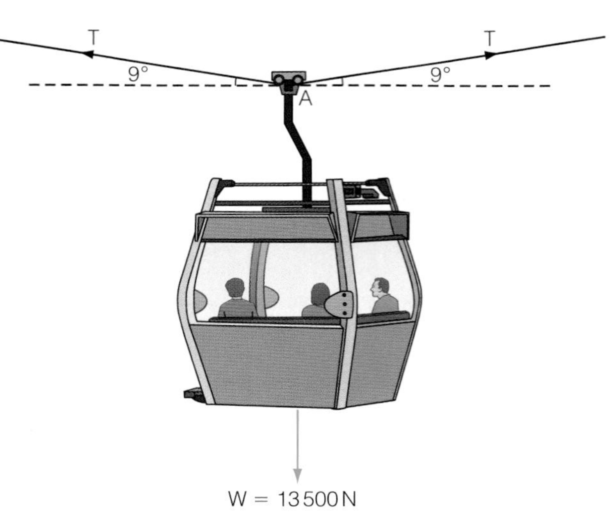

Figure 7.39

17 A shelf with a weight of 16 N is placed (without being fixed) on two supports B and D (see Figure 7.40).

a) State the size of the reaction contact forces at B and D. *(1)*

A box of weight 40 N is now placed on the shelf as shown in Figure 7.41.

b) Use the principle of moments to calculate the reaction contact forces at B and D now. *(4)*

The box is now moved to the left of B.

c) Calculate the position of the box so that the contact force at D is zero. *(2)*

d) Calculate what additional weight you would have to place above C to ensure that the box will not tip the shelf when it is placed at A or E. *(3)*

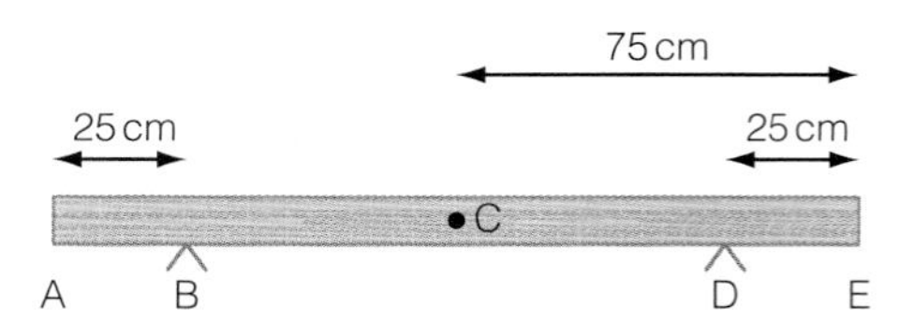

Figure 7.40

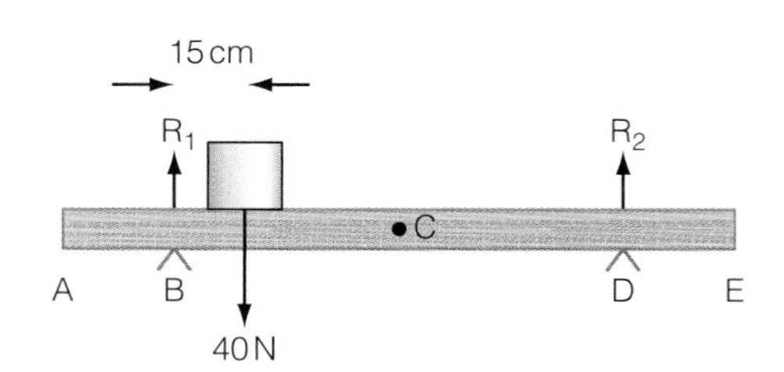

Figure 7.41

18 A pedal bin is a labour-saving device in the kitchen. Pressing on the pedal with your foot causes a series of pivots and levers to lift the lid so that you can throw rubbish into the bin. When the pedal is depressed, rod R lifts the lid (see Figure 7.42).

a) When a force of 20 N is applied to the pedal, as shown, calculate the force acting upwards at A, to lift the rod R. *(2)*

b) If 20 N is the minimum necessary force applied to the pedal to lift the lid, calculate the weight of the lid. *(2)*

c) Calculate the contact force between pivot 2 and the lid as it is lifted by the 20 N force from the pedal. *(2)*

d) Explain how a small displacement of the pedal lifts the lid a much larger distance. *(2)*

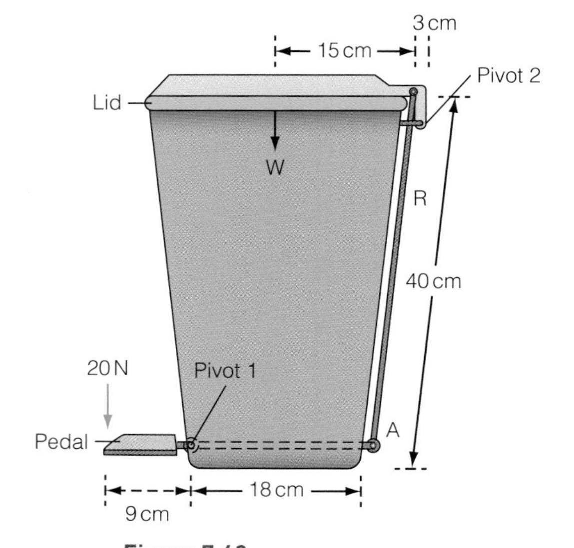

Figure 7.42

Stretch and challenge

19 The figure below shows a big wheel at a funfair. The wheel's centre of mass is at its centre of rotation, point O. In the following calculations, you may ignore any effects of friction.

At present only two of the seats on the wheel are occupied: at A two people sit with a combined weight of 1600 N; one person sits at B with a weight of 800 N.

a) Calculate the vertical force required at the embarking point C to prevent the wheel rotating:

i) in the position shown

ii) after the wheel has been rotated clockwise by 60°.

b) In what position will the wheel be in stable equilibrium?

c) In which position would you place two more passengers, and what should their combined weight be, so that the wheel has no tendency to rotate in any position?

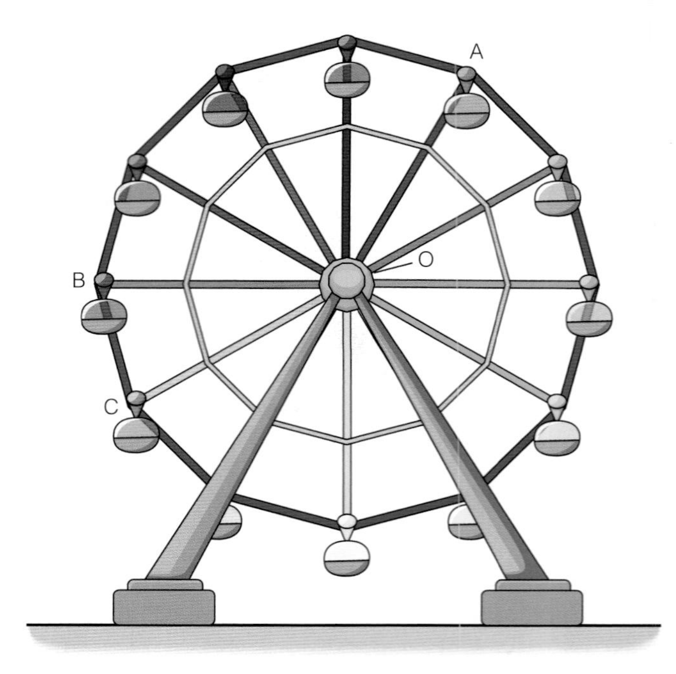

Figure 7.43

20 A light, rigid straight beam ABC is smoothly hinged at end A, on a vertical wall, and supported in a horizontal position by a wire, which is attached to the wall and the midpoint of the beam, B, at an angle of 60° to the beam.

Calculate the tension in the wire and the direction and magnitude of the force on the hinge when a mass M is suspended from point C.

8 Motion and its measurement

PRIOR KNOWLEDGE

- average speed = $\frac{\text{distance}}{\text{time}}$
- Speed is measured in $m\,s^{-1}$ or $km\,h^{-1}$.
- When we deal with the vector quantities, displacement and velocity, we write a similar equation:
  average velocity = $\frac{\text{displacement}}{\text{time}}$
- When an object speeds up it accelerates, and when it slows down it decelerates.
- Acceleration is a vector quantity as it has a direction. It is defined by this equation:
  acceleration = $\frac{\text{change of velocity}}{\text{time}}$
- The units of acceleration are m/s per second, usually written as $m\,s^{-2}$.
- The gravitational field strength, g, near a planet is defined as the gravitational force acting on each kilogram:
- On Earth, $g = 9.81\,N\,kg^{-1}$
- Newton's first law of motion: an object will continue to move in a straight line with a constant speed (or remain at rest) unless acted on by an unbalanced force.
- Newton's second law of motion: resultant force = mass × acceleration
- A velocity vector may be resolved into horizontal and vertical components.

TEST YOURSELF ON PRIOR KNOWLEDGE

1 A car travels at $40\,km\,h^{-1}$ for 2 hours then $60\,km\,h^{-1}$ for 4 hours. Calculate the car's average speed.

2 A car slows down from a speed of $30\,m\,s^{-1}$ to $12\,m\,s^{-1}$ in a time of 6 s. Calculate the car's acceleration.

3 Convert a speed of $54\,km\,h^{-1}$ to $m\,s^{-1}$.

4 A ball is falling in a direction at 35° to the horizontal with a speed of $12\,m\,s^{-1}$.
Calculate the vertical and horizontal components of its velocity.

5 An astronaut throws a ball horizontally when standing on the surface of the Moon, where there is no atmosphere.
Draw a diagram to show the direction of the force(s) acting on it. (There is no need to calculate the size of any force.)

Motion in a straight line

In October 2012 Felix Baumgartner broke the existing records for the highest manned balloon flight, the highest altitude for a parachute jump and the greatest free fall velocity. He jumped from a height of nearly 39 km and reached a speed of 377 m s^{-1} (1357 km h^{-1}). The successful completion of such a challenge depends on careful planning and design. Felix needed a balloon inflated to exactly the right pressure at sea level so that it could expand as the atmospheric pressure reduced at high altitude; he needed a pressurised space suit, capable of withstanding the shock of free fall; and he relied on the calculations of physicists using the equations of motion to predict his time of fall and knowing when it was safe to open his parachute. You will learn about these equations in the chapter that follows.

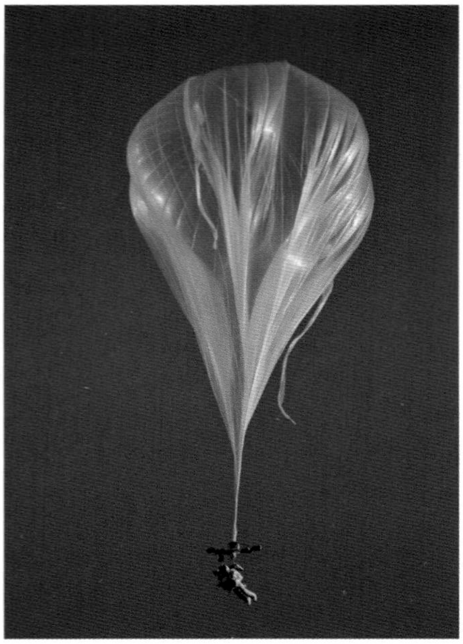

Figure 8.1 Alan Eustace on his way up to break Felix Baungartner's record. In October 2014, Alan fell from a height of 41.5 km.

Displacement-time graphs

In the work which follows we study motion in a straight line, but since the motion might be from left to right, or upwards then downwards, the direction is important. Therefore we shall use the vector displacement, s, and velocity, v.

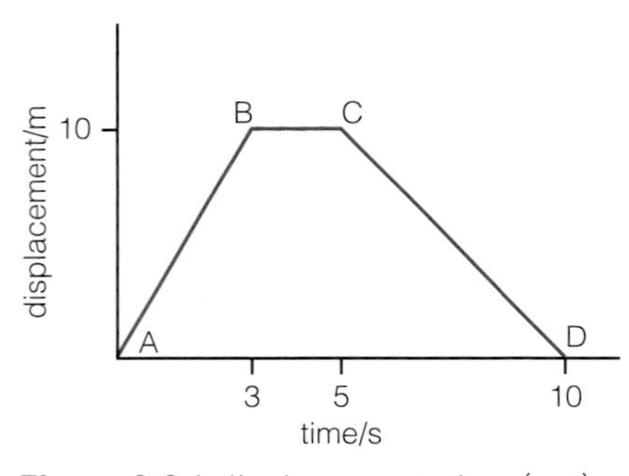

Figure 8.2 A displacement-time (s–t) graph for a walker.

Figure 8.2 shows a displacement–time (s–t) graph for a walker. In the first three seconds she walks to the right (which is chosen to be positive); then she stands still for two seconds, so her displacement does not change; then she moves back to the left for five seconds and gets back to where she started. So, after ten seconds, her displacement is zero.

The velocity of the walker can be calculated from the gradient of the graph.

Over the region AB:

$$v = \frac{(10 - 0)\,\text{m}}{3\,\text{s}} = 3.3\,\text{m}\,\text{s}^{-1}$$

Over the region BC:

$$v = 0$$

Over the region CD:

$$v = \frac{(0 - 10)\,\text{m}}{5\,\text{s}} = -2\,\text{m}\,\text{s}^{-1}$$

Because the gradient is negative over the region CD the velocity is negative; the walker is moving to the left.

In Figure 8.2 there are three constant velocities. This means that the gradients of the graph are constant, so the graphs are 'straight line'. Figure 8.3 is a displacement–time graph for a vehicle that is accelerating. As the velocity increases, the gradient increases.

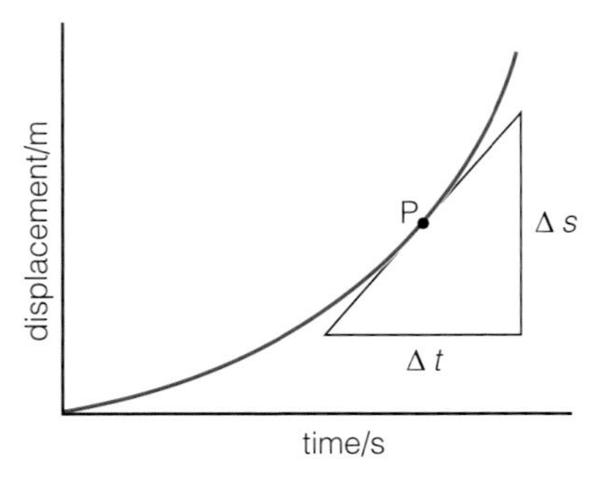

Figure 8.3 A displacement–time graph for a vehicle that is accelerating.

To calculate the velocity at point P, we measure the gradient, which we do by drawing a tangent to the curve at that point. We measure a small change in displacement, Δs, which occurred over a small time interval, Δt.

Then

$$v = \frac{\Delta s}{\Delta t}$$

TIP

The gradient of a displacement–time graph is velocity.

TEST YOURSELF

1. Figures 8.4a, b and c show displacement–time graphs for three moving objects. In each case describe the motion in as much detail as possible.
2. Draw displacement–time graphs to illustrate the motion in each of the following cases.
   a) A walker travels at a constant velocity of $3\,\mathrm{m\,s^{-1}}$ for 30 s.
   b) An express train slows down as it comes into a station and stops; five minutes later it sets off and accelerates up to its original velocity.
   c) A parachutist jumps out of a plane, accelerates to a high constant velocity before opening the parachute and coming to ground. Draw a graph with the ground as the point of zero displacement.
   d) A rubber ball falls from a height of 1 m and hits the ground. It bounces back to height of 0.5 m. Sketch a graph to show the displacement against time up to the second bounce.

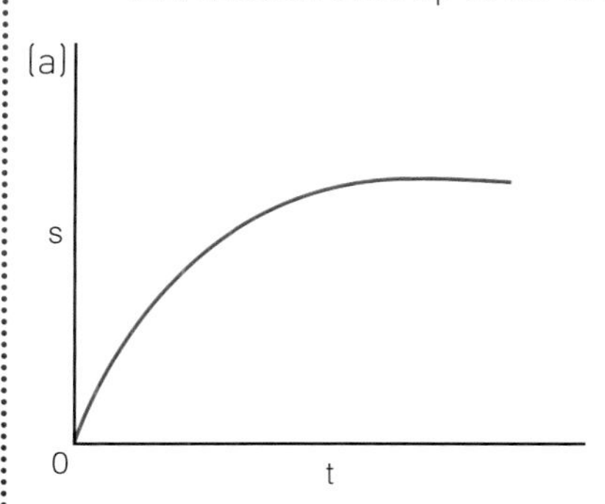

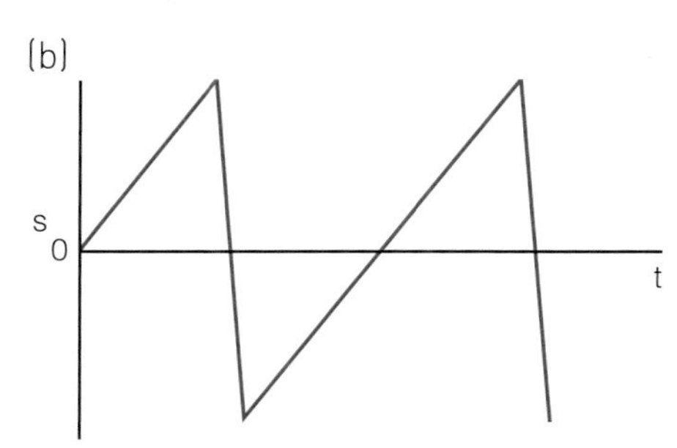

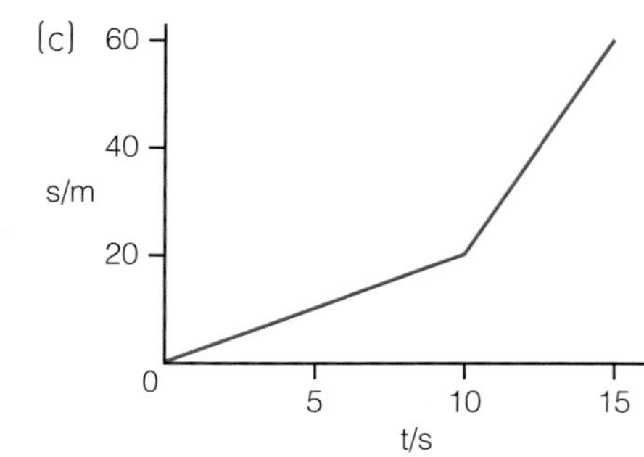

Figure 8.4 Displacement–time graphs for three moving objects.

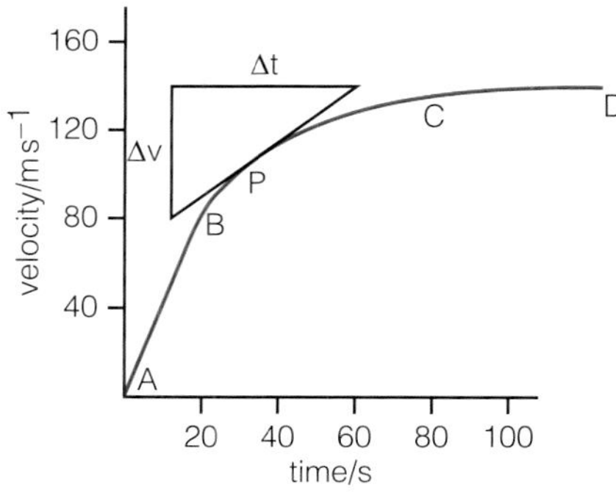

Figure 8.5 A velocity–time graph for an airplane at take off.

Velocity–time graphs

Figure 8.5 shows a velocity–time graph for an aeroplane as it takes off. Initially, the plane accelerates at a constant rate, shown on part AB of the graph. Then the acceleration slows until the plane reaches a constant velocity at point D.

Over the part AB, you can see from the graph that the acceleration is $4\,\mathrm{m\,s^{-2}}$. This is the gradient of the graph: $(80\,\mathrm{m\,s^{-1}})/(20\,\mathrm{s})$. But, over the region BC, the gradient is changing, so we now use the formula:

$$a = \frac{\Delta v}{\Delta t}$$

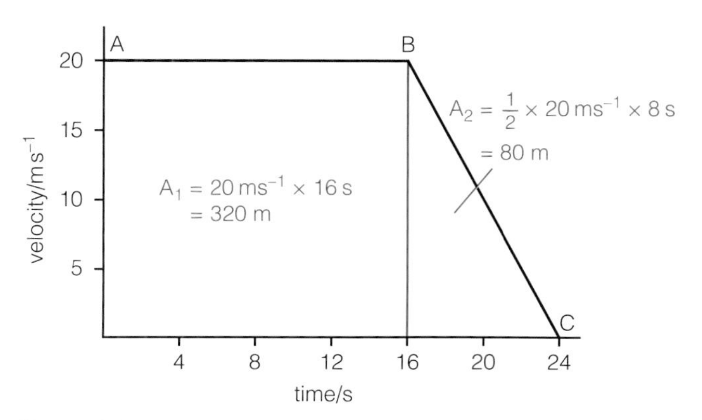

Figure 8.6 How to use a velocity–time graph to calculate the distance travelled by a motorbike.

where Δv means a small change in velocity, and Δt means a small interval of time.

At point P:

$$a = \frac{\Delta v}{\Delta t} = \frac{60\,\mathrm{m\,s^{-1}}}{48\,\mathrm{s}} = 1.25\,\mathrm{m\,s^{-2}}$$

Figure 8.6 shows how we can use a velocity–time graph to calculate the distance travelled by a motorbike. In this graph, the motorbike travels at a constant velocity over the period AB, before decelerating over the period BC. While the velocity is constant, the distance travelled is represented by area A_1, which is 320 m.

While the bike decelerates, we could use the formula;

distance = average velocity × time

But the average velocity is $10\,\text{m}\,\text{s}^{-1}$, which is the average of $0\,\text{m}\,\text{s}^{-1}$ and $20\,\text{m}\,\text{s}^{-1}$. So the distance can also be calculated using the area:

$$A_2 = \frac{1}{2} \times 20\,\text{m}\,\text{s}^{-1} \times 8\,\text{s} = 80\,\text{m}$$

Figure 8.6 also shows that when the gradient of a velocity–time graph is zero, then the velocity is constant; when the gradient is negative, the bike is decelerating.

TIP

The area under a velocity–time graph is the distance travelled. The gradient of a velocity–time graph is the acceleration.

TEST YOURSELF

3 a) Draw a velocity–time graph for a train that slows down and stops at a station, and then sets off again in the same direction, reaching the same speed.

b) Draw a second velocity–time graph for a train that comes into a terminus, stops and then returns back in the opposite direction, reaching the same speed.

4 Figure 8.7 is a velocity–time graph for the Maglev train, which travels from Longyang Station in Shanghai to Pudong International Airport.

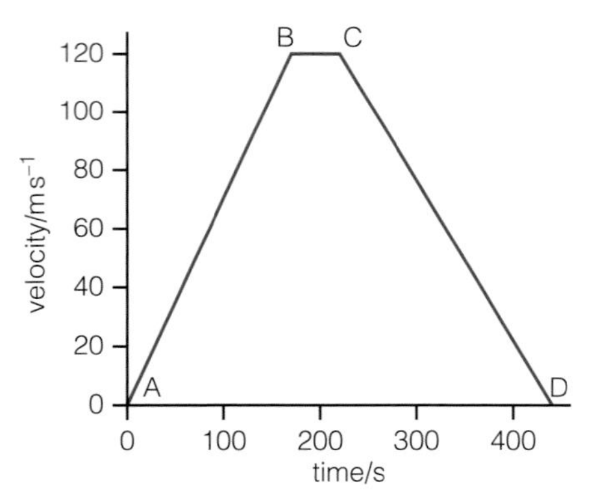

Figure 8.7 A velocity–time graph for the Maglev train.

a) Calculate the train's acceleration over the region AB.

b) Calculate the distance travelled when the train travelled at a constant velocity.

c) Calculate the distance from Longyang Station to Pudong Airport.

d) Calculate the train's average speed for the journey.

5 Figure 8.8 is a velocity–time graph for a firework rocket which takes off vertically and, when the fuel runs out, falls back to where it took off.

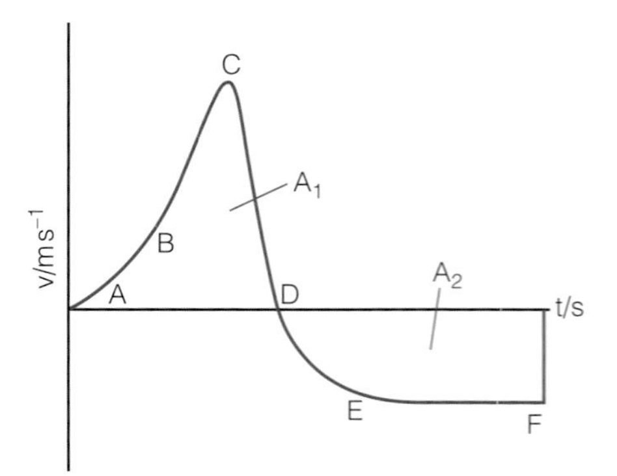

Figure 8.8 A velocity–time graph for a firework rocket.

a) Explain why the gradient gets steeper over the region AB.

b) At which point does the rocket stop burning?

c) Explain why the gradient of the graph should be about $10\,\text{m}\,\text{s}^{-2}$ over the region CD.

d) Why is the gradient zero over the region EF?

e) Explain why the areas A_1 and A_2 are the same.

6 A rubber ball falls for a time of 0.6 s before hitting the ground. It is in contact with the ground for 0.02 s, before bouncing upwards for 0.4 s; it then falls to the ground in 0.4 s. Sketch a velocity-time graph for the ball up to the second bounce.

Equations for uniform acceleration

So far you have learnt to analyse motion using graphs. We can use a velocity–time graph, where the acceleration is constant, to derive formulas to help us calculate velocity and displacement changes.

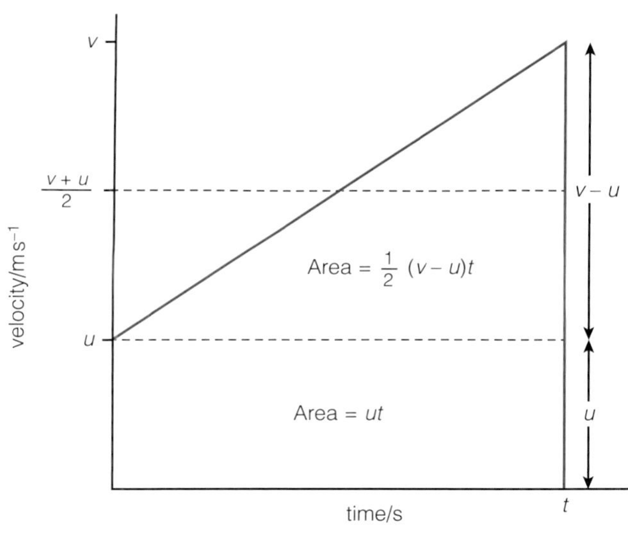

Figure 8.9 A velocity–time graph showing a vehicle increasing its velocity with a constant acceleration.

Figure 8.9 is a velocity–time graph showing a vehicle increasing its velocity with a constant acceleration, a, from an initial velocity, u, to a final velocity, v, over a time, t.

The first equation is:

$$\text{acceleration} = \frac{\text{change of velocity}}{\text{time}}$$

or

$$a = \frac{v - u}{t}$$

Rearranging the equation to make v the subject of the formula gives:

$$v = u + at \qquad \text{(Equation 1)}$$

The second equation is:

displacement = average velocity × time

The average velocity is half way between the initial and final velocities.

Substituting for average velocity using:

$$\frac{(u + v)}{2}$$

gives:

$$s = \left(\frac{u + v}{2}\right) t \qquad \text{(Equation 2)}$$

The displacement can also be calculated from the area under the velocity-time graph, which is the sum of the two areas shown in Figure 8.9.

The third equation is:

$$s = ut + \frac{1}{2}(v - u)t$$

(which is the area under the graph).

Therefore

$$s = ut + \frac{1}{2}at^2 \qquad \text{(Equation 3)}$$

because

$$(v - u) = at \qquad \text{(from Equation 1)}$$

The fourth equation, which links velocity, acceleration and displacement, is:

$$v^2 = u^2 + 2as \qquad \text{(Equation 4)}$$

There is no need to be able to derive these equations and you will find them on your formula sheet. The derivation of Equation 4 is shown in the Maths box.

MATHS BOX

The fourth equation can be derived as follows:

$$v = u + at$$
$$\rightarrow v^2 = (u + at)^2$$
$$\rightarrow v^2 = u^2 + 2uat + a^2 t^2$$
$$\rightarrow v^2 = u^2 + 2a\left(ut + \frac{1}{2}at^2\right)$$
$$\rightarrow v^2 = u^2 + 2as \qquad \text{(using Equation 3)}$$

EXAMPLE

You drop a stone off the edge of a high cliff (making sure there is no-one below). How far does it fall after 2 seconds?

Answer

Use

$$s = ut + \frac{1}{2}at^2$$

The initial velocity is zero, and the acceleration due to gravity is $9.8\,m\,s^{-2}$.

$$s = 0 + \frac{1}{2} \times 9.8\,m\,s^{-2} \times 2^2 s^2$$

$$s = 19.6\,m$$

EXAMPLE

What is the displacement after 2 seconds if you throw the stone:

a) with an downward velocity of $5.0\,m\,s^{-1}$

b) with a upward velocity of $5.0\,m\,s^{-1}$?

Answer

We need to remember that velocity and displacement are vector quantities, so we need to define a direction. We will call down a positive direction, and up a negative direction. (The choice is arbitrary; you can reverse the direction signs if you wish.)

a) Use

$$s = ut + \frac{1}{2}at^2$$

$$s = 5.0\,m\,s^{-1} \times 2\,s + \frac{1}{2} \times 9.8\,m\,s^{-2} \times 2^2 s^2$$

$$s = 29.6\,m$$

b) Use

$$s = ut + \frac{1}{2}at^2$$

$$s = -5.0\,m\,s^{-1} \times 2\,s + \frac{1}{2} \times 9.8\,m\,s^{-2} \times 2^2 s^2$$

$$s = 9.6\,m$$

EXAMPLE

A ball is thrown upwards with an initial velocity of $15\,m\,s^{-1}$ from a height of 1 m above the ground. When does it reach a height of 7 m above the ground?

Answer

The first point to appreciate about this example is that there are two times when the ball is 7 m above the ground: when it is on the way up, and when it is on the way down again.

We use the same equation again, using $s = -6\,m$ (for the distance travelled from the start); $u = -15\,m\,s^{-1}$; $g = 9.8\,m\,s^{-2}$. Here the downward direction has been defined as positive.

Use

$$s = ut + \frac{1}{2}at^2$$

$$-6 = -15t + \frac{1}{2} \times 9.8 \times t^2$$

$$4.9t^2 - 15t + 6 = 0$$

This can be solved using the standard formula for the solution of quadratics:

$$t = \frac{15 \pm \sqrt{15^2 - 4 \times 4.9 \times 6}}{2 \times 4.9}$$

$$t = \frac{15 \pm 10.3}{9.8}$$

$$t = 2.6\text{ s or }0.5\text{ s}$$

So the ball reaches a height of 7 m after 0.5 s on the way up, and passes the same height on the way down after 2.6 s.

TEST YOURSELF

7 A plane accelerates from rest along a runway at a constant rate of $1.8\,m\,s^{-2}$. The plane takes off after 40 s.
   a) Calculate the velocity of the plane at take off.
   b) Calculate the minimum length of runway required for the plane to take off.

8 A car slows down from a speed of $30\,m\,s^{-1}$ to $22\,m\,s^{-1}$ over a distance of 120 m. Calculate the deceleration of the car.

9 A Formula 1 car accelerates off the grid reaching a speed of $55\,m\,s^{-1}$ in 3.6 s.
   a) Calculate the acceleration of the car over this time.
   b) Calculate the distance travelled by the car in this time.

10 An express train travelling at $50\,m\,s^{-1}$ applies its brakes for 3 minutes before reaching a station, which it passes at $30\,m\,s^{-1}$. How far from the station was the train when the brakes were first applied?

11 A car starts from rest with a uniform acceleration of $2\,m\,s^{-2}$ along a long straight track. At the moment the car starts, a second car passes it on a parallel track travelling at a constant speed of $20\,m\,s^{-1}$. Calculate:
   a) the time at which the cars are level
   b) the velocity of the first car at that time
   c) the distance travelled by the first car when it catches the second car.

12 A ball is launched upwards from ground level with a speed of $20\,m\,s^{-1}$. At what times is the ball at a height of 10 m above the ground? (Hint: look at the example above.)

Acceleration due to gravity, g

One of the first scientists to study the acceleration of objects due to gravity was Galileo Galilei. There is story that he dropped two iron balls of different masses from the top of the Leaning Tower of Pisa. Thereby, he demonstrated that gravity accelerates all masses at the same rate, provided that air resistance is negligibly small.

By such an experiment you could calculate the acceleration due to gravity, using the equation:

$$s = \frac{1}{2}gt^2$$

which rearranges to give

$$g = \frac{2s}{t^2}$$

For example, Galileo's assistant could have timed such a fall (from the top of the tower, which is 55 m high) as three and a quarter swings of his 1-second pendulum.

$$g = \frac{2 \times 55\,\text{m}}{3.25^2}$$

$$g = 10.4\,\text{m}\,\text{s}^{-2}$$

This answer illustrates one of the problems we have if we try to measure g accurately. We need very accurate timing and, because the calculated answer depends on the value of t^2, small errors in the measurement of t become significant.

REQUIRED PRACTICAL 3

Determination of g by a free-fall method

Note: This is just one example of how you might tackle this required practical.

Figure 8.10 shows an experimental arrangement using two light gates to calculate the gravitational acceleration. A steel ball is dropped through the two gates, and the times to fall through the gates are measured and recorded in the table. The ball is guided through the light gates by a glass tube.

Experiment number	1	2	3
Diameter of the ball (cm)	1.32	1.32	1.32
Time for the ball to pass through gate A (ms)	8.3	6.7	9.7
Time for the ball to pass through gate B (ms)	2.8	3.1	5.2
Time for the ball to pass from the centre of gate A to the centre of gate B (ms)	318.0	219.7	121.2

Using the results from experiment 1, the acceleration g is calculated as follows.

The velocity of the ball as it falls through gate A:

$$v_A = \frac{0.0132\,\text{m}}{0.0083\,\text{s}} = 1.59\,\text{m}\,\text{s}^{-1}$$

The velocity of the ball as it falls through gate B:

$$v_B = \frac{0.0132\,\text{m}}{0.0028\,\text{s}} = 4.71\,\text{m}\,\text{s}^{-1}$$

$$g = \frac{v_B - v_A}{t}$$

$$= \frac{4.71\,\text{m}\,\text{s}^{-1} - 1.59\,\text{m}\,\text{s}^{-1}}{0.318\,\text{s}}$$

$$= 9.81\,\text{m}\,\text{s}^{-2}$$

An advantage of using light gates and a computer is that the experiment can be repeated several times, and the computer can be programmed to calculate the acceleration directly.

1 The experiment is repeated using different separations of the light gates. Use the data in experiments 2 and 3 to calculate the acceleration due to gravity.
2 Use the equations of motion to calculate the separations of the light gates in each of the experiments.
3 Explain why it is not necessary to know the separation of the light gates to calculate to gravitational acceleration.
4 Does it matter how far above the first light gate you release the steel ball?
5 Discuss the possible sources of error in this experiment.
6 Calculate the average value of the acceleration and estimate the percentage error in the measurements.

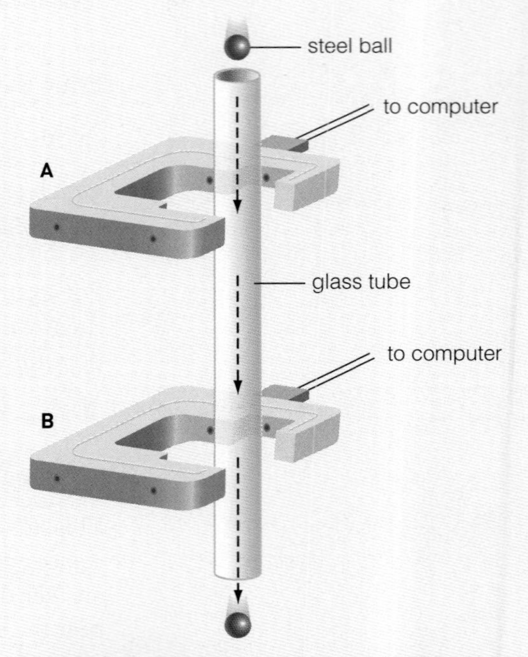

Figure 8.10 A ball is timed as it falls between two light gates.

Terminal speed is the speed reached when the weight of an object in free fall is balanced by drag forces acting upwards on the object.

Drag is the name given to resistive forces experienced by an object moving through a fluid such as air or water.

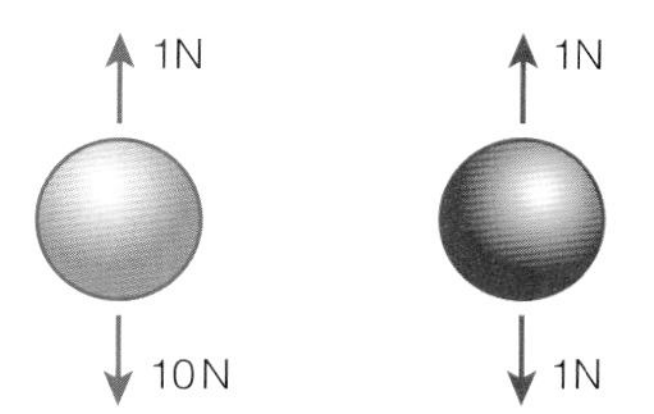

Figure 8.11 At this instant the two balls fall at the same speed; the drag is the same on each but the blue ball continues to accelerate.

Terminal speed

In everyday life we are all aware of the effect of wind resistance, or drag, when something falls to the ground. A piece of paper that has been rolled up into a ball accelerates quickly downward, but the same piece of paper left as a sheet will flutter from side to side as it falls due to a large drag acting over the larger surface area.

The size of the drag on a falling object increases with:

- speed
- surface area.

Figure 8.11 explains in more detail how drag affects the accelerations of two similar balls. They are both of identical size and shape, but the blue ball has a weight of 10 N and the red ball 1 N. In the diagram they are falling at the same speed and they both have an upward drag of 1 N. The blue ball continues to accelerate because there is a resultant downwards force acting on it; but the red ball moves with a constant speed as the pull of gravity is balanced by the drag. The red ball has reached its terminal speed.

TEST YOURSELF

13 A parachutist together with his parachute has a weight of 850 N.

a) Draw a diagram to show his weight and the drag force acting on him when he falls at his terminal speed *without* his parachute open.

b) Draw a second diagram to show the weight and drag forces when he falls at his terminal speed with his parachute open. Explain why the two terminal speeds are different.

14 A student drops a ping-pong ball and a steel ball of the same diameter from a height of 5 m.

a) Explain why the steel ball reaches the ground first.

b) Sketch graphs to show how the velocity of each ball changes with time.

15 Figure 8.12 shows another method of measuring g. A small steel ball is dropped from a height, h, and passes through a light gate. The data from the gate goes into a computer, which calculates the ball's speed, v.

Height fallen: h (m)	Speed: v (m s^{-1})	v^2 (m^{2}s^{-2})
0.10	1.40	
0.15	1.72	
0.26	2.26	
0.38	2.73	
0.55	3.28	
0.75	3.84	

a) Copy the table and complete the column to calculate value of v^2.

b) The relationship between v and h is given by the equation:

$$v^2 = 2gh$$

This relationship can be deduced from the equation of motion:

$$v^2 = u^2 + 2as.$$

Plot a suitable graph and use it to calculate the value of g.

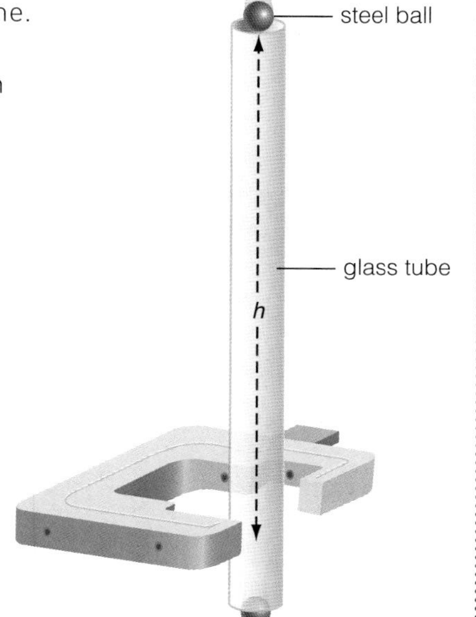

Figure 8.12 The speed is measured as the ball falls from different heights.

16 Figure 8.13 shows the velocity-time graph that described Felix Baumgartner's free fall from a height of 39 km.

a) Determine his acceleration over the region AB of the graph.

b) Explain why his acceleration decreases over the region BC of the graph.

c) Over the region CD of the graph, Felix was falling at his terminal speed. Yet this terminal speed kept decreasing. Explain why.

d) At time E Felix opened his parachute. Use the graph to estimate the distance he fell before he opened his parachute.

e) Calculate Felix's average speed over the period A to E.

f) One of the claims made about this descent was that Felix reached 'Mach 1.2': that he travelled at 1.2 times the speed of sound. From the graph, deduce the speed of sound at high altitude. Why is this speed less than the speed of sound at sea level, at 20 °C, which is $340\,m\,s^{-1}$?

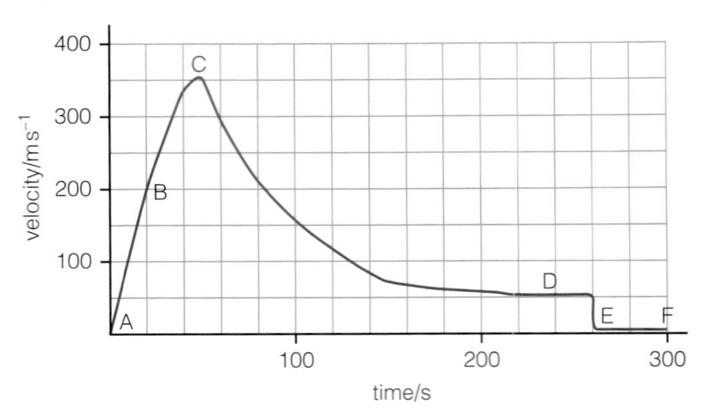

Figure 8.13 The velocity–time graph that described Felix Baumgartner's free fall from a height of 39 km.

Projectile motion – or falling sideways

Projectile motion builds on the following principles.

- Newton's first law of motion: an object will continue to move in a straight line with a constant speed (or remain at rest) unless acted on by an unbalanced force.
- Newton's second law of motion: an object accelerates in the direction of a resultant force acting on it (resultant force = mass × acceleration), see Chapter 9.
- Projectile paths may be predicted using the equations of motion.
- A vector velocity may be resolved into horizontal and vertical components.

Figure 8.14 A man jumping off a diving board into the sea.

Falling sideways

Figure 8.14 shows a diver jumping into the sea. Each image is taken 0.2 seconds apart. You can see that the diver is displaced sideways by a constant amount in each frame of the picture. This is because there is no horizontal force acting on him, so he keeps moving with a constant speed in that direction. Yet his displacement downwards increases with each frame of the picture. This is because gravity acts to accelerate him, so his downwards velocity is increasing.

An important principle is that the vertical and horizontal motions of the diver are independent of each other. When the diver jumps horizontally off the diving board, he will always reach the water in the same time because he has no initial velocity in a vertical direction. However, the faster he runs sideways, the further he will travel away from the board as he falls.

In all of the calculations that follow, we shall assume that we may ignore the effects of air resistance. However, when something is moving very quickly – a golf ball for example – the effects of air resistance need to be taken into account. The effect of air resistance on the path of a golf ball is to reduce its maximum height and to reduce the horizontal distance (range) that it travels.

EXAMPLE

In Figure 8.14, the board is 12.5 m above the sea, and the man jumps 7.0 m sideways, from the end of the board, by the time he splashes into the sea.

a) How long does it take him to fall into the sea?

b) How fast was he running when he jumped off the board?

Answer

a) Use the equation:

$$s = ut + \frac{1}{2}at^2$$

This equation is being used to consider the vertical displacement under the influence of gravity; the final displacement is 12.5 m, the initial (downward) velocity is zero, and the acceleration due to gravity, g, is $9.8\,\mathrm{m\,s^{-2}}$.

$$s = \frac{1}{2}gt^2$$

$$12.5\,\mathrm{m} = \frac{1}{2} \times 9.8\,\mathrm{m\,s^{-2}} \times t^2$$

Rearranging to find t^2:

$$t^2 = \frac{12.5}{4.9}\,\mathrm{s}^2$$

Taking square roots of each side of the equation gives:

$$t = 1.6\,\mathrm{s}$$

b) Because his horizontal velocity, v_h, remains constant:

$$v_h = \frac{\text{horizontal displacement}}{\text{time}}$$

$$v_h = \frac{7\,\mathrm{m}}{1.6\,\mathrm{s}}$$

$$= 4.4\,\mathrm{m\,s^{-1}}$$

EXAMPLE

A basketball player passes the ball at a speed of $11\,\mathrm{m\,s^{-1}}$ at an angle of 30° as shown in Figure 8.15.

Calculate

a) the greatest height the ball reaches

b) the time the ball takes to reach this height

c) the horizontal distance, x, it travels.

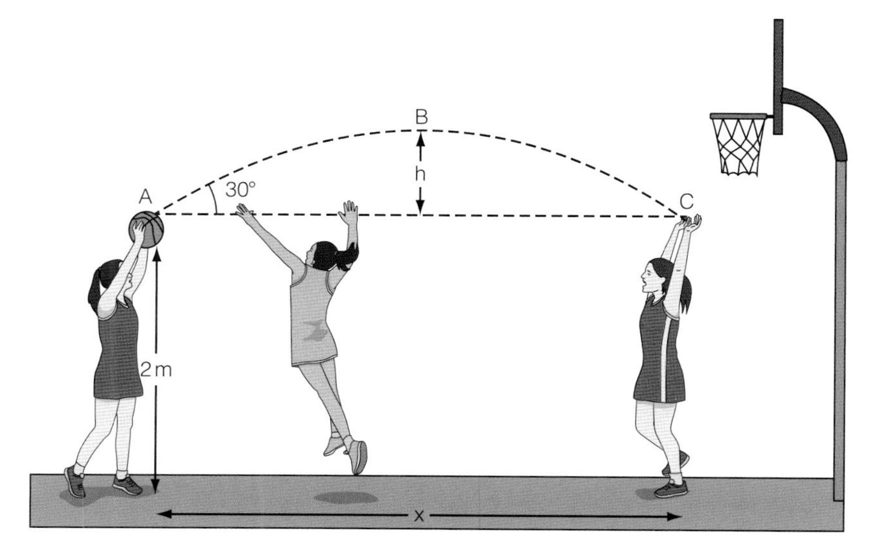

Figure 8.15 A basketball player passing a ball.

Answer

a) Figure 8.16 shows how we can resolve the velocity of the ball into vertical and horizontal components, v_v and v_h.

$$v_v = 11 \sin 30° = 5.5\,\mathrm{m\,s^{-1}}$$

$$v_h = 11 \cos 30° = 9.5\,\mathrm{m\,s^{-1}}$$

The height, h, to which the ball rises above the line AC can be calculated using the equation of motion:

$$v_v^2 = 2gh$$

$$(5.55\,\mathrm{m\,s^{-1}})^2 = 2 \times 9.81\,\mathrm{m\,s^{-2}} \times h$$

$$h = 1.5\,\mathrm{m}$$

so the greatest height the ball reaches above the ground is 3.5 m.

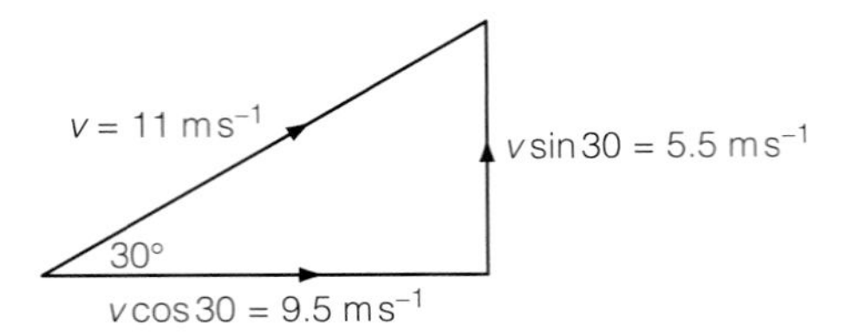

Figure 8.16 Resolving the velocity of the ball into vertical and horizontal components.

b) The time taken to reach this height is calculated using the equation of motion:

$$v = u + at$$

The final vertical velocity at the top of the ball's path (point B) is zero, the initial velocity is $-5.5\,\mathrm{m\,s^{-1}}$, and the acceleration is $+9.8\,\mathrm{ms^{-2}}$.

$$0 = -5.5\,\mathrm{m\,s^{-1}} + 9.8\,\mathrm{m\,s^{-2}} \times t$$

$$t = \frac{5.5\,\mathrm{m\,s^{-1}}}{9.8\,\mathrm{m\,s^{-2}}}$$

$$= 0.56\,\mathrm{s}$$

c) The time taken for the ball to travel from A to B is 0.56 s, so the time taken for it to travel from A to C is 2 × 0.56 s = 1.12 s.

The horizontal displacement, x, is calculated using the equation:

$$x = v_h t$$

$$x = 9.5\,\mathrm{m\,s^{-1}} \times 1.12\,\mathrm{s}$$

$$x = 10.6\,\mathrm{m}$$

ACTIVITY

Projectile paths

Figure 8.17 shows a multiple exposure photograph of two small balls that are allowed to fall freely under gravity. Ball A is dropped from rest and falls vertically from position A_1 to position A_5. Ball B is released at the same time as A, but falls with an initial sideways velocity of $1\,\mathrm{m\,s^{-1}}$; ball B falls from position B_1 to position B_5. The grid behind the balls allows you to calculate how far they travel.

All the positions of ball A – A_1 to A_5 – are shown, but positions, B_3 and B_4 of ball B have been removed.

1 A student thinks that a ball which is thrown sideways takes longer to reach the ground than a similar ball dropped from the same height. Look at Figure 8.17 and comment on this observation.

2 Use the information in the diagram to show that the time interval between each photograph is 0.1 s.

3 Either copy the diagram and mark in the positions B_3 and B_4, or write the coordinates of B_3 and B_4.

4 Calculate the velocity of ball A at position A_4 ($g = 9.8\,\mathrm{m\,s^{-2}}$).

5 Calculate the magnitude and direction of the velocity of ball B at position B_5.

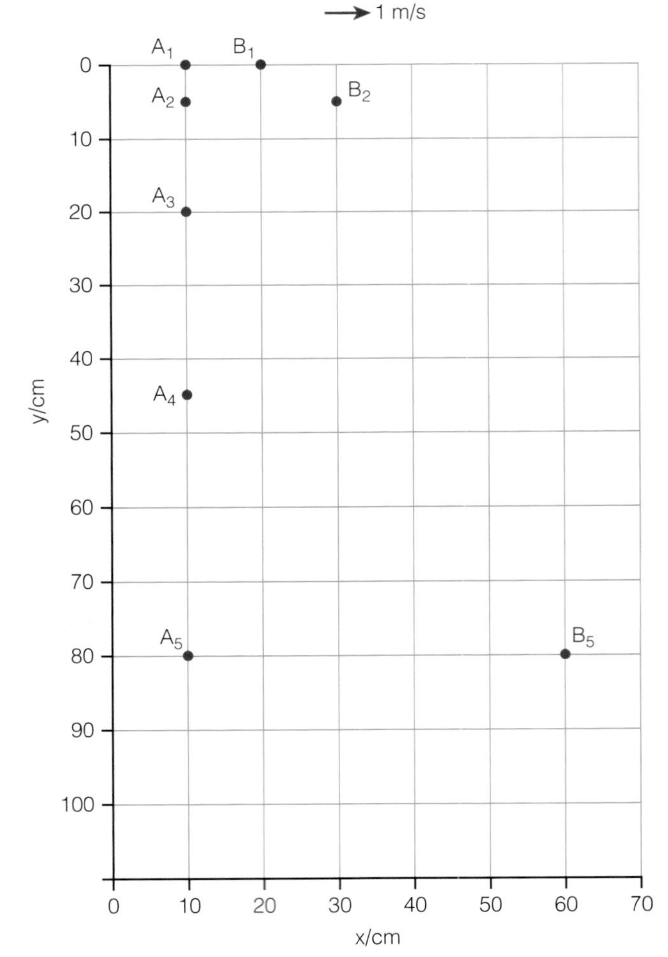

Figure 8.17 A multiple exposure photograph of two small balls.

In a second experiment balls A and B are photographed again with multiple exposures separated by 0.1 s. This time A is thrown upwards with an initial velocity of $7\,m\,s^{-1}$, while B has an initial upward velocity of $7\,m\,s^{-1}$ and a horizontal velocity of $2\,m\,s^{-1}$ to the right.

6 Copy the diagram on to graph paper and mark in the positions B_4, B_5, B_6 and B_7, or write their coordinates.

7 Position A_8 is the maximum height reached by ball A.

a) State the velocity of ball A at this highest position.

b) State the velocity of ball B at its highest position, B_8.

8 Draw diagrams to show the positions of balls A and B at intervals of 0.1 s as they fall back to the ground.

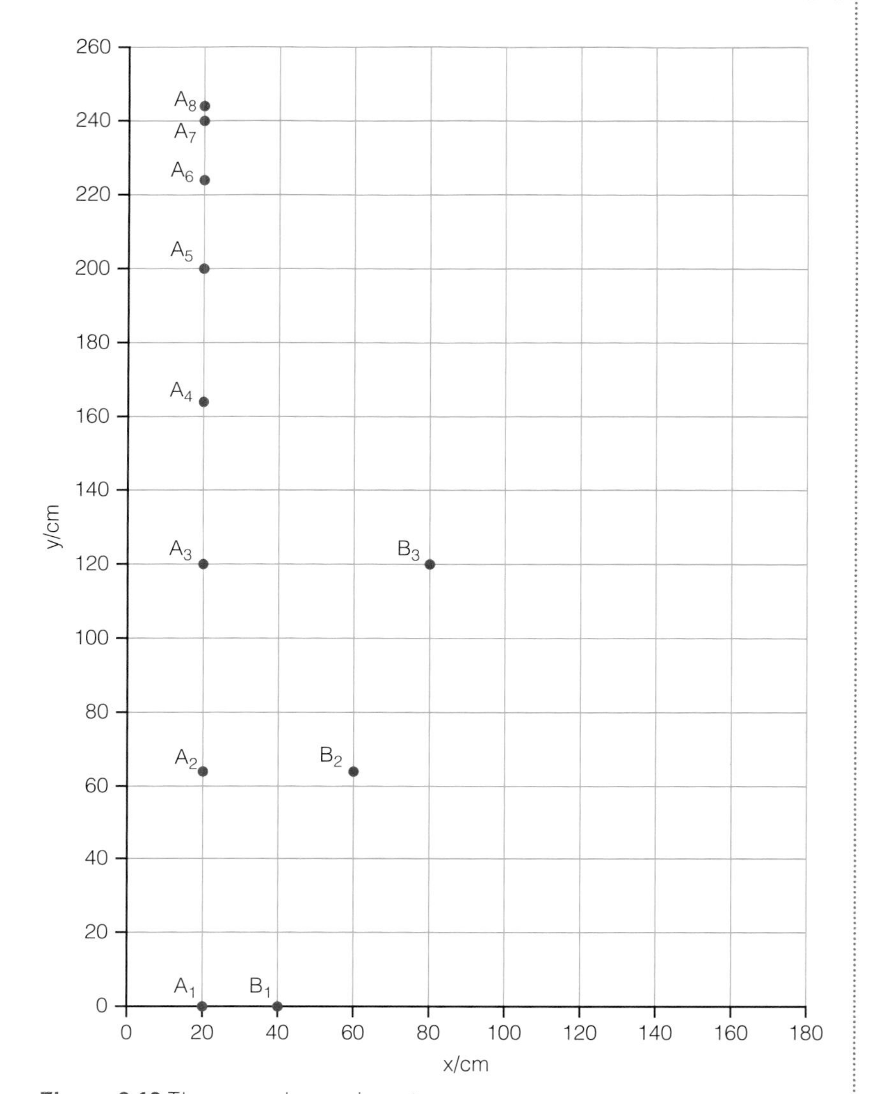

Figure 8.18 The second experiment.

TEST YOURSELF

17 Explain which of the following statements is/are true.

a) A bullet fired horizontally takes longer to fall to the ground than a bullet which is dropped from the same height.

b) When you throw a ball sideways, there is a force pushing it forwards while the ball is in mid-air.

c) The pull of gravity on an object is independent of the velocity of the object.

18 A marksman aims his rifle at a target 160 m away from him. The velocity of the bullet as it leaves the rifle horizontally is $800\,m\,s^{-1}$.

a) Calculate the time the bullet takes to reach the target.

b) How far has the bullet fallen in this time?

c) The target is now placed 480 m away from the marksman.

i) Calculate how far the bullet falls now before reaching the target.

ii) Explain why the marksman must adjust his sights before aiming at the second target.

19 A cricketer throws a ball sideways with an initial velocity of $30\,m\,s^{-1}$. She releases the ball from a height of 1.3 m. Calculate how far the ball travels before hitting the ground.

Practice questions

1 A ball is thrown vertically upwards, with an initial speed of $20\,m\,s^{-1}$ on a planet where the acceleration due to gravity is $5\,m\,s^{-2}$. Which of the following statements about the ball's velocity after 6 s is true?

A The ball's velocity is 0.

B The ball's velocity $10\,m\,s^{-1}$ upwards.

C The ball's velocity is $10\,m\,s^{-1}$ downwards.

D The ball's velocity is $5\,m\,s^{-1}$ downwards.

2 A car accelerates from rest at a rate of $3.5\,m\,s^{-2}$ for 6 s; it then travels at a constant speed for a further 8 s. Which of the following is the correct distance that the car has travelled in this time?

A 231 m

B 215 m

C 147 m

D 49 m

3 A plane accelerates from rest along a runway. It takes off after travelling a distance of 1200 m along the runway at a velocity of $60\,m\,s^{-1}$. Which of the following is the average acceleration of the plane along the runway?

A $3.00\,m\,s^{-2}$

B $1.50\,m\,s^{-2}$

C $0.67\,m\,s^{-2}$

D $0.33\,m\,s^{-2}$

4 A ball **P** of mass 2m is thrown vertically upwards with an initial speed u. At the same instant a ball **Q** of mass 3m is thrown at an angle of 30° to the horizontal with an initial speed of $2u$. Which of the following statements about the two balls is true?

A Q reaches twice the vertical height as P.

B Q reaches the ground first.

C Q and P reach the ground at the same time.

D The horizontal component of Q's velocity is u.

In each of the questions 5–7, select from the list A–D the relationship that correctly describes the connection between y and x.

A y is proportional to x^2

B y is proportional to $x^{\frac{1}{2}}$

C y is proportional to x

D y is proportional to $\frac{1}{x}$

Question	y	x
5	the speed of an object falling freely from rest	time
6	the speed of an object falling freely from rest	height fallen
7	the distance of an object falling freely from rest	time

In each of the questions 8–10, select from the graphs A–D the one that correctly shows the relationship between y and x.

Question	y	x
8	the speed of a car which decelerates at an increasing rate	time
9	the horizontal distance travelled by a projectile moving without air resistance	time
10	the speed of an object falling when air resistance acts on it	time

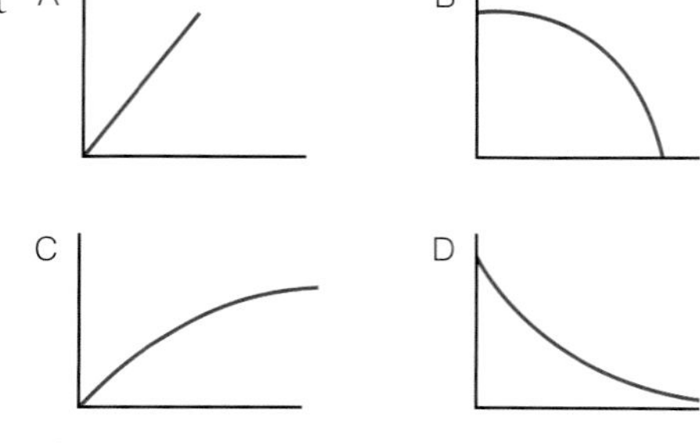

Figure 8.19 Four graphs.

11 Figure 8.20 shows the trajectory of a real golf ball (path A) and an idealised ball (path B) that is not affected by air resistance.

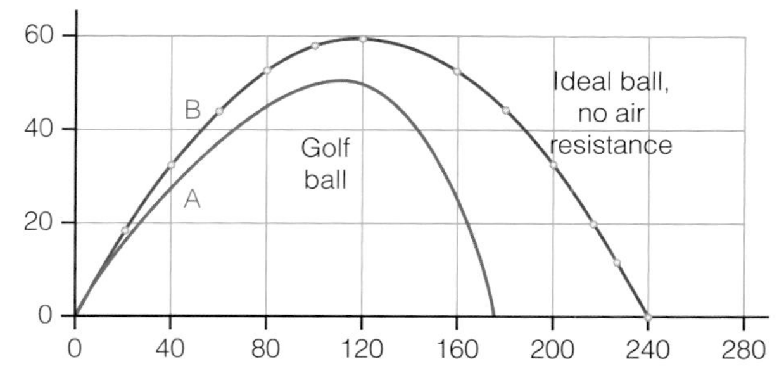

Figure 8.20 The trajectory of a real golf ball (path A) and an idealised ball (path B) that is not affected by air resistance.

a) State three differences between the two trajectories. (2)

b) Figure 8.21 shows the forces acting on the ball just after it has been struck by the club. Use this information to explain your observations in part (a). (3)

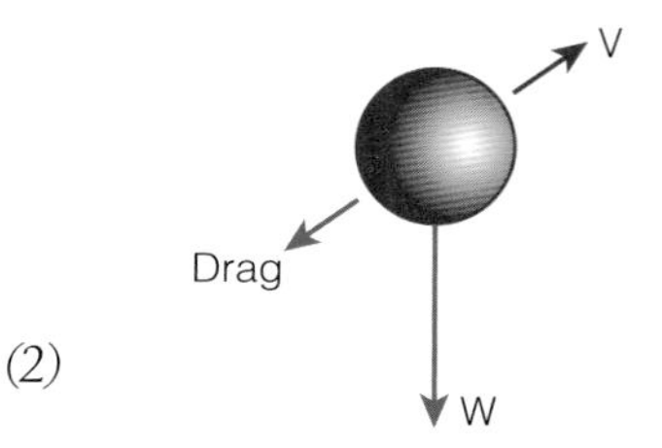

Figure 8.21 The forces acting on the ball just after it has been struck by the club.

12 A boy throws a ball vertically upwards; it rises to a maximum height of 19.0 m in 2.0 s.

a) Calculate the speed of the ball as it left the boy's hand. (2)

b) Calculate the height of the ball, above its position of release, 3.0 s after the boy threw it. (2)

c) State any assumptions you have made in these calculations. (1)

13 Figure 8.22 shows a velocity–time graph for a hard rubber ball which is dropped on to the floor before the rebounding upwards.

a) Use the graph to calculate the acceleration of the ball:

i) while it is falling (2)

ii) while the ball is in contact with the floor. (3)

In each case state the magnitude and direction of the acceleration.

b) i) Calculate the distance from which the ball was dropped. (2)

ii) Calculate the height of the first rebound. (2)

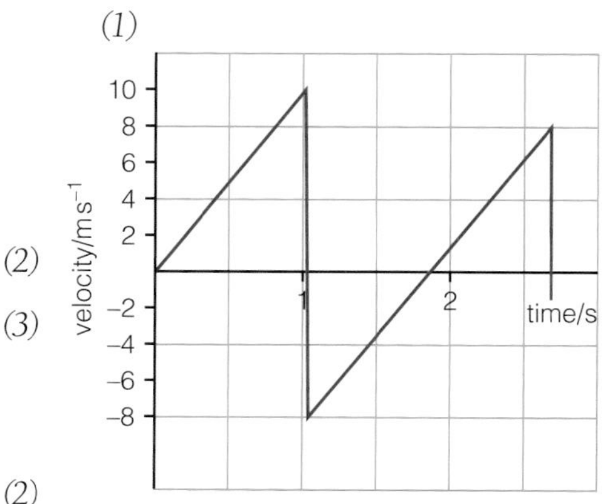

Figure 8.22 Velocity – time graph for a hard rubber ball.

14 Figure 8.23 is a velocity–time graph for a moving object.

a) Explain the significance of:

i) the positive and negative gradients of the graph (2)

ii) the areas under the graph, both above and below the time axis. (2)

b) Use your answer to part (a i) to help you draw an acceleration–time graph for the object. (2)

c) Use your answer to part (a ii) to help you to draw a displacement–time graph for the moving object. (3)

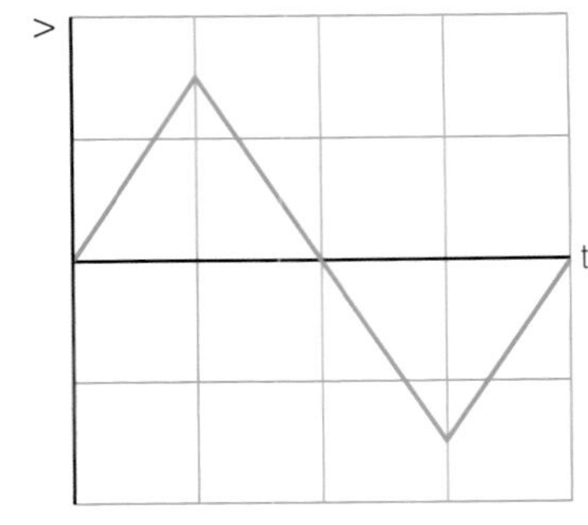

Figure 8.23 A velocity–time graph for a moving object.

15 A bungee jumper takes a plunge off a bridge. Figure 8.24 shows a velocity–time graph for the jumper as he falls from the bridge.

a) Use the graph to calculate his acceleration over the first 2 seconds. (2)

b) i) At which time does the jumper reach the lowest point of his fall? (1)

ii) Use the graph to estimate approximately how far he falls before being stopped by the bungee rope. Is the distance fallen closest to: 20 m, 40 m, or 60 m? (1)

iii) Explain how the graph shows you that the jumper does not bounce back to his original height. (1)

c) Calculate the acceleration at point B on the graph. (1)

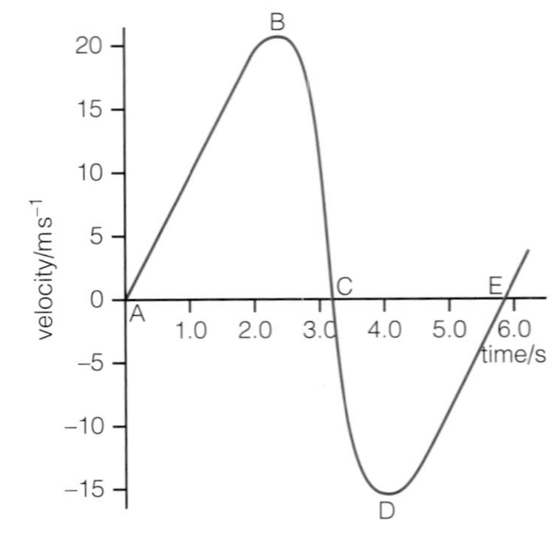

Figure 8.24 A velocity–time graph for the jumper as he falls from the bridge.

16 Figure 8.25 shows a fort that has been built to protect the coast from enemy ships. A cannon ball, which is fired horizontally from the fort, falls into the sea a distance of 1250 m from the fort.

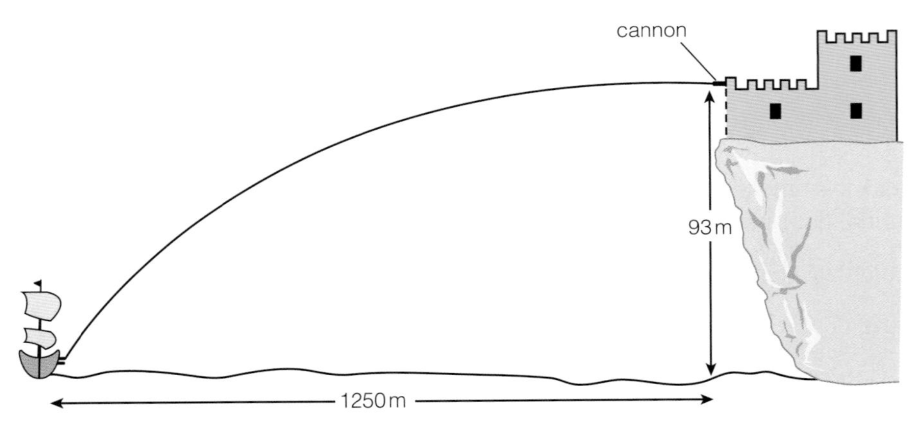

Figure 8.25 A fort that has been built to protect the coast from enemy ships.

a) Use the information in the diagram to show that the ball takes about 4.4 s to fall into the sea. (3)

b) Calculate the horizontal velocity of the cannon ball during its flight. (2)

c) Calculate the vertical velocity of the cannon ball after it has fallen for 4.4 s. (2)

d) Calculate, or determine by scale drawing, the magnitude and direction of the ball's velocity as it falls into the sea. *(3)*

e) Discuss how you would adjust the cannon to hit a ship that is further away from the fort. *(1)*

18 A student uses two light gates as shown in Figure 8.26 to measure the time taken for a small ball bearing to fall through different heights. The table below shows the results of the experiments. The ball bearing was dropped three times from each different height. To release the ball bearing the student held it just above the higher light gate.

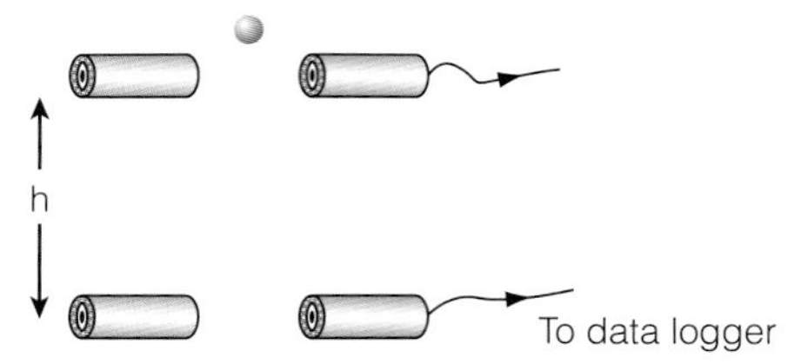

Figure 8.26 Using light gates to measure the time taken for a small ball bearing to fall through different heights.

Height (mm)	t_1 (s)	t_2 (s)	t_3 (s)
100	0.14	0.15	0.14
200	0.18	0.20	0.21
300	0.25	0.25	0.26
400	0.29	0.28	0.30
500	0.32	0.31	0.32
600	0.34	0.35	0.34

a) Calculate the average time of fall from each height. *(1)*

b) Calculate the maximum percentage error in timing when the ball bearing is dropped from a height of

i) 100 mm

ii) 600 mm. *(2)*

c) Calculate the approximate percentage error in measuring the height of 100 mm. *(1)*

d) Comment of the errors that could be introduced by the student's method of releasing the ball bearing. *(2)*

e) Plot a graph of height on the *y*-axis against t^2 on the *x*-axis. Use the graph to determine a value for *g*. Comment on the accuracy of the result. *(5)*

19 A plane travelling with a constant horizontal speed of $54\,m\,s^{-1}$ drops a food parcel to people stranded in a desert. The parcel takes 3.7 s to reach the ground.

a) Calculate the height of the plane above the ground. *(3)*

b) Calculate the horizontal distance travelled by the parcel before it hits the ground. *(2)*

c) i) Calculate the parcel's vertical component of velocity when it hits the ground. *(2)*

ii) Calculate the magnitude of the parcel's velocity as it hits the ground. *(2)*

Stretch and challenge

20 A family enjoying an outing on a river are cruising upstream in their boat. Just as they pass a bridge (point A) their sandwich box falls into the river.

They continue upstream for 10 minutes before they notice their loss. They then turn round (point B in the diagram) and travel downstream. They pick up the sandwich box 1.0 km downstream from the bridge at point C.

Calculate the velocity of the river, assuming that the boat has the same velocity relative to the water up and downstream, and that no time is wasted turning round the boat at point B.

(Hint: you do *not* need to know the velocity of the boat; you *do* need to set up some equations to calculate the time taken to catch up with the sandwich box.)

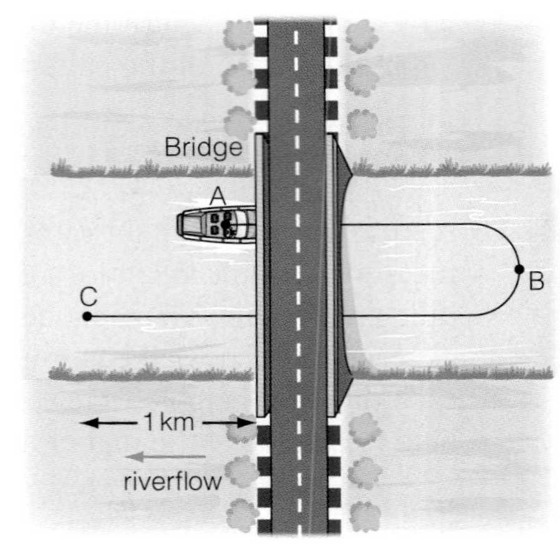

Figure 8.27 A family enjoying an outing.

21 A swimmer at point A next to the side of a swimming pool wishes to get to a friend at point B as quickly as possible. He can walk next to the pool at $3\,\text{m}\,\text{s}^{-1}$ and swim at a speed of $1\,\text{m}\,\text{s}^{-1}$.

Show that he must swim at an angle $\theta = \sin^{-1}\left(\frac{1}{3}\right)$ to get there in the shortest time.

(Hint: you need to set up an equation to calculate the time in terms of distances L, x and y, and the speeds. Then differentiation gives you the minimum time. You need to use your knowledge of Maths A-level to solve this.)

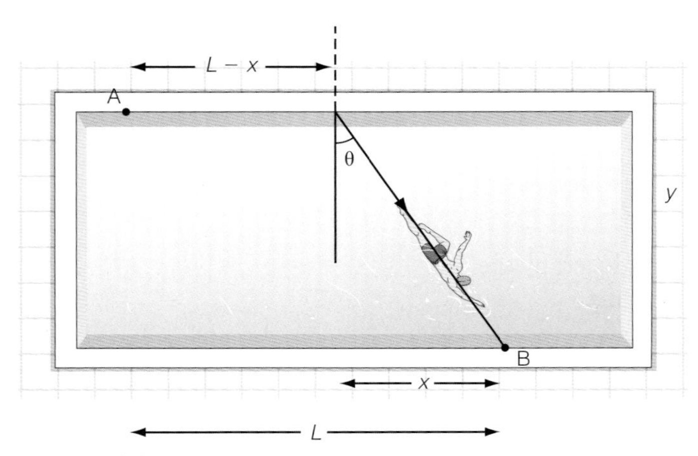

Figure 8.28 Swimming pool.

22 A supersonic aircraft leaves Heathrow to travel to Kennedy airport. You are required to use the information below to calculate the plane's position between the airports at 12:30 p.m. (Hint: sketch a velocity-time graph to help you plan your solution.)

- The plane departs from Heathrow at 10:30 a.m. sharp.
- The flight time is exactly 3 hours and 40 minutes.
- The distance between the airports is 5800 km.
- The plane flies for the first 15 minutes to a point 235 km from Heathrow. At that point it is flying at $990\,\text{km}\,\text{h}^{-1}$. It then accelerates at a constant rate for 11 minutes over a distance of 250 km to a cruising speed of $2160\,\text{km}\,\text{h}^{-1}$.
- The plane flies in a straight line.
- The deceleration into Kennedy airport is uniform.

9 Newton's laws of motion

PRIOR KNOWLEDGE

- A force is a push or a pull. Forces can be contact or non-contact forces. A contact force is exerted between bodies that touch each other; non-contact forces are exerted over a distance by gravitational, electric and magnetic fields.
- The strength of a gravitational field is measured in $N\,kg^{-1}$. On the surface of the Earth the gravitational field strength is $9.8\,N\,kg^{-1}$.
- Weight = mass × gravitational field strength
- Newton's first law of motion: a body will remain at rest or continue to move in a straight line with a constant velocity unless it is acted on by an unbalanced force.
- Newton's second law of motion: when an unbalanced (or resultant) force acts on a body, it will accelerate in the direction of that force. The size of the acceleration may be determined by using the equation F = ma.
- Newton's third law of motion: when body A exerts a force, F, on body B, body B exerts an equal and opposite force, F, on body A.

TEST YOURSELF ON PRIOR KNOWLEDGE

1 Karen has a mass of 57 kg.

a) Calculate her weight on the Earth.

b) Calculate her weight on Mars, where the gravitational field strength is $3.7\,N\,kg^{-1}$.

2 **a)** Describe what will happen to the stage of the motion of each of the bodies shown below. In each case use a suitable phrase such as: remains at rest; continues to move at a constant speed; changes direction; accelerates; decelerates.

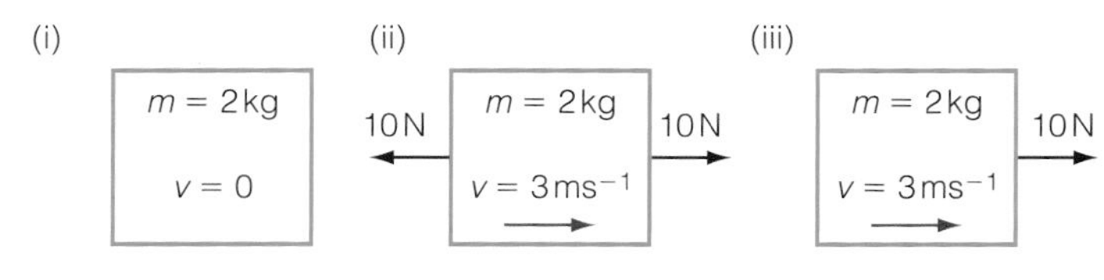

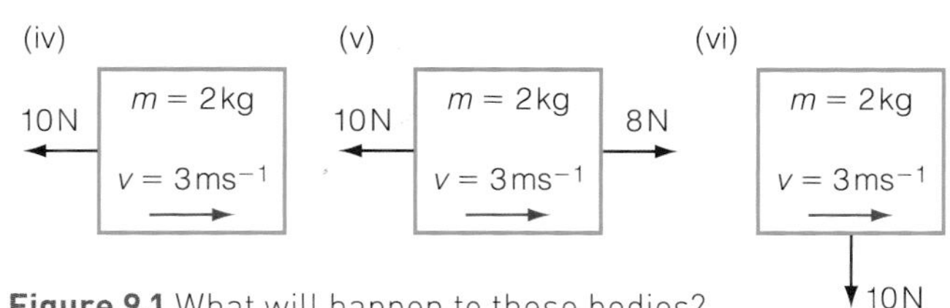

Figure 9.1 What will happen to these bodies?

b) Calculate the acceleration of the bodies in each case.

3 In Figure 9.2 you can see Tony pushing against a wall with a force of 300 N. State the magnitude and direction of the force that the wall exerts on Tony.

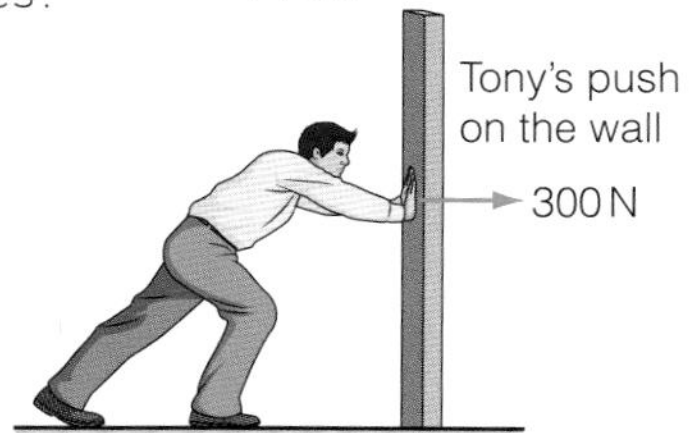

Figure 9.2 Tony pushing against a wall.

Figure 9.3 This water jet pack relies on each of Newton's three laws of motion.

Figure 9.3 shows a man enjoying a ride with a water jet pack. The pack on his back pumps water up through the yellow pipe and then forces the water out in two high-velocity jets. This recreational toy relies on each of Newton's three laws of motion. The jet pack pushes water downwards, and the escaping water pushes the jet pack back in an upwards direction. When the upwards push from the water on the jet pack balances the weight of the man and his jet pack, he remains at a constant height.

Newton's first law – free body diagrams

A body may be subjected to a number of forces, and the body may also exert forces on other bodies that come into contact with it. If we try to draw forces on more than one body, on the same diagram, it is easy to become confused over which force acts on which body. So, to understand the effect of forces on a body, we draw a free body diagram, which only shows the forces acting on a single body.

A **free body diagram** shows all the forces acting on a single body; no other body is shown in the diagram.

The diagrams below illustrate free body diagrams with balanced forces.

In Figure 9.4 a skydiver and parachute fall to the ground at a constant speed. Here the free body diagram is easy as the skydiver and parachute are not in contact with anything (except the air). The drag, *D*, upwards balances the weight, *W*, of the skydiver and parachute downwards.

Figure 9.4 Skydiver falling to the ground at a constant speed.

In Figure 9.5a, a climber is abseiling down a rock face; she has just paused for a rest and is stationary. To draw a free body diagram, we must remove the rock face. The three forces acting are shown in Figure 9.5b: the weight of the climber, *W*; the tension of the rope, *T*; the reaction from the rock face, *R*. Since the climber is stationary, these forces act through the centre of gravity of the climber. They add up to zero, as shown in Figure 9.5c.

Figure 9.5a A climber is abseiling down a rock face.

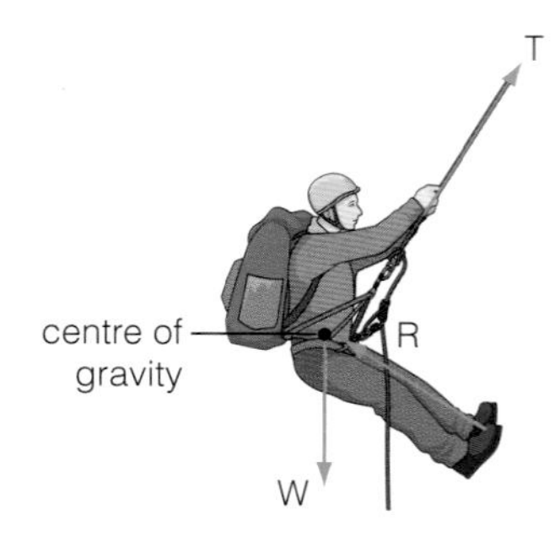

Figure 9.5b The forces acting on the climber.

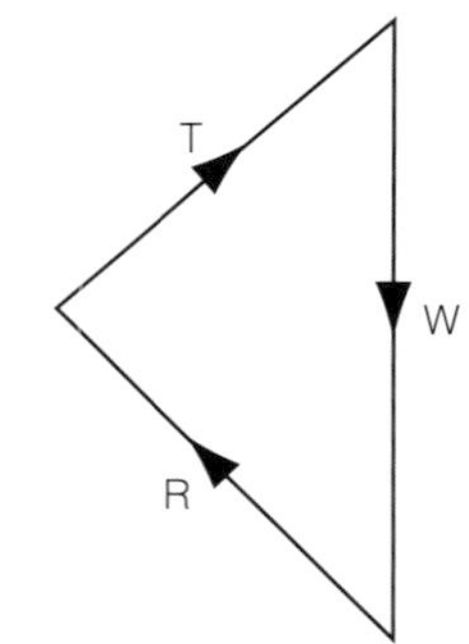

Figure 9.5c The forces add up to zero.

Figure 9.6a shows a man climbing a ladder that rests against a smooth wall. Figure 9.6b shows a free body diagram for the ladder when the man stands on it.

Figure 9.6a A man climbing a ladder.

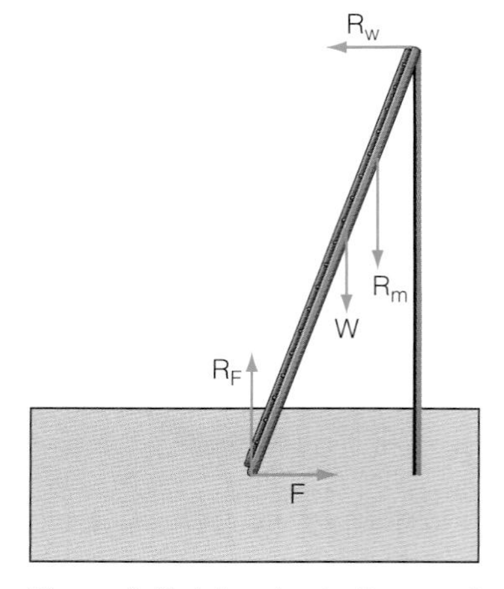

Figure 9.6b A free body diagram for the man on the ladder.

The forces acting on the ladder are:

R_W, a horizontal reaction force from the wall

R_F, a vertical reaction force from the floor

F, a horizontal frictional force from the floor

W, the weight of the ladder

R_m, a contact force from the man that is equal in size to his weight (this is *not* the man's weight, which acts on him).

Since the ladder remains stationary, the forces on it balance.

So

$$R_F = W + R_m$$

These are the forces acting vertically.

$$F = R_W$$

These are the forces acting horizontally.

TIP

To draw a free body diagram, consider only one body. Draw the forces that act on that body, not the forces that it exerts on other bodies.

Newton's second law of motion

When an unbalanced force is applied to an object, it experiences a change in velocity. An object speeds up or slows down when a force is applied along its direction of motion. When a force is applied at an angle to an object's direction of travel, it changes direction. When we calculate accelerations we use the equation:

resultant force = mass × acceleration

$$F = ma$$

Provided the force is in newtons, N, and the mass in kilograms, kg, then the acceleration is measured in $m\,s^{-2}$.

EXAMPLE

Figure 9.7 shows the forces on a cyclist accelerating along the road. The forces in the vertical direction balance, but the force pushing her along the road, F, is greater than the drag forces, D, acting on her. The mass of the cyclist and the bicycle is 100 kg. Calculate her acceleration.

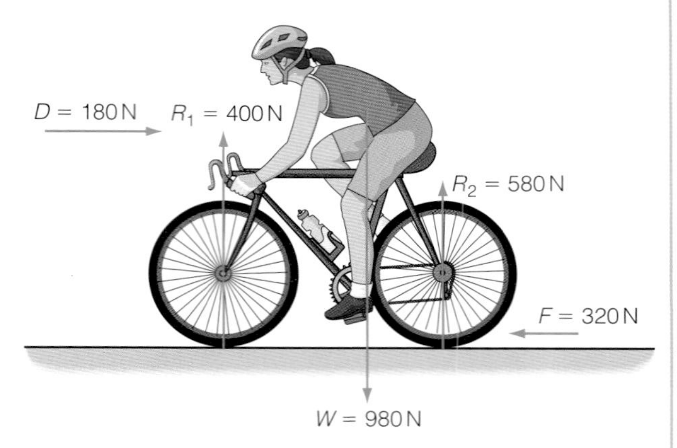

Figure 9.7 The forces on a cyclist.

Answer

Her acceleration can be calculated as follows.

$$F - D = ma$$

$$320\,\text{N} - 180\,\text{N} = 100\,\text{kg} \times a$$

$$a = \frac{140}{100}\,\text{m s}^{-2}$$

$$= 1.4\,\text{m s}^{-2}$$

EXAMPLE

A passenger travels in a lift that is accelerating upwards at a rate of $1.5\,\text{m s}^{-2}$. The passenger has a mass of 62 kg. By drawing a free body diagram to show the forces acting on the passenger, calculate the reaction force that the lift exerts on her.

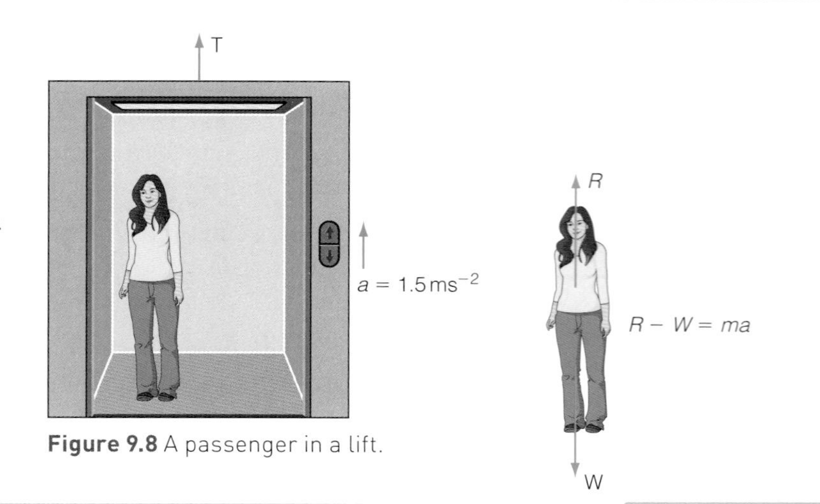

Figure 9.8 A passenger in a lift.

Answer

$$R - W = ma$$

$$R = W + ma$$

$$= 62\,\text{kg} \times 9.8\,\text{m s}^{-2} + 62\,\text{kg} \times 1.5\,\text{m s}^{-2}$$

$$= 700\,\text{N (2 sig. figs)}$$

TEST YOURSELF

1 Draw free body diagrams to show the forces acting in the following cases.

a) A car moving at a constant speed.

b) An aeroplane just at the point of take off from a runway.

c) A submarine moving at a constant speed and depth.

d) A skydiver, just after she has opened her parachute.

Illustrate your diagrams to explain whether the bodies move at constant speed, or whether they are accelerating or decelerating.

2 a) Calculate the reaction force acting on the passenger in Figure 9.8 when

i) the lift moves up at a constant speed of $3.0\,\text{m s}^{-1}$

ii) it accelerates downwards at a rate of $2.0\,\text{m s}^{-2}$.

b) Explain why the passenger feels heavier just as the lift begins to move upwards.

3 Figure 9.9 shows the horizontal forces acting on a truck that is pulling a trailer.

a) Calculate the acceleration of the truck and trailer.

b) Calculate the tension, T, in the tow bar that is pulling the trailer.

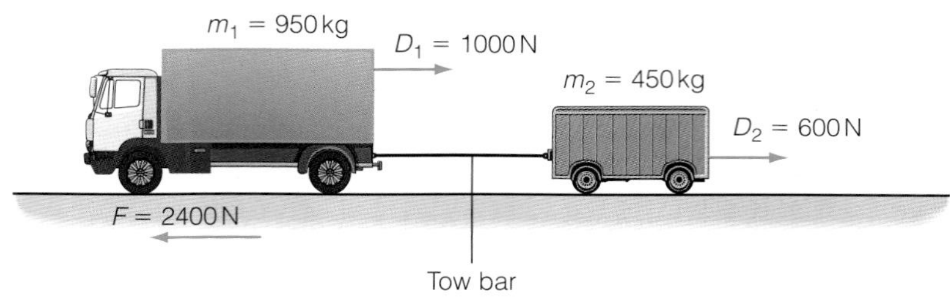

Figure 9.9 The horizontal forces on a truck.

4 A driver tests his car's acceleration. The result of this trial is shown by the velocity–time graph in Figure 9.10.

a) Use the graph to calculate the car's acceleration

i) between 0 and 2 s

ii) between 6 s and 8 s.

The car and the driver have a mass of 1200 kg.

b) Calculate the initial force that accelerates the car from rest.

c) Assuming that the same driving force acts on the car between 6 s and 8 s, calculate the drag forces acting on the car at this time using information from the graph.

d) Calculate the initial acceleration of the car when it is loaded with passengers and luggage so that its mass is 1450 kg.

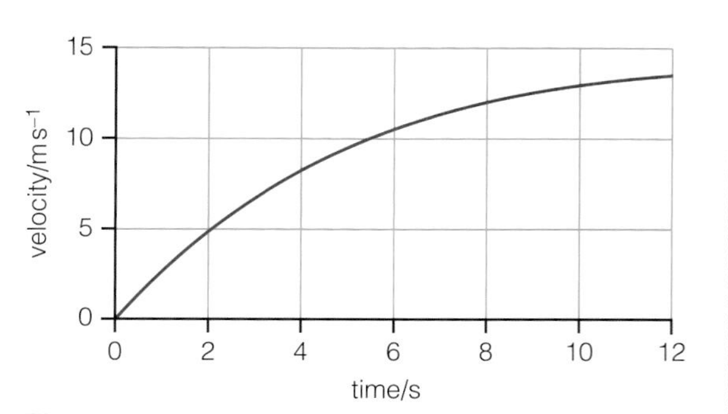

Figure 9.10 Velocity–time graph for a car.

Newton's third law of motion

This law has already been stated in the introduction but it is stated as follows in a slightly different way. Every force has a paired equal and opposite force. This law sounds easy to apply, but it requires some clear thinking. It is important to appreciate that the pairs of forces must act on *two* different bodies, and the forces must be the same type of force.

An easy way to demonstrate Newton's third law is to connect two dynamics trolleys together with a stretched rubber band, as shown in Figure 9.11.

When the trolleys are released, they travel the same distance and meet in the middle. This is because trolley A exerts a force on trolley B, and trolley B exerts a force of the same size, in the opposite direction, on trolley A. Some examples of paired forces are given in Figure 9.12.

(a)

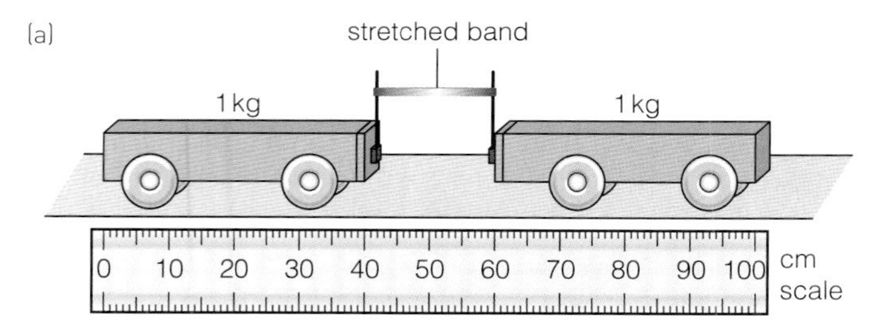

(b)

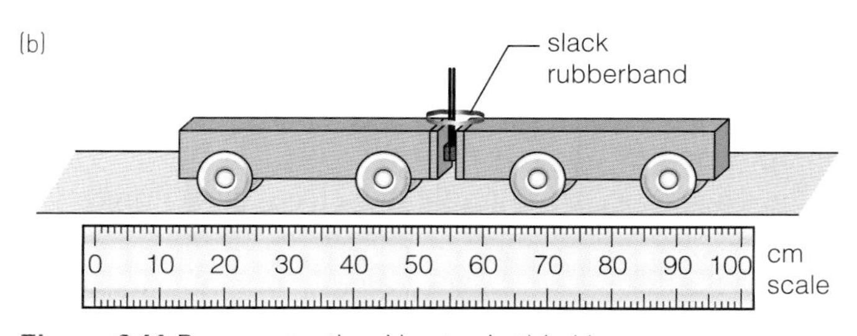

Figure 9.11 Demonstrating Newton's third law.

Figure 9.12a If I push you with a force of 100 N, you push me back with a force of 100 N.

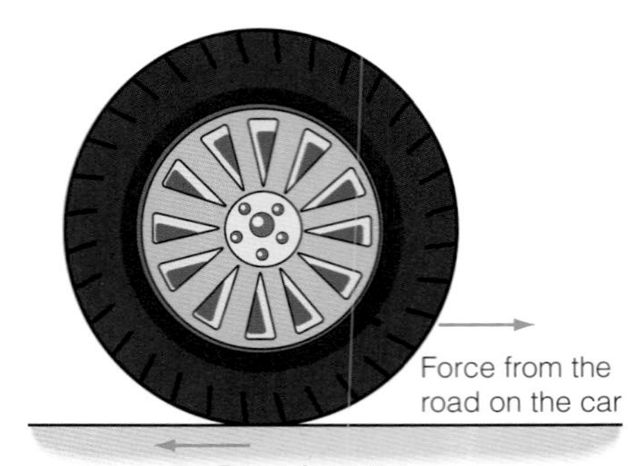

Figure 9.12b When the wheel of a car turns, it pushes the road backwards. The road pushes the wheel forwards with an equal and opposite force.

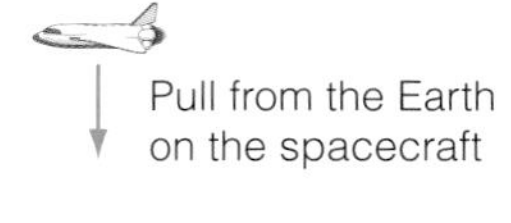

Figure 9.12c A spacecraft orbiting the Earth is pulled downwards by Earth's gravity. The spacecraft exerts an equal and opposite gravitational force on the Earth. So if the spacecraft moves towards the Earth, the Earth moves too, but because the Earth is so massive its movement is very small.

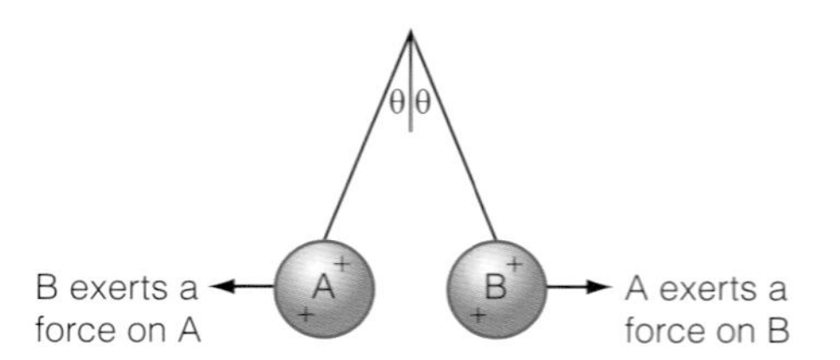

Figure 9.12d Two balloons have been charged positively. They each experience a repulsive force from the other. These forces are of the same size, so each balloon (if of the same mass) is lifted through the same angle.

All of these are examples of Newton's third law **pairs**.

Newton's third law pairs always have these properties:

- they act on two separate bodies
- they are always of the same type, for example two electrostatic forces or two contact forces
- they are of the same magnitude
- they act along the same line
- they act in opposite directions.

TIP
When applying Newton's third law, remember that the 'equal and opposite' forces act on different bodies. Do not confuse this with Newton's first law, when balanced forces keep a body at rest or at a constant speed.

A tug of war

Alan and Ben are enjoying a tug of war, as this gives them an opportunity to apply all of Newton's laws of motion. In Figure 9.13a Alan, who is stronger than Ben, is just beginning to win. Alan, Ben and the rope are accelerating to the left at $2\,\mathrm{m\,s^{-2}}$. By drawing free body diagrams for Alan (mass 80 kg), the rope (mass 2 kg) and Ben (mass 60 kg), we can analyse the forces acting on each person. Figure 9.13b shows these diagrams.

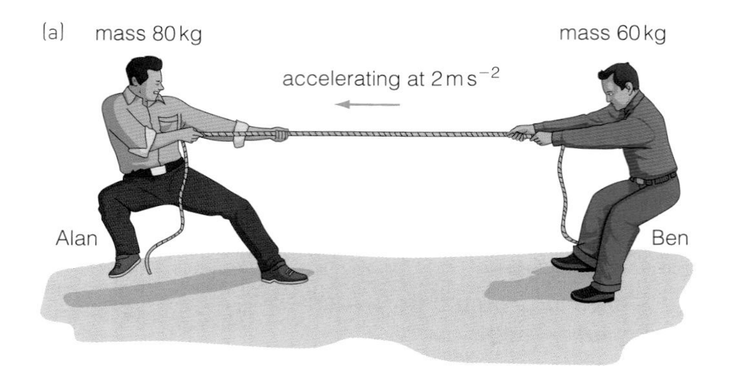

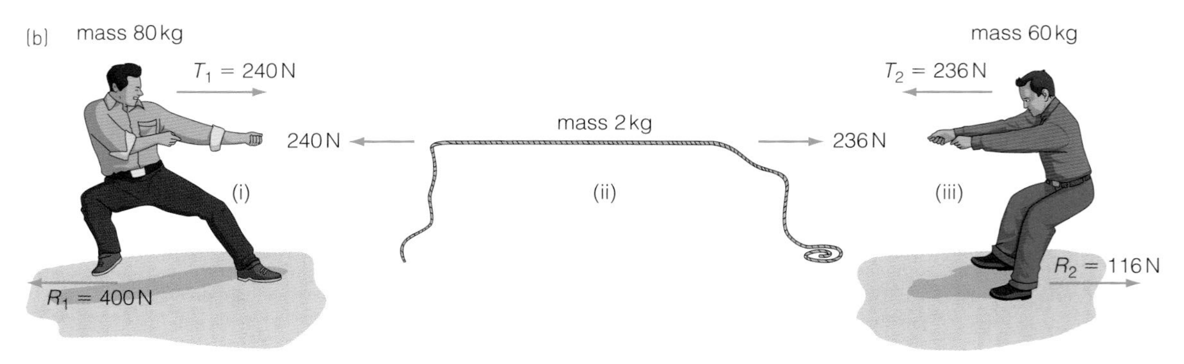

Figure 9.13 Free body diagrams for the tug of war. Exercise caution if you are repeating this experiment yourself.

For each body the resultant forces caused on acceleration of 2 m s^{-2}. These calculations use Newton's second law.

Table 9.1

	Resultant force (N)	=	Mass (kg)	×	Acceleration (m s^{-2})
Alan	(400 − 240)	=	80	×	2
Ben	(236 − 116)	=	60	×	2
Rope	(240 − 236)	=	2	×	2

Newton's third law pairs

- The Earth pushes Alan with 400 N to the left.
  Alan pushes the Earth with 400 N to the right.
- The Earth pushes Ben with 116 N to the right.
  Ben pushes the Earth with 116 N to the left.
- The rope's tension pulls Alan to the right with 240 N.
  Alan pulls the rope to the left with 240 N.
- The rope's tension pulls Ben to the left with 236 N.
  Ben pulls the rope to the left with 236 N.

TIP

To work out the acceleration of a body you must draw a free body diagram showing all the forces that act on that body. Alan and Ben accelerate because unbalanced forces act on them.

TEST YOURSELF

5 Figure 9.14 shows a dog sitting still on the ground.

a) Draw a free body diagram to show the two forces acting on it to keep it in equilibrium.

b) For the two forces above, state and explain the two forces which are their Newton third law pairs.

6 This question refers to the tug of war in Figure 9.13. Just before Alan started to win, the two men were in equilibrium. The force from the ground on Alan was 300 N.

a) Draw free body diagrams for

i) Alan

ii) the rope

iii) Ben

to show all the forces acting on them, both horizontally and vertically.

b) Count as many Newton third law pairs as you can. (Remember to include forces acting on the Earth.)

7 Use Newton's third law to explain the following.

a) You cannot walk easily on icy ground.

b) You have to push water backwards so that you can swim forwards.

8 If you blow up a balloon and then release it, the balloon flies around the room. Draw a free body diagram and explain why the balloon travels at a constant speed when air resistance balances the forwards push from the air escaping from it backwards. You need Newton's first law to explain this.

9 a) Under what circumstances would a constant force acting on a body produce zero acceleration?

b) Under what circumstances would a constant force acting on a body produce a decreasing acceleration?

c) Under what circumstances would a constant force acting on a body produce an increasing acceleration?

[Hint: for all of these three examples, there are other forces acting on the body. It is possible that the mass of a body changes while it accelerates.]

10 A body of mass 5 kg has three forces acting on it of magnitude 10 N, 7 N and 5 N.

a) Sketch a diagram to show how the body may be in equilibrium.

b) The 7 N force is removed. Calculate the body's acceleration.

11 A man, whose mass is 90 kg, jumps off a wall of height 1.5 m. While he is in free fall, what force does he exert on the Earth?

Figure 9.14 A dog sitting still on the ground.

Practical investigations

Here are several practical exercises that you can try for yourself, or you can use the data and diagrams to discuss or calculate the outcomes of these experiments. In these activities it will make your calculations more straightforward if you assume $g = 10\,\text{N kg}^{-1}$.

ACTIVITY

Investigating tension

Look at Figures 9.15a, 9.15b and 9.15c. In each case predict the reading on the forcemeter. Then check your predictions by reading the meters. Comment on your results.

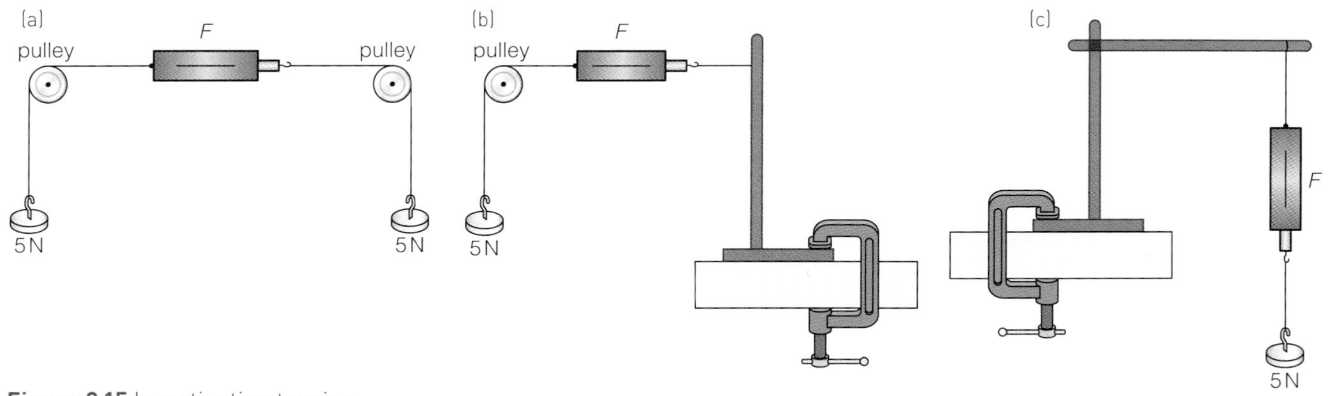

Figure 9.15 Investigating tension.

ACTIVITY

Balanced forces

Look at Figure 9.16a and predict the weight, *W*, you will have to put on the string to balance the 4 N weight that is suspended by the lower pulley block.

Now look at Figure 9.16b. Calculate the angle θ, to which you will have to tip the slope so that the 3 N weight balances the pull of gravity down the slope on the trolley.

Figure 9.16a What weight will you have to put on the string?

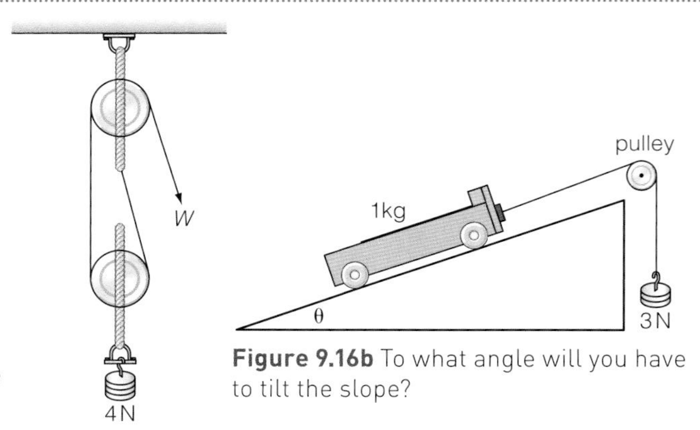

Figure 9.16b To what angle will you have to tilt the slope?

ACTIVITY

Acceleration

Refer to Figure 9.17a. A small ball is suspended from a stick attached to the trolley; it is free to move. The mass of the trolley, stick and ball is 1 kg. The trolley is held stationary by hand while a stretched spring exerts a force of 2 N on the trolley. The trolley is then released so that it accelerates to the right.

Predict what happens to the ball.

Calculate its maximum angle of deflection.

Finally, the trolley stops abruptly, when it hits a barrier.

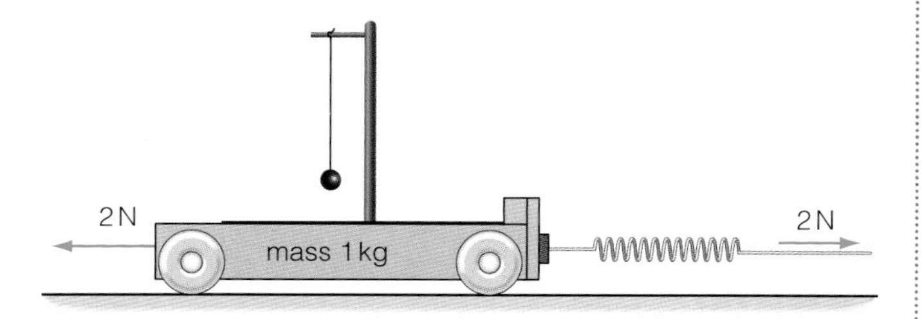

Figure 9.17a A small ball is suspended from a stick attached to the trolley. It is free to move.

Predict what happens to the ball.

Now try the experiment and check your predictions, as far as possible.

In Figure 9.17b a trolley is held in a stationary position by a weight of 2.5 N. The string holding the weight is then cut and the trolley accelerates down the slope covering the distance of 0.5 m in 0.41 s. Calculate the mass of the trolley.

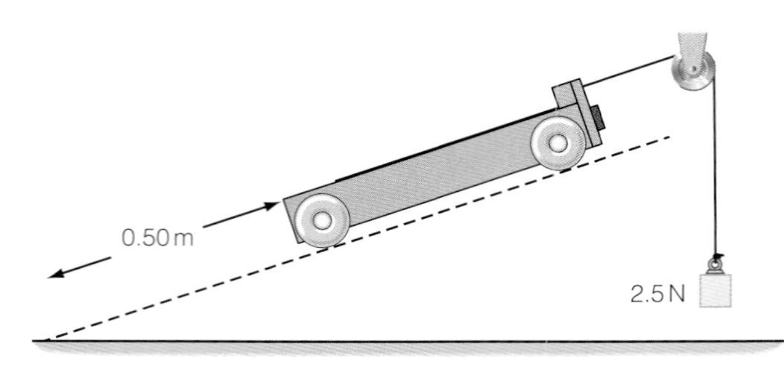

Figure 9.17b A trolley is held in a stationary position by a weight of 2.5 N.

Now check the mass of a trolley in this way with your own practical measurements.

In Figure 9.17c, three trolleys (with a mass of about 1 kg each), are held in a stationary position. The 3 N pull of the weight is being balanced by a student's hand. At the front of each trolley is a spring that has been extended by 12 cm due to the pull of the 3 N force.

The hand now releases the last trolley. Predict and explain what will happen to the extension of each spring, before you check by trying the experiment out.

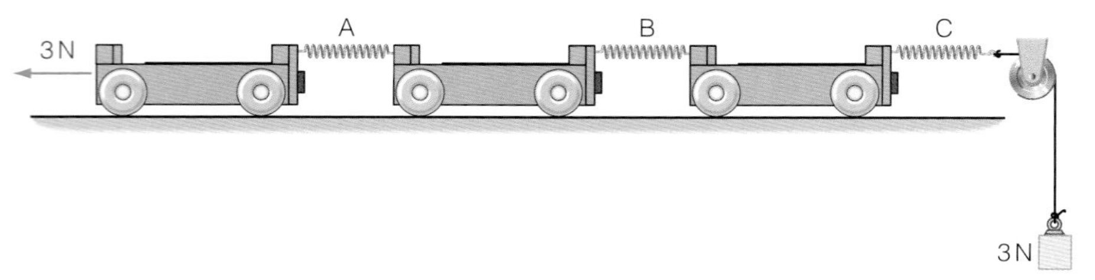

Figure 9.17c Three trolleys (with a mass of about 1 kg each), are held in a stationary position. The 3 N pull of the weight is being balanced by a student's hand.

ACTIVITY

F = *ma*

Figure 9.18 shows an arrangement of light gates that can be used to calculate accelerations directly. Two light gates measure the time it takes for a piece of card on an accelerating trolley to pass through the gates, and the time interval between each of the timed measurements.

Explain how the three time measurements allow the acceleration to be calculated. What further information is needed?

Predict the acceleration that would be calculated in this experiment.

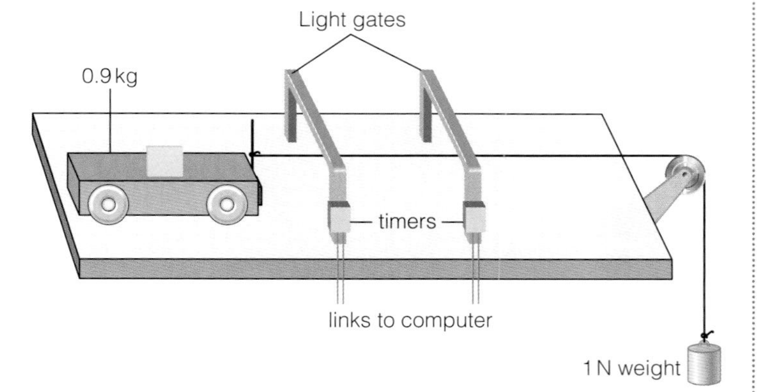

Figure 9.18 An arrangement of light gates that can be used to calculate accelerations directly.

Design an experiment to help you prove the general relationship: $F = ma$.

ACTIVITY

Newton's Third Law

In Figure 9.19a a weight is suspended from a forcemeter. It is then lowered into a beaker of water, which sits on top of an electronic balance. Use the following information to draw free body diagrams for

a) the water in the beaker
b) the suspended weight when the weight has been lowered into the water.

- When the weight is suspended above the water, the forcemeter reads 2.7 N and the balance reads 6.8 N
- When the weight is in the water the forcemeter reads 2.1 N and the balance reads 7.4 N

Name as many Newton's third law pairs as you can in this experiment.

Figure 9.19b shows a similar idea to the previous experiment. Predict what you expect to happen to the forcemeter reading and the balance reading as the magnet hanging from the forcemeter is lowered over the second magnet on the balance. Explain the significance of Newton's third law to this experiment.

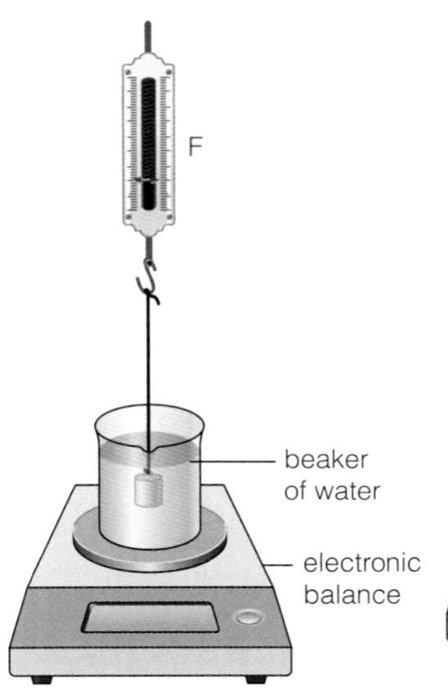

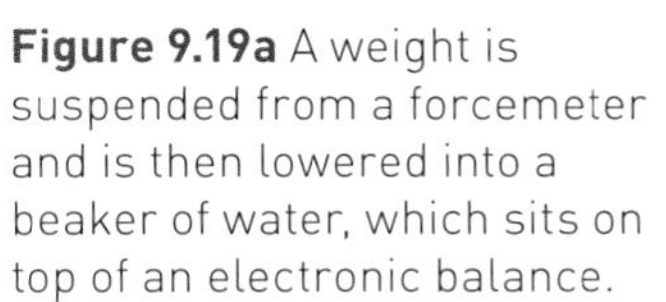

Figure 9.19a A weight is suspended from a forcemeter and is then lowered into a beaker of water, which sits on top of an electronic balance.

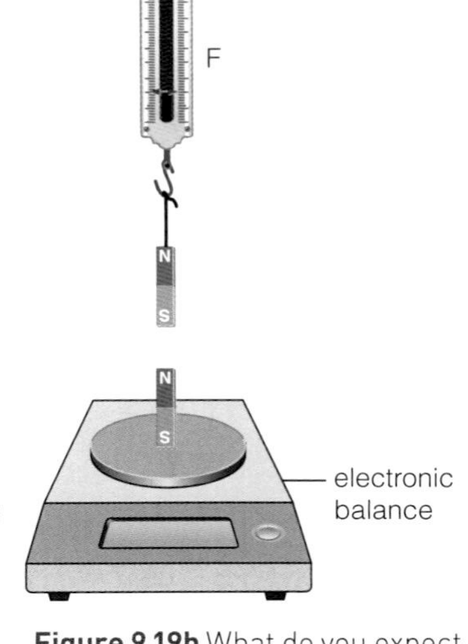

Figure 9.19b What do you expect to happen to the forcemeter reading and the balance reading as the magnet hanging from the forcemeter is lowered over the second magnet on the balance?

Practice questions

Figure 9.20 A lift and its passengers.

1 A lift and its passengers have a total mass of 2000 kg. It accelerates upwards with an acceleration of $1.1\,m\,s^{-2}$. Which of the following is the correct value for the tension in the lift cable?

A 2200 N

B 17 400 N

C 19 600 N

D 21 800 N

2 Two masses hang at rest as shown in Figure 9.21. The thread separating the two masses is burnt through.

Which of the following gives the magnitude of the accelerations of the masses after the thread burns through?

	Acceleration of A ($m\,s^{-2}$)	Acceleration of B ($m\,s^{-2}$)
A	0.5 g	g
B	g	g
C	1.5 g	1.5 g
D	1.5 g	g

2m A

m B

Figure 9.21 Two masses hang at rest.

3 Figure 9.22 shows a box at rest on the floor. The reaction force from the floor on the box equals the weight of the box. Which law predicts that these two forces are equal?

A The principle of conservation of momentum.

B The principle of the conservation of energy.

C Newton's First Law of Motion.

D Newton's Third Law of Motion.

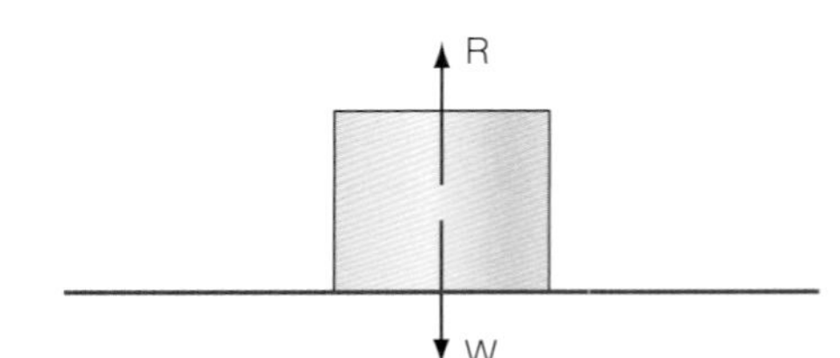

Figure 9.22 A box at rest on the floor.

4 Figure 9.23 shows a truck accelerating from rest down a slope that makes an angle of 30° to the horizontal. Which of the following is the distance travelled by the truck after 4 seconds?

A 78.4 m

B 58.8 m

C 39.2 m

D 19.6 m

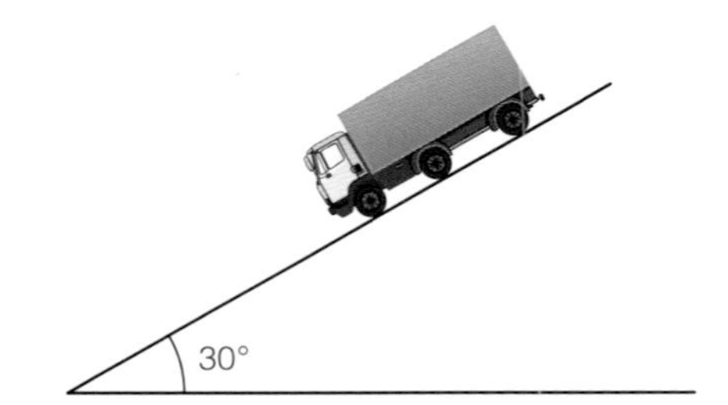

Figure 9.23 A truck accelerating from rest down a slope.

5 Figure 9.24 shows two boxes, A and B, which are connected by a light thread X. The boxes are moving at a constant speed of $2.0\,m\,s^{-1}$. A frictional force of 10 N acts on each box, and a force F is applied to a second light thread Y. Which of the following statements is true?

A F is slightly greater than 20 N.

B String X exerts a force of 10 N on B.

C X exerts a greater force on A than it does on B.

D X exerts a greater force on B than it does on A.

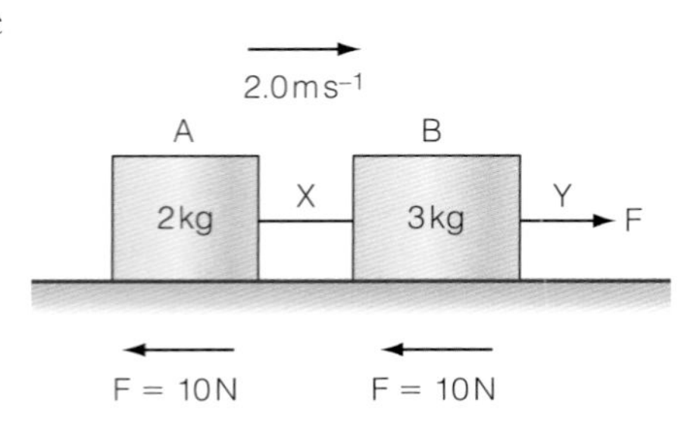

Figure 9.24 Two boxes, A and B, which are connected by a light thread X.

6 The force F in Figure 9.24 is suddenly increased to 25 N and, at the same instant, the string X breaks. Which of the following pairs of answers best describes the subsequent motion of boxes A and B?

	Motion of A	Motion of B
A	travels 0.4 m before stopping	accelerates at a rate of $3\,m\,s^{-2}$
B	stops moving in a time of 0.4 s	accelerates at a rate of $5\,m\,s^{-2}$
C	decelerates at $5\,m\,s^{-2}$	accelerates at a rate of $8\,m\,s^{-2}$
D	Stops moving in a time of 2 s	accelerates at a rate of $5\,m\,s^{-2}$

7 A car accelerates from rest and travels a distance of 20 m in 4.0 s. The driving force acting on the car remains constant at 1750 N. Which of the following is the mass of the car?

A 900 kg **C** 700 kg

B 800 kg **D** 600 kg

8 A skydiver is falling at a terminal speed of $39.8\,m\,s^{-1}$ when he opens his parachute. For the next 2 seconds the parachute exerts an average drag force of twice the skydiver's weight in an upwards direction. Which of the following statements about the skydiver's motion over the next two seconds is correct?

A The skydiver moves upwards for 2 seconds.

B After two seconds the skydiver moves at a constant speed.

C Over two seconds the skydiver falls 40 m.

D Over two seconds the skydiver falls 60 m.

9 A child pulls a trolley with a force of 15 N in the direction shown in Figure 9.25. A frictional force of 3 N acts on the trolley. Which of the following gives the acceleration of the trolley along the ground?

A $3.0\,m\,s^{-2}$ **C** $1.8\,m\,s^{-2}$

B $2.4\,m\,s^{-2}$ **D** $1.2\,m\,s^{-2}$

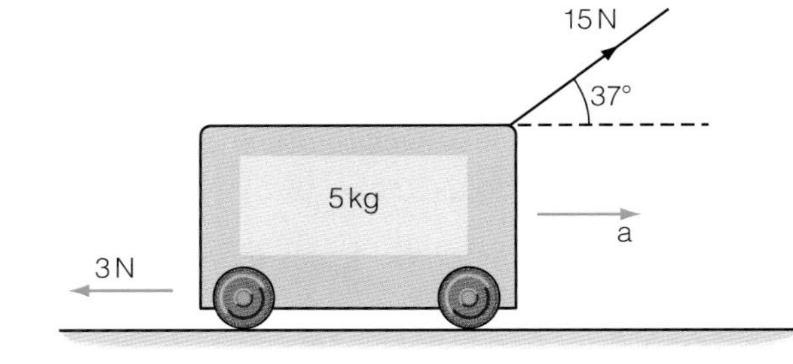

Figure 9.25 A child pulls a trolley with a force of 15 N in the direction shown.

The information below is needed for both questions 10 and 11.

A four-engine aeroplane of mass 80 000 kg takes off along a runway of length 1.0 km. The aeroplane's take-off speed is $50\,m\,s^{-1}$. The average drag forces acting against the plane during take-off are 20 kN.

10 What is the plane's minimum acceleration if it is to take-off from the runway?

A $4.00\,m\,s^{-2}$ **C** $1.50\,m\,s^{-2}$

B $2.50\,m\,s^{-2}$ **D** $1.25\,m\,s^{-2}$

11 What is the minimum thrust required from each of the four engines to achieve an acceleration of $2.0\,m\,s^{-2}$?

A 35 000 N **C** 140 000 N

B 45 000 N **D** 160 000 N

12 Figure 9.26a shows a cylinder of wood. Its density is 720 kg m^{-3}.

a) Use the information in the diagram to show that the weight of the cylinder is about 12 N. *(3)*

b) State Newton's first law of motion. *(1)*

The cylinder, Y, is now placed on a table with a second identical cylinder, X, placed on top of it. (Figure 9.26b)

Figure 9.26c is a free body diagram that shows the forces that act on cylinder Y.

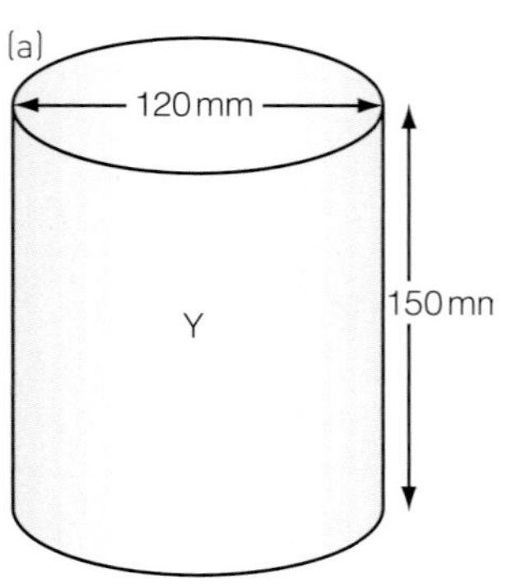

Figure 9.26a A cylinder of wood.

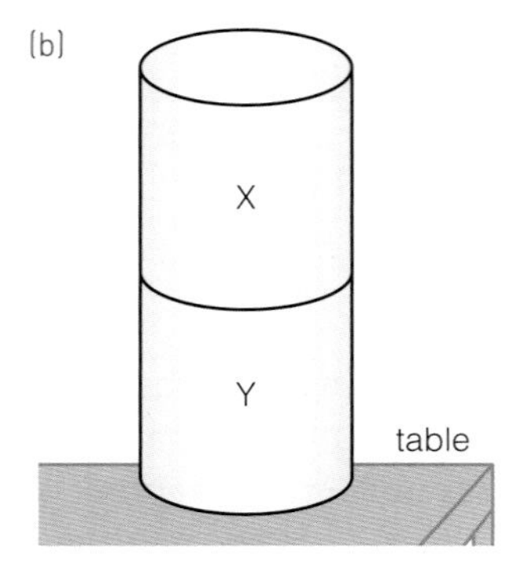

Figure 9.26b Two cylinders of wood.

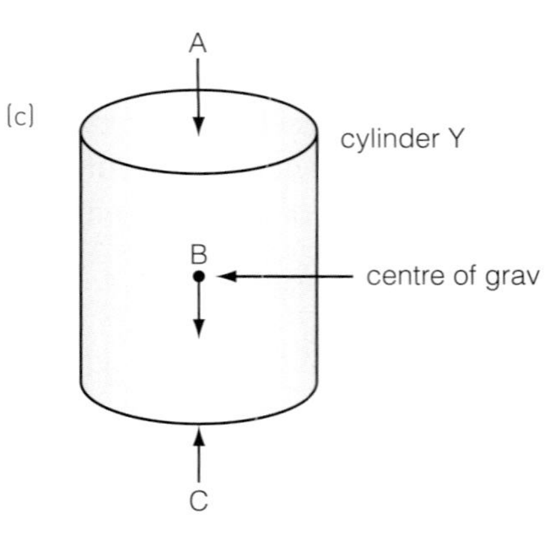

Figure 9.26c A free body diagram that shows the forces that act on cylinder Y.

c) Name and state the magnitude of the forces *A*, *B* and *C*. *(3)*

d) A force, *F*, forms a Newton's third law pair with force B.

State the following:

i) the magnitude of *F*

ii) the direction of *F*

iii) the type of force that *F* is

iv) the object on which *F* acts. *(4)*

13 A student with a weight of 580 N stands on a set of weighing scales. She holds a mass of 2 kg in her hand by her side.

a) Explain why the scales record a weight of about 600 N. *(1)*

She moves her arm upwards rapidly to lift the weight about her head. The scales change as shown in Figure 9.27.

b) Using Newton's laws of motion, explain the readings on the graph at times:

i) A

ii) B

iii) C. *(6)*

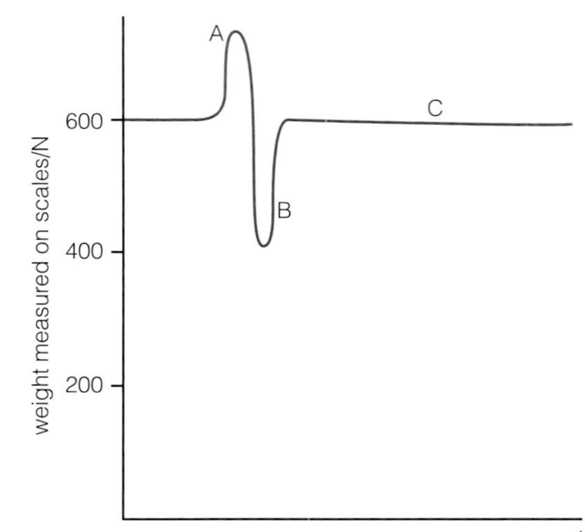

Figure 9.27 How the reading on the scales changes.

14 Two boxes are pulled with a force of 50 N, which causes them to accelerate to the right. Frictional forces of 3 N and 2 N, respectively, act on each box as shown in Figure 9.28.

a) Calculate the acceleration of the two boxes. *(3)*

b) Calculate the tension, *T*, acting in the string that connects the two boxes. *(2)*

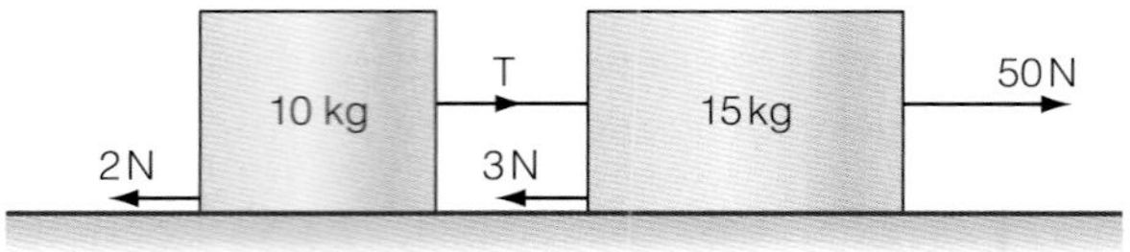

10 kg T 15 kg 50 N 2 N 3 N

Figure 9.28 Frictional forces acting on boxes.

15 Two large crates rest on the floor of a warehouse. Each crate has a mass of 48 kg. A forklift truck provides a force of 250 N and the crates accelerate to the left is shown in Figure 9.29.

Figure 9.29 Crates in a warehouse.

a) The crates accelerate at $1.2\,\text{m}\,\text{s}^{-2}$.

i) Determine the resultant force acting on the crates. *(2)*

ii) Determine the total resistive force acting on both crates. *(1)*

b) Assuming that the resistive force is spread equally between the two crates, calculate the force that crate P exerts on crate Q when the force of 250 N is applied to crate Q. Explain your answer. *(4)*

16 A toy plane is powered by a pressured water bottle. The principle is illustrated in Figure 9.30. A water bottle is pressurised with a pump and inserted into the plane. When the plane is launched, water is pushed backwards out of the plane and the plane accelerates forwards. The plane flies horizontally after it is thrown to launch it.

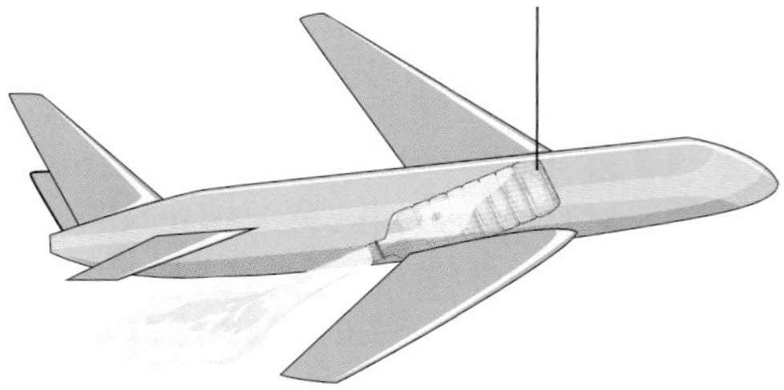

bottle of water in the plane

Figure 9.30 Principle of how a toy plane works.

a) Use Newton's second and third laws of motion to explain how the water accelerates the plane. *(4)*

The plane starts with a mass of 1.3 kg, of which 0.4 kg is the initial mass of the water in the bottle.

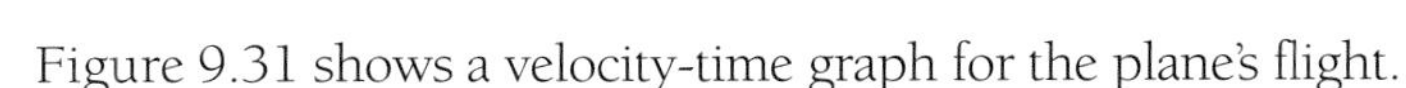

Figure 9.31 shows a velocity-time graph for the plane's flight.

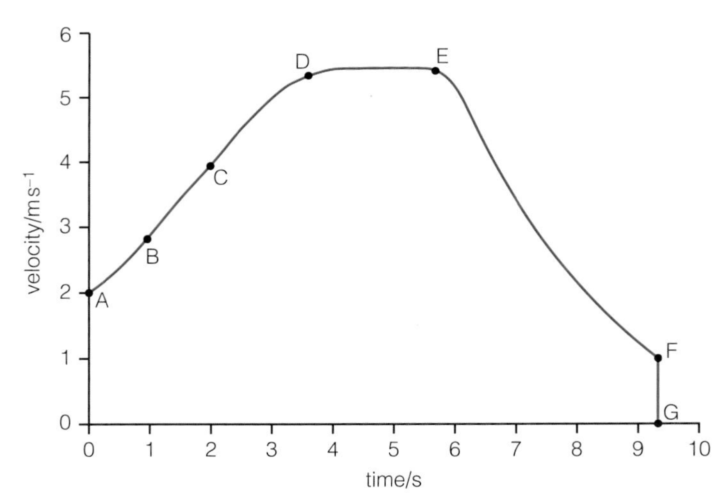

Figure 9.31 A velocity-time graph for the plane.

b) Calculate the plane's acceleration:

i) over the first second *(2)*

ii) over the time 1 second to 2 seconds on the graph. *(2)*

iii) Explain why the plane's acceleration increases during the period of 1 to 2 seconds. *(2)*

c) Calculate the resultant force on the plane at the start of its flight. *(2)*

d) At which point does the water run out? Explain your answer. *(2)*

e) Explain the shape of the velocity-time graph over the following periods of time:

i) C to D (2)

ii) D to E (1)

iii) E to F (2)

iv) F to G. (1)

f) The distance flown by the plane is closest to which of these values:

i) 15 m

ii) 35 m

iii) 55 m

iv) 75 m (2)

g) i) When the plane increases its speed, its kinetic energy increases. Explain what source provided this kinetic energy. (2)

ii) What else gains kinetic energy in this flight? (1)

17 A student designs an experiment to investigate the resistive forces that act on a small mass as it falls through a column of water. The apparatus is shown in Figure 9.32. The mass is immersed in the water and released from rest. A ticker tape attached to the mass records the motion of the mass as it falls.

The ticker timer marks the tape every 0.02 s with a small dot. The separation of the dots may be used to calculate the speed of the mass over a small interval of time. The student analyses the tape from one of their experiments and records their results in the table shown below.

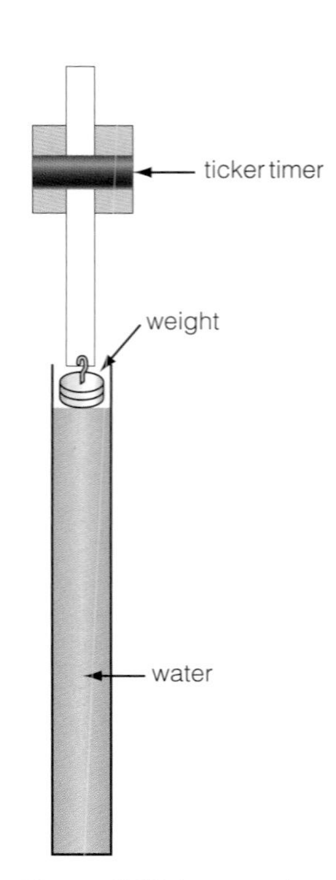

Figure 9.32 An experiment to investigate the resistive forces that act on a small mass as it falls through a column of water.

Interval number	Gap between successive dots (mm)	Interval number	Gap between successive dots (mm)
1	2.0	10	21.5
2	5.0	11	22.5
3	8.0	12	23.5
4	11.0	13	24.0
5	13.5	14	25.0
6	15.5	15	26.0
7	17.0	16	25.0
8	18.5	17	26.0
9	20.0		

a) Estimate the accuracy of these measurements. Could the results have been recorded more accurately? (2)

b) Show that the acceleration of the mass over the first four intervals is $7.5\,m\,s^{-2}$. (4)

c) Explain how the data shows that the acceleration of the mass decreases with time. (2)

d) Calculate the terminal speed of the falling mass. (2)

e) Sketch graphs to show:

i) how the velocity of the mass varies with time

ii) how the resistive forces on the mass vary with speed. (4)

f) Criticise this experiment, explaining how you would improve this method to investigate the size of the drag force on a falling object. (4)

Stretch and challenge

18 A firework rocket reaches its highest point at 100 m above the ground, where it is stationary. It then explodes into four equal parts as shown in Figure 9.33a. Each fragment has a mass of 0.1 kg.

a) Using Newton's laws explain why there should be symmetry to the explosion.

b) Calculate how long each of the fragments P, Q, R and S take to reach the ground. Ignore the effects of air resistance in this and all subsequent calculations.

c) Calculate where each fragment reaches the ground relative to the centre of gravity, C, which is 100 m above the ground.

d) Calculate the kinetic energy of each fragment as it reaches the ground.

[Hint: you might be able to solve this and the next part using your knowledge from GCSE, or you might have to read ahead into the next chapter and come back to solve this.]

e) A second rocket explodes with the four fragments moving as shown in Figure 9.33b. State the kinetic energy of each fragment when they hit the ground.

19 A man with a mass of 80 kg has designed a lift in which he can use his own body power to lift himself. The mass of the lift is 40 kg.

a) In Figure 9.34 he is stationary. By drawing free body diagrams for the man and the lift, determine the tension i n the rope and the reaction between the man and the l ift floor.

b) He pulls on the rope with a force of 648 N. Determine the acceleration of the lift now.

c) Comment on any safety issues you might have identified with this lift.

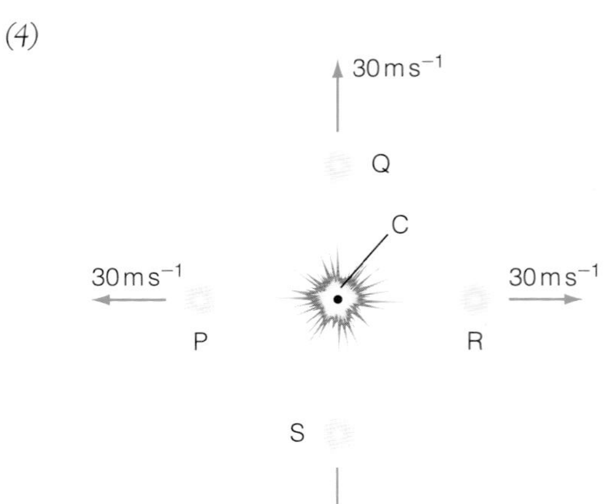

Figure 9.33a A rocket explodes into four equal parts.

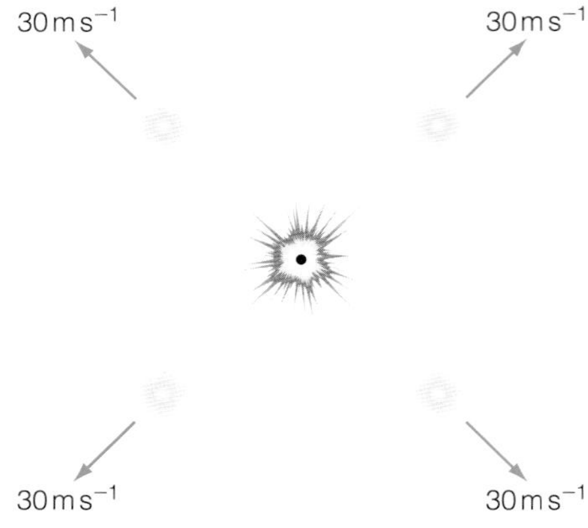

Figure 9.33b A rocket explodes into four equal parts.

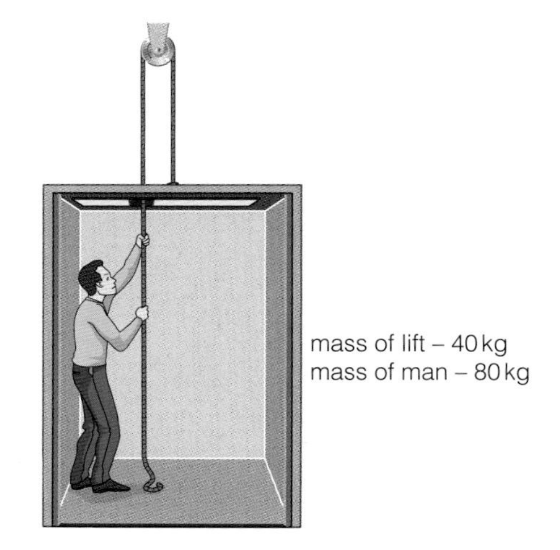

Figure 9.34 A lift in which a man can use his own body power to lift himself.

10 Work, energy and power

PRIOR KNOWLEDGE

- Work is done when a force moves an object in the direction of the force.
- Work (J) = force (N) × distance moved in the direction of the force (m)
- This equation defines the joule:

$$1\,\text{J} = 1\,\text{N} \times 1\,\text{m}$$

- Energy is the capacity to do work or to heat something up – these quantities are measured in joules.
- Power is the rate of doing work or transferring energy.

$$\text{Power} = \frac{\text{energy}}{\text{time}}$$

- This equation defines the watt:

$$1\,\text{W} = 1\,\text{J}\,\text{s}^{-1}$$

- Energy is conserved. This means that energy cannot be created or destroyed, but it can be transferred from one form to another.

TEST YOURSELF ON PRIOR KNOWLEDGE

1 Calculate the work done when you lift a mass of 24 kg through a height of 1.3 m.

2 You leave a 2 kW electrical heater on for $3\frac{1}{2}$ hours. How much electrical energy have you used?

3 In which of these examples is work being done?
  a) A weightlifter holds a weight stationary above his head.
  b) A car moves at a constant speed along a road.
  c) A spacecraft moves at a constant speed in outer space in 'zero g'.
  d) A spacecraft moves in a circular orbit around the Earth.
  e) A magnetic attraction keeps a magnet attached to a fridge door.
  f) You walk down the stairs at a steady speed.

4 Explain the energy transfers that occur in these cases.
  a) You throw a ball into the air and it lands on the ground.
  b) A car drives along a road.
  c) You use a rubber band to launch a paper pellet.

Typhoon Haiyan, which hit the Philippines in November 2013, was the most powerful storm in history to make landfall. Communities were devastated by the energy of the wind. Consider the energy transfers, which are responsible for the building up of the storm, then consider how the storm dissipates its energy when it hits land.

Work

We all know that if we expect to earn money doing a job, we must do something useful. Nobody wants to pay an employee who is not working effectively. The same idea applies to the scientific definition of work. It is possible to use energy but to do no work. Figures 10.2 and 10.3 show two examples of people using energy but not working.

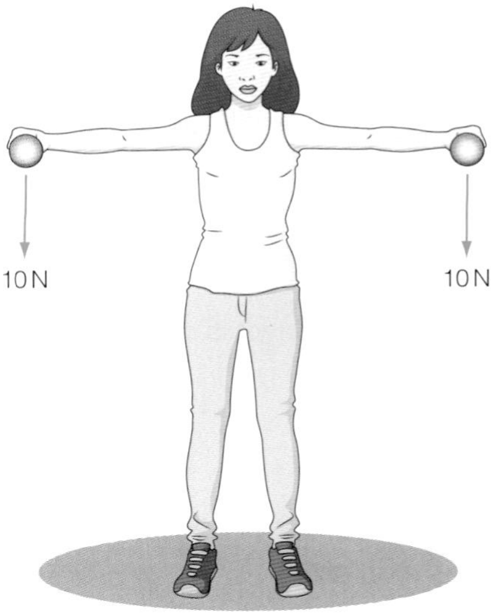

Figure 10.1 Tracy is using energy but not working.

In Figure 10.1 Tracy is doing some weight training. She is holding two weights but she is not moving them. She gets tired holding them out, and her arms convert chemical energy into heat energy, but no work is done as the weights do not move. Tracy does work when she *lifts* the weights.

In Figure 10.2 three people are helping to push-start a friend's car. Salim and Anne are pushing at the back of the car. Here Salim and Anne are working, because they are pushing along the direction in which the car is travelling. Jim does no work as he is pushing at right angles to the direction of motion.

Often an object moves in a different direction to the applied force. To calculate the work done, we resolve the force into components parallel and perpendicular to the direction of motion. Only the component parallel to the motion does work. Figure 10.4 shows a passenger at an airport wheeling her luggage. She pulls the bag with a force, F, at an angle θ to the ground.

Figure 10.2 Salim and Anne are working but Jim is not working as he is pushing at right angles to the car.

Figure 10.3 A passenger at an airport wheeling her luggage.

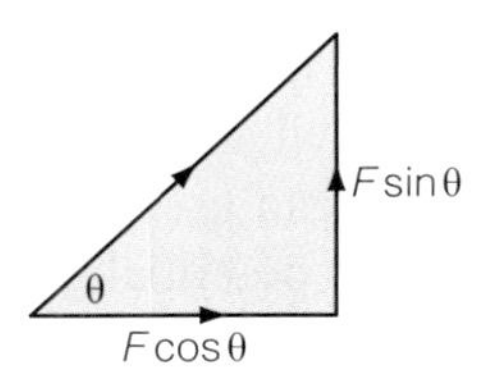

Figure 10.4 She pulls the bag with a force, F, at an angle θ to the ground.

The component of force $F\sin\theta$ does no work, but supports the bag. The component $F\cos\theta$ does work against frictional forces.

The general formula for work is

$$\text{work done} = Fs\cos\theta$$

where F is the applied force, s the displacement of the object, and θ the angle between the force and displacement.

There are times when the force we are working against does not remain constant, so the calculation of the work done is slightly more complicated. Figure 10.5 shows an example of the force increasing as an object is displaced. We calculate the work done by considering a small displacement, Δs, when the applied force is F.

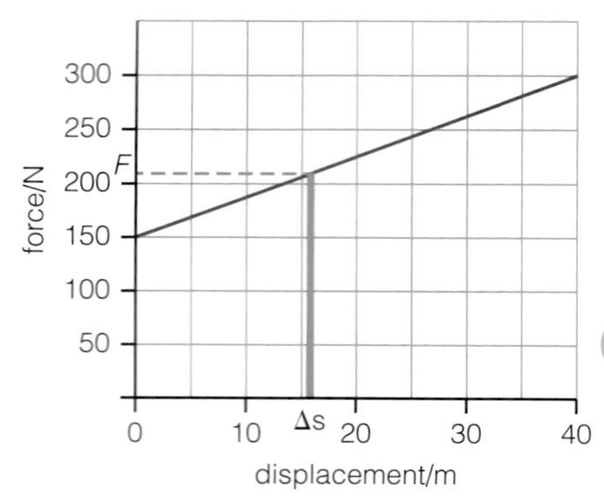

Figure 10.5 We calculate the work done by considering a small displacement, Δs, when the applied force is F.

Then the small amount of work done is:

$$\Delta W = F \times \Delta s$$

However, ΔW is the small area marked under the graph. So to calculate the work done over a bigger distance we calculate the area under the graph.

EXAMPLE

Calculate the work done from Figure 10.5 when the object is displaced 40 m from its original position.

Answer

Work done = the area under a force-displacement graph

$$= \frac{1}{2}(150\,\text{N} + 300\,\text{N}) \times 40\,\text{m}$$

$$= 9000\,\text{J}$$

TEST YOURSELF

1 A passenger pulls her suitcase, similar to the one in Figure 10.4, a distance of 400 m through an airport. The case has a mass of 20 kg, and she pulls it along with a force of 60 N at an angle of 50° to the ground.

a) Calculate the work done against resistive forces.

b) Explain why it is an advantage to pull a suitcase rather than to push it.

2 Three people help to push-start a car. The strength and direction of the forces that they apply are shown in Figure 10.6. They push the car along the road for 60 m. Calculate the work done by the three people in total.

3 A satellite, with a mass of 75 kg, is in a circular orbit of radius 7000 km around a planet. At this height the gravitational field strength is $6\,\text{N}\,\text{kg}^{-1}$. Calculate the work done by the gravitational field during one complete orbit of the planet.

4 A bag of mass 23 kg is pulled along a floor a distance of 50 m; the frictional forces acting on it are 40 N. It is then lifted on to a desk of height 0.9 m. Calculate the total work done in the process.

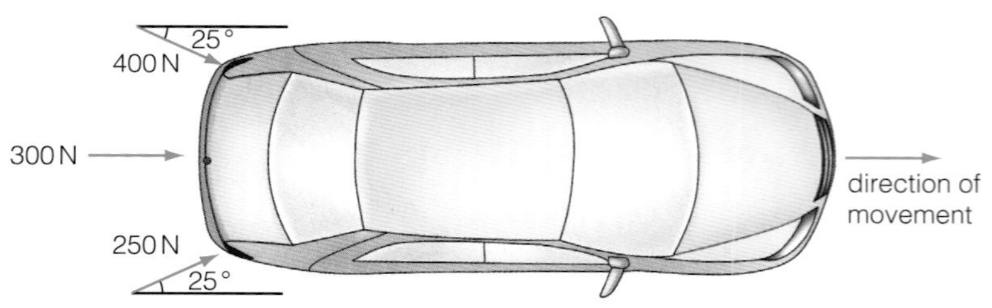

Figure 10.6 The strength and direction of applied forces.

Calculating the energy

When work is done against resistive forces, for example when the suitcase is moved at an airport, that work is transferred to heat energy. Therefore the wheels of the suitcase and their surroundings warm up by a small amount. Work can also be done to transfer energy into other forms. We can use this idea to derive formulas for gravitational potential energy, elastic potential energy and kinetic energy.

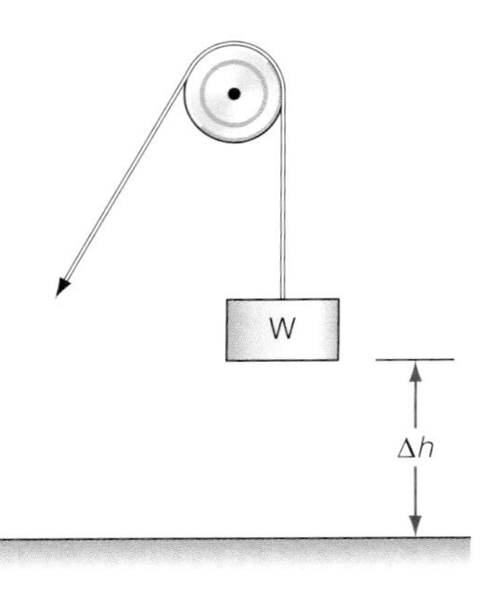

Figure 10.7 A load with a weight W lifted through a height Δh.

Gravitational potential energy

In Figure 10.7 a load with a weight W, has been lifted through a height Δh.

The work done = increase in gravitational potential energy

$$\text{work} = W \times \Delta h$$

But

$$\text{weight } W = mg$$

So the increase in potential energy, ΔE_p, is given by:

$$\Delta E_p = mg\Delta h$$

Note that we use the symbol Δ to indicate a change in energy and height. There is no defined zero point of gravitational potential energy on the Earth, so we talk about changes.

Elastic potential energy

Elastic potential energy (E_{ep}) is stored in a stretched wire or rubber band. Figure 10.8 shows how a rubber band may be stretched.

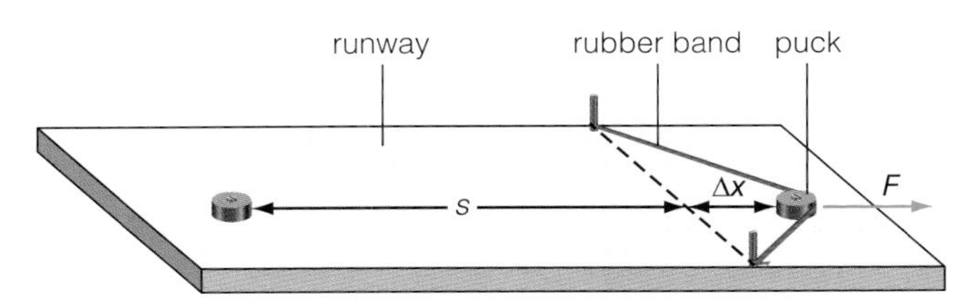

Figure 10.8 How a rubber band may be stretched.

TIP
The area under a force-displacement graph is equal to the work done.

When it is released it can convert its stored energy into the kinetic energy of a puck, which slides along a runway.

Figure 10.9 shows how the force required to stretch the band depends on the distance it is pulled back. Because the force changes we have to use the average force to calculate the energy stored.

$$\Delta(E_{ep}) = F_{av} \times s$$

In the case of Figure 10.9, where the force to stretch the band is proportional to the distance moved:

$$\text{average force} = \frac{1}{2} \times \text{final force}$$

So

$$\Delta E_{ep} = \frac{1}{2} Fs$$

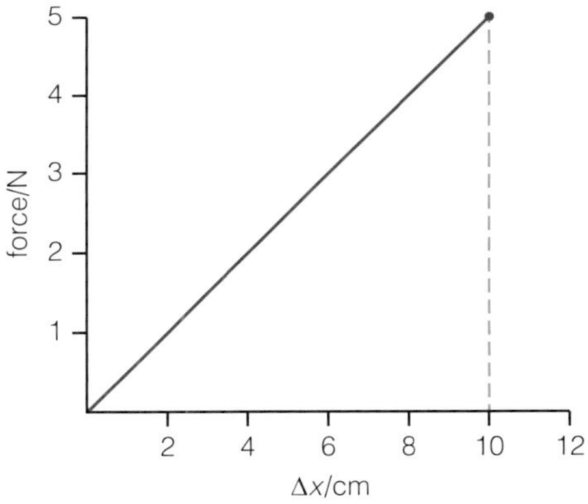

Figure 10.9 How the force required to stretch the band depends on the distance it is pulled back.

The stored elastic potential energy can also be calculated more generally using the area under the force-extension graph.

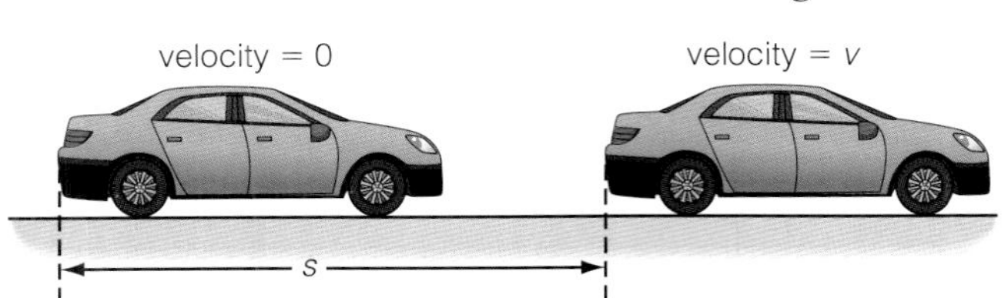

Figure 10.10 A constant force accelerates a car.

Kinetic energy

In Figure 10.10 a constant force F accelerates a car, starting at rest, over a distance s. Work is done to increase the kinetic energy of the car. We can use this idea to find a formula for kinetic energy, E_k, in terms of the car's speed and mass.

$$\Delta E_k = Fs$$

but from Newton's Second Law:

$$F = ma$$

and from the equations of motion:

$$s = \frac{1}{2}at^2$$

which gives

$$\Delta E_k = ma \times \frac{1}{2}at^2$$

$$= \frac{1}{2}ma^2t^2$$

since

$$v = at$$

$$\Delta E_k = \frac{1}{2}mv^2$$

So the kinetic energy of a body of a mass m, moving at a velocity v, is given by:

$$E_k = \frac{1}{2}mv^2.$$

Note this is a scalar quantity because v^2 has no direction.

TEST YOURSELF

5 a) i) A crane lifts a container of mass 25 000 kg though a height of 12 m above the quay next to a ship. Calculate the container's increase in gravitational potential energy.

Figure 10.11 A container being loaded on to a ship.

ii) The crane then rotates, moving the container sideways 20 m over the ship. The average frictional force acting on the crane as it moves sideways is 150 N. Calculate the work done in this movement.

iii) The container is then lowered into the ship's hold so that the container lies 8 m below the level of the quay. Calculate its potential energy now, when compared to its original potential energy.

b) A spacecraft of mass 18 000 kg falls into a black hole. The decrease in the spacecraft's potential energy as it falls 100 m near the event horizon is 5×10^{18} J. Calculate the average gravitational field strength in this region.

6 a) A Formula 1 car with a mass of 720 kg accelerates from a speed of 42 m s^{-1} to 88 m s^{-1}. Calculate its increase in kinetic energy.

b) A bullet with mass of 0.05 kg has a velocity of 300 m s^{-1}. Calculate its kinetic energy.

7 In an experiment similar to that shown in Figure 10.8, a puck of mass 0.15 kg is used. The rubber band is stretched by 0.12 m with an average force of 4.0 N.

a) Calculate the elastic potential energy stored in the rubber band when it has been pulled back a distance of 0.12 m.

b) The puck is released and it slides along the runway a distance of 0.96 m past the unstretched position of the band (the distance s in the diagram). Estimate the average frictional force acting on the puck to slow it down.

The principle of conservation of energy

You are already familiar with the principle of conservation of energy, which may be stated as follows: energy cannot be created or destroyed, but may be transferred from one form of energy to another.

In the last section you met formulas for gravitational potential energy, kinetic energy and elastic potential energy. In this section you will learn to use these formulas to calculate and predict the motion of bodies as one form of energy is transferred to another.

EXAMPLE

A boy throws a ball upwards with a speed of $16\,m\,s^{-1}$. It leaves his hand at a height of 1.5 m above the ground. Calculate the maximum height to which it rises.

Answer

The kinetic energy of the ball as it leaves the boy's hand is transferred to gravitational potential energy in the ball as it rises. At the ball's highest point it stops moving, so all its kinetic energy has been transferred to potential energy at that point.

So

$$\Delta E_k = \Delta E_p$$

$$\Rightarrow \frac{1}{2}mv^2 - 0 = mg\Delta h$$

$$\Rightarrow \Delta h = \frac{v^2}{2g} = \frac{16^2\,(m\,s^{-1})^2}{2 \times 9.8\,m\,s^{-2}} = 13\,m$$

but the total height gained is 13 m + 1.5 m = 14.5 m.

EXAMPLE

A stretched catapult stores 0.7 J of elastic potential. The catapult is used to launch a marble of mass 0.01 kg. Calculate the initial speed of the marble.

Answer

The stored elastic potential energy in the catapult is transferred to the marble's kinetic energy.

So

$$\Delta E_{ep} = \Delta E_k$$

$$0.7\,J = \frac{1}{2} \times 0.01\,kg \times v^2$$

$$v^2 = \frac{1.4}{0.01}\,(m\,s^{-1})^2$$

$$\Rightarrow v = 12\,m\,s^{-1}$$

EXAMPLE

A car of mass 1300 kg is travelling at $9\,m\,s^{-1}$. The driver applies the brakes, which exert a braking force of 4400 N on the car. Calculate the braking distance of the car.

Answer

The work done by the brakes is equal to the change of kinetic energy in the car. (The kinetic energy is transferred to heat energy in the brake discs).

$$\text{work done} = \Delta E_k$$

$$F \times s = \frac{1}{2}mv^2$$

$$4400\,N \times s = \frac{1}{2} \times 1300\,kg \times (9\,m\,s^{-1})^2$$

$$s = \frac{650\,kg \times 81\,(m\,s^{-1})^2}{4400\,N} = 12\,m$$

Efficiency

In Figure 10.12 an electric motor is being used to lift a weight. Electrical energy is being transferred to gravitational potential energy. However, the motor also transfers electrical energy into heat energy and sound energy.

In Figure 10.13 a Sankey diagram shows how 100 J of electrical energy are transferred to other types of energy. The majority of the energy is transferred into heat and sound energy, with only 30 J being transferred into *useful* gravitational potential energy.

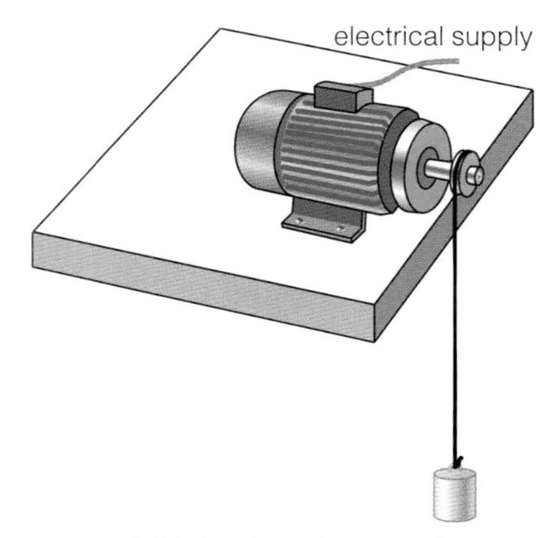

Figure 10.12 An electric motor being used to lift a weight.

Sankey diagram A specific flow diagram in which the width of the arrows shown is proportional to the quantity of the flow. In this chapter Sankey diagrams are used to show how energy is transferred in various processes.

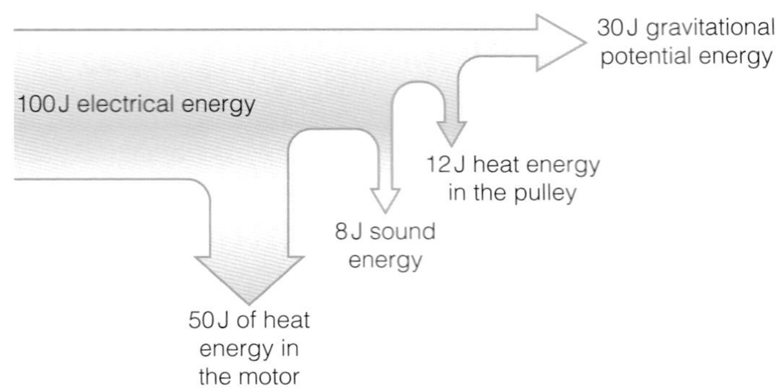

Figure 10.13 A Sankey diagram.

The efficiency of a machine is defined as follows:

$$\text{efficiency} = \frac{\text{useful energy transferred}}{\text{energy supplied}}$$

For the motor:

$$\text{efficiency} = \frac{30\,\text{J}}{100\,\text{J}}$$
$$= 0.3 \text{ or } 30\%$$

TEST YOURSELF

8 A motor similar to the one shown in Figure 10.12 uses 280 J of electrical energy to lift a mass of 2.3 kg through a vertical height of 3.5 m. Calculate the efficiency of the motor.

9 The efficiency of a car's petrol engine is 27%. This means that only 27% of the chemical energy stored in the car's fuel does useful work to drive the car forwards against the drag and frictional forces.

a) One litre of petrol in a fuel tank stores 30 MJ of chemical potential energy. Calculate the energy available in 1 litre of fuel to work against frictional forces.

b) A car travelling at a constant speed of $20\,\text{m}\,\text{s}^{-1}$ uses 1 litre of petrol to drive 7 km. Calculate the frictional forces acting on the car.

Power

Power is defined by the equation:

$$\text{power} = \frac{\text{energy transferred}}{\text{time}} \text{ or } \frac{\text{work done}}{\text{time}}$$

The unit of power is the watt, W, or $\text{J}\,\text{s}^{-1}$. This definition can be used to produce a useful formula to calculate the power transferred by moving vehicles.

$$\text{power} = \frac{\text{work done}}{\text{time}}$$
$$= \frac{F \times s}{t}$$
$$\Rightarrow \text{power} = F \times v$$

EXAMPLE

A car is moving at a constant speed of $18\,m\,s^{-1}$. The frictional forces acting against the car are 800 N in total. The car has a mass of 1200 kg.

a) Calculate the power transferred by the car on a level road.

b) The car maintains its constant speed while climbing a hill of vertical height 30 m in 16 s. Calculate the power transferred by the car now.

Answer

a) $P = F \times v$

$= 800\,N \times 18\,m\,s^{-1}$

$= 14\,kW$

b) $P = 14\,kW + \frac{mg\Delta h}{t}$

$= 14\,kW + \frac{1200\,kg \times 9.8\,N\,kg^{-1} \times 30\,m}{16\,s}$

$= 14\,kW + 22\,kW$

$= 36\,kW$

ACTIVITY

How high can you climb using the energy from a chocolate bar?

Estimate the height of a mountain that you can climb using the energy from a chocolate bar. You will need to research the efficiency of the body in turning food energy to work, and the calorific value of your chocolate bar.

The purpose of the following group of experiments is to account for energy as far as is possible. At the start of each experiment you will calculate the amount of energy in one form, and see whether you can explain how the original energy has been transformed into other forms of energy. Preferably you should do your own experiments based on these, but you can use the data provided to guide you.

ACTIVITY

Elastic potential energy transferred to kinetic energy

Here you investigate how elastic potential energy is transferred into the kinetic energy of a trolley, which is tethered by two springs. In Figure 10.14 the trolley is in equilibrium. Pull the trolley a suitable distance sideways, and measure the force required to do this. Then release the trolley and use the light gate to measure the speed directly.

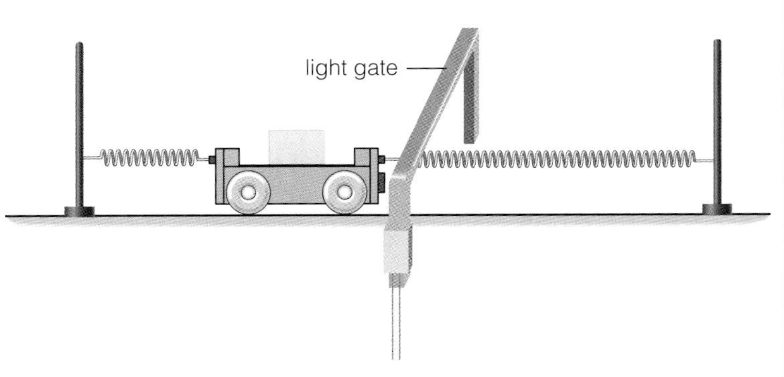

Figure 10.14 The speed of the trolley is measured as it passes through the light gate.

Here is some sample data obtained with a trolley of mass 0.85 kg.

	Extension of spring (cm)	Force to extend the spring (N)	Speed of trolley at its midpoint ($cm\,s^{-1}$)
Experiment 1	10.0	4.6	69
Experiment 2	20.0	9.3	131

Use these results to calculate the elastic potential energy stored in the springs before the release of the trolley. Use the formula:

$$\Delta E_{ep} = \frac{1}{2}\,\text{force} \times \text{extension}$$

Then calculate the trolley's kinetic energy at its fastest point of motion. Comment on your results.

ACTIVITY

Gravitational potential energy to kinetic energy

This experiment monitors the transfer of gravitational potential energy to kinetic energy. In Figure 10.15 a ball bearing is dropped through two light gates which measure the ball's speed in each position.

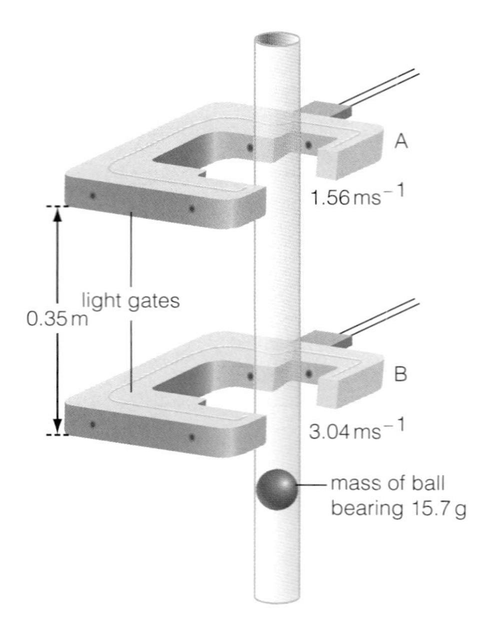

Figure 10.15 The light gates read the speed directly as the ball bearing passes through them.

Use the data provided in Figure 10.15 to calculate the change of gravitational potential energy as the ball falls from A to B, and the change in kinetic energy over the same distance. Comment on your results.

The tube is now tilted so that the ball bearing rolls down the tube rather than falling (see Figure 10.16). It falls through the same vertical height as before, 0.35 m.

When the ball bearing rolls it has two types of kinetic energy: translation kinetic energy, which is calculated by the formula:

$$\frac{1}{2}mv^2$$

and rotational kinetic energy due to the rotation of the ball about its own axis. The rotational kinetic energy can be calculated using the formula:

$$\frac{1}{5}mv^2.$$

Use this theory to predict the speed of the ball bearing after is has rolled down a vertical height of 0.35 m, to light gate B after it travelled past light gate A with a speed of 0.91 m s^{-1}. Explain which other energy transformations may take place during this process.

Check by doing the experiment for yourself.

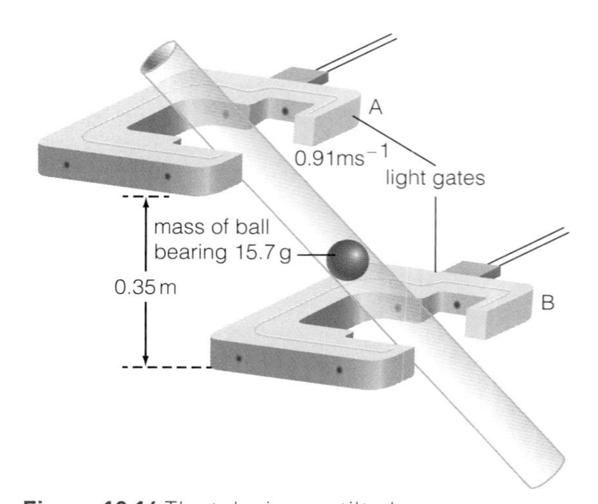

Figure 10.16 The tube is now tilted.

TEST YOURSELF

10 A ball is dropped from a height of 15 m.

a) Calculate its speed when it hits the ground below.

b) The same ball is thrown downwards from a height of 15 m with an initial speed of 10 m s^{-1}. Calculate its speed when it hits the ground.

11 A ball has a mass of 0.1 kg. It is thrown sideways with a speed of 10 m s^{-1} at a height of 20 m above the ground.

a) Calculate its kinetic energy when it is first thrown.

b) Calculate the ball's potential energy at a height 20 m above the ground.

c) Calculate the ball's kinetic energy when it hits the ground.

d) Calculate the ball's speed when it hits the ground.

12 A car of mass 1800 kg slows from a speed of $20\,m\,s^{-1}$ to $15\,m\,s^{-1}$ over a distance of 200 m. Calculate the average resistive forces acting on the car over this distance.

13 A force of F is used to accelerate, from rest, two cars over a distance, d. One car has a mass m, the second car a mass of $2m$. Which car now has the greater kinetic energy?

14 An express train is powered by diesel engines that have a maximum output of mechanical power to drive the train of 2.0 MW. Using this power the train travels at a top speed of $55\,m\,s^{-1}$.

a) Calculate the resistive forces acting on the train at its top speed.

b) The diesel fuel stores 40 MJ of chemical energy. The engines have an efficiency of 32%. Calculate the volume of fuel required for a journey of 200 km at the train's top speed.

15 A fly of mass 16 mg flies into a spider's web. The kinetic energy of the fly is converted to elastic potential energy in the web as it stretches. Figure 10.17 shows how the centre of the web is displaced as it slows the fly down.

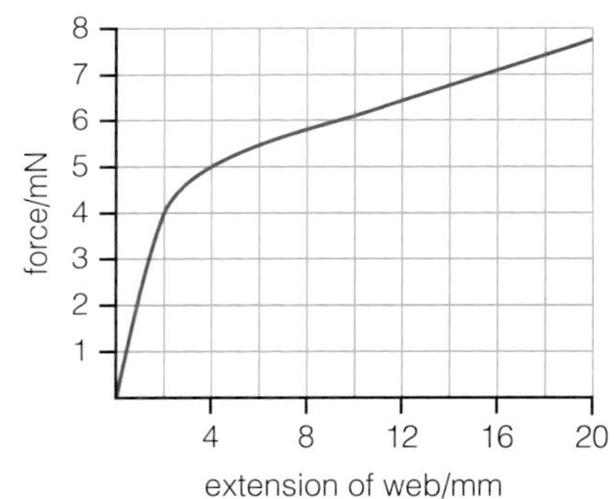

Figure 10.17 How the centre of the web is displaced as it slows the fly down.

a) The maximum stretch of the web when the fly hits it is 14 mm. Estimate the elastic potential energy stored in the web for this displacement. Give your answer in μJ.

b) Estimate the speed of the fly when it hit the web. State any assumptions you make.

16 Read this extract from a scientific article.

'The hind legs of a locust are extremely powerful. The insect takes off with a speed of $3\,m\,s^{-1}$. The jump is fast and occurs in a time of 25 milliseconds. The locust's mass is about 2.5 g.'

Use the information to answer the questions below.

a) Calculate the locust's average acceleration during take-off.

b) Calculate the average force exerted by the locust's hind legs.

c) The locust's legs extend by 5 cm during the jump. Calculate the work done by the legs.

d) Calculate the power developed in the locust's muscles.

e) Compare the power/mass ratio for the locust, with that of a high jumper. Assume the high jumper has a mass of 70 kg; he lifts his centre of gravity by 1.3 m in the jump; his take-off foot is in contact with the ground for 0.2 s.

17 A power station turns the chemical energy stored in coal into electrical energy with an efficiency of 36%. Some of the electrical energy is used to drive a tram. The tram's motor transfers electrical energy to work against resistive forces with an efficiency of 25%. What fraction of the coal's original chemical energy is used to drive the tram? What happens to the rest of the energy?

18 A hydroelectric turbo-generator is driven by water that descends 160 m from a high-level reservoir. Water passes through the turbines at a rate of $700\,m^3\,s^{-1}$. The density of water is $1000\,kg\,m^{-3}$. The efficiency of the turbo-generator is 0.8. Calculate the electrical power output from the generator. Express your answer in GW.

19 It has been suggested that a tidal barrage could be built across the River Severn to generate electricity. The idea is that, as the tide rises, water flows in through gates in the barrage into a lake. When the tide begins to turn, the gates are closed and the water is stored behind the barrage. At low tide water in the lake is then allowed to flow out through turbo-generators.

You waill be asked to carry out some calculations on the generation of electrical power. You will need the following information.

- The surface of the lake is at a height of 13 m above the turbo-generators at the start of the generation process and falls to a height of 6 m above the turbo-generators.
- The surface area of the lake is $200\,km^2$.
- The electricity is generated over a period of 5.5 hours.
- The density of sea water is $1020\,kg\,m^{-3}$.
- $g = 9.81\,N\,kg^{-1}$.
- The efficiency of the turbo-generators is 78 %.

a) Calculate the gravitational potential energy available in the lake at high tide. Express your answer in TJ.

b) Calculate the average power produced by the turbo-generators. Express you answer in MW.

c) Assuming there are two high tides a day, estimate the total electrical energy production a year.

Practice questions

1 A car exerts a tractive force of 750 N when travelling at 108 km h^{-1}. What is the work done by this force in 10 minutes?

A 10.8 MJ
B 13.5 MJ
C 15.6 MJ
D 19.6 MJ

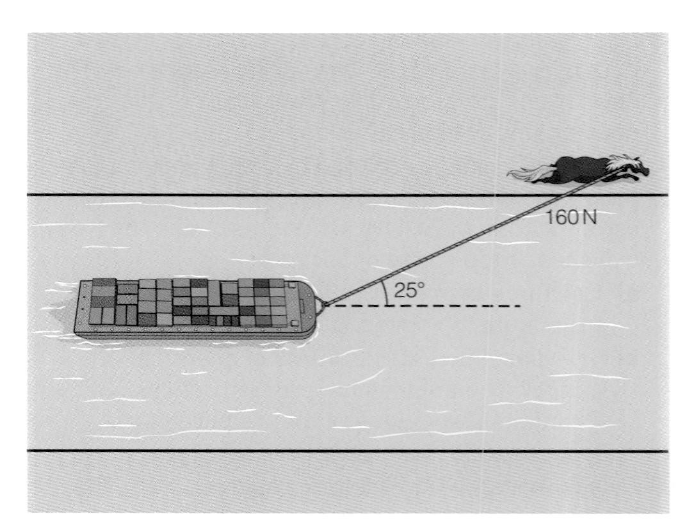

Figure 10.18 A horse pulling a barge.

2 A horse pulls a barge along a canal with a force of 160 N. The barge moves at a steady speed of 2.0 m s^{-1}. How much work does the horse do in pulling the barge forwards in 1 hour?

A 1.15 MJ
B 1.04 MJ
C 0.49 MJ
D 0.05 MJ

3 An electric motor uses 100 W to raise 10 kg vertically at a steady speed of 0.4 m s^{-1}. What is the efficiency of the system?

A 10%
B 25%
C 40%
D 100%

4 An electric motor has an efficiency of 22%. It lifts a 6.6 kg load through a height of 2 m in 4 s. What power is the motor using?

A 75 W
B 100 W
C 150 W
D 225 W

5 A car of mass 1200 kg brakes from a speed of 30 m s^{-1} to a speed of 18 m s^{-1} over a distance 240 m. What is the average braking force over this distance?

A 360 N
B 720 N
C 1080 N
D 1440 N

6 A ball is falling at a speed of 8 m s^{-1} when it is 6 m above the ground. What is the speed when it hits the ground?

A 13 m s^{-1}
B 15 m s^{-1}
C 17 m s^{-1}
D 19 m s^{-1}

The information here refers to both questions 7 and 8.

A long bow, when fully flexed, stores 75 J of elastic potential energy. The bow transfers this energy with 80% efficiency to an arrow of mass 60 g. The arrow has a range of 200 m and penetrates a target to a depth of 12 cm.

7 The speed of the arrow as it leaves the bow is:

A 50 m s^{-1}
B 45 m s^{-1}
C 40 m s^{-1}
D 35 m s^{-1}

8 The average retarding force exerted by the target on the arrow is:

A 500 N
B 200 N
C 50 N
D 5 N

9 A manufacturer states that a 50 g chocolate bar provides your body with 1080 kJ of energy. Our bodies can convert energy from food to work with an efficiency of 10%. How far up a mountain can a 60 kg student climb using the energy from this chocolate bar?

A 90 m
B 180 m
C 900 m
D 1800 m

10 A runner loses heat from her body at an average rate of 90 W when at rest. She plans to run a race that will require her to use energy at a rate of 600 W for three hours. What is the athlete's minimum energy requirement for the day?

A 6.8 MJ
B 13.3 MJ
C 16.7 MJ
D 18.0 MJ

11 A student investigates the action of a pile driver using the apparatus shown in Figure 10.19.

A steel rod is dropped on to a nail, which is embedded 4 mm into a block of wood. The student then measures the additional depth to which the nail is knocked in for each drop of the model pile driver.

Use the data below, collected by the student, to answer the following questions:

- Mass of the pile driver: 0.31 kg
- Height of the pile driver above the nail for each drop: 25 cm

Length of nail above the wood (mm)	Number of drops of the pile driver
41	0
35	1
31	2
28	3
25	4
23	5

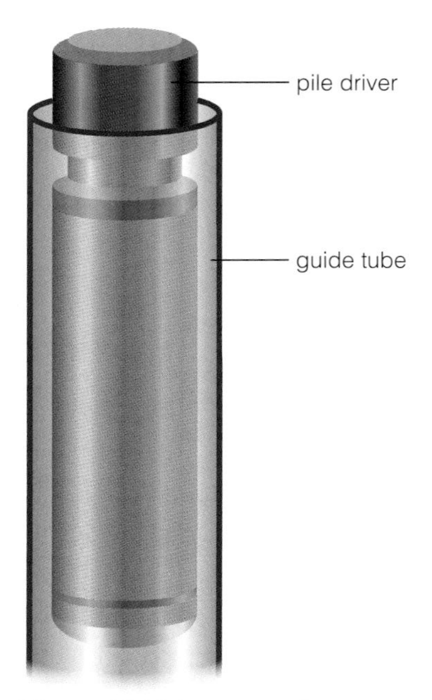

Figure 10.19 Apparatus for investigating the action of a pile driver.

a) Calculate the average force to drive the nail into the wood for the first drop of the pile driver. *(4)*

b) Use your knowledge of forces, work and energy to explain the pattern of the student's results. *(2)*

12 A conveyor belt in a power station lifts 9.6 tonnes of coal per minute to tip into the furnace.

a) Use the information in Figure 10.20 to calculate the increase in gravitational potential energy of the coal per second. *(3)*

b) The conveyor belt is powered by a 80 kW electric motor. Calculate the efficiency of the motor, ignoring frictional effects. *(2)*

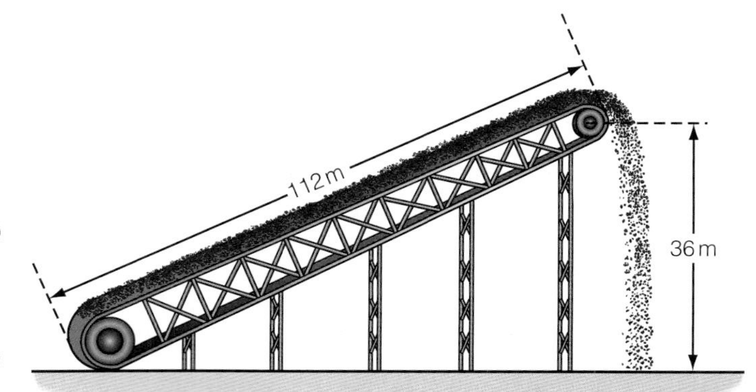

Figure 10.20 Conveyor belt lifting coal.

13 The following is an extract from a flying manual.

'The typical drag on a jet aircraft of mass 110 000 kg, flying at 240 m s^{-1} at a constant height of 12 000 m, is 92 kN. On a four-engine aircraft this means that each engine must be delivering 23 kN of thrust; and under these conditions 23 kN of thrust requires a power of just over 5.5 MW.'

a) How do you know from this information that the plane is travelling a constant speed? *(2)*

b) Calculate the lift on the aircraft. *(1)*

c) Show how the author calculated the power delivered by the engines. *(3)*

14 a) State the principle of conservation of energy. *(2)*

b) A boy of mass 32 kg travels down a slide as shown below. The distance he travels is 12 m.

Calculate his loss of gravitational potential energy. *(3)*

c) The boy reaches a speed of 9.3 m s^{-1} at the bottom. Calculate his increase in kinetic energy. *(2)*

d) The increase in the boy's kinetic energy is less than his decrease in potential energy. Explain how the principle of conservation of energy still applies here. *(2)*

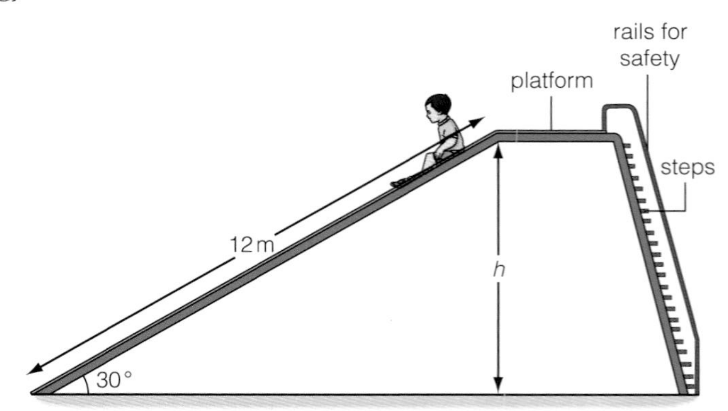

Figure 10.21 Boy sliding down slide.

e) A student suggests that the frictional force acting on the boy as he descends the slide will generate heat energy. Use this idea to calculate an average value for this frictional force. *(3)*

15 A golfer hits a golf ball so that it travels with a speed of 41 m s^{-1}. The golf club is in contact with the ball for 0.49 ms. The golf ball has a mass of 46 g.

a) i) Calculate the average acceleration of the ball while it is in contact with the club. *(2)*

ii) Calculate the force of contact that accelerated the ball. *(2)*

b) Calculate the ball's kinetic energy as it leaves the club. *(2)*

c) Calculate the distance that the club was in contact with the ball, assuming that all of the work done by the club was used to accelerate the ball. *(3)*

d) Explain in practice why the contact distance might be a little more than the distance you calculated in part (c). *(2)*

16 An attraction at a theme park is the water plunge as shown in Figure 10.22. The riders fall down a slope of 17 m in height and splash into a pool at the bottom, which slows them down to rest over a distance *d*.

The mass of a boat and its three occupants is 1300 kg.

a) Calculate the change in gravitational potential energy as the boat falls through 17 m. *(2)*

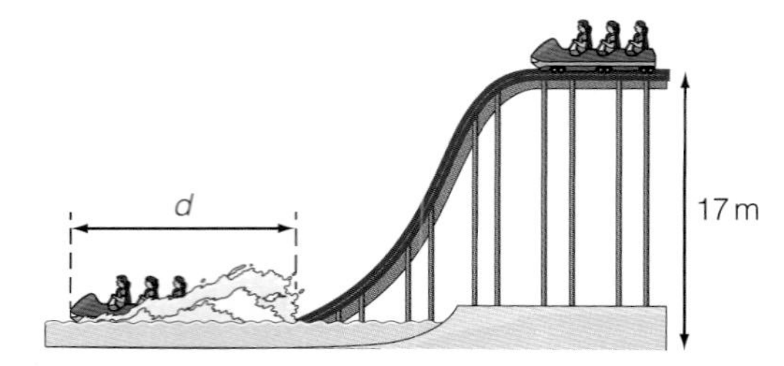

Figure 10.22 The water plunge.

b) i) Calculate the maximum speed of the boat at the bottom. (2)

ii) Explain why the speed of the boat may be less than your calculated values. (1)

The boat is now loaded with more passengers so that its mass is 1500 kg.

c) Explain carefully the effect of the larger mass on:

i) the final speed at the bottom

ii) the distance d over which the boat slows to rest. (4)

17 Figure 10.23 shows how the drag forces on a car depend on its speed.

A student observes that it will save petrol to drive the car at $20\,m\,s^{-1}$ rather than $30\,m\,s^{-1}$.

a) Explain how the graph backs up the student's comment. (3)

b) Calculate the power generated by the car to drive at a steady speed of $30\,m\,s^{-1}$. (2)

The car engine is 27% efficient when driving at a speed of $30\,m\,s^{-1}$.

c) i) Calculate the power input to the engine at $30\,m\,s^{-1}$. (2)

ii) Explain what happens to the wasted power. (1)

The petrol in the car stores 32 MJ of chemical potential energy per litre.

d) Calculate how much petrol is used in 1 hour if the car drives at a constant speed of $30\,m\,s^{-1}$. (3)

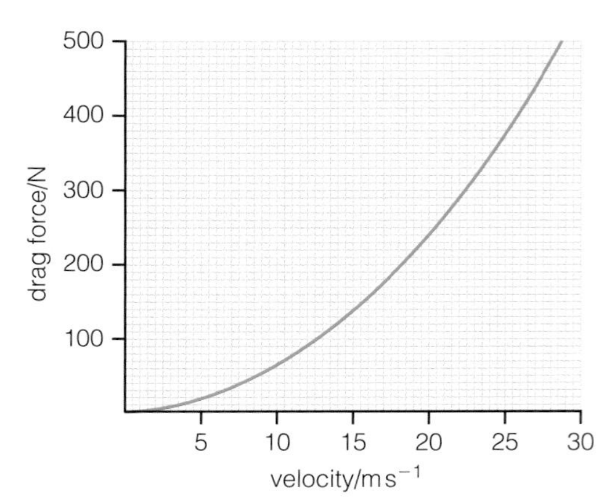

Figure 10.23 How the drag forces on a car depend on its speed.

18 The extract below is taken from a scientific article. Read it and answer the questions that follow.

'A crater in the Arizonian desert is thought to have been caused by a meteor impact about 50 000 years ago. The crater measures 1200 m across and is about 200 m deep; the volume of the crater is about $350\,000\,000\,m^3$, and the density of rock in the region is about $3000\,kg\,m^{-3}$. There is evidence from the surrounding region that the debris from the impact must have been thrown 5 km into the air before it settled back to earth. Meteors entering the Earth's atmosphere can travel as fast as $14\,km\,s^{-1}$, therefore they possess a great amount of kinetic energy.'

a) Use the data in the article to calculate the mass of rock and earth removed from the crater. (2)

b) Calculate the gravitational potential energy gained by the material as it rose 5 km above the plain. (2)

c) Calculate the minimum mass of the meteor that made this crater. (2)

d) Suggest why the meteor was probably bigger than the mass you calculated in part (c). (1)

19 A sprinter uses blocks to help him accelerate quickly after the starting gun goes. Figure 10.24 shows how the horizontal force from the blocks acts on the sprinter as he starts the race. The sprinter has a mass of 86 kg.

a) Calculate the work done by the sprinter while he is in contact with the blocks. *(3)*

b) Estimate the velocity of the sprinter just after he left the blocks. *(3)*

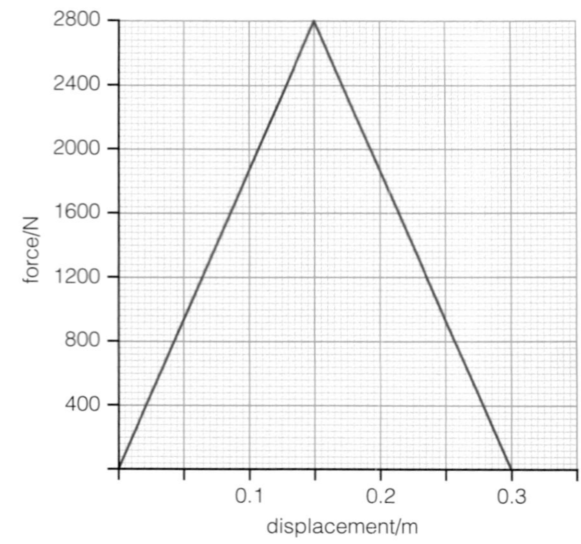

Figure 10.24 How the horizontal force from the blocks acts on the sprinter as he starts the race.

Stretch and challenge

20 A car accelerates using constant power. It takes 15 seconds to reach a speed of 40 km h^{-1}. How long does it take the car to reach a speed of 80 km h^{-1}?

21 Figure 10.25 shows how the force of attraction varies between bar magnets of length 6 cm and mass 50 g.

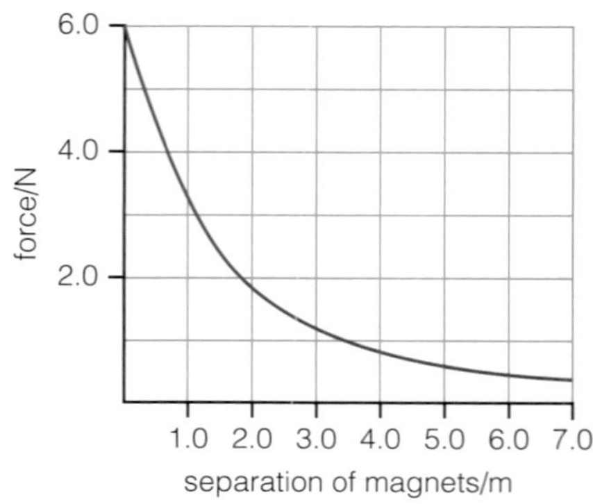

Figure 10.25 How the force of attraction varies between the bar magnets.

a) Two magnets are placed in contact. Use the graph to show that the work done to move the magnets apart to a separation of 5 cm is about 0.1 J.

b) Since work is done to pull the magnets apart, energy has been transferred. Where has this energy been transferred to?

c) While separated by 5 cm, the two magnets are released and they move back together. Each magnet experiences a constant frictional force of 0.4 N. What is the speed with which each magnet is moving, relative to the table, when they collide?

d) One magnet is now turned round and the magnets are again put into contact, so that two north poles touch each other. By using the graph and the information in the text, estimate how far the magnets move apart before they stop moving.

e) The magnets are placed 10 cm apart with a north pole facing a south pole. What is the closest distance they can be moved towards each other without the magnetic forces pulling the magnets together?

22 a) The unit of pressure is the pascal, which is equivalent to 1 N m^2. Show that the unit 1 N m^{-2} is also equivalent to 1 J m^{-2}.

b) When a champagne cork is released from a bottle, the cork (with a mass of 15 g) is launched 10 m into the air. Use the idea above to calculate the excess pressure in a 75 cl bottle.

Momentum

PRIOR KNOWLEDGE

- Momentum is defined as mass multiplied by velocity: $p = mv$.
- Momentum is a vector quantity.
- Newton's second law of motion states that resultant force is the mass multiplied by the acceleration: $F = ma$.
- Newton's third law of motion states that to every force there is an equal and opposite force. Such paired forces are of the same type, act along the same line, and act on separate bodies.
- Acceleration is the change of velocity divided by time: $a = \Delta v/\Delta t$
- Kinetic energy = $\frac{1}{2}$ mass × (velocity)^2
- The principle of conservation of energy states that energy cannot be created or destroyed, but it can be transferred from one type of energy to another.

TEST YOURSELF ON PRIOR KNOWLEDGE

1 Explain why the definition of momentum shows it must be a vector quantity.

2 Calculate the momentum of a person of mass 80 kg running with a velocity of $8\,m\,s^{-1}$. State the units of momentum.

3 A car experiences a resultant forwards force of 1470 N. It accelerates from a speed of $6\,m\,s^{-1}$ to $14\,m\,s^{-1}$ in 4 seconds. Calculate the car's mass.

4 You hold a book with a weight of 2 N in your hand. How big is the equal and opposite force to that weight, and on what body does that force act?

5 A car with a mass of 1200 kg travelling at a speed of $25\,m\,s^{-1}$ applies its brakes and comes to a halt.

a) Calculate the kinetic energy that is transferred to other types of energy.

b) Describe the energy transfers in this process.

Introducing momentum

Momentum is a useful quantity in physics because the amount of momentum in a system always remains the same provided no external forces act on that system. This principle allows us to predict what will happen in a collision or an explosion.

In the example of the exploding firework, chemical potential energy is transferred to thermal energy, light energy and kinetic energy of the exploding fragments. However, during the explosion the momentum remains the same. Momentum is a vector quantity; the momentum of

Momentum The product of mass and velocity. The unit of momentum is $kg\,m\,s^{-1}$. The symbol p is used for momentum.
$p = mv$

a fragment travelling in one direction is balanced by the momentum of a fragment travelling in the opposite direction. The same laws of physics apply equally to all masses whether they are planets, objects we meet every day or the sub-atomic particles studied by nuclear physicists.

TIP

When a ball hits a wall with momentum p and bounces back in the opposite direction with momentum $-p$, the change of momentum is $2p$.

Figure 11.1 An understanding of momentum and energy enables us to explain what is going on here.

EXAMPLE

A ball of mass 0.1 kg hits the ground with a velocity of $6\,m\,s^{-1}$ and sticks to the ground. Calculate its change of momentum.

Answer

Using the formula for momentum change:

$$\Delta p = m\Delta v$$
$$= 0.1\,kg \times 6\,m\,s^{-1}$$
$$= 0.6\,kg\,m\,s^{-1}$$

EXAMPLE

A ball of mass 0.1 kg hits the ground with a velocity of $6\,m\,s^{-1}$ and bounces back with a velocity of $4\,m\,s^{-1}$. Calculate its change of momentum.

Answer

Using the formula for momentum change, where mu is the momentum of the ball when it hits the ground and mv is its momentum when it begins to bounce back upwards:

$$\Delta p = mv - mu$$
$$= 0.1\,kg \times 4\,m\,s^{-1} - (-0.1\,kg \times 6\,m\,s^{-1})$$
$$= 1.0\,kg\,m\,s^{-1}$$

In this case the momentum before and the momentum after are in opposite directions so one of them must be defined as a negative quantity; we have defined upwards as positive and downwards as negative.

Momentum and impulse

Newton's second law can be used to link an applied force to a change of momentum:

$$F = ma$$

Substituting

$$a = \frac{\Delta v}{\Delta t}$$

gives

$$F = \frac{m\Delta v}{\Delta t}$$

or

$$F = \frac{\Delta(mv)}{\Delta t}$$

TIP
Note that momentum can be expressed in $kg\,m\,s^{-1}$ or N s.

This can be put into words as follows: force equals the rate of change of momentum. This is a more general statement of Newton's second law of motion.

The last equation may also be written in the form:

$$F\Delta t = \Delta(mv)$$

or

$$F\Delta t = mv_2 - mv_1$$

Impulse The product of force and time. The unit of impulse is N s.

where v_2 is the velocity after a force has been applied and v_1 the velocity before the force was applied. The quantity $F\Delta t$ is called the impulse.

Figure 11.2

EXAMPLE

A gymnast is practising her skills on a trampoline.

She lands on the trampoline travelling at $8\,m\,s^{-1}$ and leaves the trampoline at the same speed. Her mass is 45 kg and she is in contact with the trampoline for 0.6 s. Calculate the average force acting on the gymnast while she is in contact with the trampoline.

Answer

$$F = \frac{mv_2 - mv_1}{\Delta t}$$

$$= \frac{45\,kg \times 8\,m\,s^{-1} - (45\,kg \times -8\,m\,s^{-1})}{0.6s}$$

$$= 1200\,N$$

Gymnasts use trampolines to reach and then fall from considerable heights, for example 5 m. If you jumped from 5 m and landed on a hard floor, you would hurt yourself and might even break a bone in your foot, ankle or leg. The equation

$$F = \Delta(mv)/\Delta t$$

helps you to understand why.

When you fall you have an amount of momentum that is determined by how far you fall. The force on you when you stop moving depends on the time interval, Δt, in which you stop. On a trampoline Δt is long, so the force is small; on a hard floor Δt is short and the force much larger.

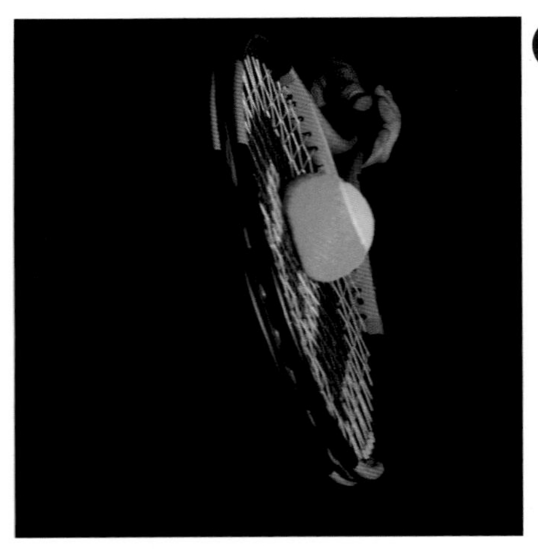

Figure 11.3 An impulsive force: the momentum of the tennis ball is changed by a large force acting for a short time.

TEST YOURSELF

1 Use the ideas of impulse and change of momentum to explain the following:
  a) why you bend your legs when you jump from a wall on to the ground
  b) why hockey players wear shin pads to protect their legs
  c) why you move your hands backwards when you catch a fast-moving ball coming towards you.

2 A gymnast of mass 45 kg jumps from a wall of height 3 m. When she lands, her legs stop her moving in 0.2 s.
  a) Calculate her momentum on landing.
  b) Calculate the force on her legs when she lands.

Car safety

The idea of impulse is vital in designing cars safely. Figure 11.4 shows two force–time graphs for passengers A and B in a high-speed car crash. Marked on the graph is a small area $F\Delta t$; this is equal to the change in momentum in that time interval, $m\Delta v$.

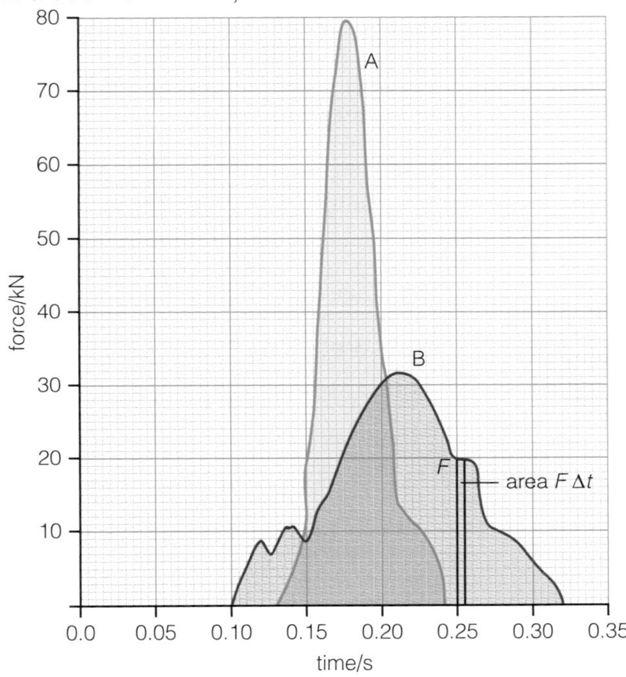

Figure 11.4 Force–time graphs for two passengers in a car crash.

So the total change of momentum of one of the passengers in the crash is the sum of all the small areas: $\Sigma F\Delta t$. Thus:

change of momentum = area under the force–time graph

The two passengers have different masses, so the areas under each graph are different. However, passenger B was strapped in by their safety belt and was stopped in the time it took the crumple zones at the front of the car to buckle. Passenger A, in the back of the car, was not restrained and was stopped as they hit the seat in front of them. Passenger A stopped in a shorter time, so the maximum force on them was much greater.

TIP
It is useful to remember that the area under a force–time graph equals the change of momentum.

Figure 11.5 In this test the dummy is protected by an airbag, and the crumple zone at the front of the car allows time for the passengers to slow down.

TEST YOURSELF

3 a) Explain why crumple zones and seat belts are vital safety measures for passengers in cars.

b) If a helicopter crashes it is most likely that the impact will take place on the bottom of the craft. Explain what safety features should be built into the seats in the helicopter.

4 In an impact, a person who experiences an acceleration of 300 m s^{-2}, of short duration, will receive moderately serious injuries. However, a much larger acceleration is likely to inflict gravely serious injuries.

a) i) Explain why rapid accelerations cause injury.

ii) Explain why injuries are likely to be more severe if a very high acceleration or deceleration acts over a longer period of time.

In the car crash described in Figure 11.4, passenger A has a mass of 82 kg and passenger B a mass of 100 kg.

b) Describe the likely injuries to both passengers A and B.

c) By using the area under graph B, show that the car crashes at a speed of about 35 m s^{-1}.

5 A karate expert can break a brick by hitting it with the side of their hand.

a) They move their hand down at a velocity of 12 m s^{-1} and the hand bounces back off the brick with a velocity of 4 m s^{-1}. Calculate the impulse delivered to the brick if the hand and forearm have a mass of 2.2 kg.

b) The time of contact between the hand and brick is found to be 64 ms. Calculate the average force exerted by the hand on the brick.

c) Calculate the average force exerted by the brick on the person's hand.

ACTIVITY

Investigating varying forces

You can investigate how forces vary with time by using a 'force plate' that is connected to a data logger. Such plates use piezoelectric crystals, which produce a p.d. that depends on the pressure applied to the crystal.

A student investigates his reaction force on the ground in various activities. He uses a force plate and a data logger, which records the changes in force with time.

In his first activity, the student steps on to the plate in a crouching position; then he stands up, before he steps off the plate again. See Figure 11.6.

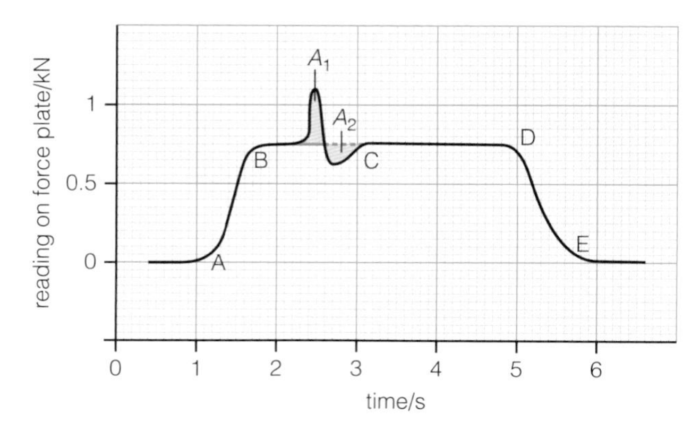

Figure 11.6

Answer the following questions, based on the student's investigation

1 a) Explain what is happening over each of the following times.
AB, BC, CD, DE
b) Explain why the areas A_1 and A_2 on the graph are equal.
c) The student now jumps on to the force plate. Sketch a graph to show how the reaction force changes now.

2 In the next activity the student measures his reaction force on the plate while running on the spot. This is recorded in Figure 11.7.
a) Use the graph to show that the student's change of momentum on each footfall is about $250\,kg\,m\,s^{-1}$.
b) Explain why his upward momentum as he takes off on one foot is $125\,kg\,m\,s^{-1}$.
c) The student's mass is 80 kg. Show that his centre of mass rises by a height of about 12 cm between each stride. (Here you will need to use the equations of motion that you learned in Chapter 8).
d) Sketch a graph to show how the horizontal component of the foot's reaction on the ground varies with time when the runner is moving at a constant speed.

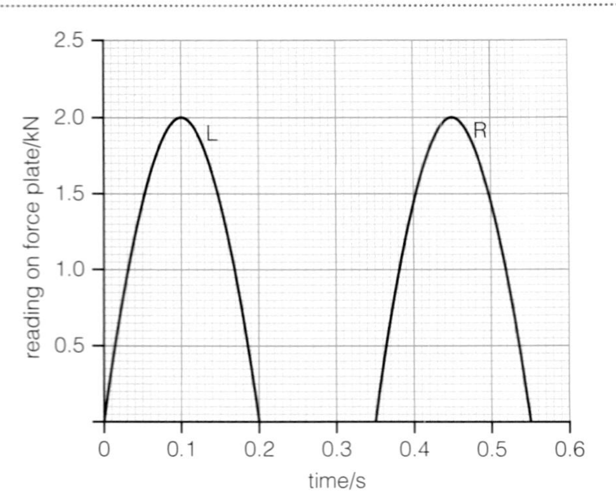

Figure 11.7

Conservation of linear momentum

Linear momentum The momentum of an object that moves only in one dimension.

We use the term **linear momentum** when we refer to collisions (or explosions) that take place in one dimension, i.e. along a straight line.

Figure 11.8 shows a demonstration of a small one-dimensional explosion. The head of a match has been wrapped tightly in aluminium foil. A second match is used to heat the foil and the head of the first match, which ignites and explodes inside the foil. The gases produced cause the foil to fly rapidly one way, and the matchstick to fly in the opposite direction. (It is necessary to use 'strike-anywhere' matches, the heads of which contain phosphorous. The experiment is safe because the match blows out as it flies through the air, but common-sense safety precautions should be taken – do this in the laboratory, not in a carpeted room at home, and dispose of the matches afterwards.)

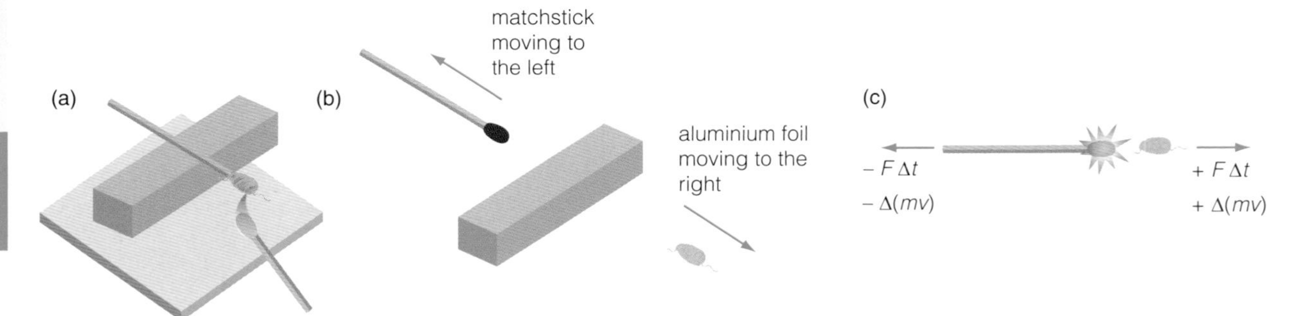

Figure 11.8 A simple experiment that demonstrates the conservation of linear momentum.

In all collisions and explosions, both total energy and momentum are conserved, but kinetic energy is not always conserved.

In this case, the chemical potential energy in the match head is transferred to the kinetic energy of the foil and matchstick, and also into thermal, light and sound energy.

As the match head explodes, Newton's third law of motion tells us that both the matchstick and foil experience equal and opposite forces, F. Since the forces act for the same interval of time, Δt, both the matchstick and foil experience equal and opposite impulses, $F\Delta t$.

Since $F\Delta t = \Delta(mv)$, it follows that the foil gains exactly the same positive momentum as the matchstick gains negative momentum.

We can now do a vector sum to find the total momentum after the explosion (see Figure 11.8c):

$$+\Delta(mv) + [-\Delta(mv)] = 0$$

So there is conservation of momentum: the total momentum of the foil and matchstick was zero before the explosion, and the combined momentum of the foil and matchstick is zero after the explosion.

Conservation The total momentum of two bodies in a collision (or explosion) is the same after the collision (or explosion) as it was before.

Collisions on an air track

Collision experiments can be carried out using gliders on linear air tracks. These can demonstrate the conservation of linear momentum. The air blowing out of small holes in the track lifts the gliders so that frictional forces are very small.

TIP

Momentum is always conserved in bodies involved in collisions and explosions provided no external forces act. This is the principle of conservation of momentum.

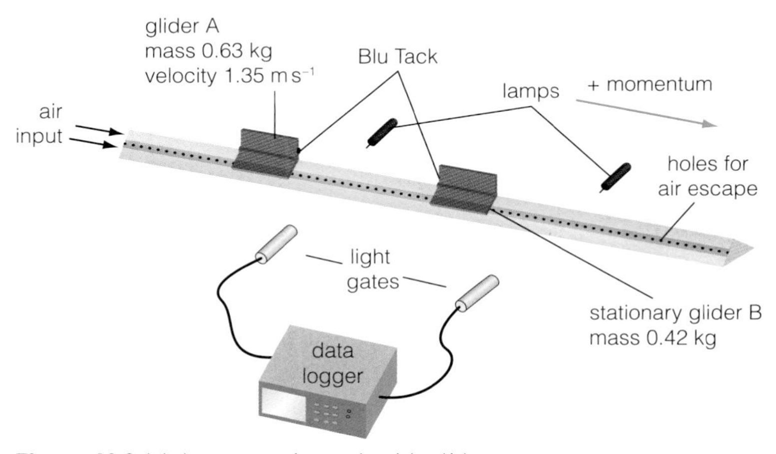

Figure 11.9 A laboratory air track with gliders.

EXAMPLE

In one air-track experiment a moving glider collides with a stationary glider. To get the gliders to stick together, a bit of BluTack can be stuck to each glider at the point of impact. We can use the principle of conservation of momentum to predict the combined velocity of the gliders after impact.

In Figure 11.9 we define positive momentum to the right.

Answer

momentum before $= m_A v_A + m_B v_B$

$= 0.63\,\text{kg} \times 1.35\,\text{m}\,\text{s}^{-1} + 0$

$= 0.85\,\text{kg}\,\text{m}\,\text{s}^{-1}$

The two gliders have the same velocity after the collision because they have stuck together.

momentum after $= (m_A + m_B)v$

$= (0.63\,\text{kg} + 0.42\,\text{kg}) \times v$

$= 1.05v\,\text{kg}\,\text{m}\,\text{s}^{-1}$

So, from conservation of momentum:

$1.05v\,\text{kg}\,\text{m}\,\text{s}^{-1} = 0.85\,\text{kg}\,\text{m}\,\text{s}^{-1}$

$v = 0.81\,\text{m}\,\text{s}^{-1}$

Both gliders move to the right at $0.81\,\text{m}\,\text{s}^{-1}$. A result such as this can be confirmed by speed measurements using the light gates and data logger.

EXAMPLE

In a second experiment, glider B is pushed to the left with a velocity of $2.7\,m\,s^{-1}$ and glider A is pushed to the right with a velocity of $1.5\,m\,s^{-1}$. What happens after this collision?

Answer

momentum before $= m_A v_A + m_B v_B$

$= 0.63\,kg \times 1.5\,m\,s^{-1} - 0.42\,kg \times 2.7\,m\,s^{-1}$

$= -0.19\,kg\,m\,s^{-1}$

Note that B has negative momentum, and that the total momentum of both gliders together is negative.

momentum after $= (m_A + m_B)v$

So

$-0.19\,kg\,m\,s^{-1} = 1.05v\,kg\,m\,s^{-1}$

$v = -0.18\,m\,s^{-1}$

So both trolleys move to the left with a velocity of $0.18\,m\,s^{-1}$.

TEST YOURSELF

6 A student decides to investigate the dynamics of the exploding match as shown in Figure 11.8. Using a data logger she discovers that the match starts to move at $7.1\,m\,s^{-1}$ immediately after the explosion.

a) The mass of the matchstick is 0.15 g and the mass of the aluminium foil is 0.06 g. Calculate the initial speed of the foil.

b) Calculate the kinetic energy of:

i) the aluminium foil

ii) the matchstick.

c) Calculate the minimum chemical potential energy stored in the match head before it explodes. Explain why the stored energy is likely to be greater than your answer.

d) The student discovers that the matchstick travels further than the foil. She expected the foil to travel further. What factors affect the distance travelled by the matchstick and the foil? (You may need to refer back to the work on drag in Chapter 8).

7 This question refers to the collisions between the two gliders on an air track as shown in, Figure 11.9. In each of the following cases, calculate the velocity of the gliders after they have collided.

a) Before the collision glider A is moving to the right at $2.0\,m\,s^{-1}$ and glider B is moving to the right at $1.0\,m\,s^{-1}$.

b) Before the collision glider A is stationary and glider B is moving to the left at $2.0\,m\,s^{-1}$.

c) Before the collision glider A is moving to the right at $1.0\,m\,s^{-1}$ and glider B is moving to the left at $1.5\,m\,s^{-1}$.

8 A student now investigates some more collisions between gliders on an air track but he makes changes to the apparatus. He replaces the BluTack in Figure 11.9 with two magnets that repel each other. Now when a collision takes place, the two gliders move independently after the collision.
Calculate the velocity of glider B after each of the following collisions.

a) Before the collision glider A is moving to the right at $2.0\,m\,s^{-1}$ and glider B is moving to the right at $1.0\,m\,s^{-1}$. After the collision glider A is moving at $1.2\,m\,s^{-1}$ to the right.

b) Before the collision glider A is stationary and glider B is moving to the left at $2.0\,m\,s^{-1}$. After the collision glider A is moving at $1.6\,m\,s^{-1}$ to the left.

c) Before the collision glider A is moving to the right at $1.0\,m\,s^{-1}$ and glider B is moving to the left at $1.5\,m\,s^{-1}$. After the collision glider A is moving at $1.0\,m\,s^{-1}$ to the left.

Momentum and energy

In **elastic** collisions kinetic energy is conserved.

In **inelastic** collisions kinetic energy is not conserved. Some or all of the kinetic energy is transferred to heat or other types of energy.

In collisions between two or more bodies, both momentum and energy are conserved. However, the total kinetic energy of the bodies does not always stay the same because the kinetic energy can be transferred to other types of energy. Collisions in which the kinetic energy of the particles is the same after the collision as it was before are described as elastic. Collisions in which kinetic energy is transferred to other forms of energy are described as inelastic. Most collisions on a large scale are inelastic, but collisions between atomic particles are often elastic.

TEST YOURSELF

To answer Questions 9 and 10 you may need to refer back to definitions
from Chapters 4 and 6.

9 Define the volt.

10 a) Calculate the speed of an electron that has been accelerated through a potential difference of 5000 V.

b) Which has the larger momentum, a proton that has been accelerated through a p.d. of 20 kV, or an electron that has been accelerated through 20 kV? ($m_p = 1.67 \times 10^{-27}$ kg; $m_e = 9.1 \times 10^{-31}$ kg; $e = 1.6 \times 10^{-19}$ C)

Some useful equations

Although momentum and kinetic energy are very different quantities, they are linked by some useful equations.

Momentum is given the symbol p.

$$p = mv$$

Kinetic energy is given the symbol E_k.

$$E_k = \tfrac{1}{2}mv^2$$

But E_k can also be written in this form:

$$E_k = \frac{m^2v^2}{2m}$$

giving

$$E_k = \frac{p^2}{2m}$$

It is important to remember that momentum is a vector quantity and kinetic energy is a scalar quantity. We give momentum a direction: for example, $+p$ to the right and $-p$ to the left. A vehicle travelling with momentum p has as much kinetic energy when travelling to the right as it does to the left.

To the right:

$$E_k = \frac{(+p)^2}{2m} = \frac{p^2}{2m}$$

To the left:

$$E_k = \frac{(-p)^2}{2m} = \frac{p^2}{2m}$$

When a vector is squared, it becomes a scalar quantity.

Elastic and inelastic collisions

Figure 11.10 shows an unfortunate situation: a van just fails to stop in a line of traffic and hits a stationary car; and they move forwards together. Is this an elastic or inelastic collision? Without doing any calculations, we know that this is an inelastic collision because the crash transfers kinetic energy to other forms – the car is dented, so work must be done to deform the metal, and there is noise. However, we can calculate the kinetic energy transferred as shown in the example below.

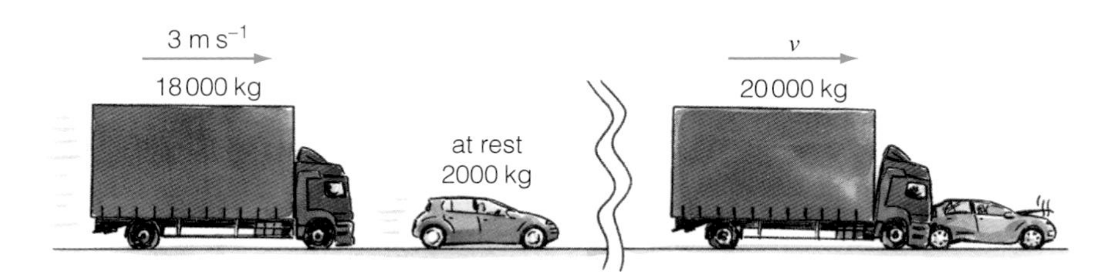

Figure 11.10 An inelastic collision.

In the world of atomic particles, collisions can be elastic because, for example, electrostatic charges can repel two atoms or nuclei without a transfer of kinetic energy to other forms.

EXAMPLE

In Figure 11.10 the momentum before the crash is equal to the momentum after the crash.

$$18\,000\,\text{kg} \times 3\,\text{m s}^{-1} + 0 = 20\,000\,\text{kg} \times v$$

$$v = 2.7\,\text{m s}^{-1}$$

Before the crash the kinetic energy of the van was:

$$E_k = \tfrac{1}{2} m_v v_v^2$$
$$= \tfrac{1}{2} \times 18\,000\,\text{kg} \times (3\,\text{m s}^{-1})^2$$
$$= 81\,\text{kJ}$$

After the collision the kinetic energy of the van and car together is:

$$E_k = \tfrac{1}{2}(m_v + m_c)v^2$$
$$= \tfrac{1}{2} \times 20\,000\,\text{kg} \times (2.7\,\text{m s}^{-1})^2$$
$$= 73\,\text{kJ}$$

So about 8 kJ of kinetic energy is transferred to other forms of energy.

TIP

In elastic collisions, kinetic energy is conserved. In inelastic collisions energy is transferred to other forms.

EXAMPLE

A helium nucleus of mass $4m$ collides head-on with a stationary proton of mass m. Use the information in Figure 11.11 to show that this is an elastic collision.

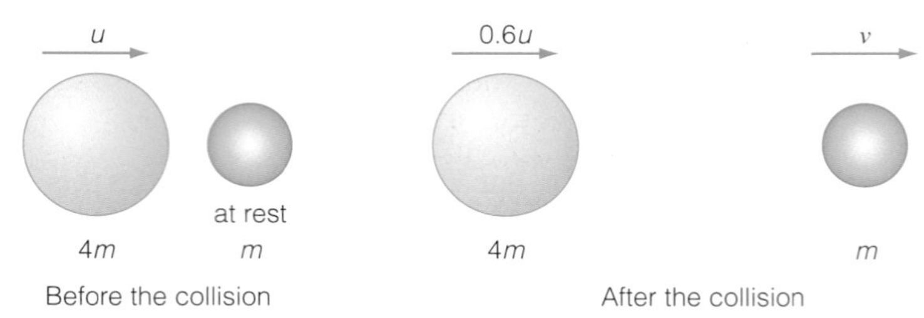

Figure 11.11

Answer

The velocity of the proton can be calculated as follows.

The momentum before the collision equals the momentum after the collision.

$$4mu + 0 = 4m \times 0.6u + mv$$
$$mv = 4mu - 2.4mu$$
$$v = 1.6u$$

The total kinetic energy before the collision was:

$$E_k = \tfrac{1}{2} \times 4mu^2$$
$$= 2mu^2$$

The total kinetic energy after the collision was:

$$E_k = \tfrac{1}{2} \times 4m(0.6u)^2 + \tfrac{1}{2} \times m(1.6u)^2$$
$$= 2m(0.36u^2) + \tfrac{1}{2}m(2.56u^2)$$
$$= 0.72mu^2 + 1.28mu^2$$
$$= 2mu^2$$

So the total kinetic energy is conserved in this collision. It is a useful rule to know that in an elastic collision the relative speeds of the two particles is the same before and after the collision.

TEST YOURSELF

11 Look at the diagram of two dodgem cars in a collision at a funfair.

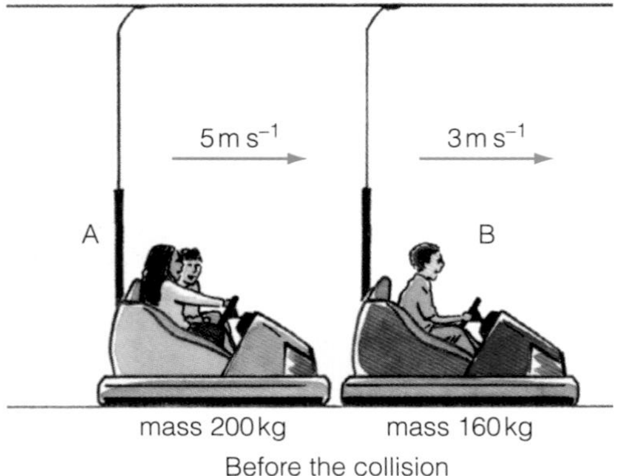

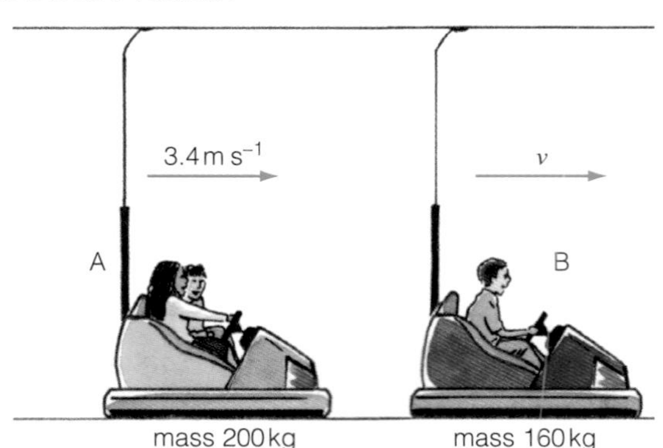

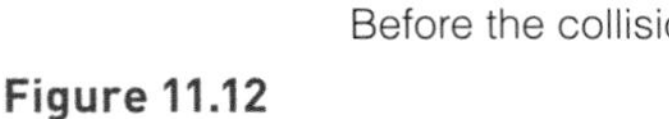

Figure 11.12

a) Use the information in Figure 11.12 to calculate the velocity of car B after the collision.

b) Explain whether this is an elastic or an inelastic collision.

12 Two ice skaters, each of mass 65 kg, skate towards each other. Each skater has a velocity of $1\,m\,s^{-1}$ relative to the ice. The skaters collide; after the collision each skater returns back along their original path with a velocity of $2\,m\,s^{-1}$.

a) Show that momentum is conserved in the collision.

b) Calculate the gain in kinetic energy of each skater during the collision.

c) Explain where the extra energy has come from.

13 Look again at Question 8. Show that each of the collisions between the two gliders is elastic.

14 A ball of mass 0.1 kg falls from a height of 5.0 m and rebounds to a height of 3.2 m. For this question assume $g = 10\,m\,s^{-2}$.

a) Calculate the velocity of the ball just before it hits the ground.

b) Calculate the velocity of the ball just after it has hit the ground.

c) The ball is in contact with the ground for 0.004 s. Calculate the average force that acts on the ball in this time.

d) The momentum of the ball has changed during this process. The principle of the conservation of momentum states that 'momentum is always conserved'. Explain how momentum is conserved in this case. [Hint: there is another body involved in this collision – what is its momentum?]

More advanced problems in momentum

The generalised form of Newton's second law of motion enables you to solve some more complicated problems.

force = rate of change of momentum

$$F = \frac{\Delta(mv)}{\Delta t}$$

EXAMPLE

Calculate the force exerted by water leaving a fire hose on the firefighter holding the hose. The water leaves with a velocity of $17\,m\,s^{-1}$; the radius of the hose is 3 cm; the density of water is $1000\,kg\,m^{-3}$.

Answer

The mass of water leaving the hose each second is $\frac{\Delta m}{\Delta t}$. This is equal to the density of water (ρ) × the volume of water, V, flowing per second:

$$\frac{\Delta m}{\Delta t} = \rho \frac{\Delta V}{\Delta t}$$

But the volume flow per second is the cross-sectional area of the hose (A) multiplied by the velocity of the water (v). So:

$$\frac{\Delta m}{\Delta t} = \rho A v$$

Thus the force, or change of momentum per second, is:

$$F = \frac{\Delta p}{\Delta t} = \rho A v \times v$$

$$= \rho A v^2$$

$$= 1000\,kg\,m^{-3} \times \pi \times (0.03)^2\,m^2 \times (17\,m\,s^{-1})^2$$

$$= 820\,N$$

This large force explains why a hose is sometimes held by two firefighters, and why large hoses have handles.

Before the collision

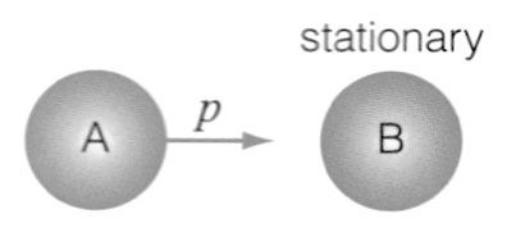

After the collision

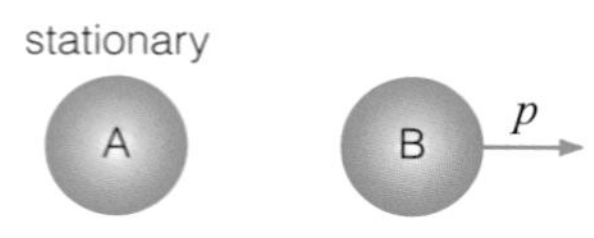

Figure 11.13 Head-on collision of two protons.

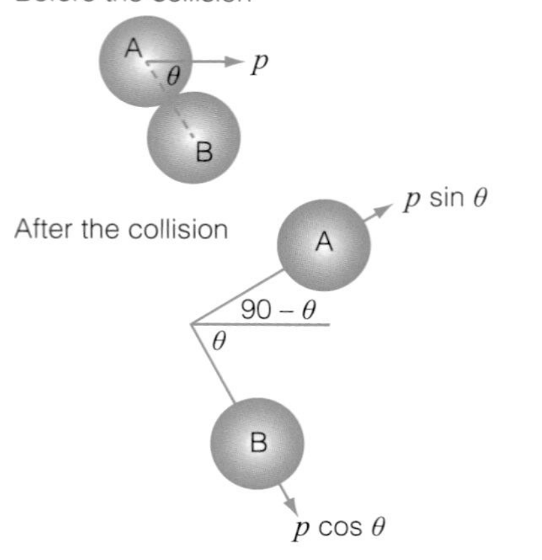

Figure 11.14 Non-head-on collision of two protons.

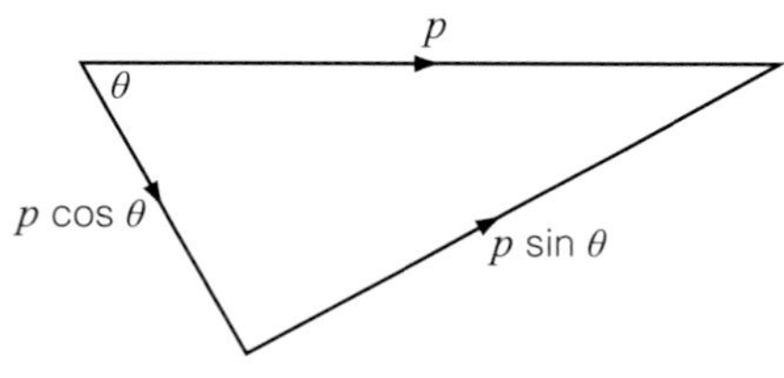

Figure 11.15 In vector terms, $p = p\cos\theta + p\sin\theta$.

Collisions in two dimensions

An interesting special result occurs when two atomic particles of the same mass collide elastically, when one of the particles is initially stationary.

If the two particles (protons for example) collide head-on, then the momentum and kinetic energy of the moving proton (A) is transferred completely to the stationary proton (B). See Figure 11.13.

In this way both momentum and kinetic energy are conserved. This only happens when the particles are of the same mass; in all other cases both particles will be moving after the collision.

What happens when two particles of the same mass collide, but the collision is not head-on, as shown in Figure 11.14? The momentum of particle A can be resolved into two components: $p\cos\theta$ along the line of collision and $p\sin\theta$ perpendicular to the line of the collision. As in Figure 11.13, all of the momentum of A along the line of the collision (here $p\cos\theta$) is transferred to particle B. This leaves particle A with the component $p\sin\theta$, which is at right angles to the line of the collision.

So, in a non-head-on elastic collision between two particles of the same mass, they always move at right angles to each other.

Figure 11.15 shows how momentum is conserved as a vector quantity.

Kinetic energy is also conserved. Before the collision the kinetic energy is:

$$\frac{p^2}{2m}$$

After the collision the kinetic energy of the two particles is:

$$\frac{p^2}{2m}\cos^2\theta + \frac{p^2}{2m}\sin^2\theta$$

However, since $\cos^2\theta + \sin^2\theta = 1$, the kinetic energy after the collision is the same as before, $\frac{p^2}{2m}$.

TEST YOURSELF

15 An alpha particle with a mass of about 4 amu is emitted from a uranium nucleus, mass about 238 amu, with a kinetic energy of 4.9 MeV.
1 amu (atomic mass unit) = 1.66×10^{-27} kg;
e = 1.6×10^{-19} C

a) Calculate the momentum of the alpha particle.

b) Calculate the momentum of the nucleus after the alpha particle has been emitted.

c) Calculate the kinetic energy of the nucleus after the alpha particle has been emitted.

d) Discuss where the kinetic energy has come from in this process.

16 A rocket of mass 400 000 kg takes off on a voyage to Mars. The rocket burns 1600 kg of fuel per second ejecting it at a speed of 2600 m s^{-1} relative to the rocket.

a) Calculate the force acting on the rocket due to the ejection of the fuel.

b) i) Calculate the acceleration of the rocket at take-off.

ii) Calculate the acceleration of the rocket 90 seconds after take-off.

Practice questions

1 Which of the following is a correct unit for momentum?

A N s **C** N s^{-1}

B kg m s^{-2} **D** kg m^{-1} s

2 A body is in free fall. Which of the following quantities is equal to the rate of increase of the body's momentum?

A kinetic energy **C** weight

B velocity **D** decrease in potential energy

3 A tennis ball of mass 60 g is travelling at 35 m s^{-1}. A player hits the ball back in the opposite direction, with a velocity of 45 m s^{-1}. What is the magnitude of the impulse which the tennis racket exert on the ball?

A 2.4 Ns **C** 0.6 Ns

B 4.8 Ns **D** 2.7 Ns

4 An alpha particle has kinetic energy of 8.0 MeV. The mass of the alpha particle is 6.8×10^{-27} kg; the charge on an electron is 1.6×10^{-19} C. Which of the following gives the correct value for the particle's momentum?

A 1.8×10^{-22} kg m s^{-1} **C** 2.6×10^{-22} kg m s^{-1}

B 1.8×10^{-19} kg m s^{-1} **D** 1.3×10^{-19} kg m s^{-1}

5 A lorry travelling at 5.5 m s^{-1} collides with a stationary car. After the collision, the lorry and car move forwards together. Use the information in the diagram below to calculate which of the following values is the correct velocity of the car and lorry after the collision.

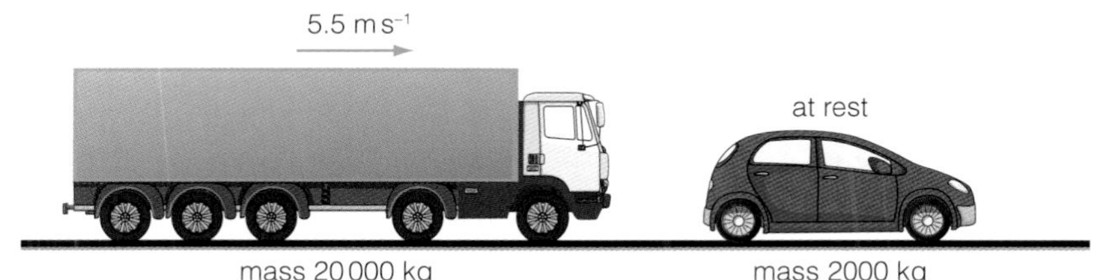

Figure 11.16

A 5.5 m s^{-1} **C** 5.0 m s^{-1}

B 2.75 m s^{-1} **D** 1.5 m s^{-1}

6 In the collision described in Question 5, some of the kinetic energy of the lorry is transferred to other forms, such as heat and sound. Which of the following gives the correct value for the change of kinetic energy during the crash?

A 300 000 J **C** 150 000 J

B 275 000 J **D** 27 500 J

7 A snooker ball of mass 160 g collides elastically with the cushion of a snooker table. Use the information in the diagram below to calculate which of the following is the correct value for the change of momentum of the ball in the direction normal (perpendicular) to the cushion.

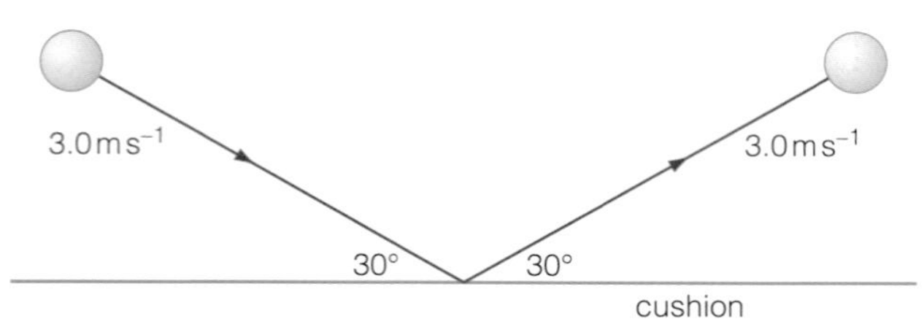

Figure 11.17

A 0 **C** 0.42 kg m s^{-1}

B 0.24 kg m s^{-1} **D** 0.48 kg m s^{-1}

8 A driver passing through a town sees a hazard and brings his car to an emergency stop. The driver has a mass of 80 kg. The graph shows how the force exerted on him by his seat belt varies while the car slows down. Use the graph to calculate which of the following gives the correct value for the initial speed of the car before the brakes were applied.

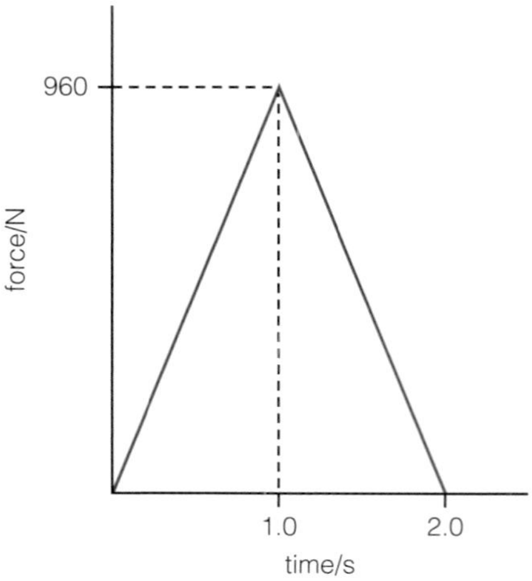

A 24 m s^{-1} **C** 12 m s^{-1}

B 16 m s^{-1} **D** 6 m s^{-1}

960
force/N
1.0
2.0
time/s

Figure 11.18

9 A rubber ball of mass 0.2 kg lands on a floor with a velocity of 6 m s^{-1}; it bounces back up with a velocity of 3 m s^{-1}. The ball is in contact with the floor for 0.06 s. Which of the following is the correct value for the average force that the floor exerts on the ball during the bounce?

A 60 N **C** 20 N

B 30 N **D** 10 N

10 Two trolleys collide as shown in the diagram below. After the collision they stick together. Use the information in the diagram to calculate which of the velocity values below is the correct value for the trolleys after the collision.

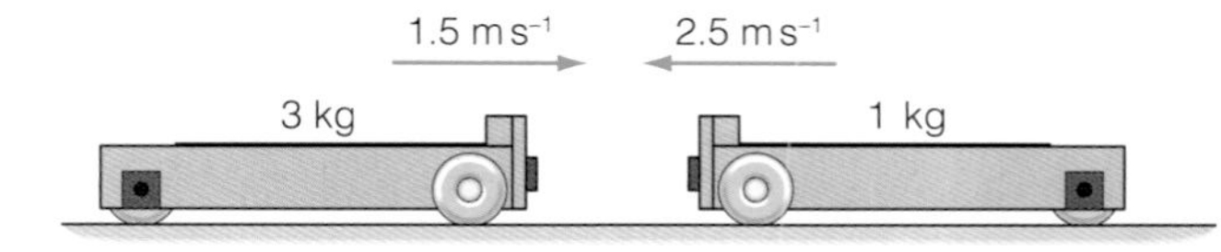

Figure 11.19

A 1.75 m s^{-1} to the right

B 1.75 m s^{-1} to the left

C 0.25 m s^{-1} to the left

D 0.50 m s^{-1} to the right

11 A car travelling at speed collides with a wall and comes to a halt. The force acting on the car during the crash is shown in the diagram.

a) Use the graph to show that the change of momentum of the car in the crash is approximately 26 000 N s. *(3)*

b) The car and its passengers have a mass of 1300 kg. Calculate the speed of the car before the crash. *(2)*

c) There were two passengers in the car. One was wearing a seat belt and the second was not. Explain why the one without the seat belt is more likely to receive serious injuries. *(3)*

d) Sketch a graph to show how the velocity of the car changed with time during the crash. *(3)*

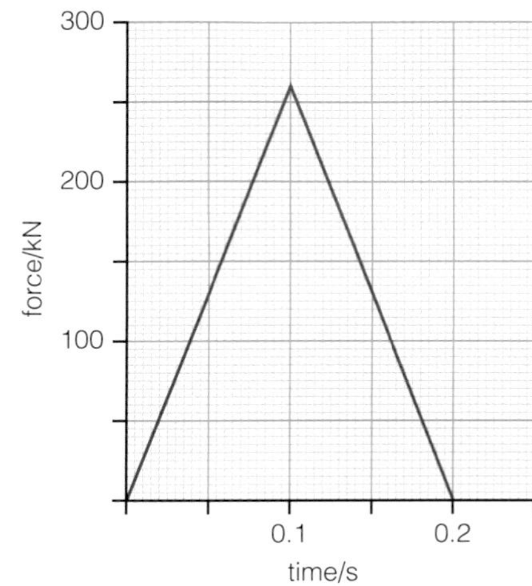

Figure 11.20

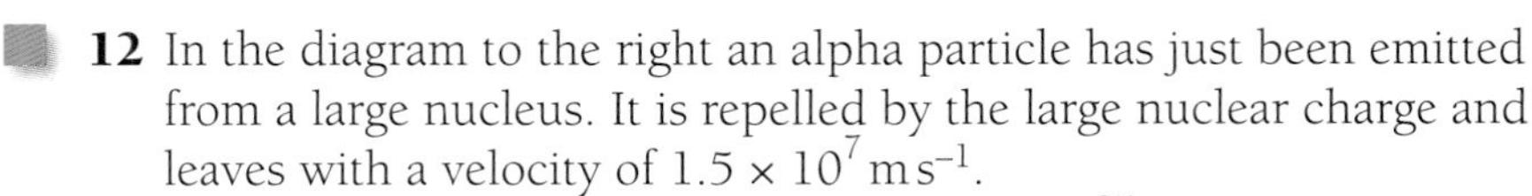

12 In the diagram to the right an alpha particle has just been emitted from a large nucleus. It is repelled by the large nuclear charge and leaves with a velocity of $1.5 \times 10^7\,\text{m s}^{-1}$.

a) The alpha particle has a mass of 6.8×10^{-27} kg and the nucleus a mass of 4.0×10^{-25} kg. Calculate the recoil velocity of the nucleus. *(3)*

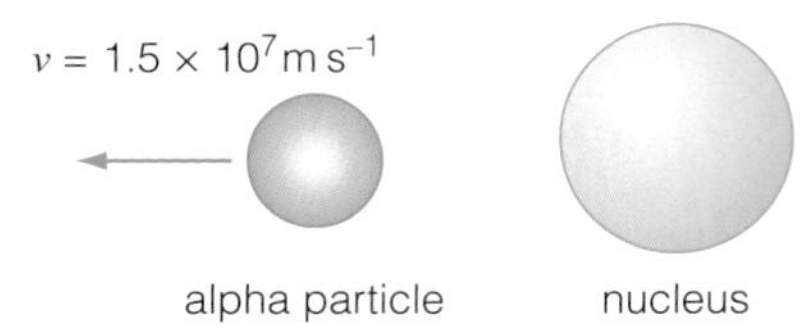

Figure 11.21

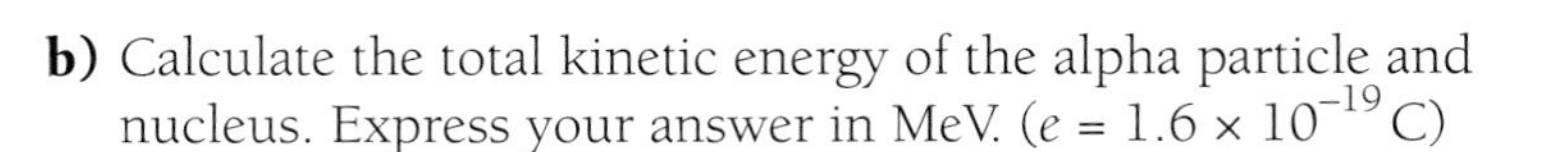

b) Calculate the total kinetic energy of the alpha particle and nucleus. Express your answer in MeV. ($e = 1.6 \times 10^{-19}$ C) *(5)*

c) Explain how momentum and energy have been conserved in this alpha decay process. *(2)*

13 The diagram below illustrates part of a ride at a funfair. One of the vehicles, with a mass of 240 kg, reaches the bottom of the slope reaching a speed of $14\,\text{m s}^{-1}$. It is slowed down to a speed of $5\,\text{m s}^{-1}$ by passing through a trough of water. The vehicle slows down over a time of 0.6 s. When it hits the water, a mass of water is thrown forwards with a velocity of $18\,\text{m s}^{-1}$.

Figure 11.22

a) Explain what is meant by the principle of conservation of momentum. *(2)*

b) Show that the mass of water thrown forwards by the vehicle is about 120 kg. *(3)*

c) Calculate the change in kinetic energy of the vehicle, and the increase in kinetic energy of the water, as the vehicle splashes into the trough. *(4)*

d) Explain how energy is conserved during this process. *(2)*

e) Calculate the average deceleration of the vehicle in the water. *(2)*

14 A soft rubber ball is allowed to fall on to the ground and bounce back up again. A data logger measures the speed as the ball falls. The graph shows how the ball's velocity changes with time.

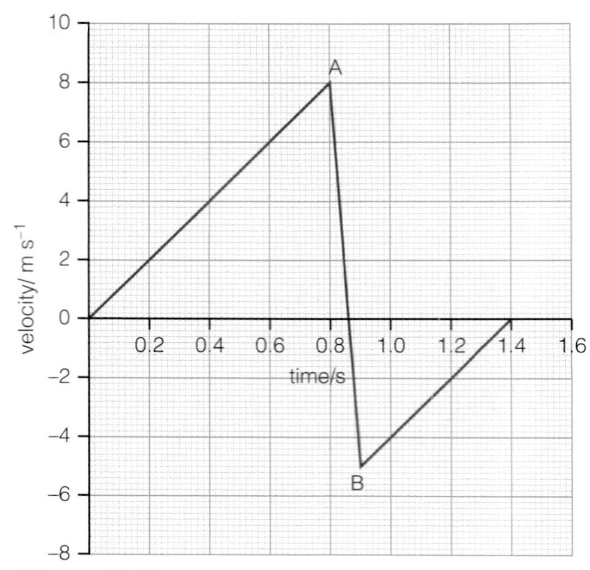

Figure 11.23

a) Use the graph to determine the time of the ball's bounce over the region AB. *(1)*

The ball has a mass of 0.08 kg.

b) Determine the momentum change of the ball as it hits the ground. *(3)*

c) Use your answer to part (b) to calculate the force exerted by the ground on the ball as it is in contact with the ground. *(2)*

d) How big is the force that the ball exerts on the ground while it is in contact? *(1)*

e) A hard rubber ball of the same mass, 0.08 kg, is now dropped from the same height. When it is in contact with the ground it exerts a larger force than the soft ball. Explain **two** changes you might see in the graph of velocity against time measured by the data logger. *(4)*

15 A steel ball of diameter 10 cm is allowed to fall from a height of 2 m so that it collides with a steel spike that is embedded in a piece of wood.

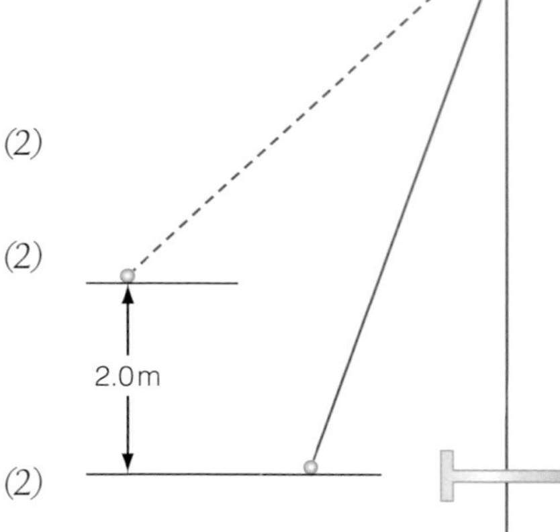

Figure 11.24

a) Steel has a density of 9000 kg m^{-3}. Show that the ball has a mass of about 4.7 kg. *(2)*

b) Determine the ball's speed as it comes into contact with the steel spike. *(2)*

The ball now moves forwards with the spike together. The spike has a mass of 2.6 kg.

c) Calculate the velocity of the ball and spike as they move forwards. *(2)*

d) By calculating the change of kinetic energy in the collision, explain whether this is an elastic or inelastic collision. *(4)*

The spike penetrates 3.5 cm into the wood before coming to rest.

e) Calculate the average force acting to slow the ball and spike. *(3)*

16 a) A neutron travelling with a velocity of 1.2×10^7 m s^{-1} is absorbed by a stationary uranium-238 nucleus. Calculate the velocity of the nucleus after the neutron has been absorbed. *(3)*

b) An alpha particle decays from a nucleus of polonium-208. The speed of the alpha particle is 1.5×10^7 m s^{-1}. Calculate the velocity of the polonium nucleus. The alpha particle has a relative atomic mass of 4. *(3)*

Stretch and challenge

17 The rotor blades of a helicopter push air vertically downwards with a speed of $5.0\,m\,s^{-1}$.

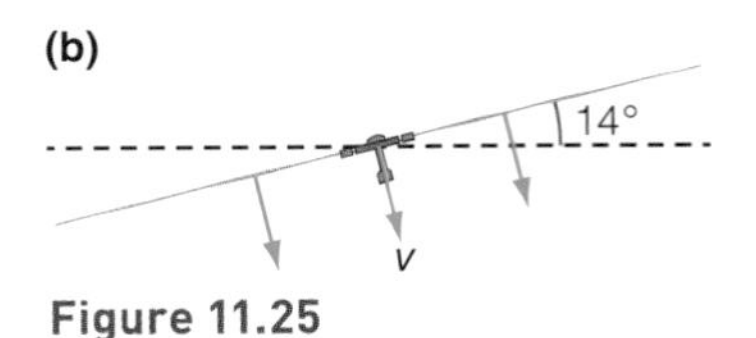

Figure 11.25

a) Use the information in diagram (a) and the value for the density of air, $1.2\,kg\,m^{-3}$, to calculate the momentum of the air pushed downwards per second.

b) When the helicopter pushes air downwards with a speed of $5.0\,m\,s^{-1}$, it hovers stationary. Calculate the helicopter's mass.

c) The helicopter's rotor blades are now tilted forwards at an angle of 14° to the horizontal as shown in diagram (b). The speed of rotation is increased so that the helicopter flies horizontally and accelerates forwards.

Calculate:

i) the speed of air being pushed away from the blades

ii) the initial horizontal acceleration of the helicopter.

18 A firework rocket is launched upwards. When it reaches its highest point, it explodes symmetrically into a spherical ball with 100 fragments, each of which has a mass of 20 g. The chemical energy stored in the explosives is 2 kJ; 80% of this energy is transferred to the kinetic energy of the fragments. Calculate the speed of each fragment.

19 The photo on the right shows a computer-enhanced image of a cloud chamber photograph. The event shown is the radioactive disintegration of a helium-6 nucleus, ${}^{6}_{2}He$, into a lithium-6 nucleus, ${}^{6}_{3}Li$, and a β-particle. The helium nucleus was originally stationary at point O. The β-particle is ejected along the thin broken red track OA, and the lithium-6 nucleus travels a short distance along the thick green track OB. Both particles travel in the plane of the paper. The β-particle's track is curved due to the presence of a strong magnetic field.

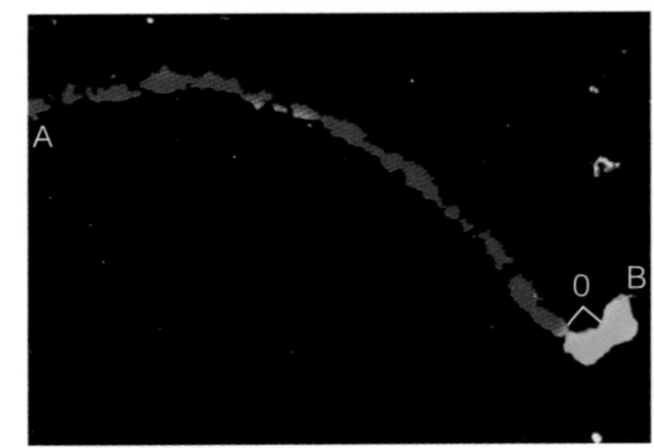

Figure 11.26

a) Deduce the direction of the magnetic field.

b) Explain why the lithium nucleus track is short and thick, and the β-particle track longer and thinner.

c) Judging from the cloud chamber tracks, momentum is not conserved in this decay process. Explain why this appears to be the case.

Since physicists accept the principle of conservation of momentum to be a universal law, it was suggested that another particle must be emitted in β-decay. This particle is the antineutrino, which is uncharged and leaves no track in the cloud chamber.

d) The β-particle has momentum p_1 and the lithium nucleus momentum p_2. Use the principle of momentum conservation to determine the momentum of the antineutrino. Express your answer in terms of p_1 and p_2, giving both a magnitude and direction.

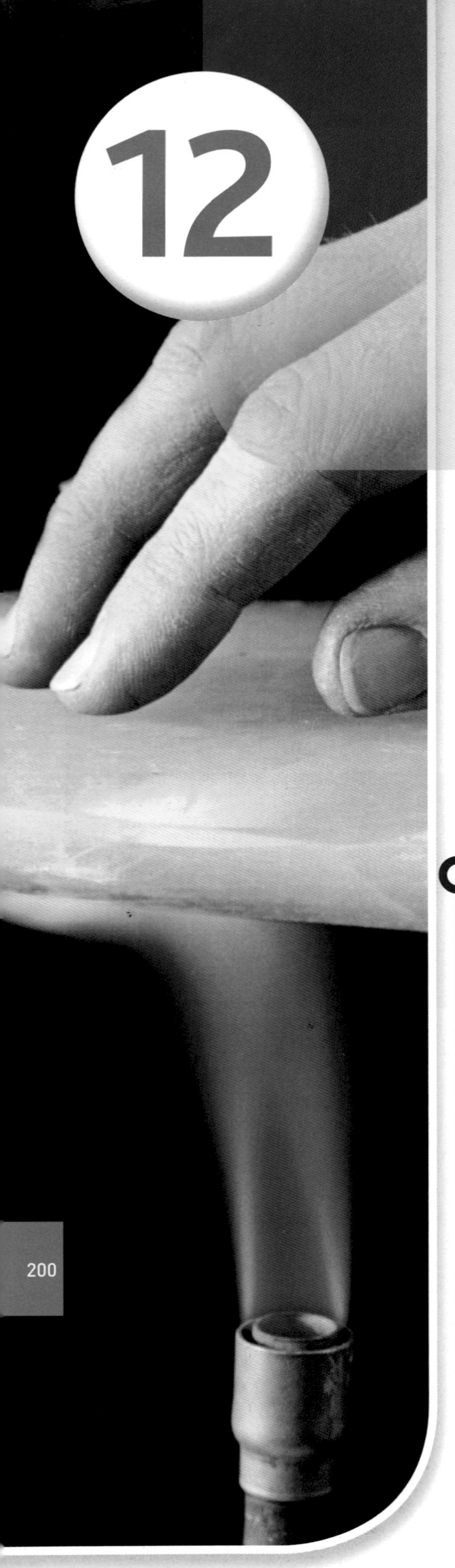

12 Properties of materials

PRIOR KNOWLEDGE

- A force acting on an object can cause it to change shape and stretch, squash or bend.
- Weight is the pull of gravity on an object and can be calculated using the formula:

  $W = mg$

  where W is the weight in N
  m is the mass of the object in kg
  g is the gravitational field strength in $\mathrm{N\,kg^{-1}}$.
- Work done (or energy transferred) by a force on an object depends on the magnitude of the force in the direction of motion of the object. This can be calculated using the formula:

  $W = Fs\cos\theta$

  where W is the work done in J
  F is the magnitude of the force in N
  s is the distance moved in the direction of the force in m
  θ is the angle between the force and the direction in which the object moves.

TEST YOURSELF ON PRIOR KNOWLEDGE

1 The *Voyager 2* spacecraft is travelling through space at a steady speed of approximately $23\,\mathrm{km\,s^{-1}}$ relative to Earth. Discuss whether work is being done by the spacecraft to travel at this speed, assuming there are no gravitational forces acting on *Voyager*.

Figure 12.1 *Voyager* 2, launched over 36 years ago and still travelling through space.

2 The Mars lander, *Curiosity*, has a mass of 899 kg. Calculate the lander's weight on Earth and on Mars. The gravitational field strength on Earth is $9.81\,N\,kg^{-1}$ and on Mars is $3.7\,N\,kg^{-1}$.

3 Alex is going on holiday. At the airport she has to lift her suitcase and carry it up a flight of stairs which has 20 steps, each 18 cm tall and 22 cm deep. Her suitcase has a mass of 20 kg.
Calculate the total amount of work Alex did in moving her suitcase (assume $g = 9.81\,N\,kg^{-1}$).

4 A girl is sledging on a snowy day. She pulls her sledge on to a path where the snow has melted. The girl does 12 000 J of work to pull the sledge 100 m along the path. Calculate the horizontal force she has used to pull the sledge along the path.

5 Figure 12.2 shows an example of a penguin toy race. Wheeled penguins are lifted by a motor up a moving staircase. When they get to the top they roll down the long slide and back to the staircase. The height of the staircase is 20 cm, and each penguin has a mass of 20 g.

a) Calculate the work done by the motor lifting one penguin from the bottom to the top of the staircase.

b) Compare the value you calculated in part (a) with the change in gravitational potential energy of the penguin as it goes down the slide.

Figure 12.2 A penguin toy race.

Bulk properties of solids

People have been using, and altering, the properties of materials since the Stone Age. Flint tools, samurai swords and glass beakers all made use of the strength, flexibility or optical properties of the materials they were created from.

Today, engineering materials researchers use their knowledge of the properties of materials to develop novel materials, and find new uses for old ones.

Figure 12.3 The pole used in pole vaulting must be lightweight, strong and have some flexibility. Vaulting poles were originally made from wood. Nowadays, they are usually made from layers of fibreglass and carbon-fibre materials.

Density

The density of a material is a measure of the mass per unit volume. It is given the symbol, ρ. Density is a useful quantity because it allows us to compare different materials.

$$\rho = \frac{m}{V}$$

where ρ is density in $kg\,m^{-3}$

m is mass in kg

V is volume in m^3

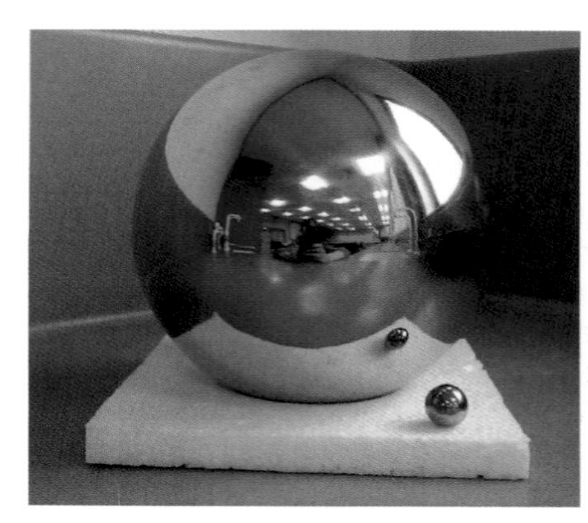

Figure 12.4 The large stainless steel globe is more massive than the small stainless steel ball bearing. The globe floats; the ball bearing doesn't. Why?

EXAMPLE

Calculate the density of the stainless steel globe shown in Figure 12.4 using this data.

Mass of globe = 0.30 kg

Diameter of globe = 0.2 m

Volume of a sphere = $\frac{4}{3}\pi r^3$

Answer

Volume of the globe = $4.19 \times 10^{-3}\,m^3$

$$\text{density} = \frac{\text{mass}}{\text{volume}}$$

$$= \frac{0.3\,kg}{4.19 \times 10^{-3}\,m^3}$$

$$= 72\,kg\,m^{-3}$$

The density of the globe depends on the density of the steel and the density of the air that is within the globe. Together, the combined (or average) density is much less than the density of water, which is $1000\,kg\,m^{-3}$; this is why the globe floats. The ball bearing, which has a density of $7800\,kg\,m^{-3}$, sinks.

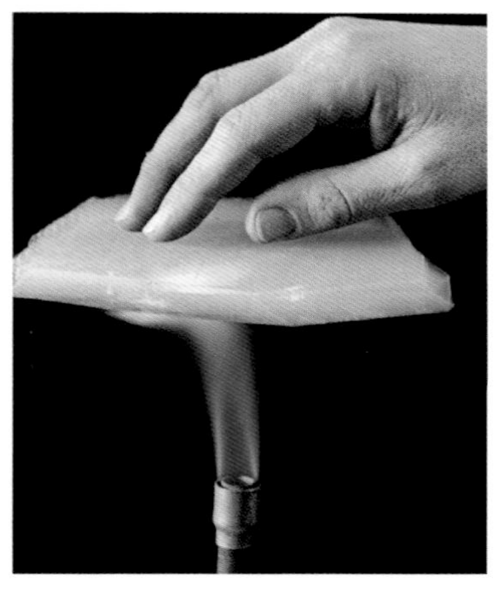

Figure 12.5 Aerogel: a very low-density material that is 99.8% air.

Aerogels have the lowest density of any solid. Table 12.1 compares the density of aerogel with other common materials.

Table 12.1 The density of some common materials

Material	Density ($kg\,m^{-3}$)
Aerogel	20
Air (at 20°C)	1.2
Water	1000
Silica glass	2200
Steel	7480–8000

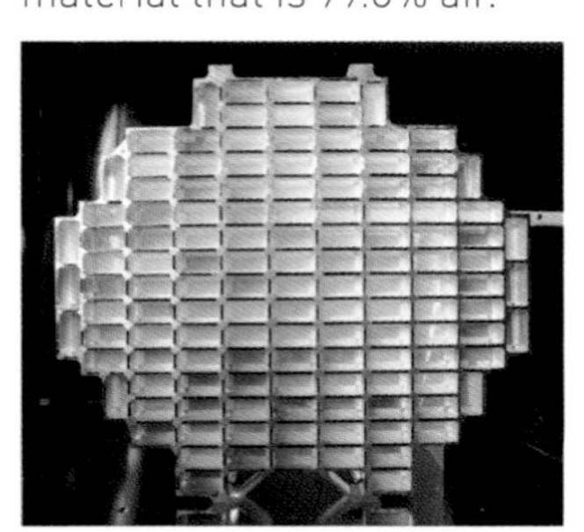

Figure 12.6 The sample collector from the *Stardust* space mission. Each rectangular block is made from aerogel.

Aerogel has been used in aerospace industries as a very low-density insulation material. Aerogel is a good insulator because it traps air and prevents it from moving. If air can circulate it will transfer heat through convection currents. NASA used aerogel on the Mars landers *Spirit* and *Opportunity* to provide thermal insulation of key components.

The NASA *Stardust* mission used aerogel to capture particles from the tail of comet Wild 2. The very low density of the aerogel ensured that the comet dust particles were slowed down without being damaged. The insulating properties of aerogel also meant that the dust particles were protected from the heat generated during re-entry of the sample collector into the Earth's atmosphere.

TEST YOURSELF

1 Suggest why different samples of steel can have different densities.

2 a) The stainless steel ball bearing shown in Figure 12.4 has a mass of 32.1 g and a diameter of 19.98 mm. Calculate the density of the stainless steel used to make the ball bearing.

b) The diameter of the ball bearing was measured using a pair of vernier calipers. These measure to an accuracy ±0.02 mm. Estimate the percentage uncertainty for the value of the diameter that was measured.

3 The globular cluster M13 contains approximately 300 000 stars. It has a diameter of 145 light years. A light year is the distance travelled by light in a year. Take the speed of light to be $3 \times 10^8\,\text{m}\,\text{s}^{-1}$.

a) If a typical star in the cluster has a mass of 2×10^{30} kg calculate the average density of the globular cluster.

b) Use your answer to part (a) to suggest why stars within the globular cluster rarely collide with each other.

4 A window made from silica glass has an area of 1.5 m × 2 m and is 3 mm thick. Calculate the mass of the window using the value for density given in Table 12.1.

5 A plumber uses a brass washer when fitting a tap. The washer is a flat ring of metal that is 0.8 mm thick. It has an internal diameter of 4 mm and an external diameter of 8.5 mm. The density of brass is $8.4\,\text{g cm}^{-3}$. Calculate the mass of the washer.

6 The aerogel sample collector from the *Stardust* mission used 132 aerogel blocks, each of which was 3 cm thick. The dust collector had a total surface area of $1039\,\text{cm}^2$ of aerogel.
Calculate the mass of aerogel used to capture the comet dust.

7 Estimate the density of a typical human body.

Hooke's law

Compressive forces are forces that tend to squeeze an object and reduce its size in the direction that the forces are applied. For example, a heavy weight placed on a column, together with the upwards force on the bottom of the column, will reduce the height of the column.

Tensile forces are forces that act to pull or stretch an object. The metal ropes holding up a lift have tensiles force acting on them due to the weight of the lift and the upwards pull from the ceiling on the end of the rope.

Materials can be characterised by the properties they show when forces are applied to them. Different materials will exhibit very different properties. When scientists are testing materials they may apply compressive forces or tensile forces.

In 1678, Robert Hooke wrote about his discovery of elasticity. He used his investigations into springs to develop a spring for use in the first portable timepiece, or a watch as they are now known.

Hooke realised that the extension of some springs shows a linear region for a range of applied forces. In other words, the extension was proportional to the force applied.

$$\text{extension} \propto \text{force}$$

$$\Delta l \propto F$$

Figure 12.7 A bungee trampoline makes use of the elastic properties of ropes. When a tensile force is applied to the ropes, they stretch. The greater the force applied, the more they stretch.

Spring constant, k, is a measure of how hard it is to bend or stretch a spring. A large spring constant means that the spring is stiff. k has units of $N\,m^{-1}$.

Extension is the length a material has stretched when a load is added. It is calculated by subtracting the original length of the material from the length when stretched.

Limit of proportionality is the endpoint of the linear section of a force–extension graph.

Elastic limit is the load above which a material is permanently deformed.

A material is said to be elastic when it returns to its original dimensions once the applied load is removed.

A material is said to be plastic when it is permanently deformed and does not return to its original dimensions once the applied load is removed.

How much a spring extends will also depend on the spring constant of the spring. This is a measure of how easy it is to stretch when a force is applied. A spring that extends a large amount for a force of 1 N is not as stiff as a spring that extends only a small amount for the same force. The spring constant, k, defines the stiffness of a spring as the force required for a unit extension of the spring.

We can therefore write Hooke's law as:

$$F = k\Delta l$$

where

F is the applied force in N

k is the spring constant in $N\,m^{-1}$

Δl is extension in m

Using the simple equipment shown in Figure 12.8, the spring constant of a spring can be measured. The spring is stretched using a tensile load. This is done by adding masses to a mass hanger. The extension of the spring is measured by noting the change in the position of the pointer against the rule.

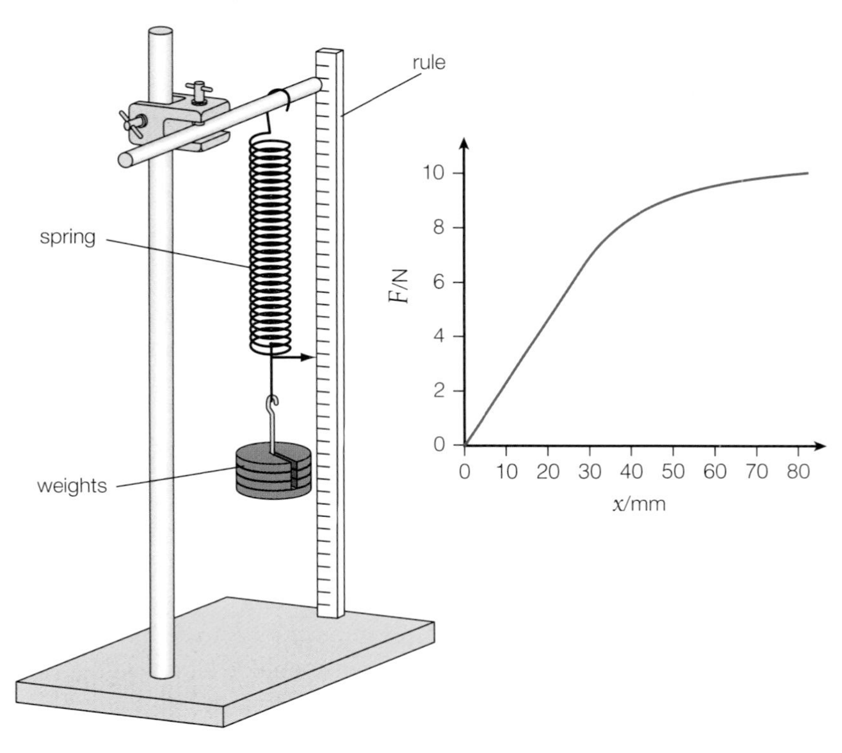

Figure 12.8 Experimental set-up for Hooke's law and graph.

When the results from the experiment are plotted, the graph initially has a linear region, as shown in Figure 12.8. Here the spring is still coiled and the extension is directly proportional to the load. When more masses are added, the spring starts to stretch more for each increase of load: the extension is no longer proportional to the load. The point at which this occurs is called the limit of proportionality.

The elastic limit of a material is the point below which the spring will return to its original length when the load is removed. The spring is showing elastic deformation. Above the elastic limit, the spring will be stretched out of shape and will not return to its original length. This is known as plastic deformation.

Figure 12.9 These two springs were the same length. One of them has gone beyond its elastic limit and will not return to its original length.

EXAMPLE

Springs are used in newton-meters to make measurements of force. One particular newton-meter is used to measure forces of up to 50 N. The length of the scale on the newton-meter is 10 cm long.

Calculate the value of the spring constant for the spring in the newton-meter.

Answer

$$F = k\Delta l$$

so

$$k = \frac{F}{\Delta l}$$

$$k = \frac{50\,\text{N}}{0.1\,\text{m}}$$

$$k = 500\,\text{N m}^{-1}$$

ACTIVITY

Extension of springs

Figure 12.10 shows two different combinations of springs: in series and in parallel. All the springs are the same length and have the same spring constant.

A student hangs a load from the bottom of each spring combination and then measures the extension. Her results are given in the table below.

Load (N)	Extension (cm)	
	Series combination	Parallel combination
0	0	0
0.5	2.5	0.7
1.0	6.2	1.5
1.5	9.5	2.6
2.0	13.6	3.4
2.5	17.5	4.4
3.0	21.4	5.3

1 Plot a graph of load against extension for the series and parallel combination of springs.

TIP

It is usual to plot the independent variable on the *x*-axis and the dependent variable on the *y*-axis. However, in this case we plot the load (independent variable) on the *y*-axis so that the spring constant, *k*, is given by the gradient.

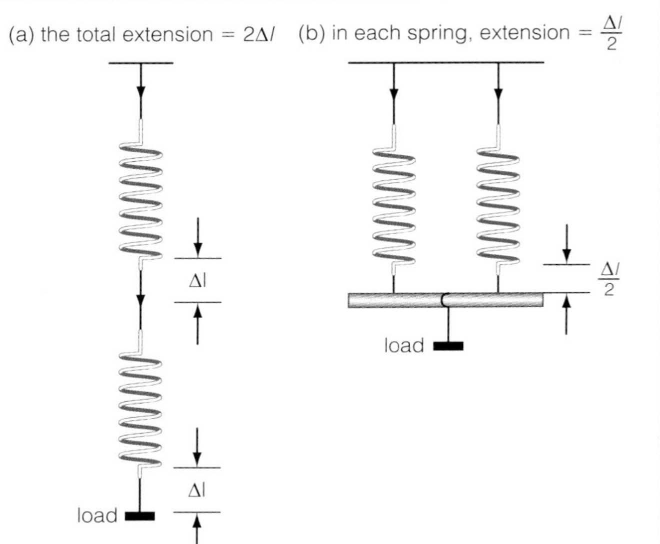

Figure 12.10 Springs in series and parallel.

2 Use your graph to calculate the effective spring constant for each combination of springs.

3 The student predicted that the effective spring constant of the series combination will be four times smaller than the effective spring constant of the parallel combination.
Use the data from your graph to show if this prediction is correct.

4 Predict the spring constant when two identical springs, which are placed in parallel, are then put in series with another pair of springs, which are also in parallel.

5 For the series combination of springs, calculate the percentage uncertainty in each measurement if the student measured the extension of the springs with an accuracy of ±0.1 cm.

TEST YOURSELF

8 The experiment shown in Figure 12.8 is repeated with a spring having double the value of spring constant. Describe, and explain, how the linear part of the graph would change.

9 A spring has a spring constant of $1\,\text{N}\,\text{cm}^{-1}$. The scale on a newton-meter is 10 cm long. Would the spring be suitable for a newton-meter used to measure forces up to 10 N?

10 An acrobat is climbing up an aerial rope that is suspended vertically from a high ceiling. The unstretched length of the rope is 5 m, and it stretches to 5.7 m when the acrobat is starts climbing up. The mass of the acrobat is 55 kg. Assuming that the rope obeys Hooke's law, calculate the spring constant of the rope.

11 Three identical springs with a spring constant k are hung in series with each other. A force, F, is applied. What is the effective spring constant, k_e, for this combination of springs in series.

12 a) A pair of springs in parallel are attached to the bottom of three springs in series. All five springs are identical and have a spring constant k. A force, F, is suspended halfway between the lower ends of the parallel springs. What is the effective spring constant, k_e, of this combination?

b) The three springs, originally in series, are now put in parallel and then attached, in series, to the pair of springs in parallel. Calculate the effective spring constant of the new combination.

Investigating wires and fibres

Nearly all materials show Hooke's law behaviour up to a point. This includes metals such as copper and steel, fibres such as cotton and silk, natural rubber and polymers.

The applied force beyond which materials no longer obey Hooke's law will be different for each material. In a school laboratory it is possible to investigate the properties of materials such as copper wire and nylon thread. These materials produce measurable extensions for an easily obtainable range of applied forces. Other materials, such as steel wire, require much larger forces and so specialist equipment is required.

ACTIVITY

Investigating the properties of copper wire

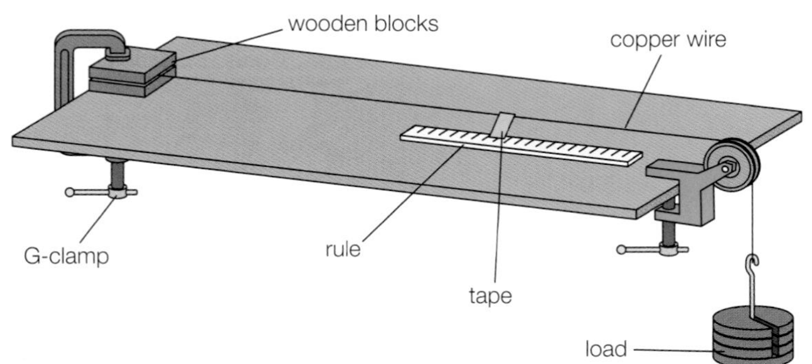

Figure 12.11 Stretching copper wire.

Figure 12.11 shows a simple experimental set-up that can be used to measure the properties of copper wire.

Wear safety glasses while carrying out the experiment. Place padding beneath the masses to break their fall.

A thin copper wire is clamped between two wooden blocks at one end of a bench. The other end of the wire is passed over a pulley.

A rule is placed on the bench next to the wire, and a strip of sticky tape is attached to the wire so that it touches the rule.

To begin with, carry out a preliminary test to determine what length of wire will produce a reasonable amount of extension as you increase the load. The tape marker must not go past the pulley. Choose a length of wire, and attach a mass hanger on to the end. Place masses on to the mass hanger and note how far the tape marker moves as you increase the load.

Once you have decided on the length of wire, set up the experiment again. Before attaching the mass hanger,

stick the tape marker on the wire so that it is at the lower end of the metre scale on the rule.

Carefully add a 100 g mass to the hanger and read the extension off the scale. Continue taking readings until the extension obtained for each increase in weight becomes much larger than before. At this point the wire has passed the elastic limit and is now behaving plastically. Continue to add masses, but allow time for the wire to continue to extend after each addition. Eventually the wire will break.

1 Plot a graph of force against extension for the copper wire. On your graph label:
   - the region where the wire obeyed Hooke's law
   - the elastic limit
   - the breaking point of the wire.
2 Calculate the spring constant for your wire in the region where the wire obeys Hooke's law.
3 Explain why using a longer wire gives a larger range of values for extension.
4 Estimate how accurately you were able to read the position of the marker on the rule. Explain what effect this will have on the value of spring constant you have calculated.

Extension

- Repeat the experiment for different thickness of copper wire and compare the results.
  Carry out a similar experiment for fibres such as cotton and nylon thread, and compare their behaviour with that of copper.

If you do not have data for this experiment then you can use these results for a copper wire.

Mass (kg)	0	0.5	1.0	1.5	2.0	2.5	3.0	3.5	4.0
Extension (mm)	0	0.8	1.6	2.3	3.2	4.2	5.3	6.8	9.6

Ductile materials can be formed into wires by stretching them. They show **ductility**.

A **brittle** material is one that shows little, or no, plastic deformation before breaking.

Wires obey Hooke's law because the bonds between the metal atoms act like springs. When the wire is stretched the bonds lengthen slightly. When the force is removed, the bonds return to their original length. However, if the force applied is too great, and the elastic limit exceeded, then the metal atoms will be able to move past one another and the wire lengthens. This is known as **ductility**, and is a very useful property as it allows metals to be formed into thin wires. The wire formed shows plastic behaviour and will not return to its original length when the force is removed. Ductile behaviour is also an example of plastic deformation.

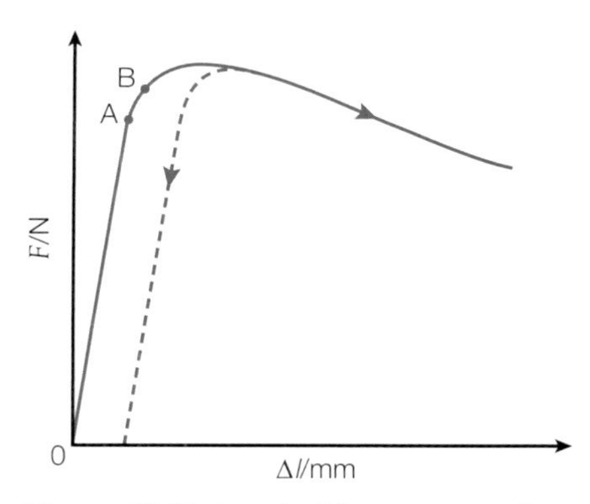

Figure 12.12 A typical force–extension graph for a copper wire. The wire shows plastic behaviour when it is stretched beyond the elastic limit.

In Figure 12.12 the dotted line represents the extension measured as the force is removed from the loaded wire. It can be seen that the wire has permanently lengthened because, even with no applied force, there is still a measurable extension.

Some materials do not show plastic behaviour but are **brittle** and break when the elastic limit is exceeded. Cast iron and glass are two examples of brittle materials. Figure 12.13 shows a typical force–extension graph for high-carbon steel, which is also a brittle material. The material fractures and breaks. It does not show plastic behaviour.

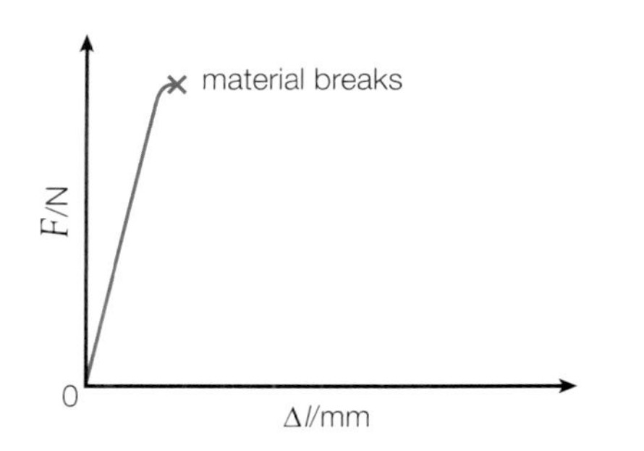

Figure 12.13 Force–extension curve for high-carbon mild steel.

The way in which ductile and brittle materials fracture is also different. In a ductile material, the sample of material will elongate and 'neck' before it breaks. On a force–extension graph, necking occurs in the plastic region of the graph. In a brittle material there is no change in the shape of the material because it does not undergo plastic behaviour. A straight break in the material is seen. Figure 12.14 shows the difference between the two types of fracture.

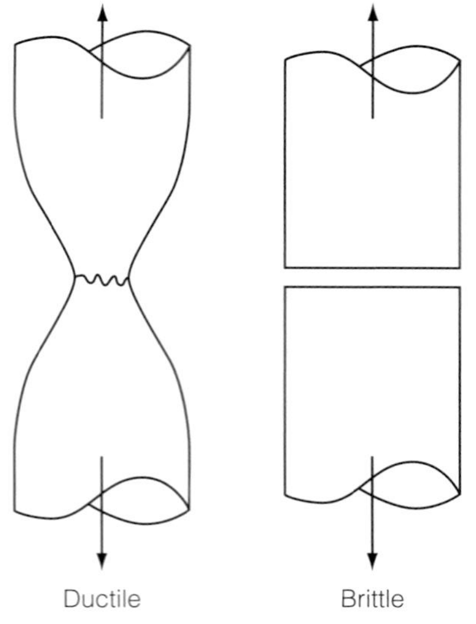

Figure 12.14 Ductile and brittle fracture.

Elastic strain energy is the energy stored by stretched materials.

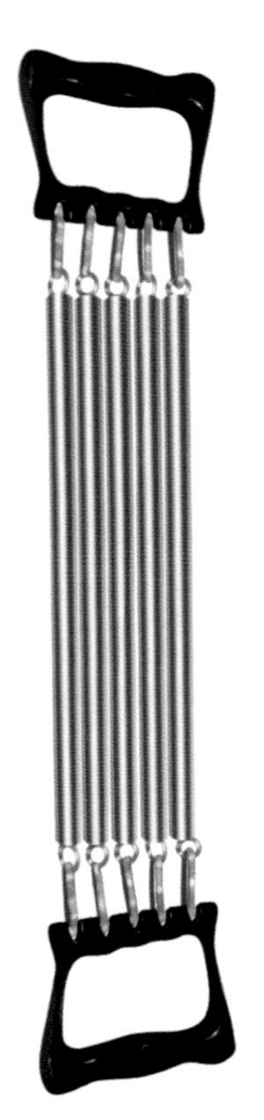

Figure 12.15 This exercise equipment uses parallel springs to strengthen arm muscles.

Elastic strain energy

The exercise equipment shown in Figure 12.15 makes use of springs. When you pull the two handles apart you increase the energy stored by the springs. We can calculate how much energy is stored as elastic strain energy in the springs.

The energy stored is equal to the work done stretching the springs. The work done depends on the average force applied and the extension of the springs.

work done = average force × extension

For a material that obeys Hooke's law, the average force is:

$$\frac{F_{max}}{2}$$

We can express the elastic strain energy using the equation:

$$\text{elastic strain energy} = \frac{1}{2} \times F \times \Delta l$$

where:

J is elastic strain energy

F is force in N

Δl is extension in m.

We can also express the elastic strain energy in terms of the spring constant:

$$\text{Elastic strain energy} = \frac{1}{2} \times F \times \Delta l$$

and

$$F = \mathrm{k}\Delta l$$

substituting for F we obtain:

$$\text{elastic strain energy} = \frac{1}{2} \times (\mathrm{k}\Delta l) \times \Delta l = \frac{1}{2} \times \mathrm{k} \times \Delta l^2$$

For a material that does not fully obey Hooke's law, we can still calculate the elastic strain energy by calculating the area under the load–extension graph.

For any graph, we can choose small changes in extension, δl, and calculate the work done by the load to produce that small extension. The total work done is then the sum of all these values.

$$W_T = \Sigma F \delta l$$

Figure 12.16 shows how this is done for a simplified force–extension graph.

On Figure 12.16 the region of the graph marked as OA represents the applied loads for which the material obeys Hooke's law. For this region the area under the graph is a triangle, so:

$$W_1 = \frac{1}{2} F_{max} \Delta l_1$$

For the second region of the graph, AB, the material no longer obeys Hooke's law. However, energy is still required to stretch the material, so work is still being done.

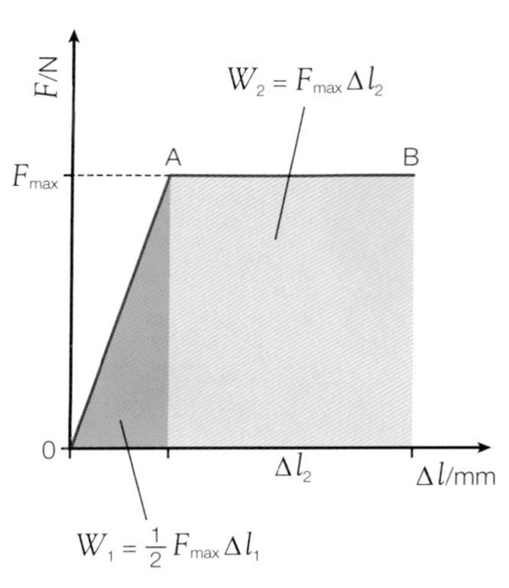

Figure 12.16 Calculating elastic strain energy.

The area of this region is that of the rectangle below the line:

$$W_2 = F_{max} \times \Delta l_2$$

which means that the total work done in stretching this material is:

$$W_T = W_1 + W_2$$

$$W_T = \frac{1}{2}F_{max}\,\Delta l_1 + F_{max}\Delta l_2$$

or

$$W_T = F_{max}\left(\frac{1}{2}\Delta l_1 + \Delta l_2\right)$$

The elastic strain energy stored is equivalent to the work done in stretching the material.

EXAMPLE

For non-linear graphs elastic strain energy is calculated by counting the number of squares under the line and multiplying this by the value of work equivalent to each square. Find the elastic strain energy in Figure 12.17.

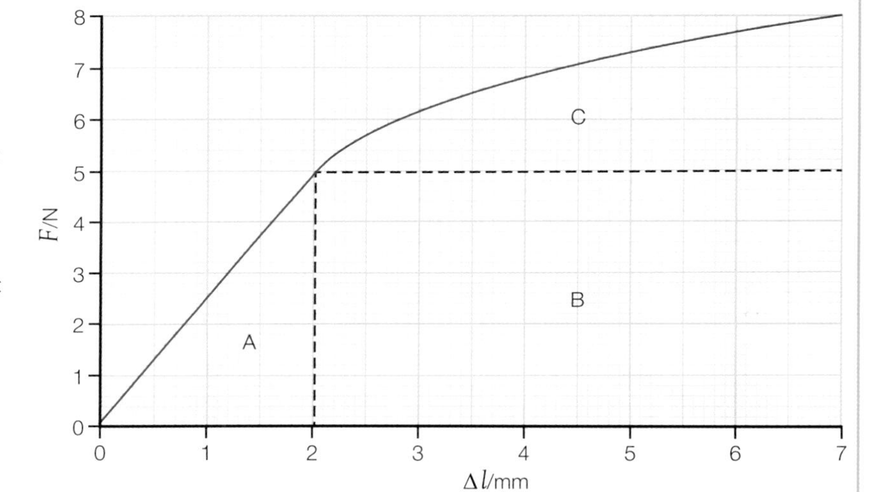

Figure 12.17 Elastic strain energy is calculated by counting the number of squares under the line and multiplying this by the value of work equivalent to each square.

Answer

Note: We could have estimated the elastic strain energy by assuming that shape C was a triangle and calculating the area of that triangle. However, that would give a less accurate answer than the method of counting squares shown here.

In Figure 12.17, the work equivalent of each small square

$= 0.2\,\text{N} \times 0.1 \times 10^{-3}\,\text{m} = 0.02 \times 10^{-3}\,\text{J}$

area of A $= \frac{1}{2} \times 5\,\text{N} \times 2 \times 10^{-3}\,\text{m}$
$= 5 \times 10^{-3}\,\text{J}$

area of B $= 5\,\text{N} \times 5 \times 10^{-3}\,\text{m}$
$= 25 \times 10^{-3}\,\text{J}$

Number of small squares in region C ≈ 448

Work equivalent $= 448 \times 0.02 \times 10^{-3}\,\text{J} = 9 \times 10^{-3}\,\text{J}$

So

elastic strain energy $= 3.9 \times 10^{-2}\,\text{J}$

TIP

Convert all measurements to standard units, for example, cm to m.

Energy and springs

When a material is stretched or compressed its elastic strain energy is altered. Figure 12.18 shows two common toys that make use of stored elastic strain energy.

EXAMPLE

The spring jumper toy shown in Figure 12.18 is compressed by pushing the suction cup on to the stand. When the suction cup releases the toy 'jumps' to a height of 65 cm. The toy has a mass of 16 g. The spring is originally 3.6 cm long, and is compressed to 0.9 cm long.

Assuming that no energy is dissipated as heat and sound when the toy jumps, calculate:

a) the elastic strain energy stored by the spring
b) the spring constant of the spring.

Figure 12.18 Each of these toys stores elastic strain energy by deforming an elastic material.

Answer

a) If there are no energy loses during the jump then the amount of gravitational potential energy gained by the toy will be equal to the amount of elastic strain energy stored by the spring.

gpe = mass × gravitational field strength × height
$= 0.016\,\text{kg} \times 9.81\,\text{N kg}^{-1} \times 0.65\,\text{m}$
$= 0.102\,\text{J}$

Therefore, the elastic strain energy = 0.102 J

b) Elastic strain energy $= \frac{1}{2}k\Delta l^2$

In this case the spring has been compressed, not stretched. The equation is the same but instead of the amount the spring has extended we need to know the amount it has compressed.

compression = initial length − final length
= 3.6 cm − 0.9 cm = 2.7 cm

Rearranging the equation for elastic strain energy we get:

$$k = \frac{2 \times \text{elastic strain energy}}{\Delta l^2}$$

$$k = \frac{2 \times 0.102\,\text{J}}{0.027\,\text{m}^2}$$

$$k = 280\,\text{N m}^{-1}$$

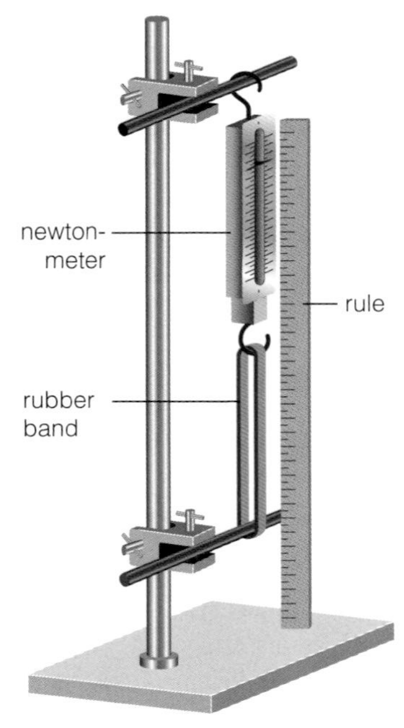

Figure 12.19 Experimental set-up for measuring the extension of a rubber band when a force is applied. Raising and lowering the clamp stand boss alters the applied force, which is measured using the newton-meter. If you are conducting this experiment, wear safety glasses.

The elastic properties of some materials, such as rubber, can be complex. When a rubber band is stretched it will return to its original length. However, the way in which it does this is very different from a metal wire. Figure 12.19 shows an experimental set-up that can be used to carry out this measurement. Figure 12.20 shows a typical force–extension curve obtained from this experiment.

Initially, there is a small amount of extension as the force is applied. Then, as more force is applied, the rubber band stretches easily. Finally, just before it breaks (not shown in Figure 12.20) it becomes harder to stretch again. If you have ever blown up a balloon, you will be familiar with this changing behaviour. Initially, balloons are hard to inflate, but become much easier once you've blown some air inside. Just before they burst, it becomes more difficult to inflate them further.

Figure 12.20 also shows that the extension for a given force is different when the rubber band is being loaded (top curve) or unloaded (bottom curve). This means that the strain energy stored when the rubber band is being loaded is greater than the strain energy released when the rubber band is being unloaded.

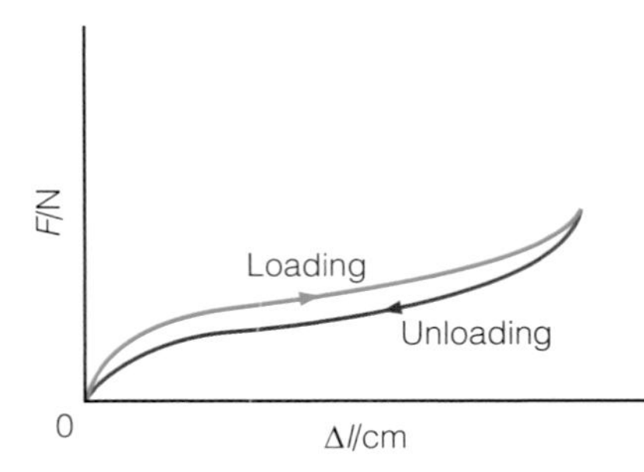

Figure 12.20 Force–extension curve for a rubber band. Notice that the loading and unloading characteristics are different. What happens to the elastic strain energy stored by the rubber band?

However, the law of energy conservation states that energy cannot be created or destroyed in a closed system. The difference in strain energy must be accounted for. In the case of the rubber band, it will become warm as it is stretched and relaxed. This is why there is a difference in energy between loading and unloading.

TEST YOURSELF

13 A weight of 35 N is attached to the bottom of a vertical wire. The wire stretches by 0.8 mm. Calculate the elastic strain energy stored in the wire.

14 A 2 m rope is suspended from a tree branch. A child of mass 30 kg hangs on to the end of the rope, which stretches to 2.05 m. Calculate the elastic strain energy stored in the rope. Take $g = 9.81\ \text{N kg}^{-1}$.

15 Guitar strings are stretched using a tension key to tune them to the correct frequency. A 750 mm long nylon guitar string is stretched so that it experiences a tension force of 27.5 N. At this tension the guitar string is 785 mm long. Calculate the elastic strain energy stored in the string.

16 The exercise equipment shown in Figure 12.15 is designed to provide a total resistive force of 1000 N when pulled to an extension of 40 cm. The effective spring constant of the five springs in parallel is $2500\ \text{N m}^{-1}$. Calculate the minimum energy that the person who is exercising needs to use to stretch the springs by this extension and explain why this is a minimum value.

17 A toy uses a compression spring to launch a small ball at a target. The ball moves horizontally. The spring constant of the spring is $1250\ \text{N m}^{-1}$. The spring is compressed by 3 cm before the ball is launched. Calculate the elastic strain energy stored by the spring before launch.

Stress and strain: the Young modulus

Look back at Figure 12.10 showing springs joined in parallel and in series.

When two identical springs were joined together in series, we saw that the springs extended twice as much as a single spring on its own. The extension depends on the original length. Similarly, if we join two springs in parallel, then the springs extend half as much as a single spring on its own. The extension depends on the area over which the force is acting.

In each case, the property of the object depends on the dimensions of the object itself.

If we want to compare materials fairly, rather than compare two different objects, then it is better to have a measurement that doesn't depend on the shape or size of the object. This comparison is made by using stress and strain instead of force and extension.

Tensile stress $= \dfrac{\text{force}}{\text{cross-sectional area}}$

Tensile stress is a measurement of the force applied over the cross-sectional area of sample of material.

$$\text{tensile stress} = \frac{\text{force}}{\text{cross-sectional area}}$$

$$\text{tensile stress} = \frac{F}{A}$$

where tensile stress is measured in Pa, or $\mathrm{N\,m^{-2}}$

F is applied force in N

A is area in $\mathrm{m^2}$

Stress is sometimes given the symbol σ, so $\sigma = \dfrac{F}{A}$

Tensile strain $= \dfrac{\text{extension}}{\text{original length}}$

Tensile strain is the ratio of the extension and original length of the sample.

$$\text{tensile strain} = \frac{\text{extension}}{\text{original length}}$$

$$\text{tensile strain} = \frac{\Delta l}{l}$$

where tensile strain is dimensionless because it is a ratio

Δl is the extension in m

l is the original length in m.

Strain may also be expressed as a percentage. It is sometimes given the symbol ε.

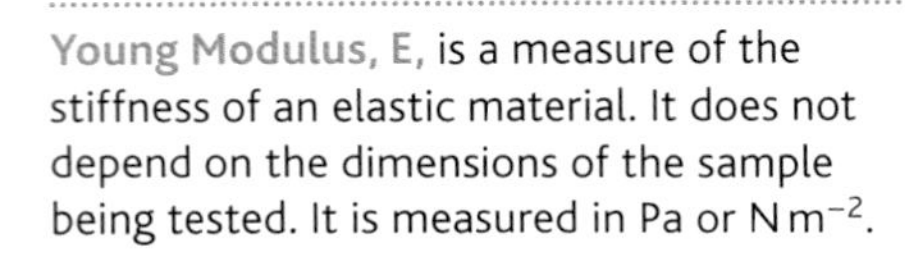
Young Modulus, E, is a measure of the stiffness of an elastic material. It does not depend on the dimensions of the sample being tested. It is measured in Pa or $\mathrm{N\,m^{-2}}$.

We can now use these quantities to calculate a measure of the stiffness of an elastic material that is independent of the shape of the sample of material. This is called the Young modulus, E, of the material.

$$\text{Young modulus} = \frac{\text{stress}}{\text{strain}}$$

$$E = \frac{\sigma}{\varepsilon}$$

$$E = \frac{\left(\frac{F}{A}\right)}{\left(\frac{\Delta l}{l}\right)}$$

$$E = \frac{Fl}{A\Delta l}$$

Young modulus is measured in Pa (or $\mathrm{N\,m^{-2}}$).

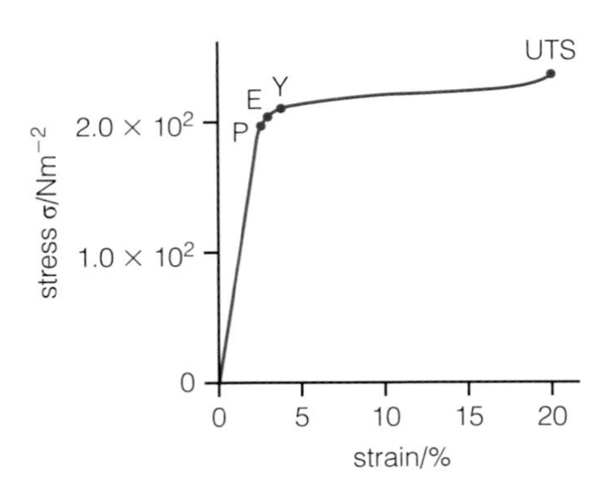

Figure 12.21 A simplified stress–strain graph for copper.

Figure 12.21 shows a stress–strain graph for copper. It looks like the force–extension graph shown earlier but this graph will be valid for any sample of copper.

O–P on the graph represents the range of tensile stress for which the copper obeys Hooke's law. The gradient of this section is the Young modulus for copper. This value will be the same no matter what the size or shape of the sample of copper being used. Point P represents the limit of proportionality for the material.

Point E on the graph represents the elastic limit. Up to point E, if the stress is removed, the sample of copper will return to its original length. Beyond this point, copper behaves plastically. It does not return to its original length.

The yield point of the material is given by Y. This is the value of stress beyond which the strain increases rapidly for small increases in stress.

TIP

Make sure that you convert all the measurements to standard units before attempting the question.

The ultimate tensile stress (UTS) of copper is sometimes called the maximum strength or strength of the wire. Just before this point the copper becomes narrower at the weakest point, known as necking. This can be seen in Figure 12.14 for a ductile sample. The UTS is the stress at which the material breaks.

Some metals will also show a phenomenon known as creep. This is when the sample continues to extend over a period of time, even though the stress applied is not increased.

EXAMPLE

Rock climbers often rely on their ropes. Some ropes are chosen so that they will stretch if the climber loses their footing on the rock and falls a short distance. This reduces the danger from the fall.

To test the suitability of ropes for climbing a weight of 800 N is suspended from the rope and the extension measured.

This test is carried out on 50 m of a rope that has a diameter of 10 mm. If the rope extends 2 m during the test, calculate the Young modulus of this rope.

Answer

First we calculate tensile stress.

$$\text{tensile stress} = \frac{\text{force}}{\text{area}}$$

The area of the rope is the area of a circle = πr^2

The radius of the rope is half the diameter (in m) = 5×10^{-3} m

$$\text{area} = \pi(5 \times 10^{-3})^2\,\text{m}^2$$
$$= 7.85 \times 10^{-5}\,\text{m}^2$$

so

$$\text{tensile stress} = \frac{800\,\text{N}}{7.85 \times 10^{-5}\,\text{m}^2}$$
$$= 10.19 \times 10^{6}\,\text{N}\,\text{m}^{-2}$$

Next we calculate tensile strain, making sure that both measurements are in the same units.

$$\text{tensile strain} = \frac{\text{extension}}{\text{original length}}$$
$$= \frac{2}{50}$$
$$= 0.04$$

$$\text{Young modulus} = \frac{\text{tensile stress}}{\text{tensile strain}}$$
$$= \frac{10.19 \times 10^{6}\,\text{N}\,\text{m}^{-2}}{0.04}$$

Young modulus of the rope = 2.5×10^{8} N m^{-2}

TIP

Sometimes the tensile stress and Young modulus will be given with the SI units of Pascal, Pa; 1 Pa is equal to 1 N m^{-2}.

REQUIRED PRACTICAL 4

Determination of the Young modulus by a simple method

Note: This is just one example of how you might tackle this required practical.

A student carried out an experiment to measure the extension of a copper wire as weights were applied. The original length of the wire was 2.5 m. The student used a micrometer screw gauge to take three measurements of the diameter of the wire at different places along the wire. The mean diameter of the wire was 0.52 mm.

The data obtained are shown in the table.

Mass (kg)	Extension (mm)	Stress ($N\,m^{-2}$)	Strain
0.5	0.5		
1.0	1.0		
1.5	1.5		
2.0	2.0		
2.5	2.5		
3.0	3.0		
3.5	4.0		
4.0	6.0		

1 Suggest why the student took three measurements of the diameter of the wire.
2 Calculate the stress and strain values for the results.
3 Plot a graph of stress against strain and calculate the Young modulus of copper from your graph.
4 Explain whether the wire returned to its original length when the weights are removed.

Extension

Measure the Young modulus of a number of different diameter copper wires and compare the values obtained.

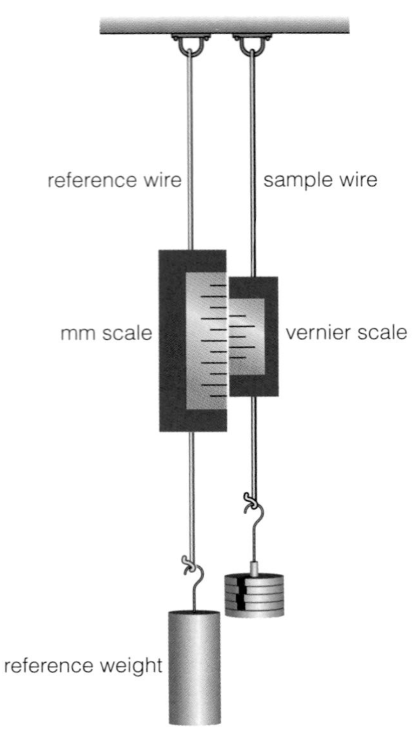

Figure 12.22 Experimental set-up to measure the Young modulus of a metal wire. If you are conducting this experiment, wear safety glasses. (Not drawn to scale.)

A simple experimental set-up for copper can be used to measure its Young modulus. However, we can use a more accurate experimental set-up as shown in Figure 12.22. This is particularly useful for materials such as steel, which generally give smaller values of strain.

In Figure 12.22, the left-hand wire is a reference wire that holds the main scale of the vernier calipers, usually calibrated in millimetres. The reference wire has a mass hung from it to keep it taut. The sample wire is hung close to the reference wire and holds the smaller scale of the vernier calipers. The two wires are made from the same material.

As weights are added to the sample wire it extends, and the small scale moves relative to the main scale. This allows the extension of the wire to be measured.

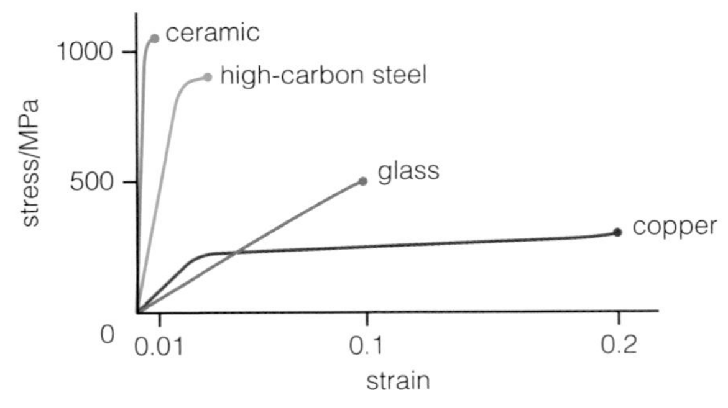

Figure 12.23 Stress–strain graphs for four different materials.

Interpreting stress–strain graphs

Stress–strain graphs allow us to describe the properties of materials, and also to predict the stresses at which changes in those properties might occur.

Figure 12.23 compares the stress–strain graphs for four different materials: ceramic, steel, glass and copper.

Ceramics are extremely strong and have very high UTS values. However, they are show very little (if any) plastic behaviour before they fracture, so they are also very brittle. Glasses have lower UTS values than ceramics and so are less strong, but they are also brittle, generally showing no plastic behaviour before they break.

Steel is made by adding different elements to iron to form an alloy. Common elements used in steel-making are carbon, manganese and chromium. Types of steel differ in the percentage composition of the various elements added to iron to create them. This affects the properties of the steels and they are generally much stiffer than ductile metals such as copper. This can be seen on the graph from the shallower gradient of copper giving a lower value of Young modulus.

The high-carbon steel shown in the graph is a strong but brittle material. It shows elastic behaviour at higher values of stress but fractures with very little plastic behaviour. This type of steel is often used in cutting tools and drill bits because it has a higher UTS value. Other types of steel may show plastic behaviour but have a lower UTS value.

Copper has a long plastic region because it is a ductile material, and this makes it ideal for forming into wires for use in electrical circuits.

Strain energy density

The strain energy density is the strain energy per unit volume of a sample. Earlier we showed that the spring constant, k, depends on the dimensions of the material sample being tested, but the Young modulus depends only on the material used. In the same way we can calculate the strain energy density, which is a measure of the energy stored in a material that does not depend on the dimensions of the sample being tested.

Strain energy density is the strain energy per unit volume of the material.

From earlier:

$$\text{strain energy} = \frac{1}{2}F\Delta l.$$

If l is the original length of the wire, and A its cross-section, then the volume of the wire = Al.

Therefore:

$$\text{strain energy per unit volume} = \frac{1}{2}F\frac{\Delta l}{Al}$$
$$= \frac{1}{2}\left(\frac{F}{A} \times \frac{\Delta l}{l}\right)$$

but $\frac{F}{A}$ = stress and $\frac{\Delta l}{l}$ = strain.

Therefore:

$$\text{strain energy per unit volume, or strain energy density} = \frac{1}{2}(\text{stress} \times \text{strain})$$

$\frac{1}{2}$(stress × strain) is the area under a linear stress–strain graph. Therefore, the area under any stress–strain graph is equal to the energy per unit volume.

EXAMPLE

Why do all animals jump the same height?

Answer

Obviously all animals don't jump exactly the same height. However, they nearly all jump somewhere between 0 and 1 m.

If all animals are the same shape and they are made of the same stuff then we can say that:

$$\text{strain energy per unit volume} = \frac{1}{2}(\text{stress} \times \text{strain})$$

Therefore:

$$\text{energy stored in muscles} = \frac{1}{2}(\text{stress} \times \text{strain} \times \text{volume of animal muscle})$$

gravitational potential energy (E_p) gained in jumping $= mgh$

We can estimate the mass of the animal (muscle) using:

$$\text{mass} = \text{density} \times \text{volume}$$

so

$$E_p = \text{density} \times \text{volume} \times g \times h$$

Assuming that all the energy stored in the animal is used to increase the height of the animal, we can say:

$$\text{energy stored in muscles} = \text{increase in gpe}$$

$$\frac{1}{2}(\text{stress} \times \text{strain} \times \text{volume of animal muscle}) = \text{density} \times \text{volume} \times g \times h$$

therefore

$$\text{height jumped} = \frac{1}{2}(\text{stress} \times \text{strain})/\text{density} \times g$$

So, if all animals are the same shape and made out of the same stuff, i.e. they all have the same density and muscle properties, then the height jumped is *independent* of the size of the animal and they will all jump the same height.

Although real animals do jump different heights, the range of heights is much smaller than the corresponding range of masses. For example, from a standing start a flea can jump about 20 cm and a human 60 cm. However, a typical flea has a mass of approximately 0.5 mg and a typical human has a mass of approximately 70 kg.

TEST YOURSELF

18 A brass wire of length 1.5 m and diameter 0.4 mm is extended by 3 mm when a tensile force of 32 N is applied. Calculate:

a) the applied stress

b) the strain on the wire

c) the Young modulus of the brass.

19 A towbar on a car is used to pull a trailer. The trailer exerts a tensile force of 23 kN on the towbar. The towbar can be modelled as a cylindrical bar of steel with a diameter of 6 cm and a length of 25 cm. The high-strength steel has a Young modulus of 200 GPa. Calculate how much the towbar will extend when it is in use.

20 The high-strength steel used in the towbar in question 19 is a brittle material with an ultimate tensile strength of 500 MPa. Estimate the maximum force the towbar can sustain before it fractures and breaks.

21 A metal component has a length of 0.5 m. The strain must not exceed 0.1% (1×10^{-3}). Calculate the maximum extension allowed.

22 A steel strut has a cross-sectional area of 0.025 m^2 and is 2.0 m long. Calculate the compressive force that will cause the strut to shorten by 0.3 mm. Assume that the Young modulus for this steel is 200 GPa.

23 Estimate the stress on the lower leg of a person who is standing still. Suggest why your value is likely to be much less than the ultimate compressive stress of bone.

24 An architect makes a scale model of a bridge she is going to build. It is made 20 times smaller than the real bridge in every dimension. The model is made from exactly the same material as the real bridge. The bridge is supported by four steel pillars. Calculate the ratio of the stresses in the support pillars, in the real bridge to model bridge.

25 Lions are approximately ten times larger than cats. Use your understanding of stress to explain why the legs of lions and cats are not the same shape.

Practice questions

Figure 12.24 shows the stress–strain graph for two materials up to their breaking points. You need this graph to answer questions 1 to 3.

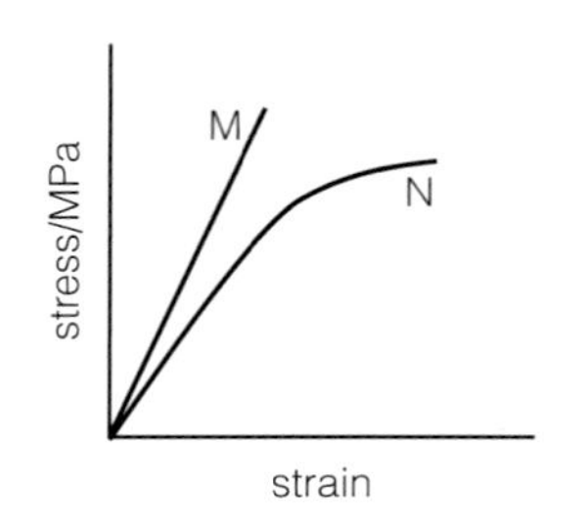

Figure 12.24 A stress–strain graph for two materials.

1 Which word best describes the behaviour shown for material M?

A brittle **C** fragile

B ductile **D** plastic

2 Which word best describes the behaviour shown for material N?

A brittle **C** plastic

B fragile **D** strong

3 Which combination of materials could be best represented by the graph?

	M	**N**
A	glass	cast iron
B	high-carbon steel	glass
C	cast iron	high-carbon steel
D	cast iron	rubber

4 A superbounce bouncy ball has a diameter of 2.5 cm and a mass of 35 g. What is the density of the ball in $kg\,m^{-3}$?

A 4.3×10^{-3} **C** 4.3×10^{3}

B 4.3 **D** 4.3×10^{6}

5 A 10 cm spring is stretched until it is 16 cm long. The force needed to stretch the spring by this amount is 30 N. What is the spring constant of the spring in $N\,m^{-1}$?

A 1.9 **C** 480

B 50 **D** 500

You need the following information for questions 6 and 7.

A 2 m long square metal bar has sides of 40 mm width. A force of 80 kN is applied to the bar and it extends by 0.046 mm.

6 What is the stress in the metal bar?

A $5.0\,N\,m^{-2}$ **C** $5.0 \times 10^{7}\,N\,m^{-2}$

B $5000\,N\,m^{-2}$ **D** $5.0 \times 10^{9}\,N\,m^{-2}$

7 What is the strain in the metal bar?

A 2.3×10^{-5} **C** 23

B 0.23 **D** 2.3×10^{5}

You need the following information for questions 8 and 9.

A vertical steel wire of length 0.8 m and radius 1.0 mm has a mass of 0.2 kg attached to its lower end. Assume that the Young modulus of steel is $2.0 \times 10^{11}\,N\,m^{-2}$ and $g = 10\,N\,kg^{-1}$.

8 What is the extension of the wire?

A 0.128×10^{-6} m **C** 0.255×10^{-3} m

B 2.55×10^{-6} m **D** 2.55 m

9 What is the energy stored in the wire when stretched?

A 2.6×10^{-6} J **C** 2.6 J

B 2.6×10^{-4} J **D** 26 J

You need the following information for questions 10 and 11.

Figure 12.25 shows the stress–strain graph for a metal wire that is stretched until it breaks.

10 What is the Young modulus of the metal?

A 16 GPa **C** 74 GPa

B 53 GPa **D** 120 GPa

11 What does the area under the graph represent?

A the total stress on the wire

B the breaking strain

C the energy stored in the wire per unit volume

D the elasticity of the wire

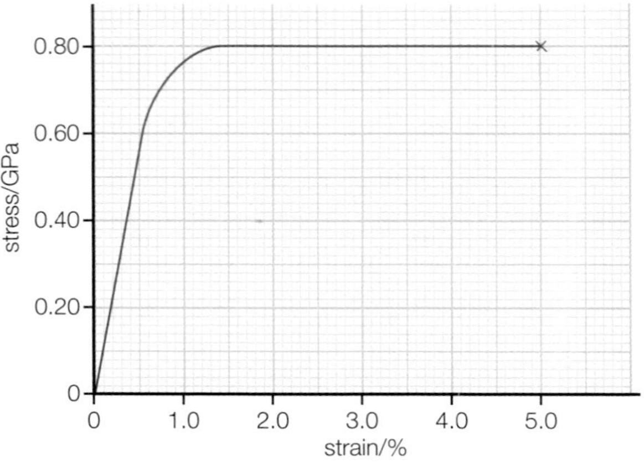

Figure 12.25 The stress–strain graph for a metal wire that is stretched until it breaks.

12 A sample of wood is tested in a tensile testing machine. The wood sample breaks when the applied force is 840 N. The cross-sectional area of the sample is 1.3×10^{-5} m^2. Calculate the ultimate tensile stress for the sample. *(1)*

13 A stone paving slab has a mass of 13 kg. Its dimensions are 600 mm × 300 mm × 35 mm.

a) Calculate the density of the stone. *(1)*

b) Calculate the maximum compressive stress that the slab could exert on the ground when resting on one of its edges. *(2)*

14 A cord made from natural rubber is initially 20 cm long. A load of 30 N is attached to it. The new length is measured as 1.0 m.

a) Calculate the strain in the rubber cord. *(1)*

b) The stress on the cord is 14 MPa. Calculate the cross-sectional area of the cord. *(1)*

15 An experiment is carried out to determine the spring constant of a metal in the form of a wire. Weights are added to the wire in steps of 5.0 N up to 25.0 N.

	Loading	Unloading
Load (N)	Extension (mm)	Extension (mm)
0.0	0.00	0.00
5.0	0.12	0.12
10.0	0.23	0.23
15.0	0.35	0.34
20.0	0.48	0.49
25.0	0.61	0.61

a) Use the results in the table (without plotting a graph) to state and explain if the deformation of the wire is plastic or elastic. *(1)*

b) Describe how the length and extension of the wire could be measured experimentally and explain what safety precautions should be taken when carrying out the measurements. *(4)*

c) Plot a graph using the results for loading the wire and use your graph to calculate the spring constant of the wire. *(3)*

d) What additional measurements would need to be made to allow calculation of the Young modulus of the wire? *(1)*

16 A lift has a mass of 500 kg. It is designed to carry five passengers with a maximum combined mass of 450 kg. The lift is moved by means of a steel cable of diameter 20 mm. When the lift is on the ground floor of the building the cable is at its maximum length of 30 m. The density of the cable is $9550\,\text{kg}\,\text{m}^{-3}$.

a) Calculate the mass of the 30 m cable. *(1)*

b) The lift is stationary on the ground floor of the building. Show that the tensile stress in the cable due to the lift and the mass of the cable is approximately 30 MPa. *(3)*

17 A student measures the extension of an elastic band when a force is applied. She then continues to measure the extension as the force is progressively removed from the elastic band. A graph of her results is shown in Figure 12.26.

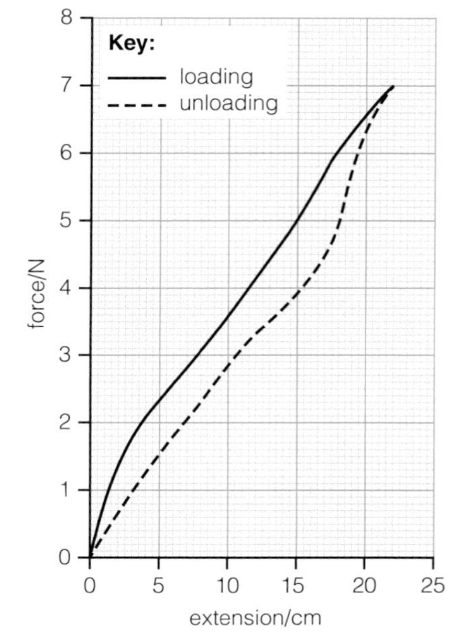

Figure 12.26 Results from loading and unloading an elastic band.

a) Describe a simple experiment that would allow measurement of the force applied and the extension of the elastic band. *(4)*

b) The student did not take repeat measurements. Suggest what effect this might have on her results. *(1)*

c) Does the elastic band show Hooke's law behaviour? Explain your answer. *(2)*

d) Explain how the student could estimate the amount of energy dissipated by the elastic band during the experiment. *(2)*

e) What physical change would occur in the elastic band if it was repeatedly stretched and released in a short space of time? *(1)*

18 Figure 12.27 shows a stress–strain graph for copper wire.

a) Explain why no units are given for strain on the *x*-axis. *(1)*

b) Describe the behaviour of copper up to a strain of 1.0×10^{-3}. *(1)*

c) State the breaking stress of this copper wire. *(1)*

d) Calculate the Young modulus for copper. *(2)*

e) A similar copper wire is loaded up to a strain of 3.5×10^{-3}. The load is then

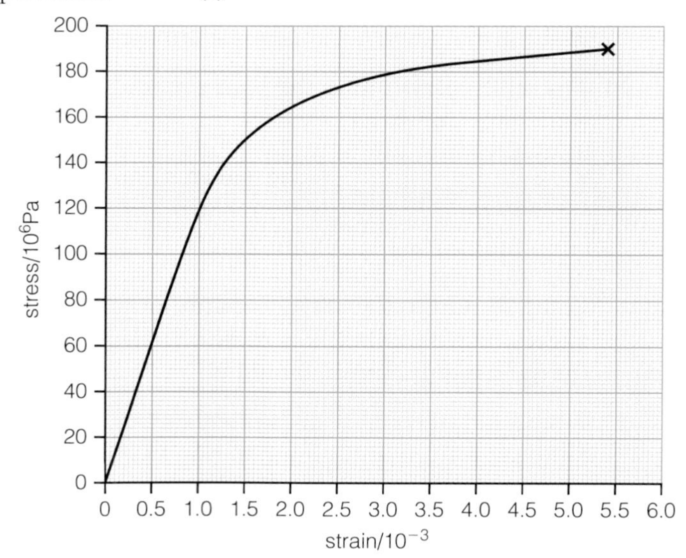

Figure 12.27 A stress–strain graph for copper wire.

removed from the wire. Sketch on the graph the stress–strain measured as the load is removed and explain the shape of the graph you have drawn. (2)

Stretch and challenge

19 Figure 12.28 shows an experiment in which two wires, each 2.0 m long, are hung vertically from a fixed bar. The ends of the wires are attached to a lightweight horizontal bar 0.1 m apart.

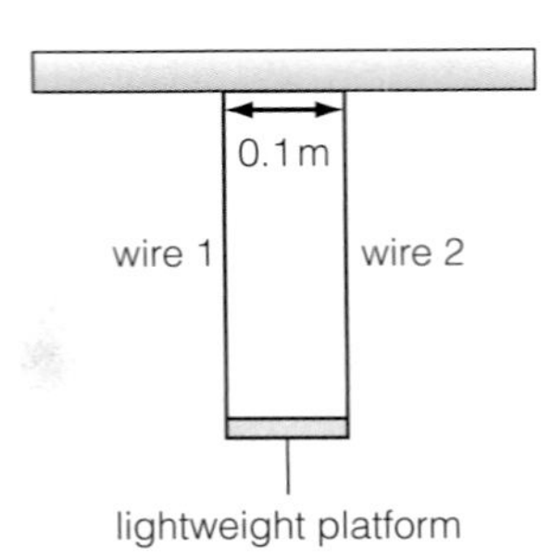

Figure 12.28 An experiment in which two wires, each 2.0 m long, are hung vertically from a fixed bar.

Wire 1 is made from steel and has a diameter of 0.8 mm. Wire 2 is made from brass and has a diameter of 0.68 mm. A force of 100 N is applied vertically downwards to the centre of the platform. The wires remain vertical but the brass wire extends more than the steel wire and the bar tilts at 1° to the horizontal.

a) Calculate the difference between the extensions of the two wires.

b) The Young modulus for steel is $2.0 \times 10^{11}\,N\,m^{-2}$; show that the extension of the steel wire is 1.0 mm.

c) Calculate the Young modulus for the brass wire.

d) Calculate the energy stored in the steel wire due to its extension.

20 A pinball machine firing mechanism is used to launch a small ball bearing, of mass 100 g, into the playing area. The design of the mechanism is shown in Figure 12.29. The playing area and the mechanism are at 8.5° to the horizontal.

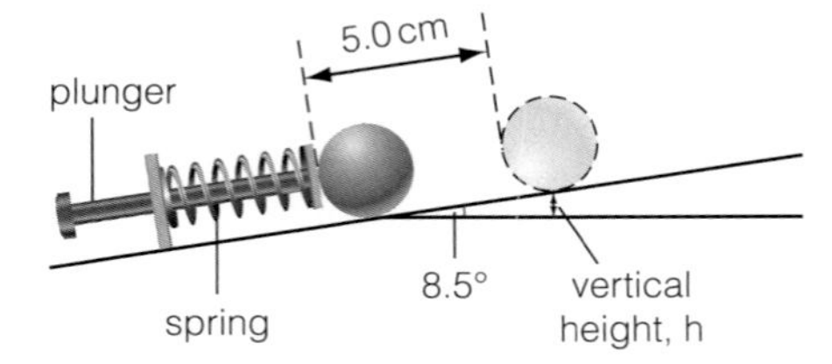

Figure 12.29 A pinball machine firing mechanism.

Initially the spring is relaxed. To fire the ball, the plunger is pulled back and compresses the spring by 5.0 cm. When the plunger is released the spring extends and the plunger applies a force to the ball. The dashed circle shows the position of the ball when it has a maximum speed, just as it loses contact with the plunger.

The ball leaves the plunger at a speed of $0.68\,m\,s^{-1}$. Use ideas about conservation of energy to calculate the spring constant of the spring. Assume that the mechanism is frictionless.

21 The table below shows the Young modulus and breaking strength of spider silk and steel.

	Young modulus (Pa)	Breaking strength (Pa)
Spider silk	1×10^{10}	1×10^{9}
High-tensile steel	2×10^{11}	2×10^{9}

Using data from the table, and considering the strain and elastic energy, suggest why spiders silk is more effective at catching a fly than steel (of the same thickness) would be.

22 A long rod of metal is suspended and stretches under its own weight. Its length is l, its density ρ, and Young Modulus E. Show that its extension under its own weight is $\rho g l^2/2E$.

13 Current electricity

PRIOR KNOWLEDGE

- Electric current is a flow of electric charge.
- The size of the electric current, I, is the rate of flow of electric charge, which is given by the equation:

$$I = \frac{Q}{t}$$

where Q is the charge in coulombs, C, and t is the time in seconds.
- Electric current is measured in amperes, A, by an ammeter connected in series with components.
- The potential difference or voltage, V, between two points in an electric circuit is the work done (energy transferred), W, per coulomb of charge that passes between the points and is given by the equation:

$$V = \frac{W}{Q}$$

- Potential difference is measured in volts, V, by a voltmeter connected in parallel with components.
- The resistance of a component, R, can be found by measuring the current, I, through and potential difference, V, across the component; resistance is defined by the equation:

$$R = \frac{V}{I}$$

- Resistance is measured in ohms, Ω.
- The current flowing though, the potential difference across and the resistance of a component are related by the equation:

$$V = I \times R.$$

For example, if the current in Figure 13.1 is 4.5 A and the resistance is 4.0 Ω, the potential difference across the resistor is 18 V.

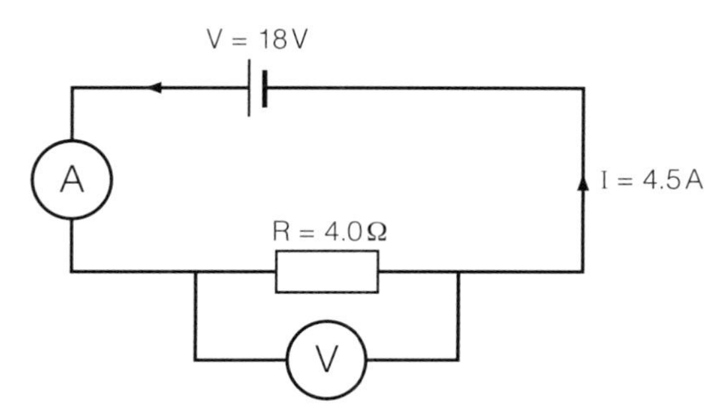

Figure 13.1 Simple electric circuit.

- The potential difference provided by cells connected in series is the sum of the potential difference of each cell (depending on the direction in which they are connected).

TEST YOURSELF ON PRIOR KNOWLEDGE

1 Draw a circuit showing how you would measure the electric current flowing through a bulb powered from a 12 V battery.
2 Write down the names and symbols of the units of:
   a) current
   b) potential difference
   c) resistance
   d) charge
3 0.6 C of charge flows through a bulb during 40 s. Calculate the current flowing through the bulb.
4 The battery provides a potential difference of 12 V across the bulb. Calculate the electrical energy transferred into heat and light by the bulb when 0.25 C of charge flows through the bulb.

Moving charge and electric current

Electric current is the rate of flow of electric charge. The unit of current is the ampere.

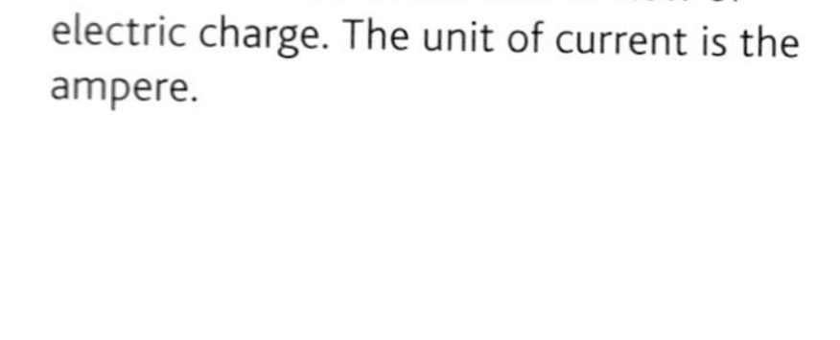

A lightning bolt is an extreme example of an electric current. During the largest cloud-to-ground lightning bolts 2×10^{21} electrons might jump from the bottom of a cloud to Earth in 0.003 s, delivering 5000 MJ of electrical energy.

An electric current is formed when the electrons move. Electric current, I, is the rate of flow of charge:

$$\text{electric current, } I \text{ (A)} = \frac{\text{amount of charge flowing, } \Delta Q \text{ (C)}}{\text{time to flow, } \Delta t \text{ (s)}}$$

$$I = \frac{\Delta Q}{\Delta t}$$

Figure 13.2 A cloud-to-ground lightning strike.

The SI unit of current is the **ampere** (symbol, A), which is nearly always abbreviated to amps. 1 amp is equal to a charge of 1 coulomb flowing in 1 second (so $1\,\text{A} = 1\,\text{C}\,\text{s}^{-1}$).

Note that here we have used the notation ΔQ and Δt, rather than Q and t. In GCSE work, you performed calculations in which the current was a constant value. At A-level you will meet questions in which the current is changing. Under those circumstances we can calculate a current by considering a small flow of charge, ΔQ, in a small time, Δt. So the equation above is a more general expression of the equation you met during your GCSE course.

EXAMPLE

The electrons moved by friction in a cloud accumulate at the bottom of the cloud producing a large negative charge. Each electron has a charge, e, where $e = -1.6 \times 10^{-19}$ coulombs (symbol, C). The total magnitude of the charge (symbol, Q) of all the electrons is about 320 C ($Q = ne = 2 \times 10^{21} \times 1.6 \times 10^{-19}\,\text{C} = 320\,\text{C}$). The bottom of the cloud is 'near' to the ground, which is 'earthed', and the electrons flow from the cloud to Earth through the air in 0.003 s, which provides a conducting pathway.

How much current is produced?

Answer

Using the current relationship for a large lightning bolt:

$$I = \frac{320\,\text{C}}{0.003\,\text{s}} = 1 \times 10^5\,\text{A (1 sf)}$$

Electronic circuits, such as those that control household appliances, operate with much smaller currents, typically milliamps (mA, 10^{-3} A), and many microelectronic circuits, such as the printed circuit boards inside many computer devices, operate with currents of the order of microamps (μA, 10^{-6} A). Even currents of the order of microamps still involve the movement of about 6×10^{12} electrons per second.

EXAMPLE

Some electric currents involve beams of electrons. Any device containing a cathode ray tube (CRT), such as an oscilloscope, involves electron beam currents.

A current of 30 μA flows in a CRT. Calculate the number of electrons flowing in the beam per second, *n*.

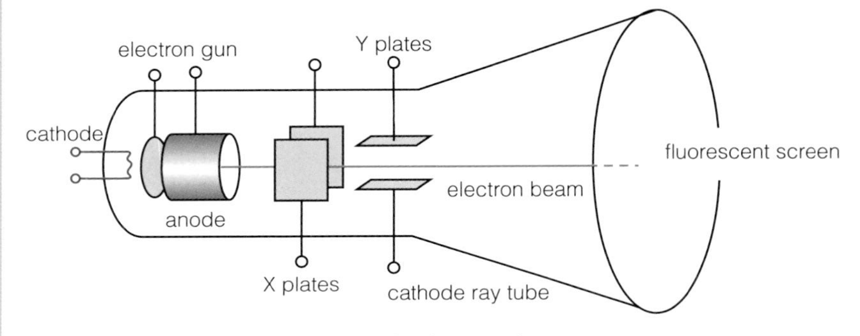

Figure 13.3 Electron beam in a cathode ray tube.

Answer

$$I = \frac{\Delta Q}{\Delta t} = \frac{n \times e}{1} = ne$$

(The charge, ΔQ, is the charge carried by each electron, e, multiplied by the number of electrons).

$$\Rightarrow n = \frac{I}{e} = \frac{30 \times 10^{-6}\text{A}}{1.6 \times 10^{-19}\text{C}} = 1.9 \times 10^{14} \text{ electrons}$$

per second

An **electrolyte** is a conducting solution, usually containing positive and negative salt ions dissolved in water.

Not all electric currents involve the flow of electrons. Charged ions in solution (called an electrolyte) can also flow and create a current. Car batteries work due to the flow of hydrogen (H^+) ions and sulfate (SO_4^{2-}) ions and can deliver currents of up to 450 A for 2.5 seconds.

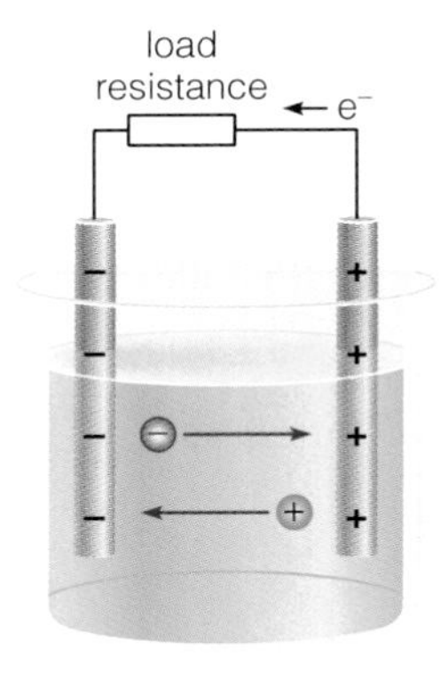

Figure 13.4 Electric current can also be due to the flow of ions.

TIP
In the examination you are assessed on your understanding of the significant figures of the numbers that you use in calculations. In the example on the right, both the pieces of data used in the calculation are to 2 significant figures (sf). As a general rule, you should state your answer to the same number of sf as the *least* significant piece of data used – 2 sf in this case.

Both of the ions in the solution contribute to the current, which can be measured externally by an ammeter connected in series with both electrodes.

EXAMPLE

If a car battery delivers 450 A for 2.5 seconds, calculate the total charge flowing, ΔQ.

Answer

$$\Delta Q = I \times \Delta t$$
$$= 450\,\text{A} \times 2.5\,\text{s}$$
$$= 1125\,\text{C}$$
$$= 1100\,\text{C or } 1.1 \times 10^3\,\text{C} \quad (2\text{ sf})$$

Variations of current with time

Figure 13.5 shows the variation of the current drawn from a cell with time. The current is drawn for a total of 10 seconds, but 5 s after turning on switch A, switch B is closed allowing current to travel through the second identical resistor:

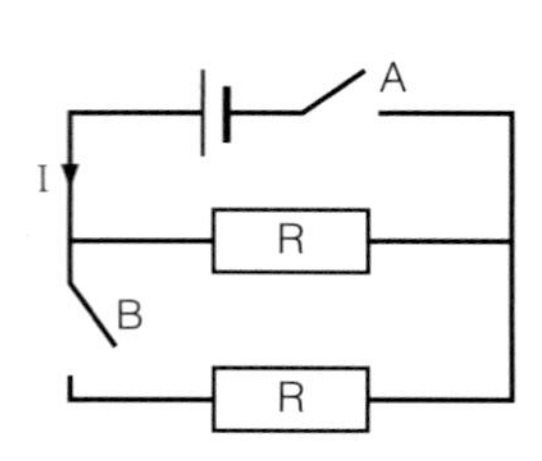

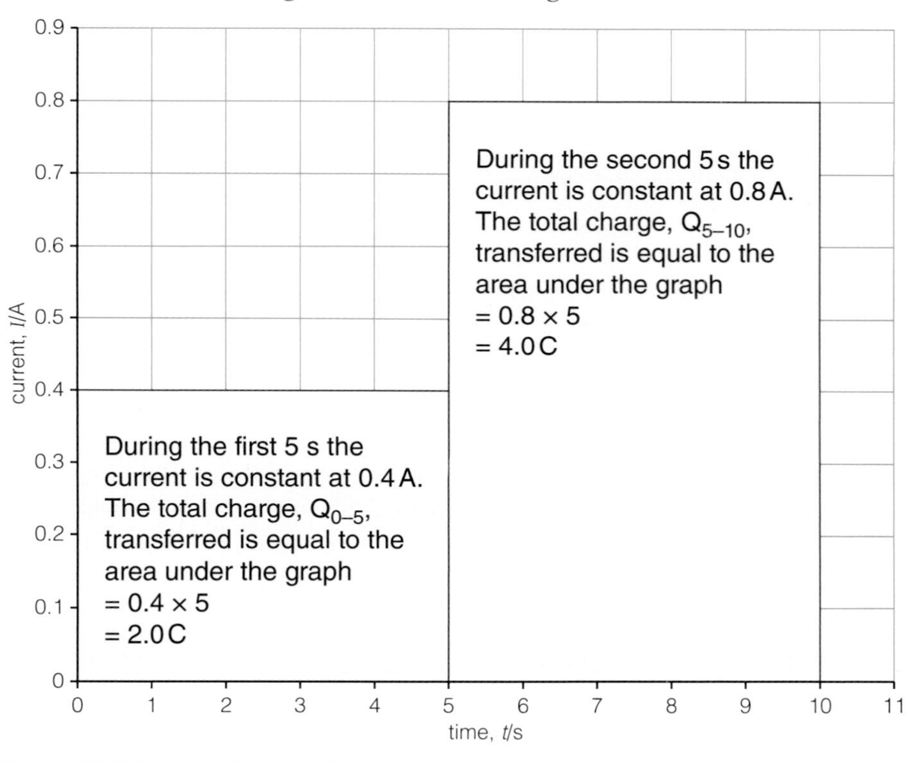

Figure 13.5 Current–time graph.

TIP
The area under a current-time graph is the charge transferred.

As

$$\Delta Q = I \times \Delta t$$

the charge transferred during the first 5 s is the area under the current–time graph from $t = 0$ s to $t = 5$ s. During the second 5 s, the current doubles. The charge transferred during this time is the area under the graph from $t = 5$ s to $t = 10$ s. The total charge ΔQ transferred during the whole 10 s period is the total area under the graph

$$\Delta Q = Q_{0-5} + Q_{5-10} = 2.0\,\text{C} + 4.0\,\text{C} = 6.0\,\text{C}$$

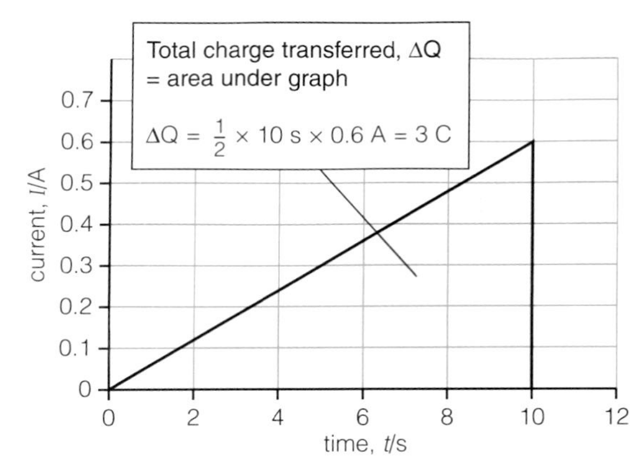

Figure 13.6 Current–time graph for a traffic light circuit.

The total charge transferred is the area under a current–time graph.

Figure 13.6 shows the variation of current with time through part of the timing circuit for a set of traffic lights.

In this case, the current varies continuously with time. The total charge transferred is still the area under the current–time graph, which is calculated to be 3 C in this case.

In a similar way, the gradient of a charge–time graph ($Q-t$) is the current. Figure 13.7 shows the current sparking off the top of a school van der Graaff generator.

The current, I, during this discharge is determined by calculating the gradient of the $Q-t$ graph as $I = \frac{\Delta Q}{\Delta t}$.

In this case,

$$I = \frac{0.5 \times 10^{-6}\text{C}}{0.1\,\text{s}} = 5 \times 10^{-6}\text{A} = 5\mu\text{A}.$$

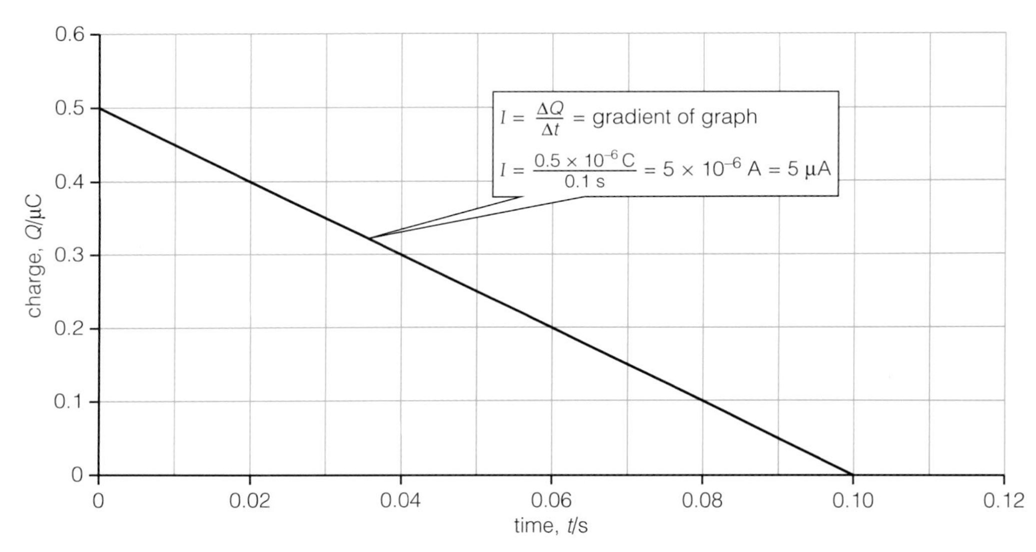

Figure 13.7 The current sparking off the top of a school van der Graaff generator.

TEST YOURSELF

1 Here is a list of electrical units:

As AV^{-1} Cs^{-1} Js^{-1} JC^{-1}

Choose the unit for:

a) electric charge
b) potential difference
c) electric current.

2 An iPhone can be charged three ways:

- via a PC USB port, which delivers 0.5 A
- via the iPhone charger, which delivers 1.0 A
- via an iPad charger, which delivers 2.1 A.

The iPhone battery capacity is 1420 mAh, which equates to a total electric charge of 5112 C.

a) Calculate the time needed to charge an empty iPhone battery from each of these charging methods.
b) Explain why 1420 mAh equals 5112 C. (A charge of 1 mAh is transferred when a current of 1 mA flows for 1 hour.)

3 Sprites are rare, large scale, very high altitude lightning bolts that often take the shape of a jellyfish.

Figure 13.8 High-altitude sprite lightning.

A sprite delivers a total charge of 285 C in 3.5 ms. Calculate the current flowing in this type of sprite.

4 A blue light emitting diode (LED) has a current of 25 mA flowing through it. Calculate the total number of electrons passing through the blue LED per second. (The charge on an electron is 1.6×10^{-19} C.)

5 A scanning tunnelling microscope works by bringing a very fine metal needle tip up very close to a surface; electrons then flow between the tip and the surface. Tiny piezoelectric crystals in the tip move the tip up and down in response to the size of the current flowing between the tip and the sample.

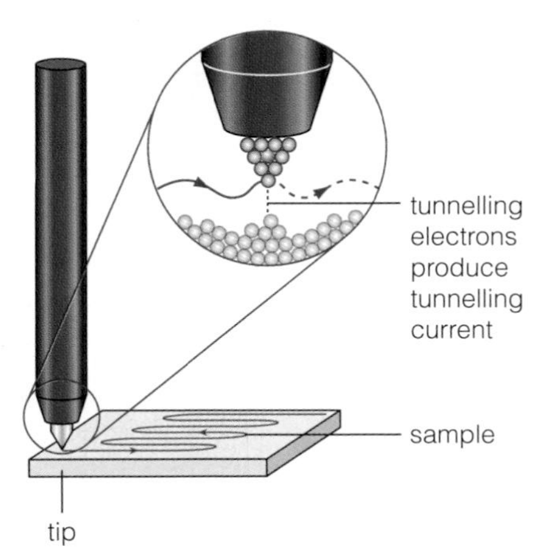

Figure 13.9 A scanning tunnelling microscope.

15 million electrons per second flow across the gap between the tip and the surface. Calculate the current in the circuit.

6 An ion beam used to construct integrated circuits delivers a charge of 80 nC in a time of 46 s. Calculate the current of the beam and the number of ions passing per second. Assume that each ion has a charge equivalent to the charge on the electron.

7 This question requires you to apply your knowledge of currents in a more complicated situation. A car battery has a low voltage, typically of the order of a few volts, but it can produce a very large current. The large current results from rapid chemical reactions involving large numbers of ions. You need to use the following data:

- magnitude of $e = 1.6 \times 10^{-19}$ C
- Avogadro number, the number of particles in one mole, $N_A = 6.0 \times 10^{23}$ particles/mole
- molar mass of lead, $M_{Pb} = 207.2$ g.
- In a lead chemical cell, the energy transferred by the chemical reactions between the lead electrodes and the electrolyte is about 24 kJ per mole of lead.

a) Calculate the charge carried by a mole of electrons (known as a Faraday of charge).

b) Show that the electrical potential energy transferred to each coulomb of charge (i.e. the potential difference, V) is 0.125 V. (Each lead atom loses two electrons to become a Pb^{2+} ion.)

c) How many lead chemical cells would be required to produce the same potential difference as a 12 V car battery?

d) When driving at night the battery delivers a current of 15 A to power the lights; how many electrons flow from the battery per second?

e) Calculate the number of moles of electrons flowing per second.

f) Chemical cells can produce large charge flows for substantial times (i.e large currents) before their chemicals have all reacted and the cell 'runs out'. In a lead chemical cell, what mass of lead is reacting per second to produce a current of 15 A? (Remember that each lead atom loses two electrons to become a Pb^{2+} ion.)

g) If the cell contains 640 g of reactable lead, for how long could it generate this current before running out? (Note that we protect a car's battery by running the lights off the car's alternator when the engine is running; the alternator generates electricity using energy from the engine.)

8 A mobile phone charger normally operates at 0.70 A. During a particular charging cycle, the charger operates normally for 1.5 hours before it suddenly develops a fault and outputs a current of 0.12 A for a further two hours.

a) Plot an I–t graph for this charging cycle.

b) Calculate the total charge transferred during the whole cycle.

Potential difference and electromotive force

Electric current only measures the rate at which charged particles (usually electrons) flow around a circuit. It tells us nothing about the electrical energy involved with circuits. The quantity used to describe the electrical energy in circuits is potential difference (pd), symbol V, measured in volts, V. A potential difference across an electrical component is measured by putting a voltmeter across the component, in parallel with it.

Potential difference is the amount of electrical work done per unit charge flowing through a component.

Electrical energy in circuits is defined in terms of the electrical work done by the electric charge flowing through the circuit. Potential difference (pd) is defined as the electrical work done per unit (coulomb) of charge flowing through components such as bulbs, motors, resistors, etc. This electrical energy is transferred into heat, light and other more useful forms of energy by the components.

$$\text{potential difference, } V\ (\text{V}) = \frac{\text{electrical work done by the charge, } W\ (\text{J})}{\text{charge flow, } Q\ (\text{C})}$$

$$V = \frac{W}{Q}$$

$$1\,\text{V} = 1\,\text{J}\,\text{C}^{-1}$$

Potential difference, however, cannot be used to describe the energy changes involved with power supplies such as cells, generators and mains power supply units. These devices transfer other forms of energy, such as chemical energy into electrical energy. To make a distinction between these different energy transfers we define another quantity, electromotive force (emf), symbol ε. Both emf and pd are measured in volts, symbol V, using a voltmeter.

Electromotive force is the amount of electrical work done per unit charge by a power supply such as a cell in an electrical circuit. The power supply transfers the other forms of energy, such as chemical energy into electrical energy. The unit is the volt, V.

Electromotive force (emf) is defined as the electrical work done per unit (coulomb) of charge as it flows through a source of electrical energy such as a cell, generator or power supply unit (psu). The sources of electrical energy transfer other forms of energy such as kinetic or light energy into electrical energy.

$$\text{electromotive force, } \varepsilon\ (\text{V}) = \frac{\text{electrical work done on the charge, } E\ (\text{J})}{\text{charge flow, } Q\ (\text{C})}$$

$$\varepsilon = \frac{E}{Q}$$

The law of conservation of energy can now be written in terms of emf and pd. In a series circuit, where the components are connected one after another in a complete loop, the total electrical energy per coulomb transferring into the circuit (the sum of the emfs in the circuit) must equal the energy per coulomb transferring into other forms of energy (the sum of the pds). (There is more about this in Chapter 14 on electrical circuits.)

Here is an example. The following circuit is set up:

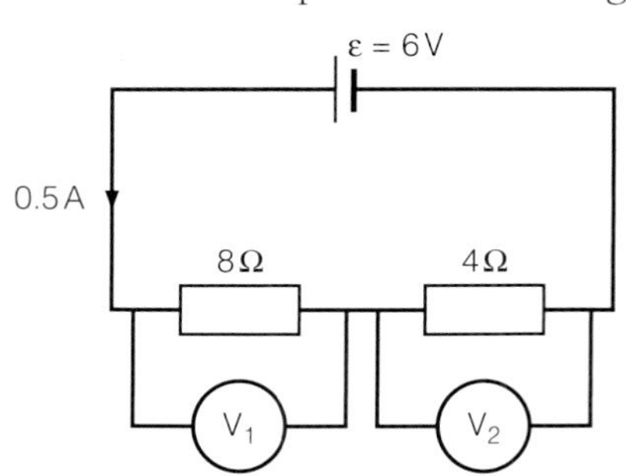

Figure 13.10 Circuit diagram.

The emf of the cell, ε, transfers 6 J C^{-1} (V) of chemical energy into electrical energy (if the cell is 100% efficient).

The 6 J C^{-1} of electrical energy is shared between the two resistors. This energy is shared in the same ratio as the resistance of the resistors (8 : 4 or 2 : 1). The potential difference, V_1, across the 8 Ω resistor is therefore 4 J C^{-1} (V) and the potential difference, V_2, across the 4 Ω resistor is therefore 2 J C^{-1} (V). Note that the law of conservation of energy still holds here as 6 J C^{-1} (V) is transferred from chemical energy into electrical energy in the cell and 6 J C^{-1} (V) in total is transferred from electrical energy into heat energy within the resistors.

ACTIVITY

Investigating the ability of a travel mug heating element to boil water

In an experiment to investigate the ability of a travel mug heating element to boil water in a mug for a cup of tea, a student connected the heating element to a car battery.

Figure 13.11 Travel heating element.

The student put 180 g of water into a mug and measured the temperature of the water every minute for 8 minutes. Here are the results:

Time, *t* (minutes)	Temperature, θ (°C)
0.0	15
1.0	26
2.0	37
3.0	48
4.0	59
5.0	70
6.0	81
7.0	92
8.0	100

One gram of water requires 4200 J of heat energy to raise the temperature of the water by 1 °C. This is called the specific heat capacity of water. An ammeter measures the current in the heater to be 11.1 A during the heating.

1 Convert the times to seconds and plot a graph of temperature, θ/°C against time, *t*/s.
2 Plot a best-fit line and calculate the gradient of the line.
3 What quantity is represented by the gradient of the line?
4 The heat energy, ΔQ, required to heat the water is given by the equation:

$$\Delta Q = mc\Delta\theta$$

where *m* is the mass of the water, *c* is the specific heat capacity of the water and $\Delta\theta$ is the change in temperature. How much energy heats 180 g by one degree? (Note: The heat energy ΔQ should not be confused with electrical charge.)
5 How much heat has been supplied therefore after two minutes?
6 How much charge is transferred in 2 minutes?
7 Assuming that the electrical energy supplied to the heater is transferred into heat energy of the water, use your answers to questions 5 and 6 to calculate *V*, the potential difference across the heater.
8 State two hazards involved with this experiment and describe the risks and control measures that you would use.
9 This experiment involves three measurements, the volume and temperature of water, and the time. What are the sources of uncertainty in these measurements? For each one, state a way of reducing this uncertainty?

Resistance

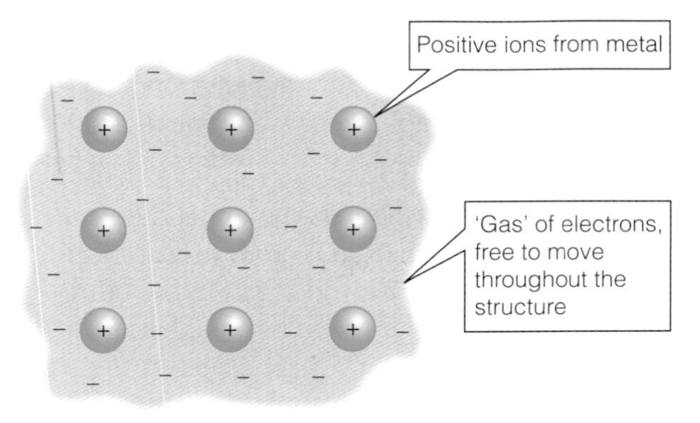

Figure 13.12 The structure of a metal.

When current flows through the material of a circuit, such as the metal of the connecting wires, the material of the circuit gets in the way of the flow of the charge. On a microscopic level, as the electrons flow through the metal they collide with the vibrating positive ion cores of the metal structure. The collisions between the electrons and the positive ion cores transfer electrical energy from the electrons to the structure of the metal, causing the metal ion cores to vibrate more, thus heating up the wire.

The opposition of a component to the flow of electric current through it is called resistance, symbol R, unit ohms, Ω.

The **resistance** of a conductor is the opposition of the conductor to electric current flowing through it. Unit is the ohm, Ω.

A **superconductor** is a material whose resistance drops to zero below a specific temperature, called the critical temperature.

The model of a metal shown above explains what happens to the resistance of the metal when it is heated. The resistance increases as the temperature increases. The vibrating, positively charged, ion cores move around much more as the temperature increases, thus getting in the way of electron flow. This opposes the flow of the gas of electrons moving through the structure.

Components with very high resistances let very little current through them, and are considered to be electrical insulators. Some materials, at very low temperatures, have zero resistance. These materials are called superconductors.

Figure 13.13 Magnetically levitating superconductor.

Superconductors

A current set up in a loop of superconducting material carries on flowing indefinitely. Superconductors also exclude magnetic fields inside them. This allows a strong permanent magnet to be repelled and held above the superconductor. Superconducting materials completely lose their resistance below a temperature called the critical temperature, T_c. Different superconducting materials have different critical temperatures; the metallic element tungsten, W, for example, has the lowest known critical temperature of 0.015 K, (−273.135 °C); and a type of ceramic copper oxide, called mercury barium thallium copper oxide, has the highest known critical temperature of 139 K (−134.15 °C). This is much higher than the boiling point of liquid nitrogen (77 K).

Figure 13.14 One of the Shanghai Maglev trains, where superconducting magnets support the train, enabling it to travel at speeds up to 270 mph (430 km/h).

As new materials are developed, so the superconducting critical temperature has risen substantially. The ultimate goal is to develop superconductors with critical temperatures that are around room temperature. Imagine the world of possibilities with zero-resistance superconductors. There could be, for example, electronic devices and computer units that don't generate any heat and don't need cooling fans; batteries that last for an extremely long time on one charge; cheap magnetic levitation; portable MRI scanners; super-strong electromagnets and electrical power transmission lines that don't waste any energy.

Current/potential difference characteristics and Ohm's law

An **electrical characteristic** is a graph (usually $I - V$) that illustrates the electrical behaviour of the component.

Most of the properties of an electrical component can be determined by plotting the current – potential difference graph of the component. This is called an electrical characteristic.

Electrical characteristics are graphs that show how the potential difference and current vary when the component is connected in both forward and reverse bias (when the current passes one way and then in the other direction). The most straightforward electrical characteristic is that of a fixed resistor.

You can see that the graph can be drawn with each variable on either axis.

> **TIP**
> I–V graphs can be drawn with I or V on either axis to illustrate the relationship between a current through a component and the potential difference across a component. I–V graphs (with V on the x-axis) are useful because the current through a component depends on the potential difference across it. This type of graph is particularly useful for people who are designing circuits for devices where they need to know how a component will behave with different applied potential differences. V–I graphs (with I on the x-axis) are useful for illustrating the resistance of a component, and the characteristics of cells and batteries are usually plotted this way.

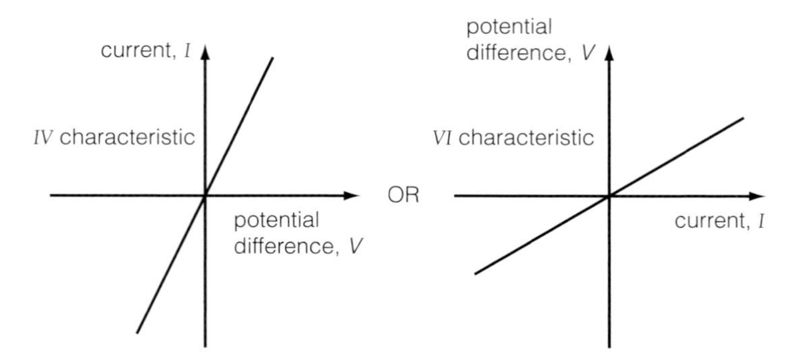

Figure 13.15 Electrical characteristic graphs of a fixed resistor.

Ohm's law

In 1827, the German physicist Georg Ohm performed potential difference – current experiments on metal wires at constant temperature. Ohm discovered that the current, I, flowing through the wire was proportional to the potential difference, V, across the wire, provided that the temperature (and other physical variables) remained constant. If this is shown on an electrical characteristic graph, then the graph will be linear. The relationship between the current and the potential difference became known as Ohm's law, which can be written mathematically by:

current, $I \propto$ potential difference, V

$$I \propto V$$

Ohm's law applies to some conductors under some circumstances. It is generally a special case applied to metal wires at constant temperature. But, using the concept of resistance, a convenient, familiar mathematical equation can be written that uses the relationship between I and V to define the unit of resistance, the ohm (Ω):

$$R = \frac{V}{I}$$

In other words, if a potential difference of 1 V produces a current of 1 A flowing through a component, then its resistance must be 1 Ω. This defines resistance in ohms. This equation applies in all cases and also makes it clear that for a fixed pd, V, a small measured current, I, implies that the component has a large resistance, R, and vice versa. Similarly, for a fixed current, I, we need a large pd, V, to drive the current through a large resistance, R.

> **TIP**
> The equation:
> $$R = \frac{V}{I}$$
> defines resistance. Some sources on the web incorrectly call the equation Ohm's Law. It is not.

> **TIP**
> Ohm's Law states that for some conductors
> $$I \propto V$$

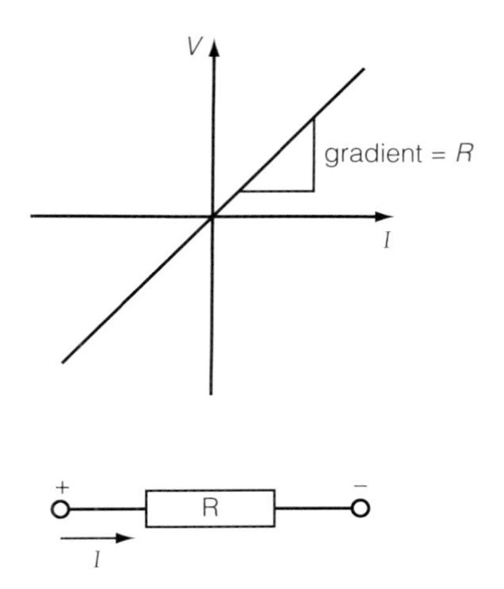

Figure 13.16 An electrical characteristic of a fixed resistor showing Ohm's law.

The current flowing through an **ohmic conductor** is proportional to the pd applied across it.

One use of the electrical characteristic graph is that if the current is plotted on the x-axis, for component's following Ohm's law then the gradient of the line is the same value as the resistance of the component. As fixed resistors and metal wires at constant temperature have a fixed resistance, then the electrical characteristic graph will be a straight line. On this type of graph, high-resistance components have large, steep gradients, and low-resistance components have small, shallow gradients. Remember though, electrical characteristics can be drawn with either variable on either axis.

Fixed resistors and metal wires at constant temperature obey Ohm's law across their current range and their electrical characteristics are linear. Components like this are said to be **ohmic conductors** – i.e. they obey Ohm's law.

Other electrical characteristics

A standard tungsten filament lamp transfers electrical energy into light and heat as the current flows through it. As the current increases, so the frequency of electron collisions with the positive ion cores of the tungsten lattice increases, transferring more kinetic energy. The positive ion cores vibrate with greater amplitude and so the resistance increases. A higher current leads to a higher temperature, which in turn leads to a higher resistance. The electrical characteristic of a filament lamp is shown in Figure 13.17:

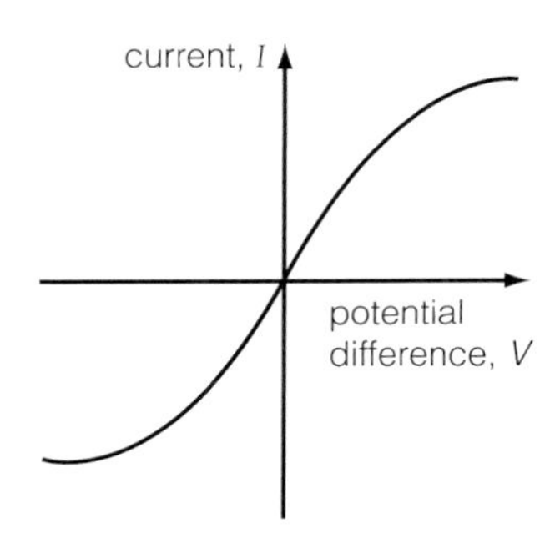

Figure 13.17 Electrical characteristic of a filament lamp.

A **non-ohmic** component is a component that does not obey Ohm's law; i.e. current is not proportional to the potential difference applied across it.

The electrical characteristic shows the ratio V/I increasing and therefore the resistance increasing. Components like this, whose electrical characteristics are non-linear, are said to be **non-ohmic** conductors.

Components such as fixed resistors and filament lamps have the same characteristics independent of the direction of the current flowing through them. Their V–I graphs produce the same shapes in forward and reverse bias. Components such as semiconductor **diodes** do not behave in this way. Diodes are generally found in electronic circuits where they act as one-way gates, preventing the current from flowing back through the circuit. They are particularly useful in mains power supplies where they can be used in circuits to convert alternating current (ac) into direct current (dc).

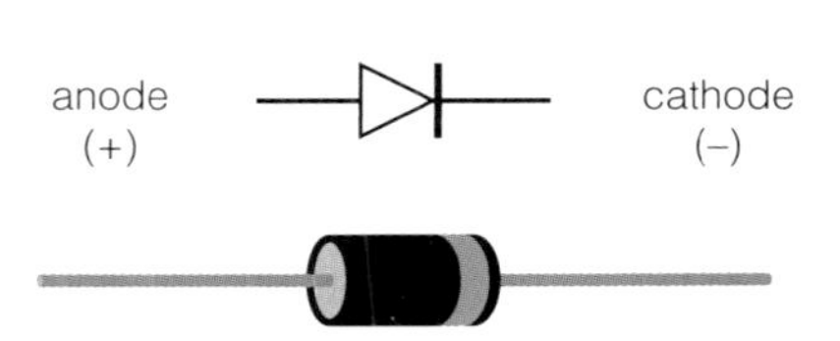

Figure 13.18 A semiconductor diode – symbol and picture. The current will only flow in the direction of the arrow (anode to cathode).

Diodes only conduct in forward bias (in the direction of the arrow on the symbol); their circuit symbol shows this direction using the arrow, and the component itself normally has a different coloured ring at the forward bias end.

Diodes do not conduct in reverse bias; this means that the resistance of the diode in reverse bias is infinite. Diodes have very low resistance in forward bias. The electrical characteristics of diodes are usually drawn in current (y-axis)–potential difference (x-axis) format:

> **TIP**
> Make sure that you are aware of how the electrical characteristic is plotted. Carefully check which quantity is on each axis. Also, be careful to ensure that you know what the units of each axis are.

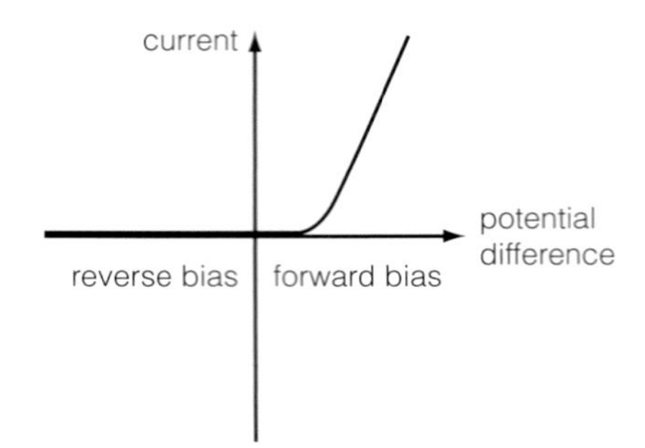

Figure 13.19 Electrical characteristic of a semiconductor diode.

ACTIVITY

Plotting electrical characteristics

The electrical characteristics of a small bulb and a fixed resistor were compared. The circuit diagrams used are shown in Figure 13.20.

The table shows data taken from the experiment. (When the components were connected in reverse bias, the current and potential difference values were numerically the same but with a negative sign.)

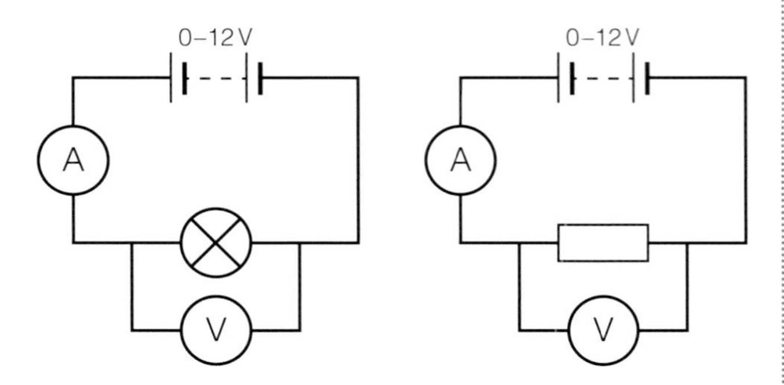

Figure 13.20 Electrical circuit diagrams.

Current, I (A)	Bulb potential difference, V_{bulb} (V)	Fixed resistor potential difference, $V_{resistor}$ (V)
0.00	0.00	0.00
0.25	0.44	1.04
0.50	1.02	2.07
0.75	1.85	3.10
1.00	3.20	4.12
1.25	4.64	5.14
1.50	6.30	6.15
1.75	8.90	7.20
2.00	11.43	8.20

1. Plot an I–V electrical characteristic illustrating this data (pd on the x-axis); show both sets of data on the same graph.
2. Use your graph to determine and label which component obeys Ohm's law.
3. Plot a V–I electrical characteristic illustrating this data (current on the x-axis); show both sets of data on the same graph.
4. Explain why the gradient of the line showing the results for the fixed resistor is the resistance (in ohms) of the resistor.
5. Use the graph to calculate the resistance of the fixed resistor.
6. At what current value is the resistance of each component the same?
7. The resistance of the bulb is not constant. Construct a copy of the table but only include the data from the bulb (in forward bias only) and add an extra column showing the resistance of the bulb at each current value.
8. Plot a graph of bulb resistance against current.
9. Explain why the resistance of the bulb varies in this way.
10. What do you think happens to R as I tends to 0 (zero)?
11. Write a method detailing how you would collect similar data for a diode. Include a circuit diagram in your description.

Thermistors

A **thermistor** is a component whose resistance varies with temperature. Many thermistors used in electrical circuits have a negative temperature coefficient (ntc), meaning that their resistance decreases with increasing temperature.

Thermistors are a type of resistor that change their resistance with temperature. Thermistors have the following circuit symbol:

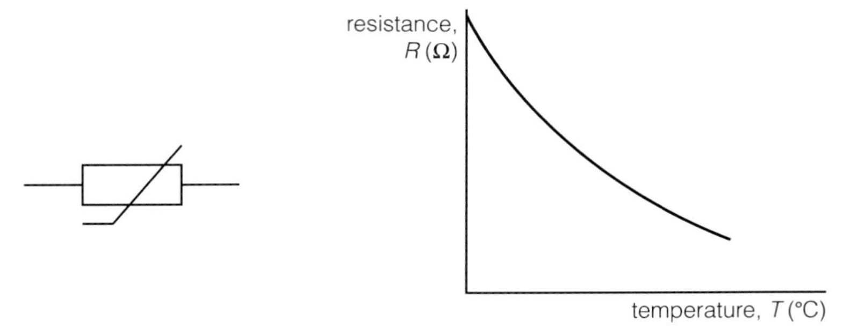

Figure 13.21 Thermistor circuit symbol and resistance – temperature graph.

Most thermistors are negative temperature coefficient (ntc) components. This means that their resistance decreases with increasing temperature. The vast majority of ntc thermistors also have a non-linear response to changing temperature; in other words, a graph of resistance against temperature is a curve.

Thermistors are widely used in circuits to sense temperature changes, and then to control devices; the change in the resistance of the thermistor affects a current, which can be used to switch devices on or off. They can be found in thermal cut-out circuits to prevent devices from overheating (such as a hairdryer), central heating circuits, digital thermometers and engine-management circuits.

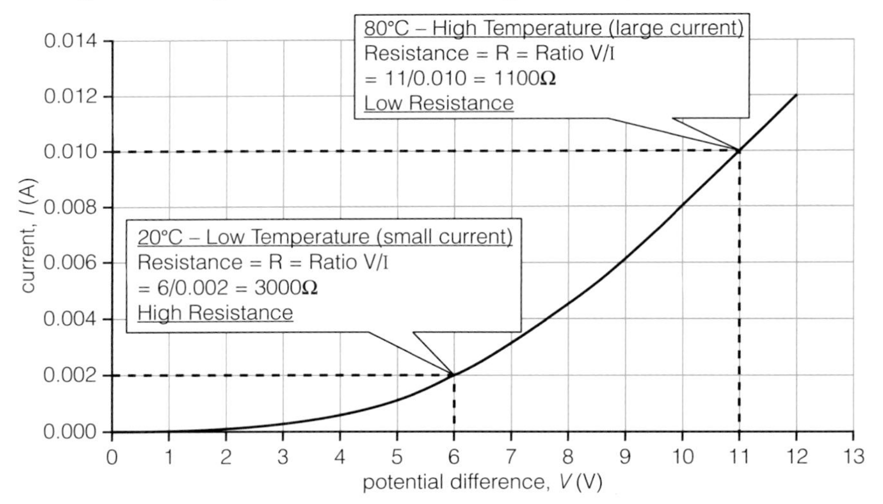

Figure 13.22 Thermistor electrical characteristic example.

At a low temperature of 20 °C, the resistance of the thermistor is high and so the current (and the V/R ratio) is low:

$$R = \frac{V}{I}$$

$$= \frac{6\,V}{0.002\,A}$$

$$= 3000\,\Omega$$

A higher temperature of 80 °C, the resistance is lower and the V/R ratio is higher:

$$R = \frac{V}{I}$$

$$= \frac{11\,V}{0.010\,A}$$

$$= 1100\,\Omega$$

the current flowing through the thermistor is therefore high.

ACTIVITY

The resistance of a thermistor

Design an experiment to determine how the resistance of a thermistor varies with temperature.

You should include:

- a description of the independent, dependent and control variables. For each variable you should state appropriate ranges, intervals, values and suitable measuring equipment
- a diagram of your experiment including a suitable circuit diagram
- a list of equipment including appropriate capacities, ranges and settings
- a statement of how you will minimise the errors on the measurements that you will take
- a numbered method stating how the experiment will be performed
- a suitable risk assessment
- a table that could be used to record your primary data
- a description of any calculations that need to be carried out, including how you could calculate the uncertainty in your measurements.

TEST YOURSELF

9 Sprites are a relatively newly discovered phenomena. The first recorded visual observation was in 1886, but the first photograph of a sprite was only taken in 1989. Jellyfish sprites occur at altitudes of up to 90 km and they can cover an area of 50 × 50 km. If a sprite lightning strike is generated by a potential difference of 200 000 000 V, and 325 C of charge is transferred, calculate the total energy transferred during the strike.

10 A small neon indicator lamp on a cooker circuit requires 95 V in order to conduct. It then draws a current of 0.8 mA.

a) Calculate the resistance of the lamp at this current and voltage.

b) Calculate the number of electrons moving through the lamp each second.

11 Figure 13.23 shows the electrical characteristic graphs of three different electrical components: A, B and C.

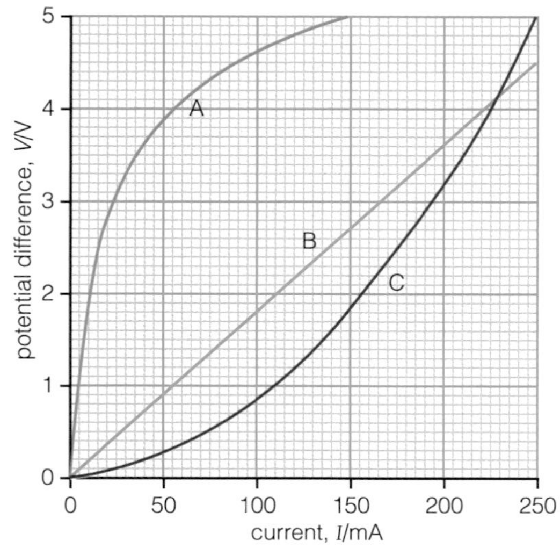

Figure 13.23 Electrical characteristic graphs of three different electrical components.

a) Which component, A, B or C, obeys Ohm's law?

b) Use values from the graph to determine which component has an increasing resistance at higher current.

12 A red-coloured LED starts to conduct electrical current when the potential difference across it is greater than 1.5 V. Figure 13.24 shows the circuit used to run the LED from a 6.0 V battery that has negligible internal resistance.

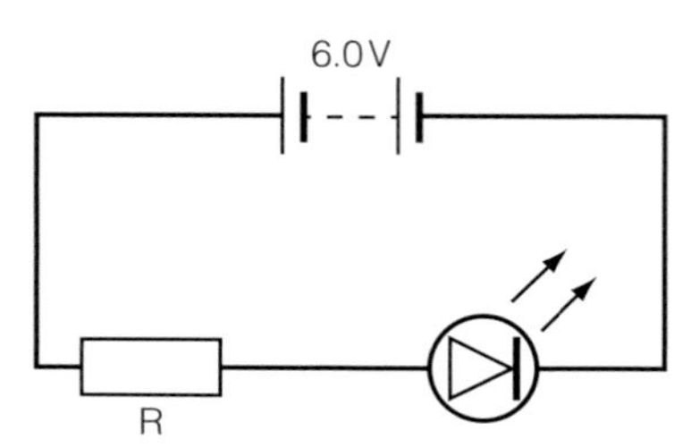

Figure 13.24 Electrical circuit diagram.

a) Explain the reason for putting the resistor, R, into the circuit.

b) When operating at its normal current of 20 mA the potential difference across the LED is 2.2 V. Calculate the value of the resistor, R, for operation of the LED at 20 mA from the 6.0 V battery. (Remember, the sum of the emfs is equal to the sum of the pds in a series circuit.)

13 Figure 13.25 shows how the potential difference varies with current for a red and a yellow LED.

a) Calculate the resistance of each LED at each of the following currents:

i) 30 mA

ii) 10 mA

b) Describe one other difference between the behaviour of the two LEDs.

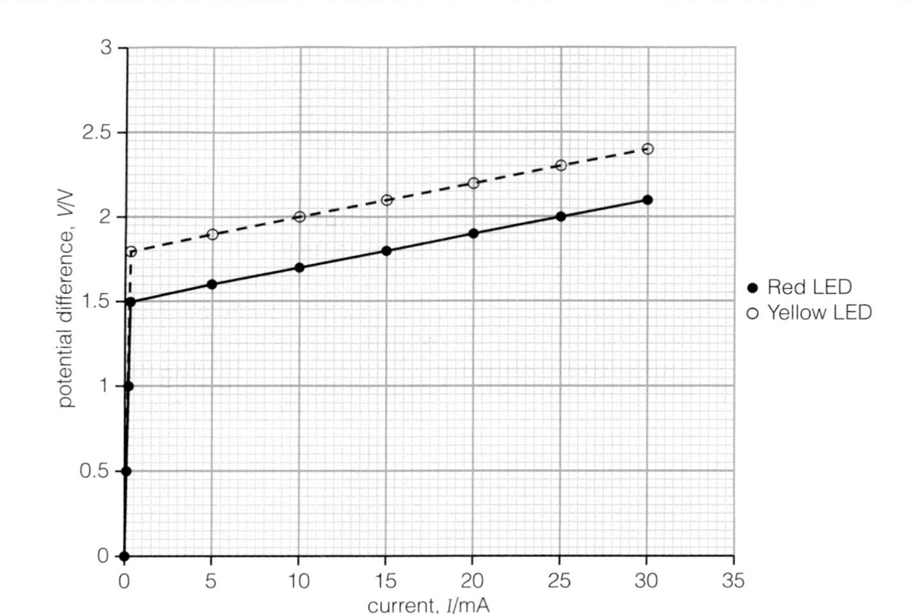

Figure 13.25 How the potential difference varies with current for a red and a yellow LED.

14 Figure 13.26 illustrates how the resistance of a negative temperature coefficient thermistor varies with temperature.

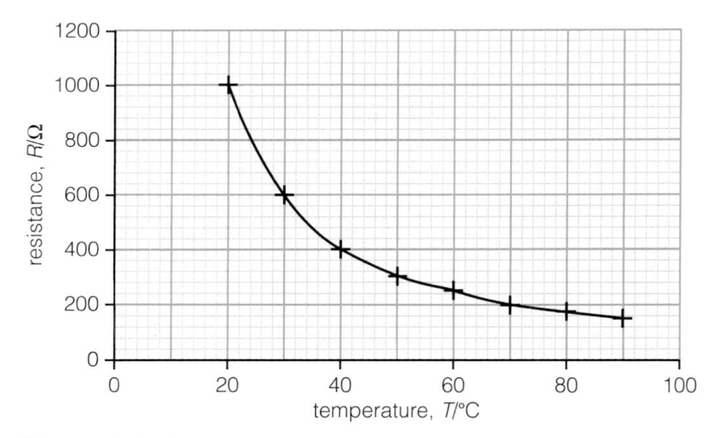

Figure 13.26 Resistance–temperature graph.

a) Use the graph to estimate the rate of change of resistance with temperature at:

i) 24 °C

ii) 84 °C.

b) The thermistor is to be used as a resistance thermometer. At which temperature is it more sensitive, 84 °C or 24 °C? Explain your answer.

15 Figure 13.27 shows the thermistor from Question 14 together with a resistor in a temperature-sensing circuit, similar to circuits found in electronic thermometers.

a) A voltmeter is connected to the circuit to indicate an increasing potential difference when the sensor detects an increasing temperature. Copy and complete the diagram showing the circuit connections for a voltmeter to measure a potential difference that increases with increasing temperature.

b) The value of the fixed resistor in the circuit diagram is 250 Ω. The thermistor is at 25 °C. Using data from the graph in Figure 13.26, calculate the current drawn from the 6.0 V supply. See Question 12 on the previous page.

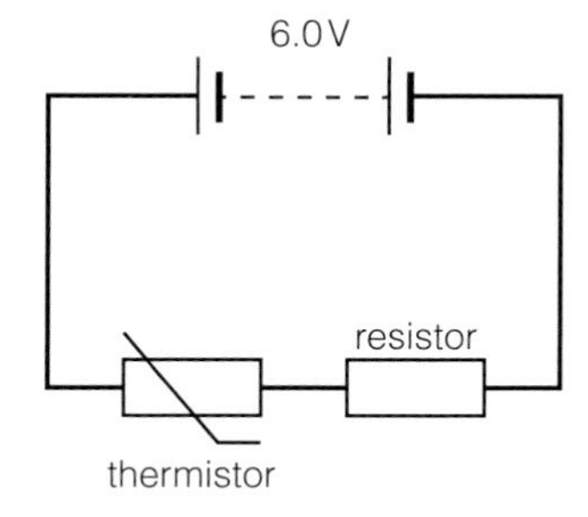

Figure 13.27 A thermistor together with a resistor in a temperature-sensing circuit.

Resistivity

An **intrinsic** property of a material is a property of the material itself, independent of other factors. The resistance of a piece of wire, for example, depends on the dimensions of the wire, in addition to the material that it is made of – resistivity is the intrinsic 'resistance' of the material, independent of its dimensions.

How is the intrinsic 'resistance' of different materials compared? How do you decide which material would be best for the internal connections of a mains plug, or for using as an insulator on a high-voltage line? The resistance of a metal wire, for example, depends on the length and area of the wire, as well as the material that the wire is made from. Consider Figure 13.28, showing a metal conductor.

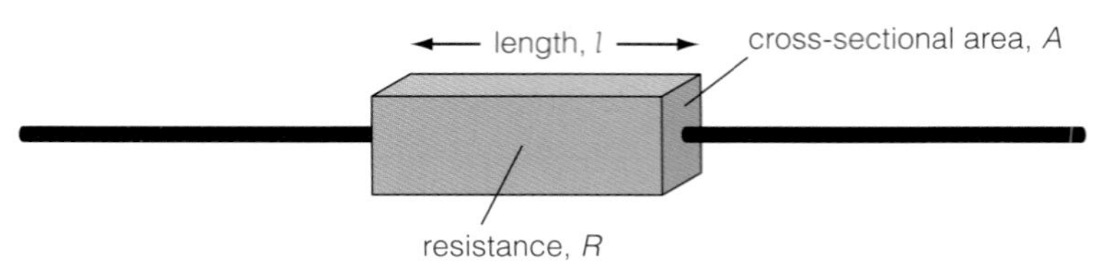

Figure 13.28 The resistance of a wire depends on its dimensions, but resistivity depends on what the material is made of.

The resistance of the conductor, R, increases with increasing length, l (there are more positive ion cores in the way of the gas of electrons moving through the conductor). In fact, if the length, l, doubles, then the resistance, R, doubles. This means that $R \propto l$.

The resistance of the conductor decreases with increasing cross-sectional area, A. (There are more conducting pathways through the conductor for the electrons to move through.) In this case, if the cross-sectional area, A, doubles,

then the resistance, R, halves. This means that $R \propto \frac{1}{A}$. Combining both of these proportionality statements together:

$$R \propto \frac{l}{A}$$

and replacing the proportionality sign and adding a constant of proportionality, ρ, we have:

$$R = \frac{\rho \times l}{A}$$

The **resistivity** of a substance is the intrinsic resistance of the material that the substance is made from; it is independent of the dimensions of the substance. Resistivity also varies with temperature, although the resistivity of metals changes only gradually with temperature.

where ρ ('rho') is called the electrical resistivity of the material. Resistivity is the property that gives the intrinsic resistance of the material independent of its physical dimensions, such as length and cross-sectional area. Resistivity has the units of ohm metres, Ω m, and is defined by the rearranged form of the equation:

$$\rho = \frac{RA}{l}$$

where R is the resistance (measured in ohms, Ω), A is the cross-sectional area (measured in metres squared, m^2) and l is the length (measured in metres, m).

The resistivity of a material depends on some intrinsic properties of the material. In particular, it relates directly to the number of free, conducting electrons that can flow through the structure and the mobility of these electrons to flow through the structure. The arrangement of the atoms in the conductor and any distribution of impurities affects this mobility, as does the temperature of the material.

At room temperature (20 °C), good insulating materials, such as ABS plastic (the material that most mains plugs are now made from), have extremely high resistivity (ABS has an electrical resistivity of 1×10^{15} Ω m at 20 °C).

Good metallic conductors have very low resistivity; the resistivity of copper, for example, is $1.68 \times 10^{-8}\,\Omega\,m$ at 20 °C. Superconductors have zero resistivity below their critical temperature.

We can use a water analogy to help understand resistivity. A hose pipe full of sand is like a high resistivity conductor, as water trickles through slowly when pressure is applied. However, that same pressure pushes water quickly through an empty hose pipe, which is like a low resistivity conductor. Remember though that the ability of the water to pass through the hosepipe also depends on the length of the hosepipe and its cross-sectional area.

Resistivity is also dependent on temperature. The resistivity of metals increases with increasing temperature, and the resistivity of many semiconductors, such as silicon and germanium, decreases with increasing temperature. The resistivity of a superconducting material decreases with decreasing temperature above its critical temperature (like a metal), but its resistivity drops to zero below the critical temperature.

REQUIRED PRACTICAL 5

Determination of resistivity of a wire using a micrometer, ammeter and voltmeter

Note: This is only one example of how you might tackle this required practical.

A 2.0 m length of 36 swg (standard wire gauge) constantan is clamped to a laboratory bench. A suitable measuring instrument is used to measure the diameter of the wire ten times at random distances along the length of the wire. A fixed 6.0 V power supply is then connected to one end of the wire and an ammeter and a flying lead and then used to connect in series to different lengths of the wire. The distance between the fixed end of the wire and the flying lead is then varied from 0.00 m to 2.00 m in 0.25 m steps.

The diameter measurements are (in mm):

0.19; 0.19; 0.20; 0.19; 0.18; 0.19; 0.20; 0.18; 0.20; 0.19

1. Draw a circuit diagram for this experiment.
2. State the name of a suitable measuring instrument for measuring the diameter of the wire.
3. Calculate the average diameter of the wire and determine an uncertainty for this measurement.
4. The following data was collected from this experiment. Copy this table and add a third row showing the resistance of each length of constantan.

Length of wire (m)	0.00	0.25	0.50	0.75	1.00	1.25	1.50	1.75	2.00
Current (A)	Short circuit	1.34	0.84	0.51	0.35	0.31	0.27	0.22	0.18

5. Plot a graph of resistance, R/Ω (y-axis) against length of wire, l/m (x-axis).
6. The ammeter introduces an error of 5% on each measurement of the current and therefore an error of 5% on each resistance measurement. Use this information to plot error bars on your graph.
7. Use a best-fit line technique to determine the gradient of the graph.
8. Use the gradient calculated in step 7 to determine the resistivity of the constantan wire.
9. Kaye and Laby, the National Physical Laboratory Tables of Physical and Chemical Constants (**www.kayelaby.npl.co.uk/general_physics/2_6/2_6_1.html**) gives the electrical resistivity of constantan to be $49 \times 10^{-8}\,\Omega\,m$ at 0 °C. Determine the percentage error in your answer compared to the value given in Kaye and Laby.
10. The temperature coefficient of resistivity for constantan is $8 \times 10^{-6}\,°C^{-1}$ (this coefficient is defined as the fractional change in resistivity per °C). Explain why the value from Kaye and Laby, quoted at 0 °C, is likely to be very similar to the value at 20 °C.

Extension

Use the spread of the error bars and a suitable graphical technique to calculate the uncertainty in the measurement of the resistivity of constantan and compare it to the value from Kaye and Laby.

TEST YOURSELF

16 A square semiconductor chip made from selenium is shown in Figure 13.29. The resistance of the chip is about 100 Ω. Current is passed in and out of the chip by the metallic connectors shown.

Figure 13.29 Selenium chip.

a) Draw a circuit diagram showing how you could obtain the electrical measurements required on this chip to determine the electrical resistivity of selenium.

b) Make a list of *all* the measurements needed to determine the resistivity, and explain how you would make them.

c) Describe how all of the data can be used to calculate a value for the resistivity of selenium.

d) Suggest *one* way in which you could improve the uncertainty in the measurement of resistivity by reducing the error in one of the measurements.

17 Table 11.7 below shows the thermal conductivity and electrical resistivity (at room temperature) of five pure metals.

Table 11.7

	Thermal conductivity, χ ($W m^{-1} K^{-1}$)	Electrical resistivity, ρ (Ωm)
Copper	385	1.7×10^{-8}
Gold	310	2.4×10^{-8}
Aluminium	240	2.7×10^{-8}
Magnesium	150	4.0×10^{-8}
Zinc	110	5.9×10^{-8}

a) Plot the data on a suitable graph with thermal conductivity on the *x*-axis.

b) Using figure 13.12, describe and explain the trend shown by the graph in terms of the microscopic structure of metals.

18 The resistance, R, of a cylindrical wire is given by the following equation.

$$R = \frac{\rho L}{A} = \frac{\rho L}{\pi r^2}$$

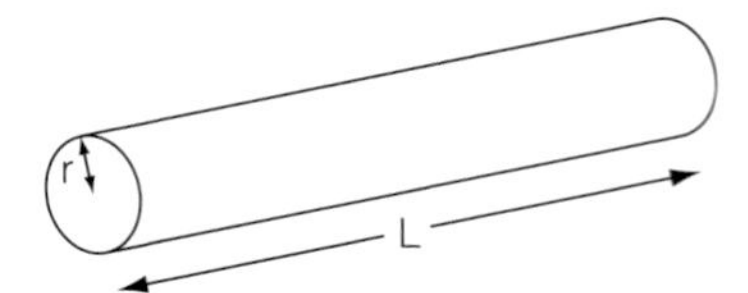

Figure 13.30 Cylinder showing dimensions of a wire.

The dimensions are shown on Figure 13.30. Here is a list of multiplying factors.

$$\times \frac{1}{4} \quad \times \frac{1}{2} \quad \times 2 \quad \times 4$$

Which of the multiplying factors above best describes the following changes:

a) The length L of the wire is doubled. The resistance R will change by a factor of...

b) The radius r of the wire is halved. The resistance R will change by a factor of...

19 Figure 13.31 shows a copper connector on the surface of a mobile phone chip. The connector has a cross-sectional area $A = 4.0 \times 10^{-10}\,m^2$, a length $L = 17.0 \times 10^{-4}\,m$ and the resistivity, ρ, of copper $= 1.7 \times 10^{-8}\,\Omega\,m$. Calculate the resistance R of this connector.

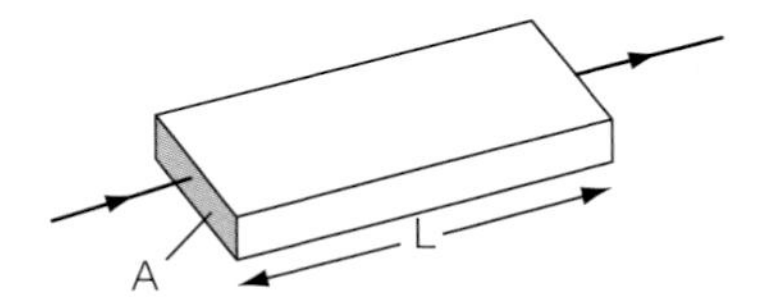

Figure 13.31 Copper connector.

Practice questions

1 The resistivity of a copper wire at room temperature is $1.7 \times 10^{-8}\,\Omega\,m$. The wire has a cross-sectional area of $3.1 \times 10^{-12}\,m^2$. The resistance of a length of 0.1 m of this wire is:

A 0.002 Ω **C** 548 Ω

B 14 Ω **D** 5484 Ω

2 A platinum wire of length 1.6 m has a constant diameter of 0.001 mm. The resistivity of platinum is $1.06 \times 10^{-7}\,\Omega\,m$. The resistance of the wire is:

A 220 mΩ **C** 2 kΩ

B 200 Ω **D** 2 MΩ

3 The current through a wire made from tin must not exceed 4.0 A. The wire has a cross-sectional area of $7.8 \times 10^{-9}\,m^2$. If a pd of 2.0 V is connected across the wire, what is the minimum length of wire needed to ensure the current remains below 4.0 A? The resistivity of tin is $1.1 \times 10^{-7}\,\Omega\,m$.

A 0.035 m **C** 0.140 m

B 0.070 m **D** 0.350 m

4 How many electrons are passing through a wire each second if the current flowing through it is 1.00 mA?

A 6.3×10^{13}

B 6.3×10^{15}

C 6.3×10^{18}

D 6.3×10^{21}

5 The heating element for a small electric heater is made from a wire of resistance *R*. It is replaced by a wire made from the same material and with the same diameter, but with half the length. The resistance of this second wire is:

A $\frac{1}{4}R$ **C** $2R$

B $\frac{1}{2}R$ **D** $4R$

6 A resistor is connected to a cell. In a time, *t*, an amount of charge, *Q*, passes through the resistor. During this time the electrical energy dissipated by the resistor is *W*. What is the current in the resistor and the emf of the cell?

	Current	emf
A	Q/t	W/Q
B	Qt	W/Q
C	Qt	W/Q
D	Q/t	W/Q

7 The wires shown in Figure 13.32 are made from the same material but have different dimensions. Which wire will have the smallest value of resistance?

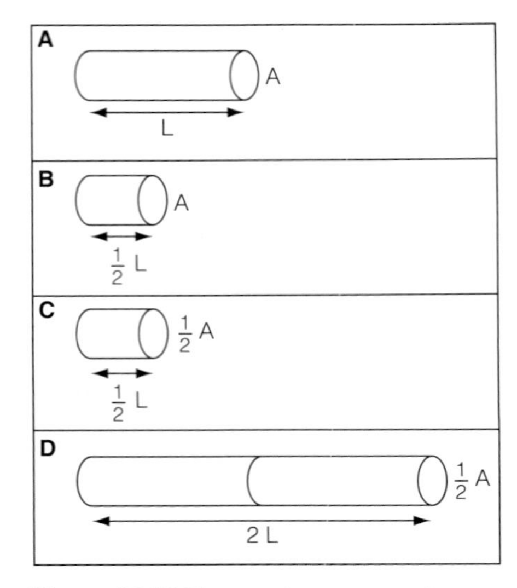

Figure 13.32 These wires are made from the same material but have different dimensions.

8 The unit of electric current, the ampere, is equivalent to:

A Cs **B** JC^{-1} **C** Js **D** Cs^{-1}

9 An electric eel can store charge in specialised cells in its body. It then discharges these cells to protect itself from predators. The discharge of an electric eel can transfer a charge of 2.0 mC in approximately 2.0 ms. What current does the eel deliver?

A 1 mA **B** 10 mA **C** 100 mA **D** 1 A

10 A car battery delivers a current of 400 A for 0.5 s. What is the charge flowing?

A 100 C **C** 400 C

B 200 C **D** 800 C

11 When a tungsten filament lamp is switched on it takes about 500 ms for the filament to heat up and reach its normal operating temperature. The variation of the current with time is shown in Figure 13.33.

a) What is the maximum current passed through the filament? (1)

b) This filament bulb is connected to a 6 V battery. Calculate the normal operating resistance of the bulb when it has reached its standard operating temperature. (2)

c) Explain why the current through the lamp increases rapidly between 0 and 50 ms before dropping to a steady value. (2)

d) Use the graph to estimate the total charge that has passed through the filament bulb in the first 500 ms of operation. (2)

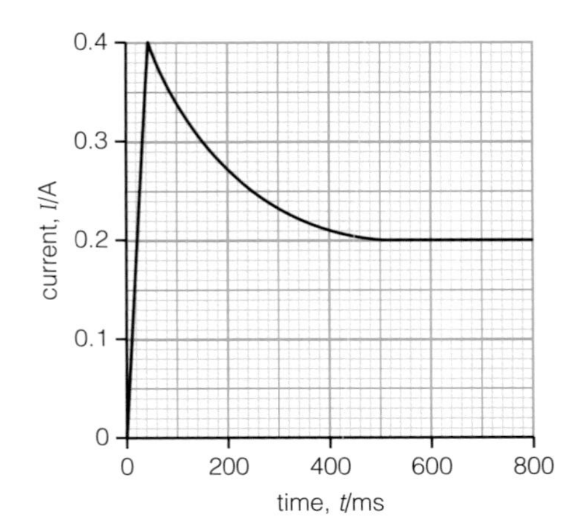

Figure 13.33 Filament lamp current – time graph.

12 The resistance of a metal wire changes with temperature. You are given a length of nichrome wire and asked to determine how the resistance of the nichrome wire changes between 0 °C and 100 °C.

a) Draw a labelled diagram of the experimental set-up that would allow you to perform this experiment. (2)

b) Write a suitable, numbered method that would allow you to obtain accurate and reliable measurements of the resistance of the wire over a range of temperatures between 0 °C and 100 °C. (3)

c) The metallic element aluminium has a critical temperature of 1.2 K. Explain what is meant by the critical temperature of an electrical conductor. (2)

13 A student wished to perform an experiment on an electrical component to determine if the component was an ohmic conductor.

a) State what is meant by the term 'ohmic conductor'. *(1)*

b) Draw a circuit diagram for this experiment. *(2)*

c) For the experimental circuit diagram that you have drawn, write an account of a suitable experiment. Your account should include:

- what measurements you would take
- how you would use your measurements
- how you would reach a conclusion. *(6)*

14 A semiconducting diode and a filament lamp are both examples of non-ohmic components.

a) Copy and complete a sketch on the axes shown in Figure 13.34 of the current–voltage characteristics for **both** components. *(2)*

b) Describe, using the current–voltage characteristic that you have drawn, how the resistance of the **filament** lamp changes as the potential difference across it changes. *(2)*

c) Draw a suitable diagram of the circuit that would enable you to collect data so that you could plot the *I–V* curve for the semiconductor diode. *(2)*

d) Write a method that you could follow in order to obtain this data for a semiconductor diode. *(4)*

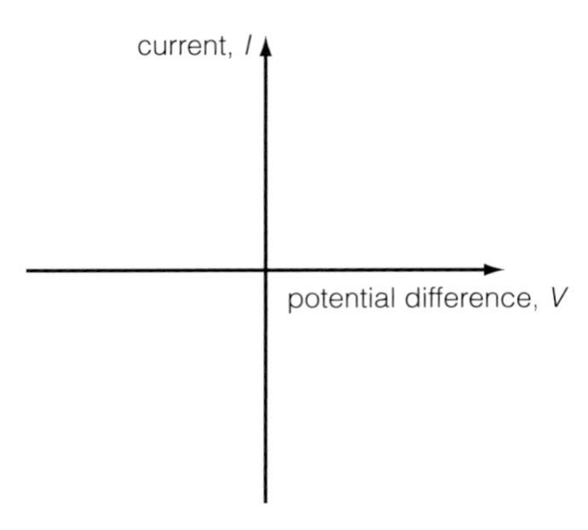

Figure 13.34 Electrical characteristic axes.

15 Figure 13.35 shows a thermistor connected in series with a resistor, *R*, and a 6.0 V battery.

When the temperature is 40 °C the resistance of the thermistor is 1.4 kΩ. The voltmeter connected across *R* reads 1.8 V.

a) Calculate the potential difference across the thermistor. *(1)*

b) Calculate the current flowing through the thermistor. *(2)*

c) Calculate the resistance of *R*. *(2)*

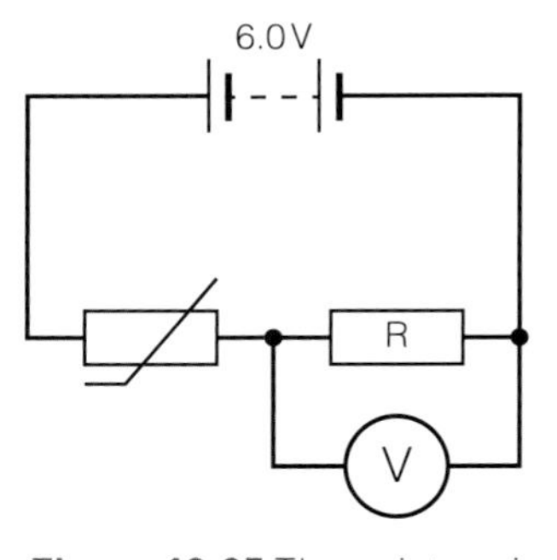

Figure 13.35 Thermistor circuit.

d) The battery is now replaced with a power supply unit with an internal resistance of 10 Ω. (This is equivalent to adding an extra 10 Ω to the circuit, but the resistance is inside the battery.) Without calculation, state and explain the effect on the voltmeter reading. *(2)*

16 Carbon-conducting assembly paste is used by electronics manufaturers to ensure good connections between components and printed circuit boards. A cylindrical length of this paste is laid out, 8.0×10^{-2} m long with a radius of 1.4×10^{-2} m. The resistivity of the paste is 0.82 Ω m.

a) Calculate the resistance between the ends of the cylinder of the assembly paste. Give your answer to a suitable number of significant figures. *(2)*

b) The paste is now reshaped into a cylinder with half the radius and a length that is four times as great. Calculate the new resistance. *(3)*

17 A plastic-clad copper connecting wire used in school electricity experiments is 0.60 m long and the copper inside the wire has a cross-sectional area of $1.3 \times 10^{-7}\,m^2$. The resistivity of copper is $= 1.7 \times 10^{-8}\,\Omega\,m$.

a) Calculate the resistance of the connecting wire. (2)

A filament lamp is connected to a power supply using two of the connecting wires; an ammeter connected in series measures a current of 2.2 A. The potential difference across the filament lamp is 12.0 V.

b) Calculate the potential difference across each of the wires. (2)

c) Calculate the potential difference across the terminals of the power supply. (2)

Stretch and challenge

All of the following questions are provided with permission of the British Physics Olympiad

18 A cell that produces a potential E (called an emf) is shown in Figure 13.36 and is connected to two resistors in series: a fixed resistor R_1 and a variable resistor R_2. The current, I, in the circuit is measured by the ammeter, A, and the potential difference, V, across resistor R_2 is measured with the voltmeter, V.

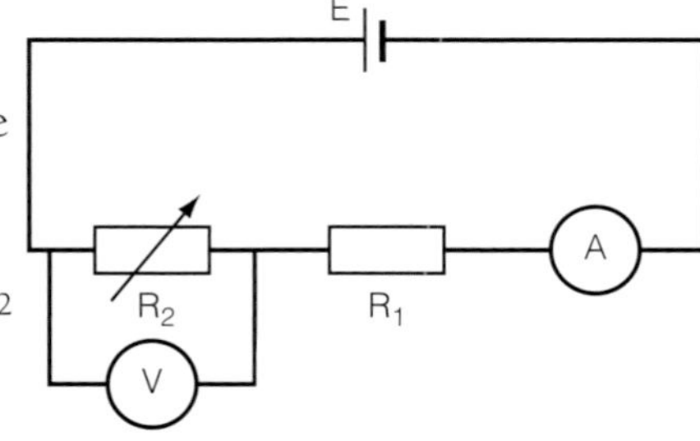

Figure 13.36 Circuit diagram.

The relation between the potential E and the current I is given by:

$$E = IR_1 + IR_2$$

Which of these graphs would produce a straight-line fit?

A V against I

B V against $1/I$

C $1/V$ against $1/I$

D I against $1/V$

19 Two students decide to calibrate a thermistor in order to measure variations in the temperature of a room. They connect a small thermistor across the terminals of a 5 V power supply and in series with a 1 A ammeter. The resistance of the thermistor is 120 Ω at room temperature. See figure 13.37.

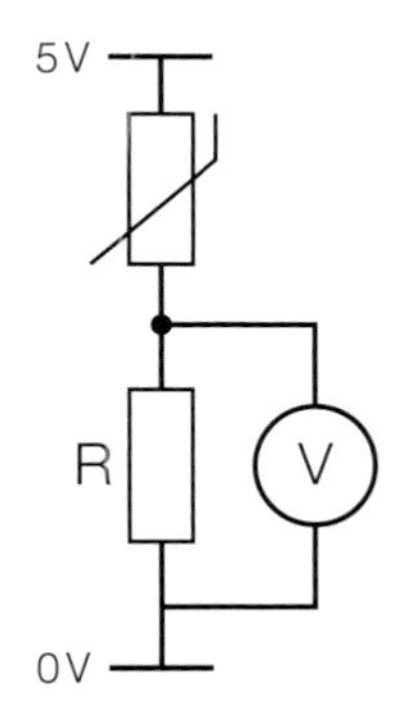

Figure 13.37 Circuit diagram.

a) Instead of showing small variations in room temperature, the thermistor is likely to go up in smoke. Explain why.

b) In the light of this experience, they decide to redesign their simple circuit as shown in Figure 13.37. They have a few values of resistor R to choose from; 5 kΩ, 500 Ω, 50 Ω.

State which value of R would give the biggest variation of V with temperature. Explain your choice.

c) State which value of R would be most likely to cause the same problem as in part (a). Again, explain your choice.

20 The resistance of a wire is proportional to its length and inversely proportional to its cross-sectional area. The resistance of a wire of length l and cross-sectional area A is given by:

$$R = \frac{\rho l}{A}$$

where ρ is a constant that depends on the material of the wire.

Some metals are ductile, which means that they can be drawn into long thin wires. In doing so, the volume V remains constant while the length increases and the cross-sectional area of the wire decreases. A wire of length 32 m has a resistance of 2.7 Ω. We wish to calculate the resistance of a wire formed from the same volume of metal but which has a length of 120 m instead.

a) Write down the relationship between V, A and l. Obtain an expression to show how R depends on the length l of the wire and its volume V.

b) Rewrite the equation with the constants ρ and V on one side and the variables we are changing, R and l, on the other.

c) Calculate the resistance of the longer wire.

14 Electrical circuits

PRIOR KNOWLEDGE

- The rate at which energy is transferred by an appliance from one energy store to another is called the power:

$$\text{power} = \frac{\text{energy transferred}}{\text{time}}$$

or

$$P = \frac{E}{t}$$

- Power, P, potential difference, V, and current, I, are related by the equation:

$$P = I \times V$$

- The potential difference provided by cells connected in series is the sum of the potential difference of each cell (depending on the direction in which they are connected) – see Figure 14.1.

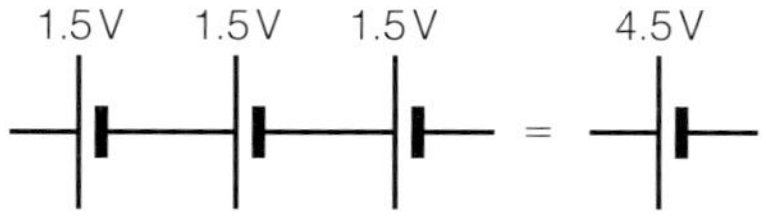

Figure 14.1 Circuit diagram showing cells and a battery.

- For components connected in series:
  - the total resistance is the sum of the resistance of each component (see Figure 14.2).

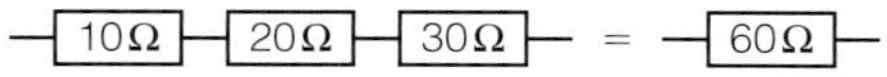

Figure 14.2 Circuit diagram showing resistance in series.

  - there is the same current through each component (see Figure 14.3).

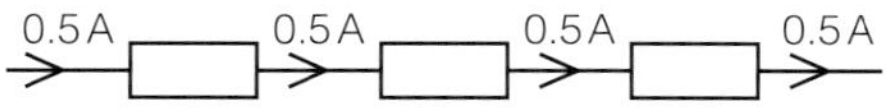

Figure 14.3 Circuit diagram showing current in series circuits.

  - the total potential difference of the supply is shared between the components (see Figure 14.4).

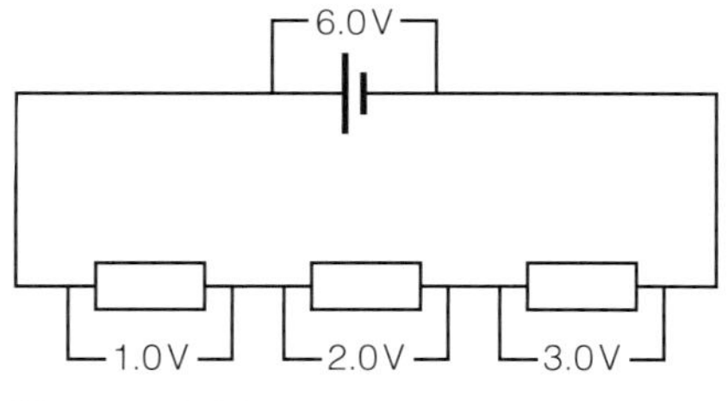

Figure 14.4 Circuit diagram showing pds in series.

- For components connected in parallel:
  - the potential difference across each component is the same (see Figure 14.5)

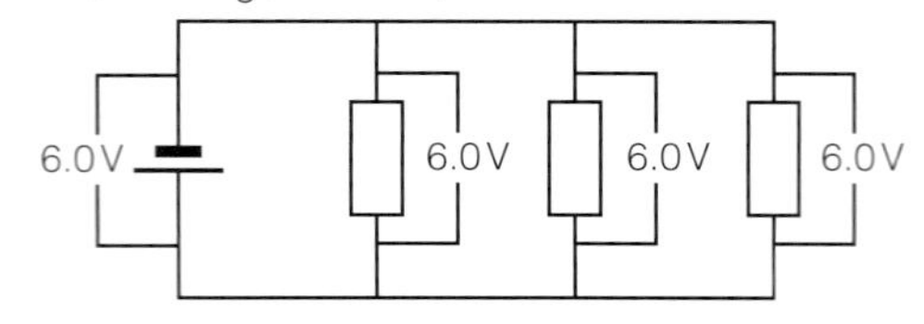

Figure 14.5 Circuit diagram sharing pds in parallel circuits.

- the total current through the whole circuit is the sum of the currents through the separate components (see Figure 14.6).

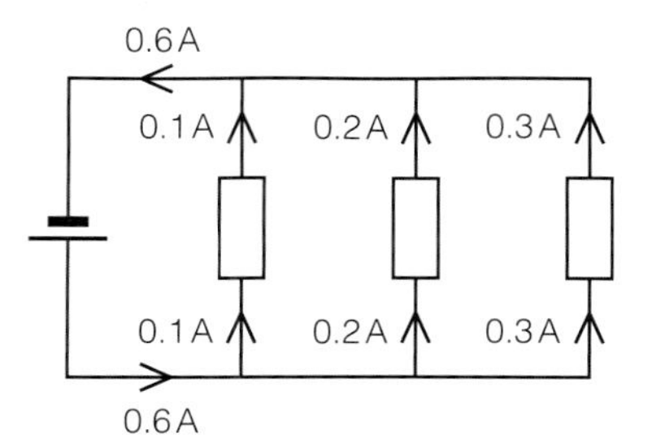

Figure 14.6 Circuit diagram sharing current in parallel circuits.

- The resistance of a light-dependent resistor (LDR) decreases as light intensity increases.
- The resistance of a thermistor decreases as the temperature increases.

TEST YOURSELF ON PRIOR KNOWLEDGE

1 An electric kettle takes 3 minutes to transfer 360 kJ of electrical energy into thermal energy stored in the water. Calculate the power of the kettle.

2 The kettle in the previous question is mains powered from a supply with a potential difference of 230 V. Calculate the current flowing through the kettle element.

3 A lab technician puts three used D cells into a battery pack. All three are connected the same way around. The measured emfs of the cells are: 1.45 V; 1.39 V and 1.47 V respectively. Calculate all the possible emfs that can be delivered from this battery pack.

4 The circuit in Figure 14.7 uses a 1.5 V cell, with negligible internal resistance, to feed current through two fixed resistors:

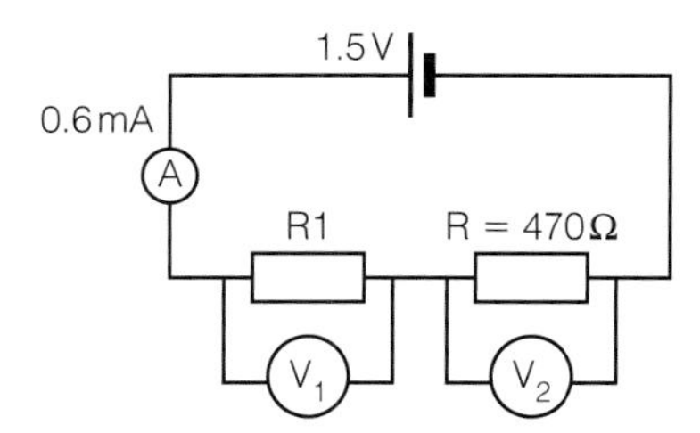

Figure 14.7 Circuit diagram for question 4.

a) Calculate the value of V_2.
b) Use the value of V_2 to calculate the value of V_1.
c) Now calculate the value of R_1.
d) State the total resistance of the circuit.

5 In Figure 14.8 a 9.0 V cell of negligible internal resistance delivers current to three resistors, R_1, R_2 and R_3, connected in parallel:

a) Calculate the total current, I, delivered by the cell.

b) Now calculate the values of R_1, R_2 and R_3.

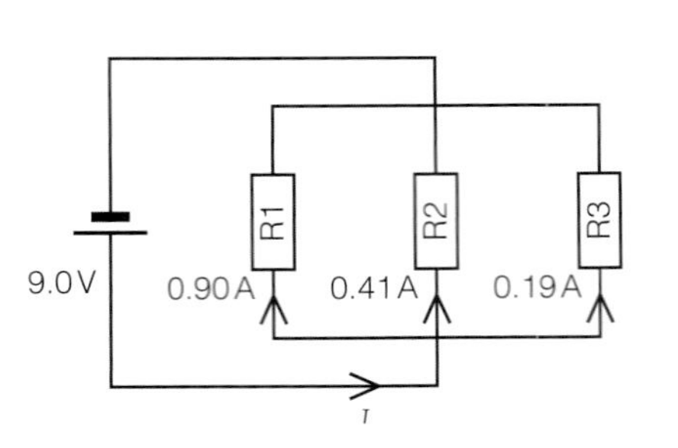

Figure 14.8 Circuit diagram for question 5.

Electrical power in circuits

Figure 14.9 The iPhone 5s A7 processor.

The integrated circuit of an iPhone is an extremely complex piece of electronic engineering. It involves the interconnection of many thousands of miniature electronic components. These are designed to work together controlling the many functions of the phone. Although the circuits are intricate, they are all based on some simple basic circuit principles and laws. Integrated circuits excel at doing simple, single-step things. They just do them very quickly in predefined sequences.

One of the most fundamental properties that electronic engineers need to consider when designing electrical devices is the power transferred by the circuit in the device. High power consumption requires specialist cooling systems and, if the device is battery powered, high battery capacity. Minimising power consumption reduces costs and bulk.

The electrical power transferred by a circuit is the sum of all the power transferred by the individual electrical components in the circuit. The electrical power transferred by a component is related to the current flowing through it and the potential difference across it. The general definition of power is:

power = rate of doing work or rate of energy transfer

or

$$P = \frac{\Delta W}{\Delta t}$$

So, in an electrical context, electrical power can be defined as the rate of doing electrical work. The electrical power transferred by a component can be determined by using the current, potential difference and resistance of the component, and there are several equations that can be used to calculate this power. These equations are produced by rearranging the basic equations for electric current, potential difference and the definition of resistance.

$$\text{current} = \frac{\text{charge}}{\text{time}}$$

in terms of symbols:

$$I = \frac{Q}{t}$$

rearranging:

$$\Rightarrow Q = I\,t \qquad \text{Equation 1}$$

$$\text{potential difference} = \frac{\text{electrical energy}}{\text{charge}}$$

in terms of symbols:

$$V = \frac{W}{Q}$$

rearranging:

$$\Rightarrow Q = \frac{W}{V} \qquad \text{Equation 2}$$

Equating Equation 1 with Equation 2:

$$It = \frac{W}{V}$$

rearranging:

$$\frac{W}{t} = IV = P$$

or

$$P = IV$$

and using the definition of resistance

$$V = IR$$

$$P = I \times (IR) = I^2R$$

or

$$P = \left(\frac{V}{R}\right) \times V$$

$$= \frac{V^2}{R}$$

The relationship $P = IV$ is a really useful one as it allows the calculation of electrical power using two easily measured and monitored quantities, current and potential difference. Data-logging ammeters and voltmeters allow for the real-time monitoring of the electrical power consumption of a circuit, which then allows battery-powered devices such as smartphones, tablets and laptops to display the amount of energy left in the battery and predict the amount of usage time remaining before charging is required.

TEST YOURSELF

1 A mains electric iron operates at 230V and is rated at 2.2kW. Calculate the resistance of the iron when operating normally.

2 Look at this list of electrical units:

A As
B VA^{-1}
C Js^{-1}
D Cs^{-1}
E VAs

Choose the unit for:

a) electric charge
b) electrical power
c) electrical resistance
d) electrical energy

3 A high-power resistor used inside electric guitar amplifiers has a resistance of 330Ω and a maximum power of 25W. Calculate the potential difference across the power resistor when it is running at full maximum power.

4 A mobile phone battery operates with a potential difference of 3.8V across its circuits and a capacity (the product of the current and time that the current can be maintained) of 1560mA h (milliampere-hours).

a) While playing a high-graphic usage game the power drawn from the battery is 0.30W. Calculate the current drawn from the battery.

b) Calculate the total charge that can be drawn from the battery.

c) Calculate the total available game-play time.

5 Two relationships for electrical components are:

$V = IR$ and $P = IV$.

a) Use these two relationships to produce an equation for electrical power in terms of V and R only.

b) The power dissipated from a 33Ω resistor operating at 1.5V is 0.068W. The pd, V, is now doubled and the resistance is halved. What is the effect on the power dissipated by the resistor?

ACTIVITY

Comparing the power of a 12V bulb and a rheostat

A student carried out a simple experiment to compare the power of a 12V bulb and a rheostat (acting as a fixed resistor). The circuits that she used are shown in Figure 14.10.

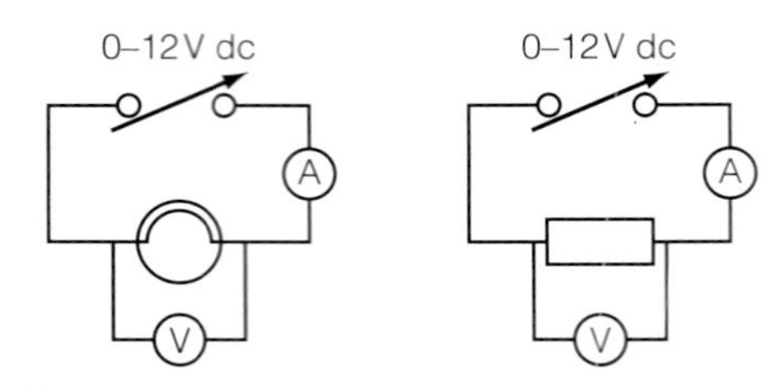

Figure 14.10 Circuit diagram.

Here are her measurements.

Potential difference, V (V) ±0.01 V	Current, I (A), ±0.01 A		Power, P (W)	
	Rheostat	Bulb	Rheostat	Bulb
0.00	0.00	0.00		
2.00	0.08	0.02		
4.00	0.16	0.09		
6.00	0.24	0.21		
8.00	0.32	0.37		
10.00	0.40	0.58		
12.00	0.48	0.83		

1 Copy and complete the table, calculating the power of each component at each potential difference.
2 Plot a graph of electrical power, P, against potential difference, V, plotting both the rheostat and the bulb on the same graph.
3 Describe in detail the pattern in the variation of power with potential difference for both components.
4 Use the data to determine the resistance of the rheostat.
5 The student's teacher suggests that the relationship between the power, P, and the potential difference, V, for the bulb is given by $P = kV^3$, where k is a constant. Use the data in the table to draw a suitable graph to enable you to determine if the teacher was correct and, if so, to calculate a value for k.
6 Explain why the power produced by the bulb varies in a different way to the rheostat.

Extension

The potential difference is measured to the nearest 0.01 V, the current to the nearest 0.01 A. Use this information to calculate a suitable maximum uncertainty for the measurement of the resistance of the rheostat.

Circuit calculations

Electronic engineers also need to know the simple ways that current and potential difference act in circuits. The 'rules' for these are universal and apply to simple circuits as well as the intricate circuits found in integrated circuit chips. These simple rules were first worked out by a German physicist called Gustav Kirchoff in 1845 and have become known as Kirchoff's First and Second Circuit Laws.

Kirchoff's First Circuit Law – the law of current

At a circuit junction, the sum of the currents flowing into the junction equals the sum of the currents flowing out of the junction.

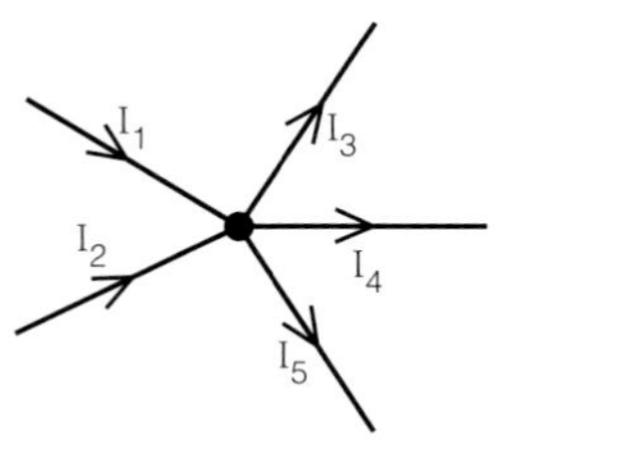

Figure 14.11 A circuit junction.

Figure 14.11 shows a circuit junction with two currents (I_1 and I_2) flowing into the junction and three currents flowing out of the junction (I_3, I_4 and I_5).

Kirchoff's First Circuit Law states:

$$I_1 + I_2 = I_3 + I_4 + I_5$$

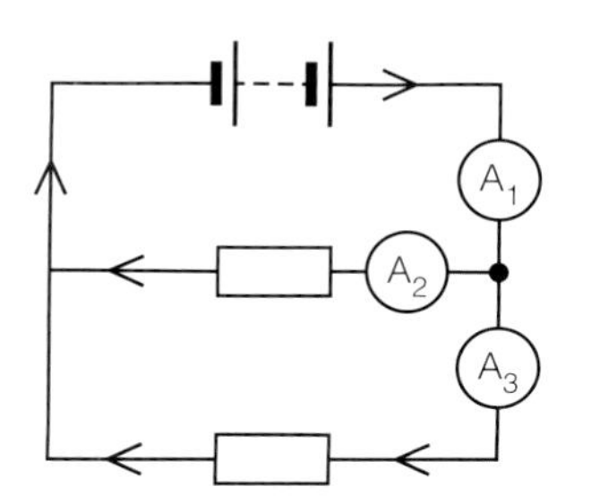

Figure 14.12 Circuit diagram showing Kirchoff's First Circuit Law.

Figure 14.12 shows Kirchoff's First Circuit Law in action in a real circuit involving ammeters.

Conventional current flows from positive to negative, so the current flowing into the junction indicated on the diagram is measured by ammeter A_1. Current splits or recombines at a junction, so the current flowing out of the junction is measured by A_2 and A_3. Kirchoff's First Circuit Law tells you that:

$$A_1 = A_2 + A_3.$$

Written more generally in mathematical notation the law can be summarised by:

$$\sum I_{\text{into junction}} = \sum I_{\text{out of junction}}$$

or, in other words, current is conserved at junctions.

Looked at from a slightly different perspective, as current is the rate of flow of charge, or

$$I = \frac{\Delta Q}{\Delta t}$$

then Kirchoff's First Circuit Law could be written in terms of charge:

At a circuit junction, the sum of the charge flowing into the junction equals the sum of the charge flowing out of the junction (per second).

$$\sum Q_{\text{into junction}} = \sum Q_{\text{out of junction}}$$

(per unit time).

Kirchoff's Second Circuit Law – the law of voltages

In a closed circuit loop, the sum of the potential differences is equal to the sum of the electromotive forces.

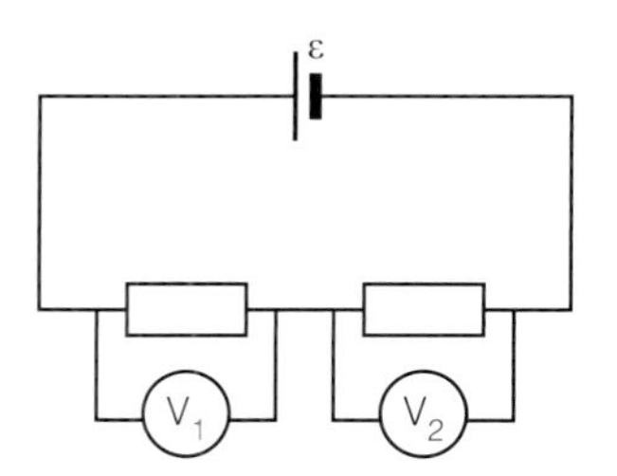

Figure 14.13 Circuit diagram showing Kirchoff's Second Circuit Law.

Figure 14.13 shows a single closed loop series circuit. In this case, there is one emf and two pds, and Kirchoff's Second Circuit Law says that:

$$\varepsilon = V_1 + V_2$$

Or more generally, using mathematical notation, for any closed circuit loop:

$$\sum \varepsilon = \sum V$$

If the circuit is extended to make it a parallel circuit such as Figure 14.14:

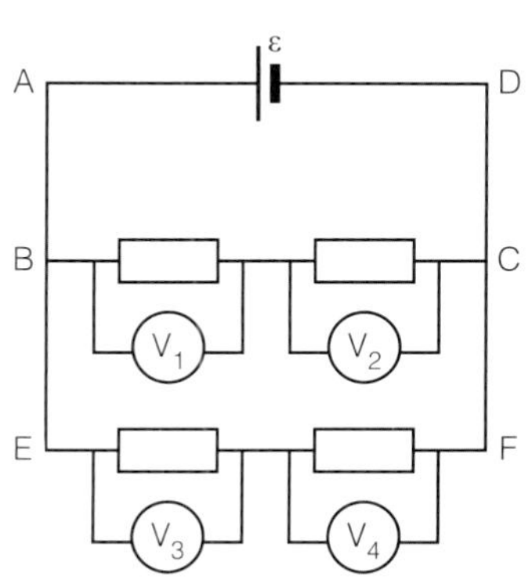

Figure 14.14 Circuit diagram showing Kirchoff's Second Circuit Law in a parallel circuit.

This parallel circuit is effectively made up of two series circuits: ABCD and AEFD, so:

$$\varepsilon = V_1 + V_2$$

and

$$\varepsilon = V_3 + V_4$$

or

$$(V_1 + V_2) = (V_3 + V_4)$$

Kirchoff's Second Circuit Law applies to any circuit but, in the case of parallel circuits, the circuit must be considered as a succession of individual series circuits with the same power supply. This law is based on the conservation of energy: the energy per coulomb transferred to the charge by the battery, ε, is then transferred to other forms of energy by the charge as it flows through the circuit components.

Figure 14.15 An AA battery.

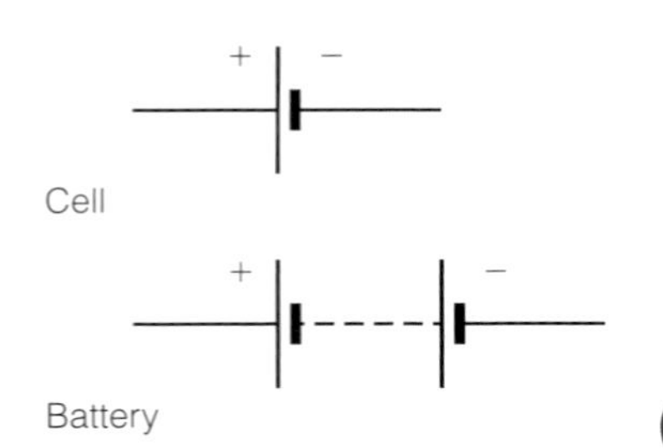

Figure 14.16 Circuit symbols for cell and battery.

TIP
Unless a question explicitly states that the power supply in the question has an internal resistance, you should ignore it.

Batteries and cells

The standard AA battery is a common battery for all manner of hand-held devices such as games controllers and remote controls. The word *battery* is actually a misnomer (a word or term that suggests a meaning that is known to be wrong): the device shown in the picture is actually a **cell**. Several cells joined in series or in parallel are called a battery. The standard electrical symbols are shown in Figure 14.16.

When identical cells are connected in parallel to form a battery, the emf of the resultant battery is just the emf of the individual cells. Connecting cells in parallel is generally not a good idea as they are rarely identical, and one cell will force current back into the other, causing damage to the cells. Cells in parallel have a higher capacity for storing and transferring electrical energy than single cells, but is it is better to just buy and use a bigger single cell.

Internal resistance and electromotive force

Real power supplies, such as batteries and laboratory power packs, always have an internal resistance, r. As the current flows through the power supply, the internal resistance creates a potential difference that leads to electrical energy being transferred to thermal energy inside the power supply. This is one of the reasons why hand-held devices such as tablets warm up when they are in use for a long period of time.

The internal resistance of most batteries and power supplies is very low: a typical AA battery has an internal resistance of about 0.2 Ω and, as such, can effectively be ignored when measuring the electrical properties of most circuits.

The internal resistance of a power supply cannot be measured directly as it is 'inside' the power supply. The only way to measure it is to use its electrical characteristic. A circuit that can be used to do this is shown Figure 14.18.

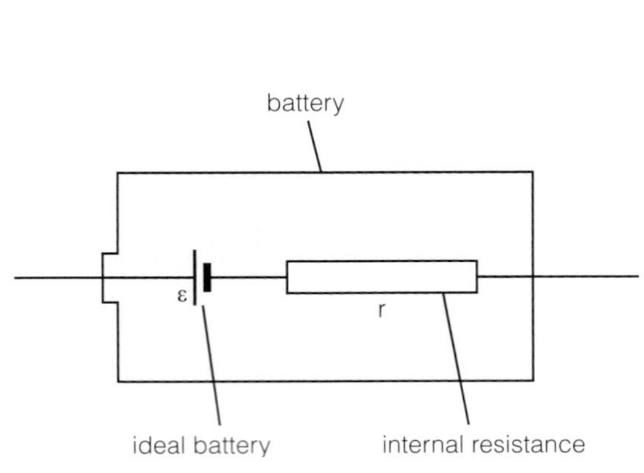

Figure 14.17 A real power supply

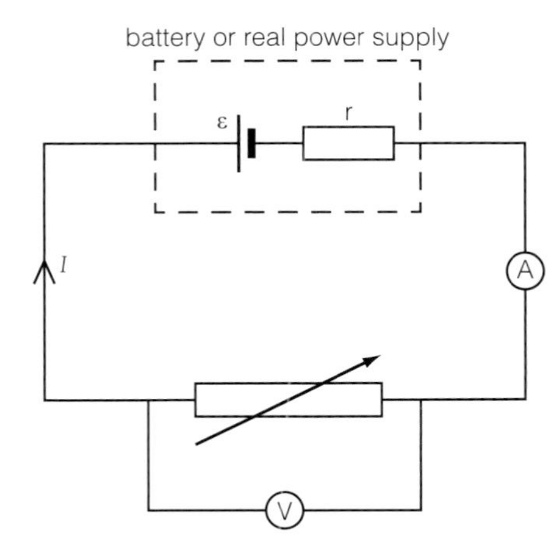

Figure 14.18 A circuit used to measure the internal resistance and electromotive force of a real power supply.

Circuit calculations can be used to determine the internal resistance and the electromotive force.

The electromotive force, ε, must be the same as the sum of the potential differences in the circuit (Kirchoff's Second Circuit Law). There are two potential differences in this circuit: one is across the external variable resistor, V, and the other is the pd across the internal resistor. This cannot be measured directly, but the pd across this resistor is equal to Ir. This means that:

$$\varepsilon = V + Ir$$

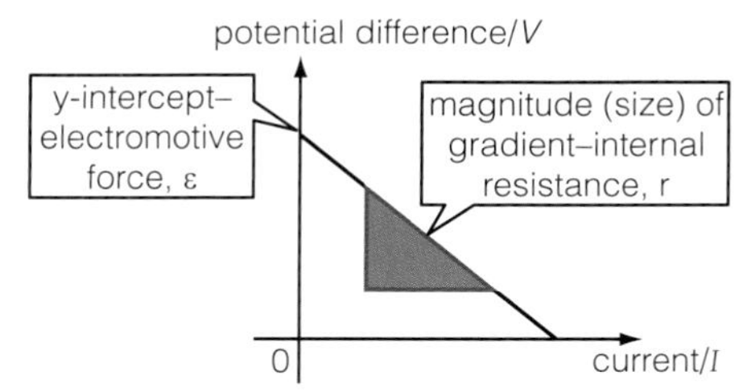

Figure 14.19 Electrical characteristic of a real power supply, such as a battery.

The current, I, can be measured directly using an ammeter; ε and r are both constants, so the equation can be rewritten as:

$$V = \varepsilon - Ir \qquad \text{or} \qquad V = -rI + \varepsilon$$

which is the equation of a straight line of the form, $y = mx + c$, where the gradient is negative. If an electrical characteristic is drawn using values of V and I from various values of R (the external load resistance) then the y-intercept of the graph is the electromotive force, ε, and the gradient is negative and equal to $-r$, the internal resistance.

TEST YOURSELF

6 Figure 14.20 shows an AAA battery with an emf, ε, of 1.5 V and an internal resistance, r, of 0.20 Ω. The cell delivers a current, I, of 300 mA into a fixed-load resistor, R, of 4.8 Ω.

a) Use the data on the circuit diagram to calculate the potential difference, V, across the fixed-load resistor, R.

b) Explain why the potential difference, V, across the load resistance is always less than the emf, ε, of the cell.

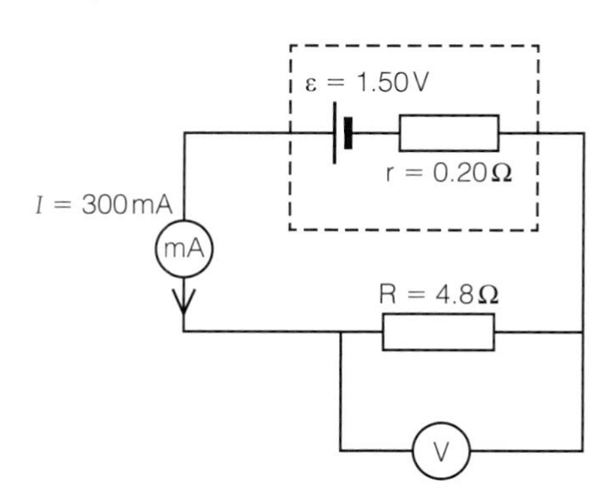

Figure 14.20 Circuit diagram for question 6.

The AAA battery is left connected to the 4.8 Ω fixed-load resistor for a few hours. Figure 14.21 shows how the current from the AAA battery varies with time.

c) Using data from the graph, describe in detail how the current varies with time.

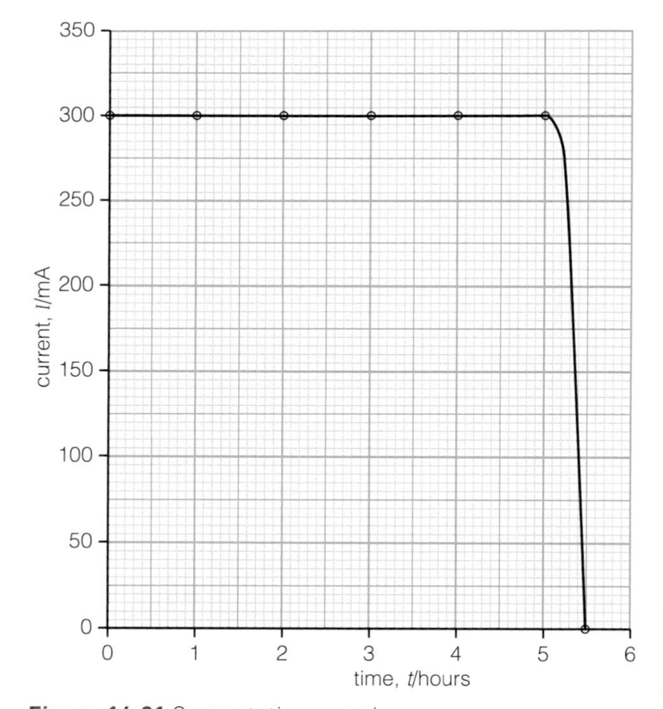

Figure 14.21 Current–time graph.

d) Give reasons why the current varies with time in this way.

e) Use the graph to estimate the total charge delivered by the AAA battery during its discharge.

7 A radio-controlled model car can be modelled with the circuit in Figure 14.22: a 9.0 V emf PP9 battery with an internal resistance r is connected across a variable-load resistor R (representing the motor).

Figure 14.23 shows the electrical characteristic for this circuit.
Use the graph to calculate the internal resistance, r, of the PP9 battery.

Figure 14.22 Circuit diagram for question 7.

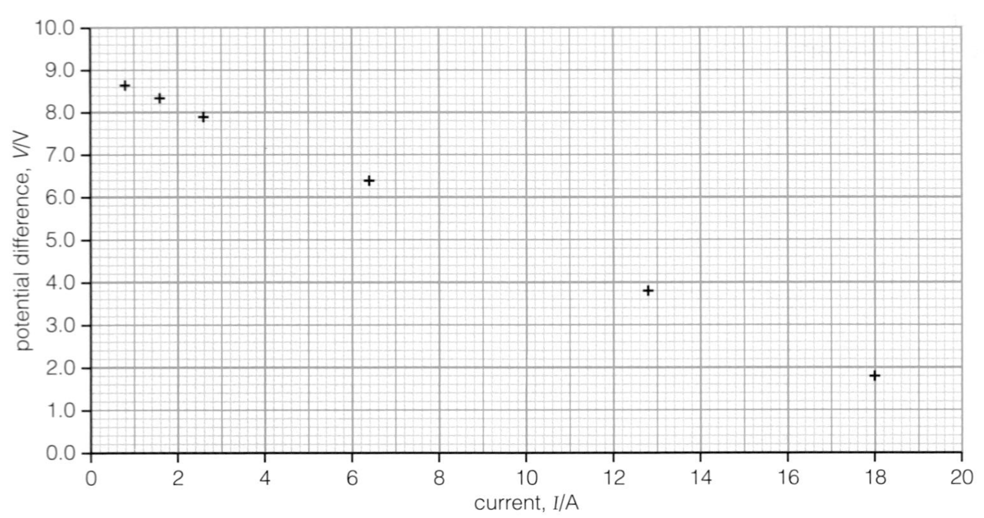

Figure 14.23 Electrical characteristic of a PP9 battery.

8 Three cells, X, Y and Z, are connected consecutively into the circuit shown in Figure 14.24.

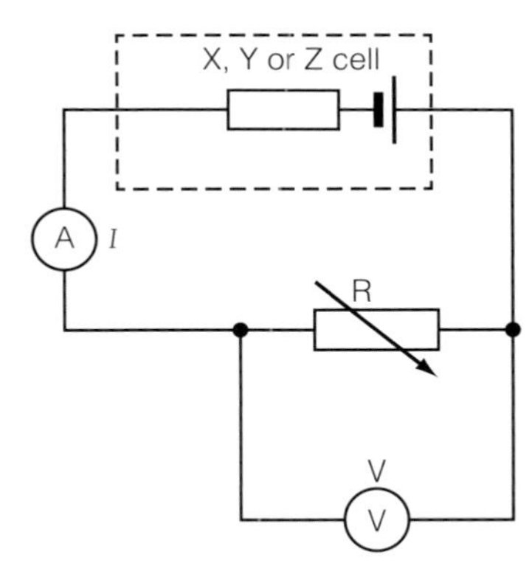

Figure 14.24 Circuit diagram for question 8.

R is varied and values of V and I are collected for each cell; a composite electrical characteristic is drawn and shown in Figure 14.25.

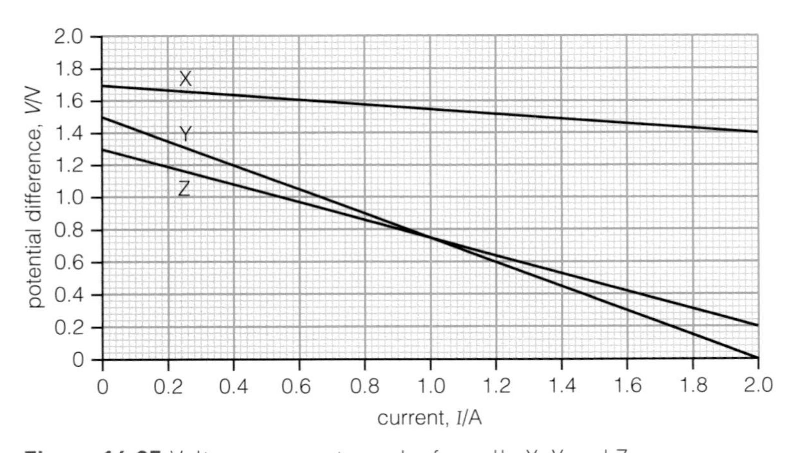

Figure 14.25 Voltage–current graphs for cells X, Y and Z.

State which of the cells X, Y or Z:

a) has the largest emf

b) will deliver the most electrical power at a current of 0.5 A

c) has the largest internal resistance

d) can deliver an electrical power of 2.8 W.

REQUIRED PRACTICAL 6

Investigation of the emf and internal resistance of electric cells and batteries by measuring the variation of the terminal pd of the cell with current in it

Note: This is just one example of how you might tackle this required practical.

Lemons contain citric acid. This acid can undergo electrochemical reactions when metal electrodes made from copper and zinc (for example) are pushed into the lemon. The electrochemical reactions generate an electromotive force that produces a current high enough to activate an LCD clock screen. Lemons, however, have a very high internal resistance which limits the current and makes lemons a poor choice for most electronic circuits. It would take about two and a half years, for example, for a lemon to charge an iPhone battery from empty.

Figure 14.26 A lemon clock.

Although having a high internal resistance is a problem for fruity powered real devices, it makes them an ideal choice for school laboratory experiments to measure emf and internal resistance. A circuit that can be used to measure these values is shown in Figure 14.27.

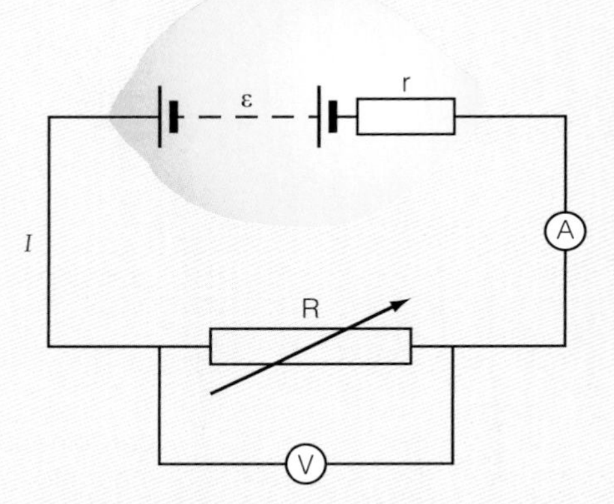

Figure 14.27 A circuit to measure the emf and internal resistance of a lemon.

In this experiment, a variable resistor is used as an external load; the current flowing through the circuit is measured with a suitable milliammeter and the potential difference across the variable resistor is measured with a voltmeter for a range of resistance values. One such experiment recorded the following data:

Resistance, R (Ω)	Current, I (mA)				Potential difference, V (V)			
	1	2	3	Av	1	2	3	Av
100	0.632	0.623	0.644		0.059	0.058	0.068	
220	0.573	0.566	0.581		0.125	0.123	0.126	
470	0.488	0.482	0.500		0.229	0.226	0.232	
1000	0.340	0.336	0.343		0.419	0.414	0.422	
2200	0.244	0.242	0.246		0.544	0.540	0.548	
4700	0.147	0.146	0.148		0.676	0.671	0.679	
10000	0.078	0.077	0.078		0.774	0.769	0.800	
22000	0.038	0.038	0.038		0.833	0.828	0.834	
47000	0.018	0.018	0.018		0.865	0.861	0.865	
100000	0.008	0.008	0.008		0.881	0.877	0.882	
220000	0.004	0.004	0.004		0.890	0.886	0.890	
470000	0.002	0.002	0.002		0.894	0.890	0.895	

1 Copy and complete the table by calculating the mean average of the currents and the potential difference.
2 Plot an electrical characteristic graph of this data with potential difference on the y-axis. Remember to include a best-fit line.
3 The formula for the internal resistance and emf of a real power supply is given by the formula $\varepsilon = I(R + r)$.

By using a suitable substitution, rearrange this formula into the equation of a straight line with V, the pd, across the variable resistor, as the subject.

4 Use the graph to measure directly the emf, ε, of the lemon.

5 Use the graph to calculate the internal resistance of the lemon.

Extension

The ammeter measures with an uncertainty of ±0.001 mA, and the voltmeter measures with an uncertainty of ±0.001 V. Use the instrument uncertainties, the spread of the data and a suitable graphical technique to determine uncertainty values for the internal resistance, r, and the emf, ε.

Voltmeters and ammeters

Ammeters are always put into circuits in series with other components. This is because they measure the flow of the charge (the electrons) through the circuit. Older, analogue ammeters involve passing the current through a coil of wire sitting inside a shaped permanent magnet.

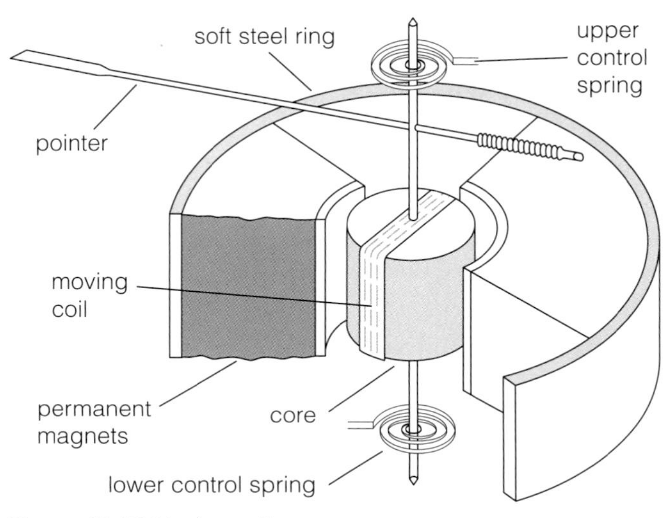

Figure 14.28 Moving coil ammeter.

The current in the coil generates a magnetic field that interacts with a permanent magnetic field, causing the coil to turn; the current is then measured by a pointer on an analogue scale. Modern, digital ammeters use an integrated circuit within the meter to measure the current, which is then displayed on a numerical display. Both designs, however, will always affect the size of the current in some way, as any device put into the circuit in series will have a resistance. The extra resistance of the ammeter will therefore reduce the current in the circuit. This effect cannot be overcome, so modern ammeters are designed to have very low resistances and are calibrated to take into account the reduction of the current due to the resistance of the meter.

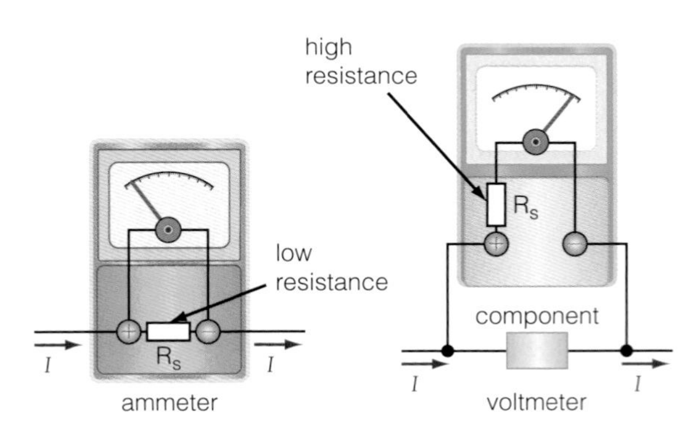

Figure 14.29 Ammeters have low resistance, and voltmeters have high resistance.

Voltmeters are always connected into circuits in parallel with other components. Both analogue and digital voltmeters work in very similar ways to ammeters, but a small current is drawn from the circuit that passes through a set, known, very high-resistance resistor so that the current is proportional to the potential difference. High-quality voltmeters therefore have very high resistance.

TEST YOURSELF

9 Look at the following list of units:

A Ω
B W
C J
D V
E A

Which of the above units is equivalent to the following:

a) $J\,s^{-1}$
b) V^2W^{-1}
c) $C\,s^{-1}$
d) WA^{-1}
e) VA^{-1}

10 A student is comparing the brightness of car bulbs when they are connected to a 12 V car battery. She connects two bulbs, P and Q, in parallel with the battery as shown in Figure 14.30.
Bulb Q is brighter than bulb P.
Copy and complete the following sentences using one of the phrases listed below:

less than   greater than   the same as

a) The current in bulb P is..................................that in bulb Q.
b) The pd across lamp P is..................................that in bulb Q.
c) The electrical power of bulb Q is..........................that in bulb P.
d) The electrical resistance of bulb Q is..........................that in bulb P.

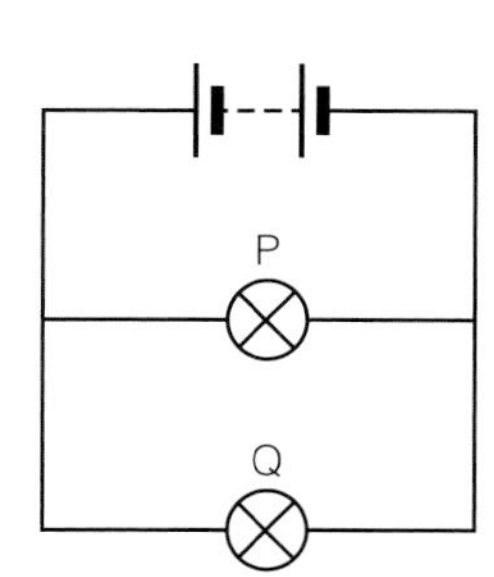

Figure 14.30 Circuit diagram for question 10.

Resistor networks

When components are connected together to make useful circuits, each component adds a certain resistance to the circuit. The effect of the added resistance depends on how the component is connected into the circuit – in **series** with other components or in **parallel** (or *both* in the case of circuits containing many hundreds of components). There are some simple rules for calculating the overall resistance of components connected in series or in parallel.

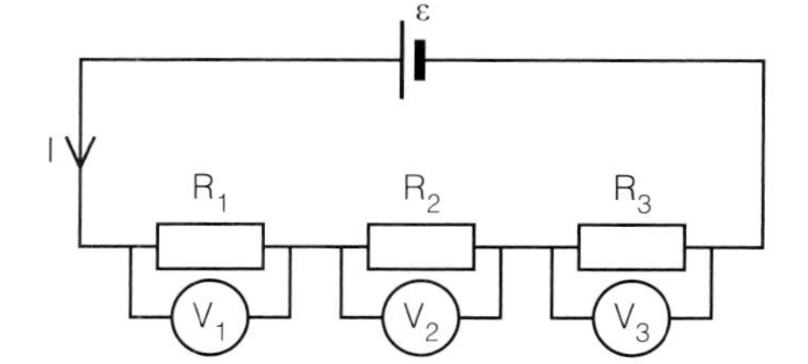

Figure 14.31a Circuit diagram showing resistor combinations in series.

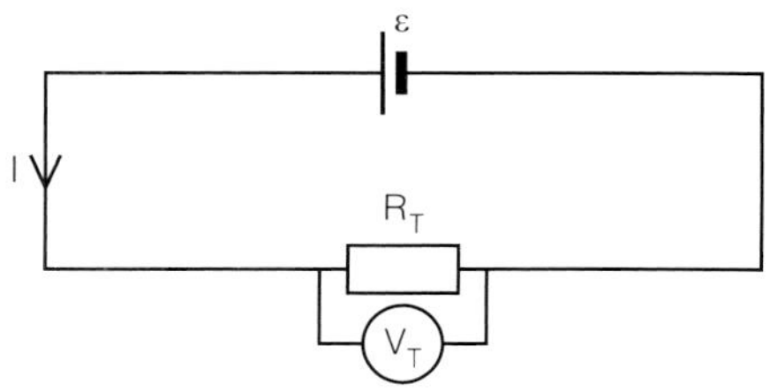

Figure 14.31b Circuit diagram showing equivalent resistor.

Resistors connected in series

Consider the resistor network shown in Figure 14.31.

Figure 14.31b represents the single resistor that could replace the three resistors in series in Figure 14.31a.

Using Kirchoff's Circuit laws and Ohm's law leads to:

$$\varepsilon = V_1 + V_2 + V_3$$

and

$$\varepsilon = V_T$$

where

$$V_T = V_1 + V_2 + V_3$$

but because

$$V = IR$$

and, as the current is the same throughout a series circuit:

$$IR_T = IR_1 + IR_2 + IR_3 = I(R_1 + R_2 + R_3)$$

so

$$R_T = R_1 + R_2 + R_3$$

and for a series network of n resistors:

$$R_T = R_1 + R_2 + R_3 + \cdots + R_n$$

or, using sigma notation

$$R_T = \sum_{i-1}^{n} R_i$$

Resistors connected in parallel

Consider the following circuit:

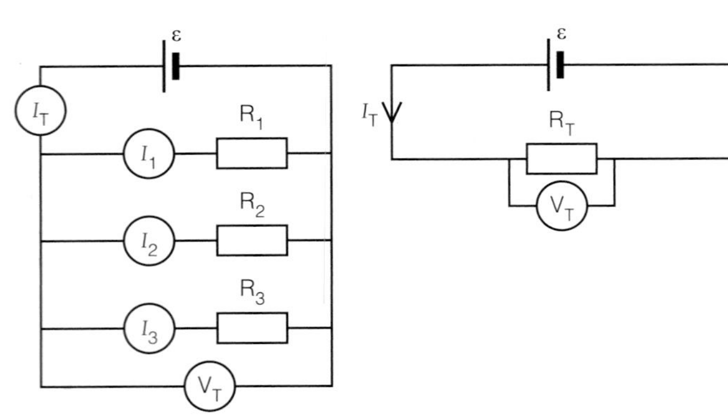

Figure 14.32 Circuit diagram showing resistors connected in parallel.

In the right-hand circuit one resistor, R_T, has been used to replace all three resistors arranged in parallel in the left-hand circuit. Again, using Kirchoff's Circuit laws and Again, using Kirchoff's Circuit laws and the definition of resistance, V = IR:

Kirchoff's First Circuit law says:

$$I_T = I_1 + I_2 + I_3$$

and

$$I = \frac{V}{R}$$

So as the potential difference, V_T, is the same across all of the resistors:

$$I_T = \frac{V_T}{R_1} + \frac{V_T}{R_2} + \frac{V_T}{R_3} = V_T\left(\frac{1}{R_1} + \frac{1}{R_2} + \frac{1}{R_3}\right)$$

Rearranging gives:

$$\frac{I_T}{V_T} = \frac{1}{R_1} + \frac{1}{R_2} + \frac{1}{R_3}$$

So:

$$\frac{I_T}{V_T} = \frac{1}{R_T}$$

where:

$$\frac{1}{R_T} = \frac{1}{R_1} + \frac{1}{R_2} + \frac{1}{R_3}$$

For a network of n resistors connected in parallel:

$$\frac{1}{R_T} = \frac{1}{R_1} + \frac{1}{R_2} + \ldots + \frac{1}{R_n}$$

or using sigma notation:

$$\frac{1}{R_T} = \sum_{i=1}^{n} \frac{1}{R_i}$$

In summary, for resistors in series the overall resistance is the sum of the individual resistances. For resistors in parallel the reciprocal of the overall resistance is the sum of all the reciprocal of the individual resistances. Resistors connected in parallel always have a resistance less than any of the resistors in that combination.

TIP

Resistors in parallel always have a resistance less than any of the individual resistors. This is a useful check when you have done a calculation.

EXAMPLE

Three resistors are connected in parallel to a power supply as shown in Figure 14.33. Calculate the value of the one resistor that could replace all three resistors.

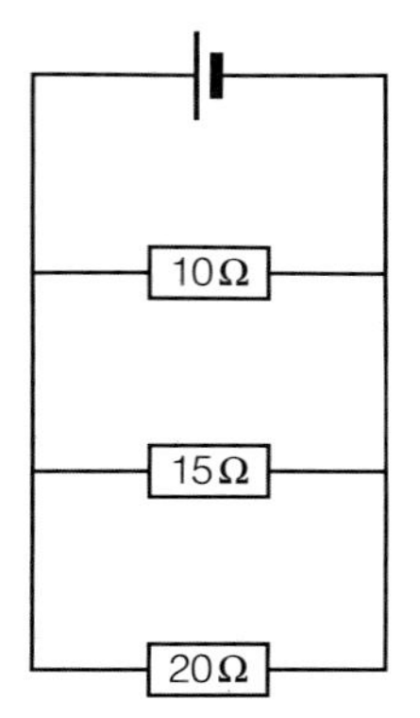

Figure 14.33 Circuit diagram.

Answer

R_T is the value of the one resistor that can replace all three so:

$$\frac{1}{R_T} = \frac{1}{R_1} + \frac{1}{R_2} + \frac{1}{R_3}$$

substitute the numbers:

$$\frac{1}{R_T} = \frac{1}{10\,\Omega} + \frac{1}{15\,\Omega} + \frac{1}{20\,\Omega}$$

$$\frac{1}{R_T} = 0.10 + 0.07 + 0.05 = 0.217\,\Omega^{-1}$$

So

$$R_T = \frac{1}{0.217\,\Omega^{-1}} = 4.6\,\Omega \qquad \text{(2 sf)}$$

When calculating the overall resistance of a circuit containing a mixture of resistors in series and in parallel, it is best to calculate the overall resistance of each of the sections arranged in parallel first, and then calculate the overall resistance by summing all the effective resistances in series.

EXAMPLE

Look at the combination circuit in Figure 14.34, containing resistors connected in series and in parallel.

What is the overall resistance of this circuit?

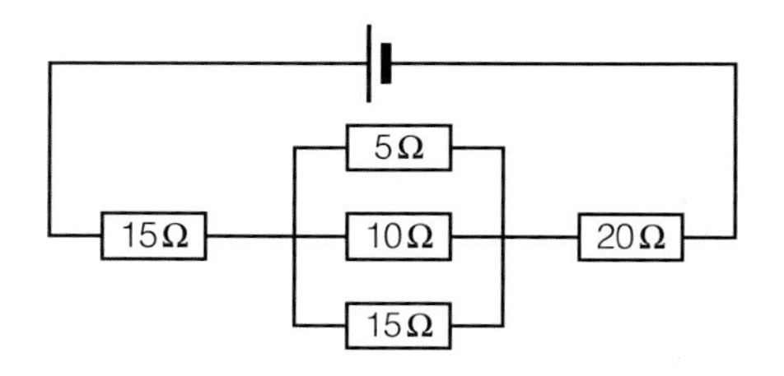

Figure 14.34 Circuit diagram.

Answer

To calculate the overall resistance, R_T, of this circuit, the first step is to calculate the effective resistance, R, of the three resistors in parallel:

$$\frac{1}{R} = \frac{1}{5\,\Omega} + \frac{1}{10\,\Omega} + \frac{1}{15\,\Omega}$$

$$\frac{1}{R} = 0.20 + 0.10 + 0.067 = 0.367\,\Omega^{-1}$$

$$R = \frac{1}{0.367\,\Omega^{-1}} = 2.7\,\Omega$$

$$R_T = 15\,\Omega + 2.7\,\Omega + 20\,\Omega$$
$$= 37.7\,\Omega$$
$$= 38\,\Omega \qquad \text{(2 sf)}$$

TEST YOURSELF

11 Three resistors, each of resistance 3.3 kΩ, are connected in parallel. Calculate the overall resistance of this network.

12 Two resistors are connected in parallel as shown in Figure 14.35.

a) Calculate the overall resistance of this circuit.

b) The resistors are connected to a 12 V battery with negligible internal resistance. Calculate the total current drawn from the 12 V battery.

13 Three 47 kΩ resistors are connected to a 6.0 V battery with negligible internal resistance as shown in Figure 14.36.

a) Calculate the total resistance of the circuit.

b) Calculate the current flowing through *each* resistor.

14 A teacher is thinking about the resistances of ammeters and voltmeters. A battery with an emf of 6.0 V and negligible internal resistance is connected first to a 2.2 Ω fixed resistor as shown in Circuit 1 in Figure 14.37.

a) Calculate the current flowing in Circuit 1.

An ammeter with a resistance of 0.3 Ω is connected into the circuit in series with the resistor as shown by Circuit 2.

b) Calculate the total resistance of Circuit 2.

c) Calculate the current flowing in Circuit 2.

d) Explain why the current in Circuit 2 is lower than the current flowing in Circuit 1.

e) Suggest how the manufacturer of the ammeter could alter the design of the ammeter to take its resistance into account.

f) The best quality ammeters, including most digital ammeters, alter the current in the circuit by as little as possible. Suggest a value for the resistance of an ammeter to make it 'perfect'.

The same battery is now connected across two fixed resistors as shown in Figure 14.38 by Circuit 3.

g) Calculate the pd across points X and Y by Circuit 3.

A voltmeter with the same resistance as the fixed resistors is now connected in parallel with X and Y as shown by Circuit 4.

h) Calculate the combined resistance across points X and Y in Circuit 4.

i) Calculate the pd across points X and Y in Circuit 4.

j) Explain why the pd across X and Y in Circuit 3 is larger than the pd across X and Y in Circuit 4.

k) Explain why the best voltmeter to connect across X and Y would be one with an extremely high resistance.

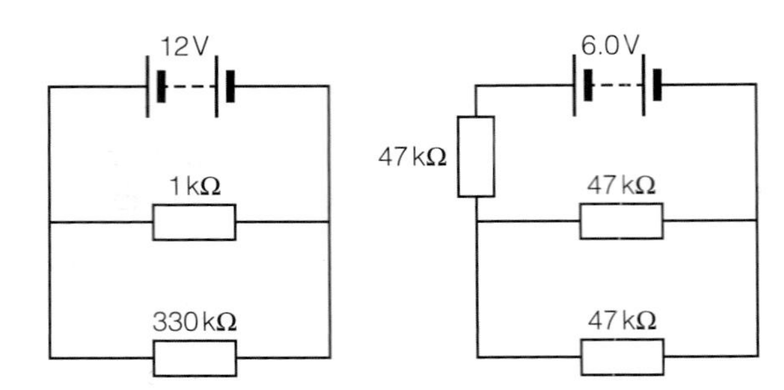

Figure 14.35 Circuit diagram for question 12.

Figure 14.36 Circuit diagram for question 13.

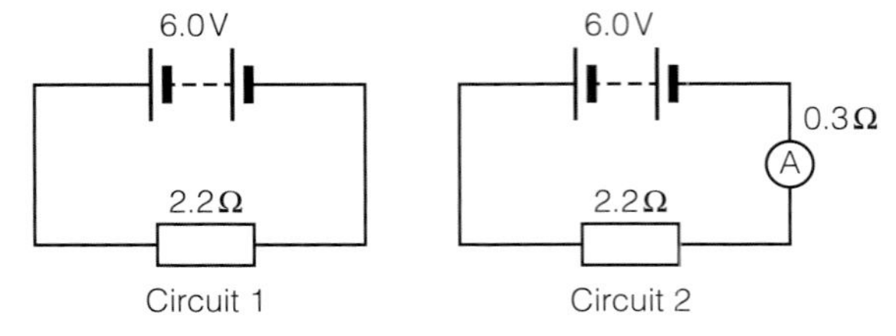

Figure 14.37 Circuit diagrams for questions 14 a) to f).

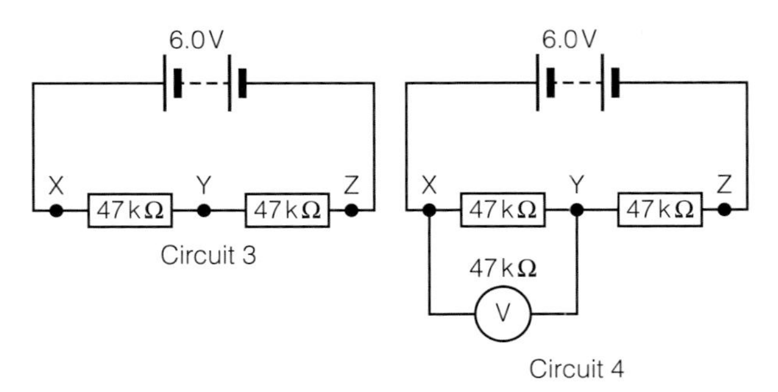

Figure 14.38 Circuit diagram for questions 14 g) to l).

Potential dividers

Potential dividers are simple, three-component circuits designed to control the potential difference in a circuit.

They consist of a power supply (such as a cell); a fixed resistor; and a third resistive component, whose resistance can be fixed (fixed resistor) or variable (variable resistors, thermistors, light-dependent resistors, etc.).

All of these components are connected in series and the emf of the power supply is shared across the two resistive components.

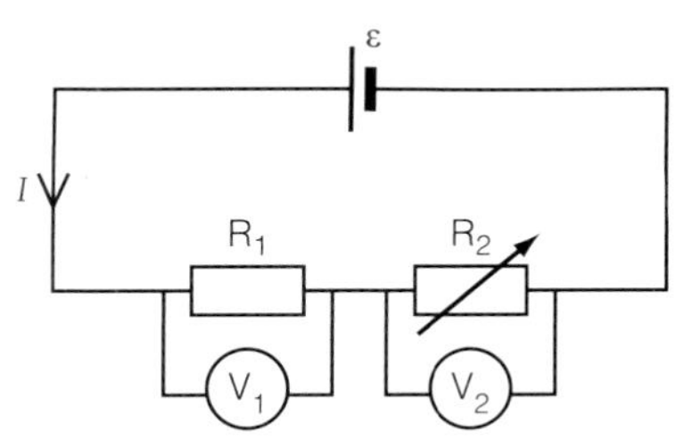

Figure 14.39 Circuit diagram of a potential divider.

The clue to how potential dividers work is in their name, *potential dividers*, i.e. something that splits up potential difference. Look at the potential divider circuit in Figure 14.39.

Generally we are trying to vary the potential difference V_1 (across R_1) by varying R_2.

Assuming that the cell has negligible internal resistance, then the total resistance of the circuit R_T is given by:

$$R_T = R_1 + R_2$$

The current, I, flowing through the circuit is given by Ohm's law:

$$I = \frac{\varepsilon}{R_T} = \frac{\varepsilon}{(R_1 + R_2)}$$

Then considering the resistor R_1 we can write:

$$V_1 = IR_1$$

Substituting for I from above:

$$V_1 = \frac{\varepsilon}{(R_1 + R_2)} \times R_1$$

so

$$V_1 = \frac{\varepsilon R_1}{(R_1 + R_2)}$$

From this equation it can be seen that if ε and R_1 are fixed, then V_1 only depends on R_2. In fact, as R_2 increases, V_1 decreases, and vice versa.

Figure 14.40 A wire potentiometer.

ACTIVITY

Analysing the results from a wire potentiometer

A potential divider can be made out of a length of high-resistance metal wire such as the metal alloy nichrome. The wire is stretched out along a ruler and a metal probe is used to make contact with the wire along its length. This piece of apparatus is called a potentiometer, as shown in Figure 14.40.

The circuit is shown in Figure 14.41:

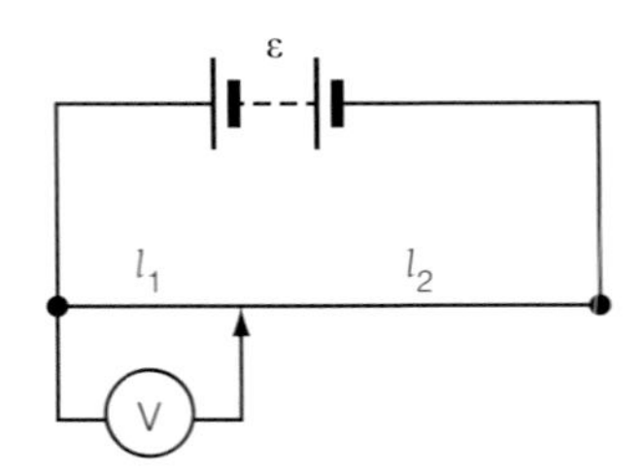

Figure 14.41 Circuit for a potentiometer.

A particular potentiometer has a maximum wire length of 1.0 m. The potentiometer is connected to a 6.0 V battery of negligible internal resistance. A student varies the distance, l_1, between the fixed connection end of the voltmeter and the sliding contact, and measures the pd for a range of different wire lengths up to 1.0 m.

Here are her results:

Length of wire, l_1 (m), ±0.002 m	pd across wire, V (V), ±0.01 V				
	1	2	3	Average	Uncertainty
0.000	0.00	0.00	0.00		
0.100	0.60	0.80	0.61		
0.200	1.20	1.04	1.02		
0.300	1.80	1.87	1.86		
0.400	2.40	2.37	2.46		
0.500	3.00	2.84	2.97		
0.600	3.60	3.52	3.56		
0.700	4.20	4.19	4.37		
0.800	4.80	4.77	4.84		
0.900	5.40	5.48	5.24		
1.000	6.00	6.00	5.87		

1. Make a copy of the table.
2. Check the data for any anomalous results. In this experiment these are normally due to pressing too hard on the wire with the slider. Deal with these anomalies in the usual way (ignore them) and calculate the average pd across the wire for each length.
3. Use the spread of the data for each length to assign an uncertainty to the pd of each length.
4. Plot a graph of average pd (y-axis) against length (x-axis), using the uncertainties and the precisions to plot vertical and horizontal error bars for each point.
5. Plot a best-fit line through the points. Should (0, 0) be a known fixed point on this best-fit line?
6. Use the best-fit line to determine an equation for the relationship between V and l.
7. The error bars that you have drawn will give you a confidence about the best-fit line and, therefore, the relationship between the two variables, V and l. Explain how the spread of the data and their error bars can help you to decide the position of the best-fit line.
8. State and explain how her original results would differ if she used a lower-resistance analogue voltmeter.

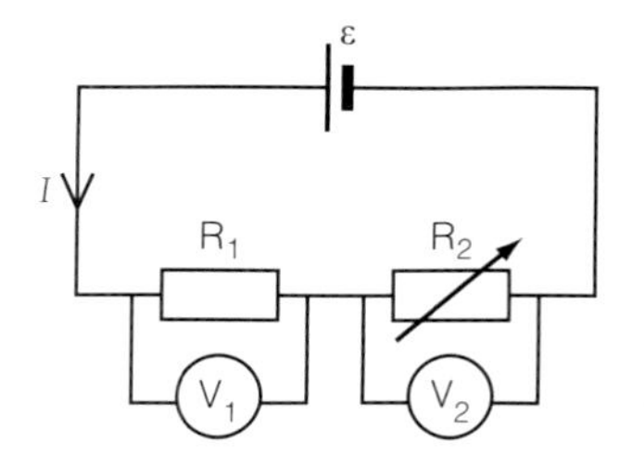

Figure 14.42 Circuit diagram for a potential divider.

Potential dividers as sensors

Another use of the potential divider is in sensor circuits. If the variable resistor, R_2, in the potential divider circuit shown in Figure 14.42 is replaced with a component whose resistance varies with an external physical variable such as temperature or light intensity, then the potential difference across the fixed resistor, R_1, will also vary with the external physical variable.

The voltmeter connected across R_1 can then be calibrated in terms of the value of the external physical variable. The circuit can then be made to act as an electronic thermometer or an electronic light sensor.

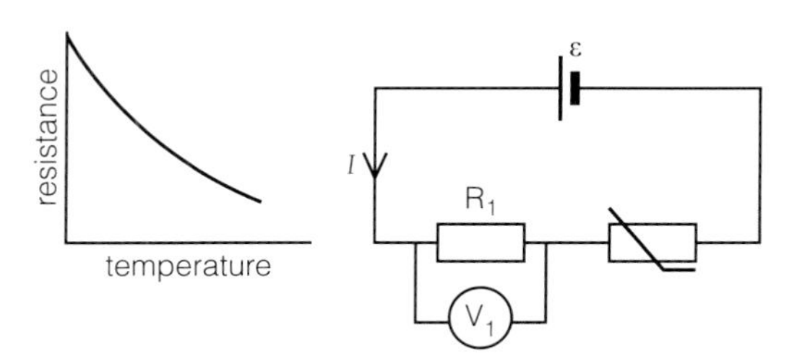

Figure 14.43 A thermistor potential divider circuit and the resistance–temperature graph for an ntc thermistor.

With an electronic thermometer, a thermistor is connected in place of R_2. Most thermistors are ntc (negative temperature coefficient) thermistors. This means that as the temperature increases, their resistance decreases, as shown in Figure 14.43, which also shows a circuit diagram showing how the components are connected:

The effect on the potential difference across the fixed resistor R_1 is that as the temperature increases, so the resistance of the thermistor, R_2, drops and V_1 rises. Thus increasing temperature produces increasing pd. If the voltmeter is connected across the thermistor, the opposite will happen: increasing the temperature will produce a decrease in pd. In most cases, applications require the pd to increase with temperature, so the voltmeter is usually connected across the fixed resistor.

EXAMPLE

The resistance–temperature graph for an ntc thermistor is shown in Figure 14.44.

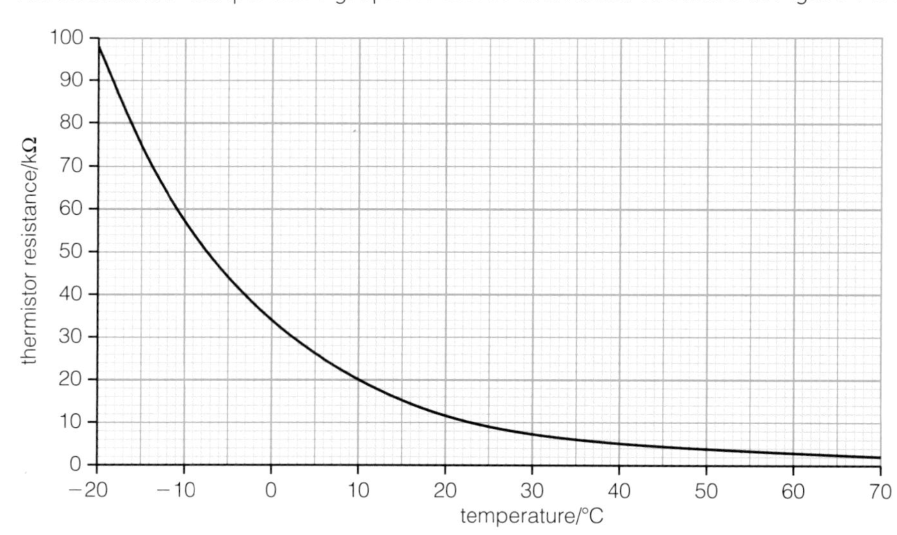

Figure 14.44 Resistance–temperature graph for an ntc thermistor.

The thermistor is connected into a potential divider circuit with a 6.0 V battery of negligible internal resistance and 45 kΩ fixed resistor.

At −10 °C the resistance of the thermistor is 55 kΩ. At 40 °C the resistance of the thermistor is 5 kΩ.

a) Calculate the voltage across the fixed resistor at −10 °C.
b) Calculate the voltage across the fixed resistor at 40 °C.
c) What effect does increasing the temperature from −10 °C to 40 °C have on the voltage?

Answer

a) $V_1 = \frac{\varepsilon R_1}{(R_1 + R_2)} = \frac{6.0\,\text{V} \times 45\,\text{k}\Omega}{(45 + 55)\,\text{k}\Omega} = 2.7\,\text{V}$

b) $V_1 = \frac{\varepsilon R_1}{(R_1 + R_2)} = \frac{6.0\,\text{V} \times 45\,\text{k}\Omega}{(45 + 5)\,\text{k}\Omega} = 5.4\,\text{V}$

c) Increasing the temperature of the thermistor from −10 °C to 40 °C doubles the voltage (in this case).

A similar effect involves the use of a light-dependent resistor (LDR). These are components that change their resistance with light intensity, which makes them incredibly useful as light sensors. LDRs are made of semi-conducting materials; as the light shines on the material, electrons are freed from the structure and can flow through it, reducing the resistance of the LDR.

In the dark, LDRs can have a very high resistance, of the order of mega-ohms, and yet in the light their resistance can drop as low as a few hundred ohms. If an LDR is connected into a potential divider in place of R_2 in the potential divider circuit then, as the light intensity increases, then so will the output pd V_1 across the fixed resistor.

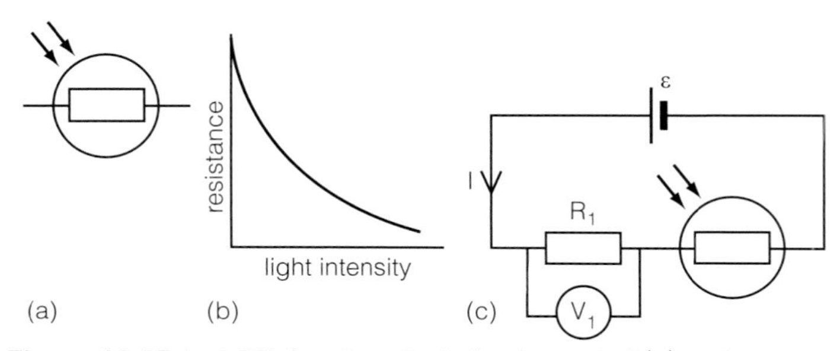

Figure 14.45 An LDR: its electrical circuit symbol (a) resistance–light intensity graph (b) and its use in a potential divider circuit (c).

EXAMPLE

A particular LDR inside a digital camera has a resistance of 150 kΩ in low light levels and a resistance of 720 Ω in bright sunlight. The LDR is connected to a 55 kΩ fixed resistor and a 3.0 V battery.

Calculate the pd across the fixed resistor in:

a) low light levels b) bright sunlight.

Answer

a) In low light levels the pd across the fixed 55 kΩ resistor is:

$$V_1 = \frac{\varepsilon R_1}{(R_1 + R_2)} = \frac{3.0\,\text{V} \times 55\,\text{k}\Omega}{(150 + 55)\,\text{k}\Omega} = 0.80\,\text{V}$$

b) In bright sunlight the pd across the fixed 55 kΩ resistor is:

$$V_1 = \frac{\varepsilon R_1}{(R_1 + R_2)}$$

$$= \frac{3.0\,\text{V} \times 55\,\text{k}\Omega}{(0.72 + 55)\,\text{k}\Omega}$$

$$= 3.0\,\text{V}$$ (2 sf)

TEST YOURSELF

15 Figure 14.46 shows part of the automatic light sensor for a digital camera. The camera battery has an emf of 6.0 V and has negligible internal resistance. The voltmeter is a digital type with a very high resistance.

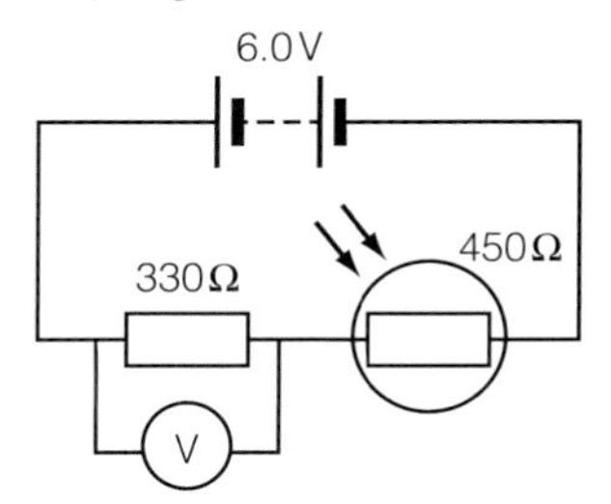

Figure 14.46 Circuit diagram for question 15.

In bright daylight the resistance of the LDR is 450 Ω.

a) Calculate the ratio

$$\frac{\text{pd across resistor}}{\text{pd across LDR}}$$

b) Calculate the voltmeter reading in bright daylight.

At low light levels, the resistance of the LDR increases to 470 kΩ.

c) Recalculate the ratio

$$\frac{\text{pd across resistor}}{\text{pd across LDR}}$$

at this lower light level.

d) Calculate the voltmeter reading in low light.

16 An ntc thermistor is connected in series with a 660 Ω fixed resistor and a 6.0 V battery forming a potential divider circuit for an electric thermometer. A high-resistance voltmeter is connected across the fixed resistor.

a) Draw a circuit diagram for this circuit.

b) At 25 °C the thermistor has a resistance of 1.5 kΩ. Calculate the current in the circuit.

c) Calculate the pd across the 660 Ω resistor.

d) The thermistor is heated to 90 °C and its resistance drops to 220 Ω. State and explain what would happen to the voltmeter reading as the temperature changes from 25 °C to 90 °C.

e) At 90 °C, when the thermistor has a resistance of 220 Ω the thermometer develops a fault and the current suddenly rises to 0.4 A. Calculate the power loss in the thermistor immediately after the fault.

f) State and explain what is likely to happen to the thermistor in the circuit.

Practice questions

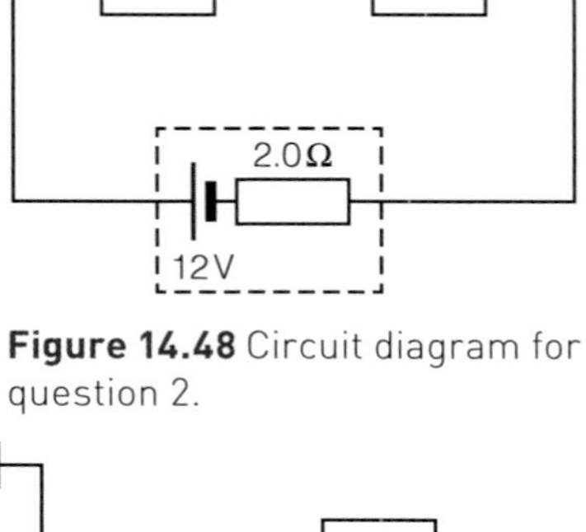

Figure 14.47 Circuit diagram for question 1.

1 A battery of emf 12 V and with negligible internal resistance is connected to the resistor network shown in Figure 14.47.

What is the current through the 47 Ω resistor?

A 0.1 A
B 0.2 A
C 0.3 A
D 0.4 A

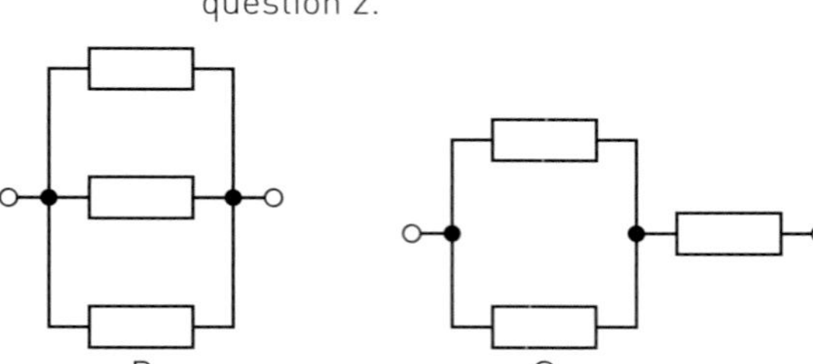

Figure 14.48 Circuit diagram for question 2.

2 A battery has an emf of 12 V and an internal resistance of 2.0 Ω. Calculate the total current in the circuit shown in Figure 14.48.

A 0.19 A
B 0.20 A
C 0.71 A
D 5.2 A

3 A student connects together three resistors together in the arrangement shown in P and Q in Figure 14.49.

Each resistor has a resistance of 75 Ω. Which line represents the equivalent resistance for each arrangement of resistors?

	P (Ω)	Q (Ω)
A	25	50
B	0.04	75.04
C	25	112.5
D	225	50

Figure 14.49 Circuit diagram for question 3.

You need the following information for questions 4 and 5.

A resistor and a thermistor are being used as part of a temperature-sensing circuit. They are connected in series with a 12 V battery of negligible internal resistance as shown in Figure 14.50.

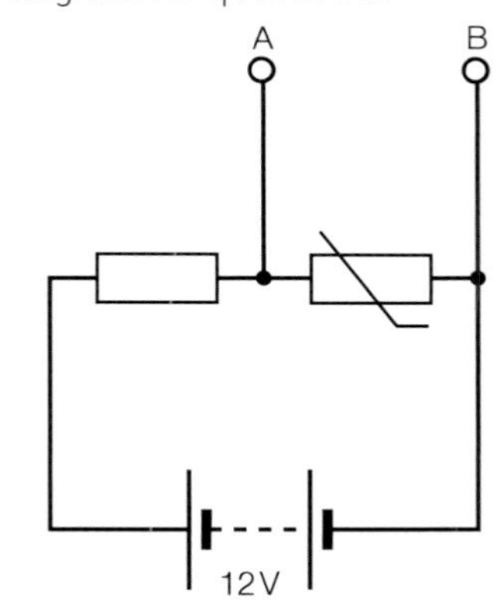

Figure 14.50 Circuit diagram for question 4.

4 At 200 °C the resistance of the thermistor is 18 Ω and the resistance of the resistor is 130 Ω.

What is the voltage across the terminals AB at this temperature?

A 1.13 V
B 1.46 V
C 10.54 V
D 10.86 V

5 What is the power dissipated by the resistor at 200 °C?

A 0.08 W
B 0.12 W
C 0.85 W
D 0.97 W

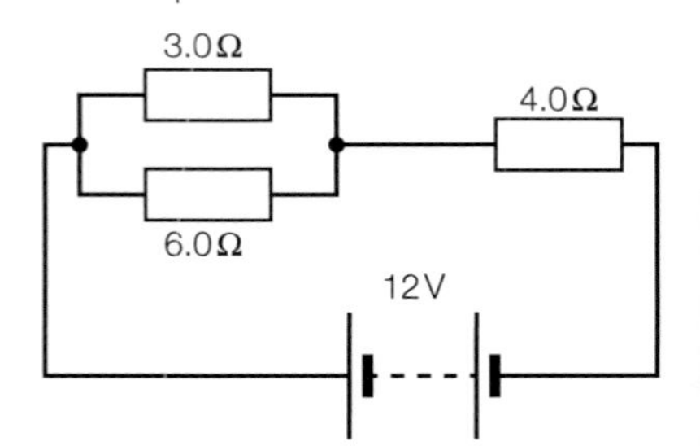

Figure 14.51 Circuit diagram for question 6.

6 Three resistors are connected to a battery of emf 12 V and a negligible internal resistance as shown in Figure 14.51.

What is the voltage across the 6.0 Ω resistor?

A 2 V
B 4 V
C 8 V
D 10 V

7 A torch bulb is connected to a battery of negligible internal resistance. The battery supplies a steady current of 0.25 A for 20 hours. In this time the energy transferred by the bulb is 9.0×10^4 J.

What is the power of the bulb?

A 1.3 W **C** 4500 W

B 21 W **D** 5625 W

8 Electric blankets have wires in the centre. The energy dissipated by the wires when a current flows through them heats the blanket and the bed. One blanket has a power rating of 63 W and is plugged into mains electricity (230 V).

What is the total resistance of the wire in the blanket?

A 4 Ω **C** 840 Ω

B 17 Ω **D** 15 000 Ω

9 A cell of emf 1.5 V and internal resistance 1.0 Ω is connected to a 5.0 Ω resistor to form a complete circuit. Calculate the current in this circuit.

A 0.25 A **C** 0.60 A

B 0.30 A **D** 1.5 A

10 An electrical generator produces 100 kW of power at a potential difference of 10 kV. The power is transmitted through cables of total resistance 5 Ω.

What is the power loss in the cables?

A 50 W **C** 500 W

B 250 W **D** 1000 W

11 Car batteries have a typical emf of 12 V and a very low internal resistance of 5.0 mΩ.

a) Explain what is meant by the terms 'emf' and 'internal resistance'. (2)

b) One such battery delivers a current of 500 A to the starter motor of a car. Calculate the potential difference across the starter motor. (2)

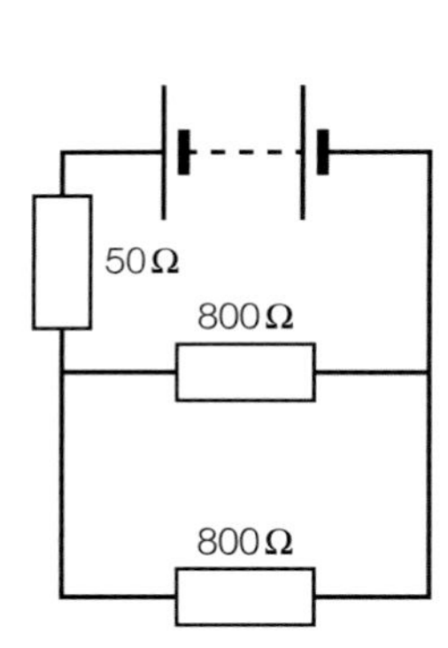

Figure 14.52 Circuit diagram for question 12.

12 Figure 14.52 shows three resistors connected to a battery of negligible internal resistance.

a) Calculate the total resistance of the circuit. (2)

b) The power dissipated by each of the 800 Ω resistors is 2.0 W. Calculate the pd across the two 800 Ω resistors. (2)

c) Calculate the current through the 50 Ω resistor. (2)

d) Calculate the emf of the battery. (1)

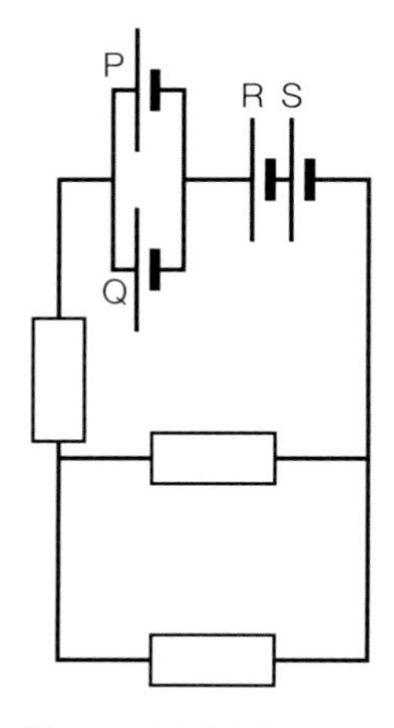

Figure 14.53 Circuit diagram for question 13.

13 Figure 14.53 shows four identical new cells, each with an emf of 1.2 V and negligible internal resistance, connected to three identical resistors each with a resistance of 5 Ω.

a) Calculate the total resistance of the circuit. (2)

b) Calculate the total emf of the cells. (1)

c) Calculate the current passing through cell Q. (2)

d) Calculate the total charge passing through cell Q in 1 second. (2)

e) Each of the cells shown in the circuit diagram stores the same amount of chemical energy, and hence can transfer the same amount of electrical energy. State and explain which two cells in the circuit would transfer electrical energy for the longest time period. (2)

14 A car engine temperature-sensor circuit consists of a 12 V car battery connected in series with a low-resistance ammeter, a 330 Ω resistor and then connected to an ntc (negative temperature coefficient) thermistor and a 1 kΩ resistor connected in parallel to each other before returning to the battery as shown in Figure 14.54.

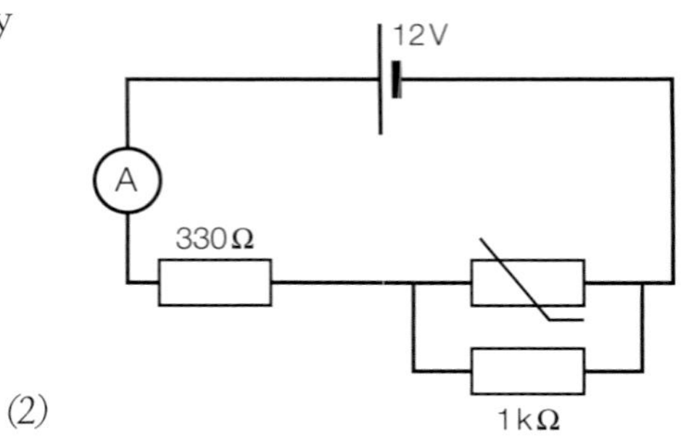

Figure 14.54 Circuit diagram for question 14.

At its normal working temperature, the current reading on the ammeter is 14.0 mA.

a) Calculate the pd across the 330 Ω resistor. (2)

b) Calculate the pd across the 1 kΩ resistor. (2)

c) Calculate the combined resistance of the thermistor and 1 kΩ resistor parallel combination. (2)

d) Calculate the resistance of the thermistor. (3)

e) The engine starts to overheat and the temperature of the thermistor starts to rise. State and explain the effect that this has on the current measured by the ammeter. (2)

15 A thermistor is connected to a combination of fixed resistors and a 6 V battery of negligible internal resistance, as shown in Figure 14.55.

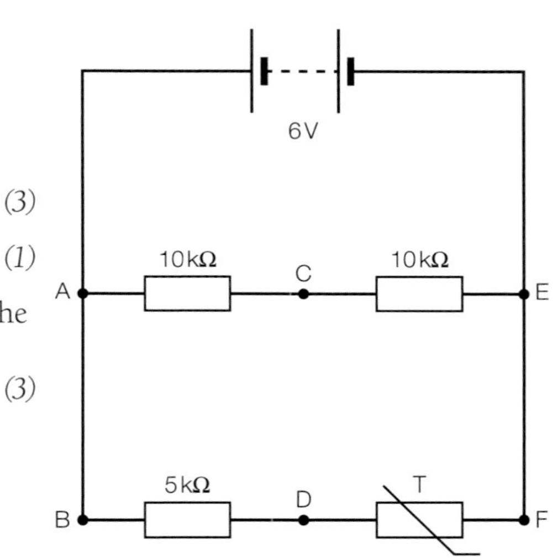

Figure 14.55 Circuit diagram for question 15.

a) At room temperature the resistance of the thermistor, T, is 2.5 kΩ. Calculate the total resistance of the circuit. (3)

b) Calculate the current flowing through the 6 V battery. (1)

c) A very high-resistance digital voltmeter is used to measure the pd across different points in the circuit. Copy and complete the table below by calculating the relevant pds. (3)

Position of the voltmeter	pd, *V* (V)
C-E	
D-F	
C-D	

d) The thermistor in the circuit is heated and its resistance decreases. State and explain the effect that this has on the voltmeter reading in the following positions: (4)

i) C-E ii) D-F.

16 An LDR is connected in series with a fixed resistor, R, and a 3.0 V battery with negligible internal resistance. A high-resistance voltmeter is connected in parallel across the fixed resistor.

a) Draw a circuit diagram of this circuit. (2)

b) On a cloudy day when the light intensity is low, the resistance of the LDR is 1.5 kΩ and the voltmeter reads 0.80 V. Calculate the pd across the thermistor. (2)

c) Calculate the current flowing through the battery. (2)

d) Calculate the value of the resistance, *R*, of the fixed resistor. (2)

e) On a bright sunny day the resistance of the LDR falls to 150 Ω. State and explain what effect this has on the reading on the voltmeter. (2)

f) State and explain the effect on the current flowing though the battery if the resistance of the voltmeter was the same as the resistance of the fixed resistor. (2)

17 Four resistors are connected into a circuit together with a 12 V car battery of negligible internal resistance. The current as measured by the low-resistance ammeter is 2.2 A. The circuit is shown in Figure 14.56.

a) Calculate the total effective resistance of the circuit. (2)

b) All the resistors have exactly the same resistance. Calculate the resistance of resistor P. (3)

c) Calculate the current flowing through resistor R. (3)

d) Calculate the power dissipated by resistor P. (2)

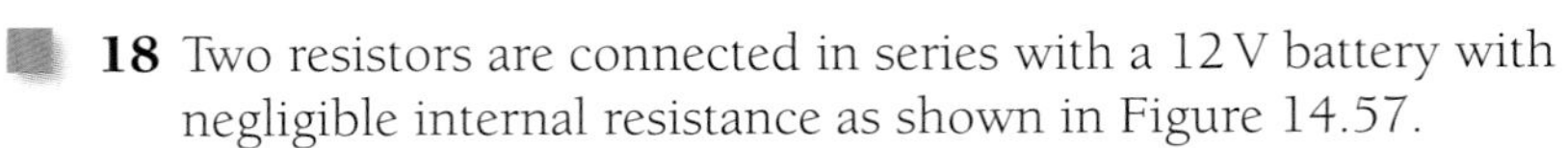

Figure 14.56 Circuit diagram for question 17.

18 Two resistors are connected in series with a 12 V battery with negligible internal resistance as shown in Figure 14.57.

a) The resistance R_2 is 150 Ω and the voltmeter reads 7.5 V. Calculate the current in the circuit. (1)

b) Calculate the power dissipated by resistor R_2. (2)

c) Calculate the resistance R_1. (2)

d) Resistor R_2 is replaced with an ntc (negative temperature coefficient) thermistor. Explain why initially the voltmeter reading remains constant, but slowly it changes and then reads another constant value. (2)

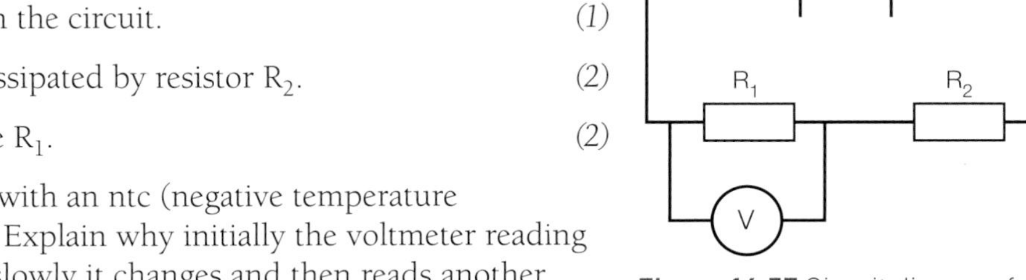

Figure 14.57 Circuit diagram for question 18.

19 An LDR (light-dependent resistor), a fixed resistor R and a variable resistor are connected in series with a 9.0 V battery with negligible internal resistance. A high-resistance voltmeter is connected in parallel with the fixed resistor, R, as shown in Figure 14.58 (overleaf).

a) In the dark, the resistance of the LDR is 120 kΩ, the resistance of the fixed resistor is 8.0 kΩ and the variable resistor has a resistance of 42 kΩ. Calculate the current in the circuit flowing through the battery. (2)

b) Calculate the reading of the voltmeter. (2)

c) The circuit is then taken out into the daylight. The resistance of the LDR drops quickly to a small value. State and explain the effect on the reading on the voltmeter. (2)

d) This light-sensing circuit forms part of an automatic light cut-off for a greenhouse. At a predetermined light intensity, the pd across the fixed resistor needs to be 0.90 V. At this light intensity, the resistance of the LDR is 12 kΩ. Calculate the required resistance of the variable resistor for this particular light level. *(2)*

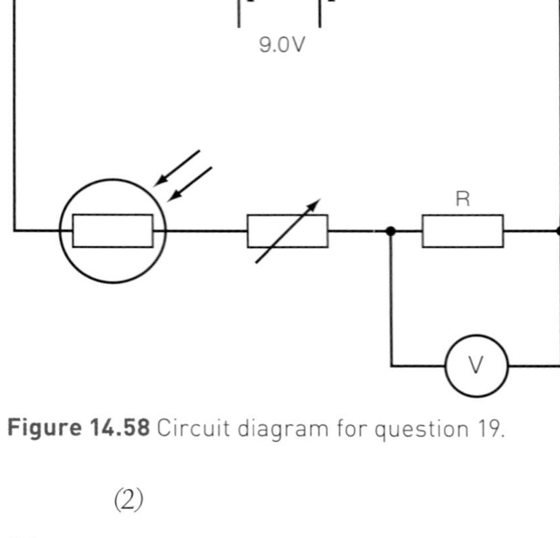

Figure 14.58 Circuit diagram for question 19.

20 The two sidelight bulbs of a car are rated as 12 V, 32 W. The bulbs are connected in parallel to a 12 V car battery of negligible internal resistance; an ammeter is connected in series with the battery to measure the total current drawn from the battery.

a) Draw a circuit diagram of this circuit. *(2)*

b) Calculate the reading of the ammeter. *(2)*

c) Calculate the resistance of the bulbs. *(2)*

d) State and explain the effect on the brightness of the bulbs if the ammeter has a higher resistance. *(2)*

Both of the bulbs are now connected to the battery in series with the ammeter.

e) State and explain the change of brightness of the two bulbs in the series circuit. *(2)*

f) Apart from the brightness of the bulbs, suggest another reason why connecting the bulbs in parallel would be preferable to connecting them in series. *(1)*

21 The internal lights on a helicopter consists of two bulbs connected in parallel to a 24 V battery of negligible internal resistance. The bulb in the cockpit needs to be dimmer than the one in the cabin. Two resistors are put into the circuit to ensure that the bulbs are both operating at their working voltage, as shown in Figure 14.59.

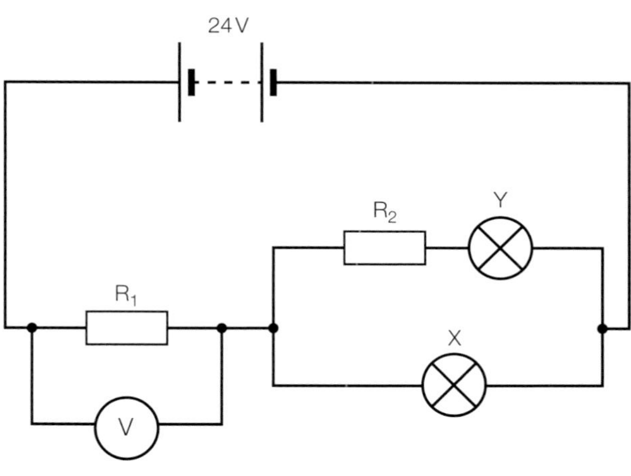

Figure 14.59 Circuit diagram for question 21.

a) Bulb X is rated at 12 V, 36 W and bulb Y at 4.5 V, 2.0 W. Use this data to calculate the current flowing though each bulb when it is operating at its working pd. *(3)*

b) Calculate the pd across resistor R_1. *(2)*

c) Calculate the current flowing through R_1. *(1)*

d) Calculate the resistance of R_1. *(1)*

e) The pd across bulb Y must be 4.5 V. Calculate the pd across resistor R_2. *(1)*

f) Calculate the resistance of R_2. *(2)*

g) Bulb Y breaks. No current runs through bulb Y and resistor R_2. Explain, without using a calculation, the effect on the voltmeter reading across R_1. *(2)*

h) State and explain what happens to bulb X. *(2)*

22 The circuit in Figure 14.60 can be used to determine the emf and internal resistance of a battery.

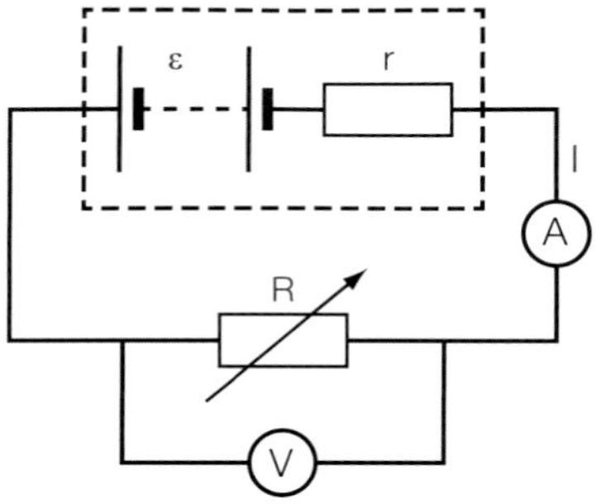

Figure 14.61 shows the results from an experiment involving a D-type cell.

Figure 14.60 Circuit diagram for question 22.

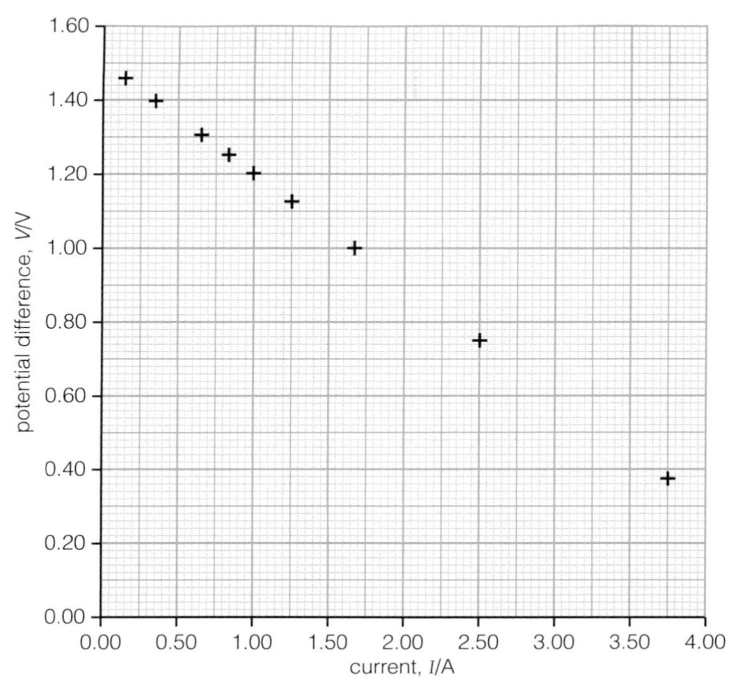

Figure 14.61 Electrical characteristic of a D-type cell.

a) Explain why the potential difference decreases as the current increases. *(2)*

b) Use the graph to determine the emf, ε, and the internal resistance, r, of the D-type cell. *(3)*

c) On a suitable sketch of the graph, draw on a line that illustrates the results obtained from another battery with the same emf but double the internal resistance of the first battery. Label this line X. *(2)*

d) On the same sketch draw another line that illustrates the results obtained by a third battery with the same emf as the other two, but with negligible internal resistance. Label this line Y. *(2)*

Stretch and challenge

All of the following questions are provided with permission of the British Physics Olympiad

23 Twelve 1 ohm resistors are connected in a cubic network, as shown in Figure 14.62.

a) Draw a two-dimensional circuit diagram of this network.

b) Use your circuit diagram to determine the overall resistance of this network between points A and B.

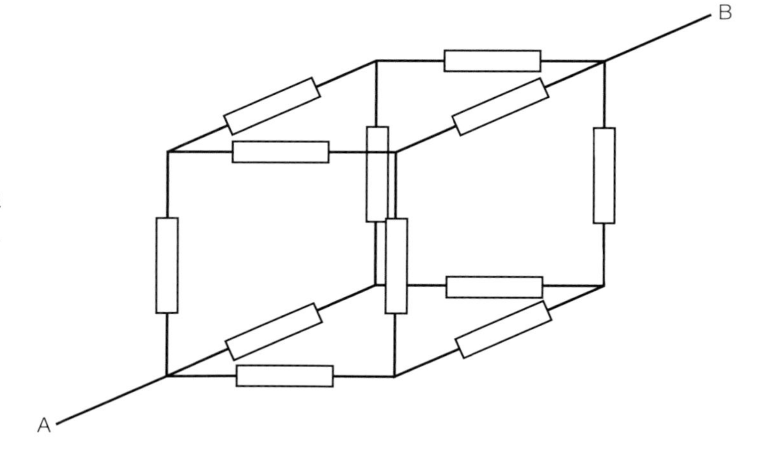

Figure 14.62 Circuit diagram for question 23.

24 If the potential difference between A and B in Figure 14.63 is V, then what is the current between A and B? All three resistors are identical.

A $\frac{V}{3R}$ **C** $\frac{3V}{2R}$

B $\frac{2V}{3R}$ **D** $\frac{3VR}{2}$

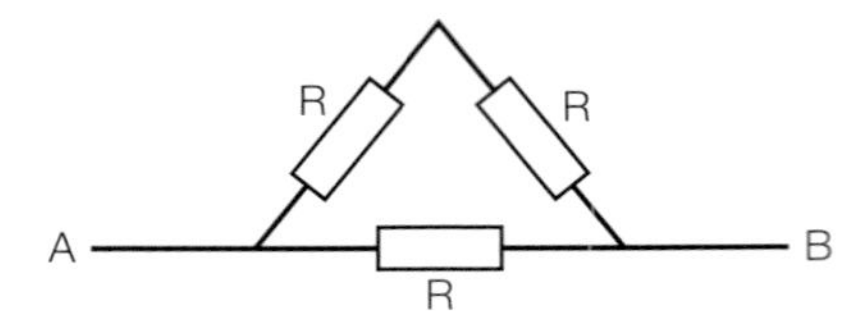

Figure 14.63 Circuit diagram for question 24 .

25 A 10 V battery with negligible internal resistance is connected to two resistors of resistance 250 Ω and 400 Ω, and to component Z, as shown in Figure 14.64.

Z is a device that has the property of maintaining a potential difference of 5 V across the 400 Ω resistor. The current through Z is:

A 2.9 mA **C** 12.5 mA

B 7.5 mA **D** 15.4 mA

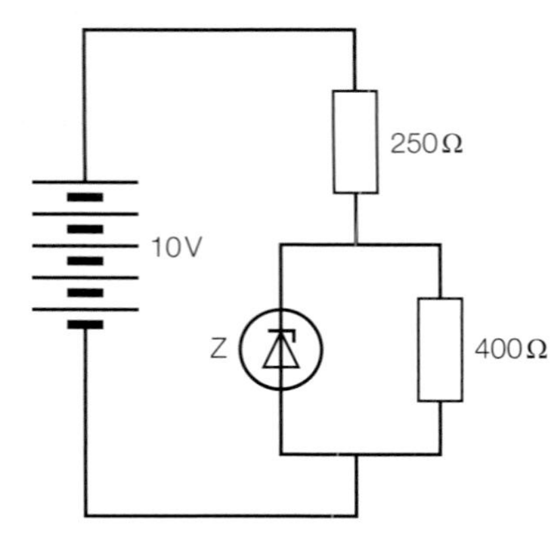

Figure 14.64 Circuit diagram for question 25.

26 In the circuit shown in Figure 14.65 the resistors have identical resistances. If the power converted in R_1 is P, what is the power dissipated in R_2?

A P/4 **C** P

B P/2 **D** 2P

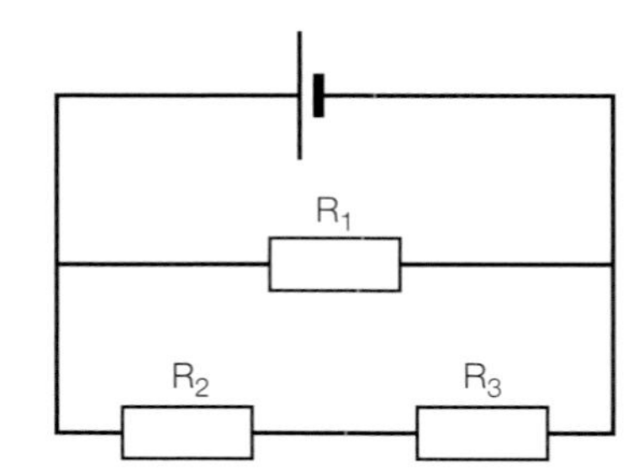

Figure 14.65 Circuit diagram for question 26.

27 Three identical voltmeters each have a fixed resistance, R, which allows a small current to flow through them when they measure a potential difference in a circuit. The voltmeters, V_1, V_2, V_3 are connected in the circuit as shown in Figure 14.66.

The voltage–current characteristics of the device D are unknown. If V_2 reads 2 V and V_3 reads 3 V, what is the reading on V_1?

A 1 V **C** 3 V

B 2.5 V **D** 5 V

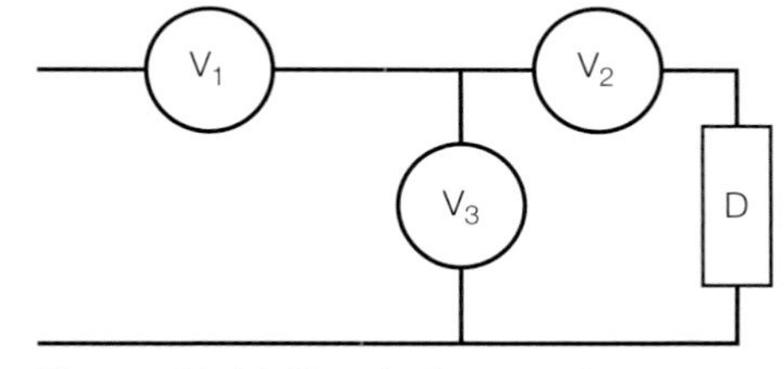

Figure 14.66 Circuit diagram for question 27.

28 A cell that produces a potential E (called an emf) is shown in Figure 14.67 and is connected to two resistors in series: a fixed resistor R_1 and a variable resistor R_2.

The current, *I*, in the circuit is measured by the ammeter, A, and the potential difference, *V*, across resistor R_2 is measured with the voltmeter, V. The relationship between the potential, *E*, and the current, *I*, is given by:

$\varepsilon = IR_1 + IR_2$

Which of these graphs would produce a straight-line fit?

A *V* against *I* **C** 1/*V* against 1/*I*

B *V* against 1/*I* **D** *I* against 1/*V*

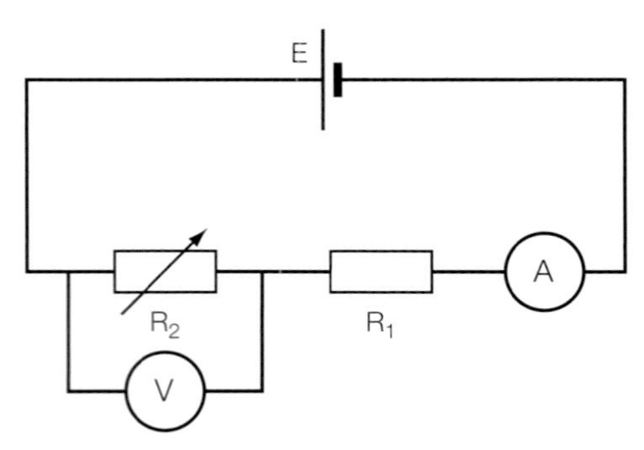

Figure 14.67 Circuit diagram for question 28.

29 A combination of resistors shown in Figure 14.68 represents a pair of transmission lines with a fault in the insulation between them. The wires have a uniform resistance but do not have the same resistance as each other. The following procedure is used to find the value of the resistor R_5.

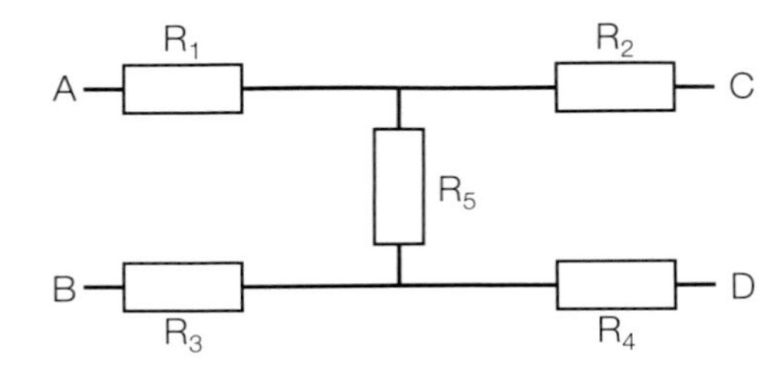

Figure 14.68 Circuit diagram for question 29.

A potential difference of 1.5 V is connected in turn across various points in the arrangement.

With 1.5 V applied across terminals AC a current of 37.5 mA flows.

With 1.5 V applied across terminals BD a current of 25 mA flows.

With 1.5 V applied across terminals AB a current of 30 mA flows.

With 1.5 V applied across terminals CD a current of 15 mA flows.

a) Write down four equations relating the potential difference, the resistor values and the currents.

b) Determine the value of resistor R_5.

c) If the ends C and D are connected together, what would be the resistance measured between A and B?

d) If the length AC (and also BD) is 60 metres, how far from A (or C) does the fault occur?

30 A student decides to calibrate a thermistor in order to measure variations in the temperature of a room. He connects a small bead-sized thermistor across the terminals of a 5 V power supply and in series with a 1 A ammeter. The resistance of the thermistor is 120 Ω at room temperature.

a) Instead of showing small variations in room temperature, the thermistor is likely to go up in smoke. Explain why.

In the light of his experience, he decides to redesign his simple circuit as shown in Figure 14.69. He has a few values of resistor R to choose from: 5 kΩ, 500 kΩ and 50 Ω.

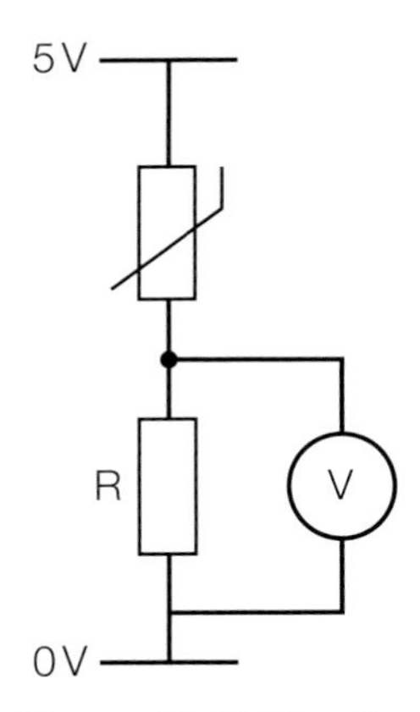

Figure 14.69 Circuit diagram for question 30.

b) State which value of R would give the biggest variation of V with temperature. Explain your choice.

c) State which value of R would be most likely to cause the same problem as in part (a). Again, explain your choice.

31 A single uniform underground cable linking A to B, 50 km long, has a fault in it at distance d km from end A, as shown in Figure 14.70. This is caused by a break in the insulation at X so that there is a flow of current through a fixed resistance, R, into the ground. The ground can be taken to be a very low-resistance conductor. Potential differences are all measured with respect to the ground, which is taken to be at 0 V.

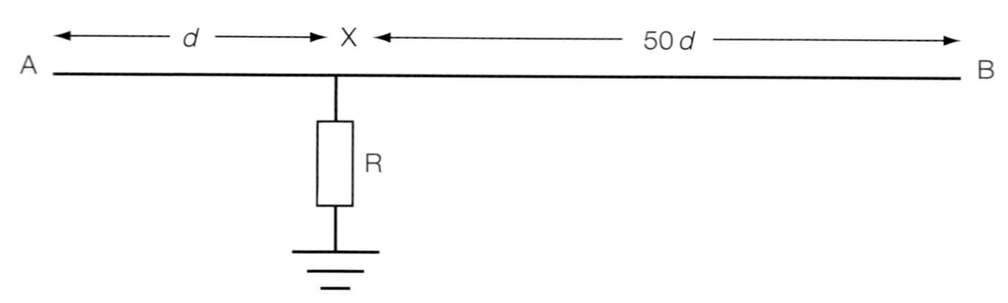

Figure 14.70 Circuit diagram.

In order to locate the fault, the following procedure is used. A potential difference of 200 V is applied to end A of the cable. End B is insulated from the ground, and it is measured to be at a potential of 40 V.

a) What is the potential at X? Explain your reasoning.

b) What is:

i) the potential difference between A and X?

ii) the potential gradient along the cable from A to X (i.e. the volts/km)?

The potential applied to end A is now removed and A is insulated from the ground instead. The potential at end B is raised to 300 V, at which point the potential at A is measured to be 40 V.

c) What is the potential at X now?

d) Having measured 40 V at end B initially, why is it that 40 V has also been required at end A for the second measurement?

e) What is the potential gradient along the cable from B to X?

f) The potential gradient from A to X is equal to the potential gradient from B to X.

i) Explain why this is true.

ii) From the two potential gradients that you obtained earlier, deduce the value of d.

32 This question is about heating elements.

a) The power dissipated as heat in a resistor in a circuit is given by $P = VI$. Show that this may also be expressed as $P = I^2R$ and $P = \frac{V^2}{R}$.

b) A student goes out to purchase an electric heater for their flat. The sales person says that to get more heat they should purchase a heater with a high resistance because $P = I^2R$, but the student thinks that a low resistance would be best because $P = \frac{V^2}{R}$. Explain who is correct.

c) Copper is a better conductor than iron. Equal lengths of copper and iron wire, of the same diameter, are connected first in parallel and then in series as shown by Figure 14.71. A potential difference is applied across the ends of each arrangement in turn, and the pd is gradually increased from a small value until, in each case, one of the wires begins to glow. Explain this, and state which wire will glow first in each case.

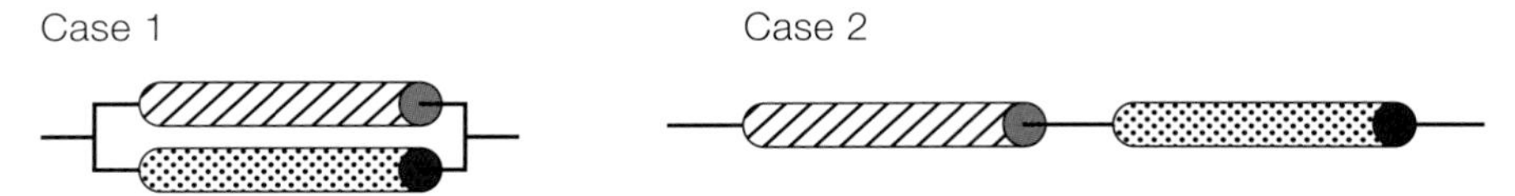

Figure 14.71 Copper and iron wire connected in different arrangements.

d) A surge suppressor is a device for preventing sudden excessive flows of current in a circuit. It is made of a material whose conducting properties are such that the current flowing through it is directly proportional to the fourth power of the potential difference across it. If the suppressor transfers energy at a rate of 6 W when the applied potential difference is 230 V, what is the power dissipated when the potential difference rises to 1200 V?

15 Maths in physics

Physicists use maths to explore how quantities are related, and to predict how they depend on each other as changes are made. Maths is used instead of words to describe how quantities relate to each other in different situations using equations. An equation can be rearranged to investigate different quantities, or values can be substituted to predict unknown quantities.

In this chapter we will look at maths skills such as handling data, algebra, graphs, and geometry and trigonometry.

Maths skills

In physics, all measurements include a number and a unit, for example $3.00 \times 10^8\,\mathrm{m\,s^{-1}}$, 0.510 999 MeV and 4.562 J. The unit gives the number a context: for example, time measured in hours is not the same as time measured in seconds.

Base units

A system of base units is used to create an internationally standardised system so that measurements made in different countries are directly comparable. These base units are:

- length in metres (m)
- time in seconds (s)
- mass in kilograms (kg)
- electric current in amperes (A)
- temperature in kelvins (K)
- amount of substance in moles (mol).

These units can be combined to give other units, such as those used for frequency (Hz, or $\mathrm{s^{-1}}$), acceleration ($\mathrm{m\,s^{-2}}$) and force (N or $\mathrm{kg\,m\,s^{-2}}$).

Converting between units

Prefixes are used for very small or very large measurements. For example, distances can be measured in km or in mm depending on their value. There are some standard prefixes you need to be able to recognise and use, which are shown in Table 15.1.

Table 15.1 Standard prefixes

Factor of 10	Name and symbol	Example
10^{-15}	femto, f	fm – femtometre
10^{-12}	pico, p	ps – picosecond
10^{-9}	nano, n	nm – nanometre
10^{-6}	micro, μ	μg – microgram
10^{-3}	milli, m	mm – millimetre
10^{-2}	centi, c	cl – centilitre
10^{3}	kilo, k	kg – kilogram
10^{6}	mega, M	MJ – megajoule
10^{9}	giga, G	GW – gigawatt
10^{12}	tera, T	TW – terawatt

To convert between units with different prefixes:

- convert the prefix into appropriate powers of ten, for example 900 nm becomes 900×10^{-9} m
- complete the calculation, combining powers of ten.

Here is an example.

$$P = I^2R$$

$$= 300\,\text{mA} \times 300\,\text{mA} \times 800\,\text{k}\Omega \text{ (substituting values)}$$

$$= 300 \times 10^{-3}\,\text{A} \times 300 \times 10^{-3}\,\text{A} \times 800 \times 10^{3}\,\Omega \text{ (converting prefixes)}$$

$$= 72\,000\,000 \times 10^{(-3-3+3)} \text{ (gathering terms)}$$

$$= 7.2 \times 10^{4}\,\text{W, or } 72\,\text{kW}$$

TIP

When calculating cross-sectional area, the prefix terms will be squared. For example, the cross-sectional area of a wire 0.3 mm in radius is $\pi \times 0.3^2 \times (10^{-3})^2\,\text{m}^2 = \pi \times 0.09 \times 10^{-6}\,\text{m}^2$, or $2.83 \times 10^{-7}\,\text{m}^2$.

Decimal and standard form

Quantities such as 3420.7 may be expressed in standard form. In this case, write down the number between 1 and 10, and add a term to indicate the correct power of ten, for example 3.4207×10^3. The number of decimal places you use indicates the precision of the number. For example, 3.4207×10^3 is more precise than 3.4×10^3.

EXAMPLE

What happens to the gravitational force between two masses, M and m, when the distance, r, separating them is doubled?

Answer

Use

$$F_1 = \frac{GMm}{r^2}$$

$$F_2 = \frac{GMm}{(2r)^2} = \frac{GMm}{4r^2} = \frac{1}{4}F_1$$

Ratios, fractions and percentages

Ratios compare one quantity with another. Normally ratios are shown with the two quantities separated by a colon. A ratio of five neutrons to six protons in a nucleus is shown as 5:6. It is best to reduce the ratio to its simplest form. For example, 5:15 can be written as 1:3.

EXAMPLE

What is the ratio of the volume of two cubes of edge length a and $3a$?

Answer

$$\frac{V_2}{V_1} = \frac{(3a)^3}{a^3} = 27$$

Fractions represent the proportion of a whole unit. Fractions are written showing the number of parts there are divided by the total number of parts. For example $\frac{3}{4}$ means there are three parts out of a total of four parts. It is best to show fractions reduced to their simplest form. For example $\frac{9}{12}$ can be written as $\frac{3}{4}$.

Percentages mean parts out of a hundred. For example, if the efficiency of a device is 56%, 56 joules are usefully transferred for every 100 joules supplied to the device. When taking measurements using a metre ruler marked in millimetres, a student states her reading as 35 mm ±1 mm. The percentage uncertainty is $\frac{1}{35} \times 100\% = 2.9\%$.

To measure percentage change, divide the actual change by the original value and multiply by 100%. For example, if the potential difference drops from 8.8 V to 6.5 V, the actual change is 2.3 V and the % change is $\frac{23}{88} \times 100\% = 26\%$.

Estimating results

You may have to estimate the effect of changing variables on your measurements or calculations.

EXAMPLE

Estimate the effect on the current in a circuit if the diameter of the wire doubles assuming the pd across the wire remains the same.

Answer

The relevant relationships are: $R = \frac{\rho l}{A}$ and $I = \frac{V}{R}$. Since resistance is proportional to 1/cross-sectional area, doubling the diameter reduces resistance by a factor of four. Reducing resistance by a factor of four increases the current four-fold if the voltage is constant.

Written mathematically:

$$R_1 = \frac{\rho l}{A_1}$$

$$R_2 = \frac{\rho l}{A_2} = \frac{\rho l}{4A_1} = \frac{1}{4}R_1$$

So

$$I_1 = \frac{V}{R_1}$$

$$I_2 = \frac{V}{R_2} = \frac{V}{\frac{1}{4}R_1} = 4I_1$$

Using calculators

Calculators usually use the BODMAS rule: brackets, orders (i.e. powers and roots), divide, multiply, add, subtract. If you don't use brackets carefully or apply operators in a certain order, operations may not apply as you expect. For example (3 + 4)/(5 × 7) = 7/35 = 0.2, could be entered as:

- (3 + 4)/(5 × 7)
- (3 + 4) ÷ 5 ÷ 7
- 3 + 4 ÷ 5 ÷ 7, which gives the wrong answer

To raise numbers to a certain power, use the [exp] button on the calculator. For example, for a number like 3.00×10^8, type in 3[exp]8. There is no need to type in the final zeros. For 6.63×10^{-34}, type in 6.63[exp]–34, although some older calculators require: 6.63[exp]34–. Check your own calculator and practise.

To check if your calculator is set to degrees or radians, press the [mode] button until this choice is displayed, and press the number corresponding to the setting you want.

To calculate sine, cosine or tangent, type in the function you want followed by the value. For example, find $\cos 45°$ by pressing [cos]45 while the calculator is in the degrees mode.

Usually π is entered by pressing the [shift] button followed by the [exp] button. To find $\sin \pi$, set the calculator to the radians mode, then type in [sin][shift][exp].

To find degrees or radians when you have been given sine, cosine or tangent, check the calculator is set to degrees or radians as required. If you are asked for $\cos^{-1} 45°$, type in [shift][cos][45]=

TEST YOURSELF

1 Write down the correct unit for these quantities:
  - **a)** velocity
  - **b)** resistivity
  - **c)** density
  - **d)** frequency.

2 Convert between these units:
  - **a)** 2 hours expressed in seconds
  - **b)** 300 mm^3 expressed in m^3
  - **c)** 2 eV expressed in joules.

3 Write these numbers in standard form:
  - **a)** 0.5109
  - **b)** 3600
  - **c)** 300 000 000.00
  - **d)** 0.009 354

4 Calculate these quantities:
  - **a)** the efficiency of a power station generating 600 MW output from a power input of 1500 MW
  - **b)** the volume of a sphere with radius 3.4 mm (volume of a sphere is $\frac{4}{3}\pi r^3$).

5 Estimate the effect of changing the named variable:
  - **a)** the effect on kinetic energy of doubling velocity when $KE = \frac{1}{2}mv^2$
  - **b)** the effect of doubling the diameter of a wire on extension, Δl, when the extension is inversely proportional to the cross-sectional area of the wire
  - **c)** the effect of halving the wavelength, λ, on fringe spacing, w, when:

$$w = \frac{\lambda D}{s}$$

  (D and s remain constant).

6 Enter these numbers in your calculator. Write down the result.
  - **a)** $6.63 \times 10^{-34} \times 3.00 \times 10^8/400$
  - **b)** $(341)^{1/2}$
  - **c)** $\tan 34°$
  - **d)** $\sin \pi/4$

Handling data

In physics, it is important that you understand the significance of numbers you may calculate or measure.

Significant figures

The results of calculations can include some numbers that do not tell us anything important. We can choose to use a particular number of significant figures so that we only include numbers that tell us something useful. For example, the mass of an electron is often given as $9.10938291 \times 10^{-31}$ kg, which is an approximate value. You only need to quote the mass of an electron to three significant figures: 9.11×10^{-31} kg.

To work out the number of significant figures, count up the number of digits remembering that:

- zeros between non-zero numbers are significant (e.g. 3405 has four significant places)
- zeros after the decimal place, or with the decimal place shown, are significant (e.g. 34.50, 3450. and 34.05 have four significant figures)
- leading zeros are not significant (e.g. 034.5 has three significant figures, and 0.00136 has three significant figures.)
- trailing zeros with no decimal place shown are significant (e.g. 3450 has four significant figures and 0.04500 has five significant figures). However, the significance of trailing zeros in a number not containing a decimal point can be ambiguous. For example, it is not always clear whether a number such as 6500 is precise to the nearest unit (and just happens to be an exact multiple of a hundred) or whether some rounding to the nearest hundred has taken place. The use of standard form, 6.5×10^3, gets around the ambiguity; alternatively you can just write 6500 (3 sf).

To reduce the number of significant figures, you should round up if the final digit is five or more, and round down if the final digit is four or less. For example,

- 3405 has four significant figures and 3410 still has four significant figures. So, to write the number to three significant figures you need to write 3410 or $3.41\,\text{sf} \times 10^3$.
- 3.46 has three significant figures but 3.5 has two significant figures.

Significant figures in calculations

The choice of significant figures in experiments is important as it gives us information about the precision of the equipment. There is more information about significant figures in Chapter 16.

When you carry out a calculation, your answer may have more significant figures than the values given in the questions. For example, $\frac{12}{3.2} = 3.75$. The answer has three significant figures but the data only has two significant figures, so you should give your answer to two significant figures, 3.8. This is because increasing the numbers of significant figures suggests the precision has improved.

TIP
Wait till you have your final answer before you round down to the correct number of significant figures. If you round down too early, your answer may be wrong. Too many significant figures overstates the precision.

Arithmetic means

To calculate an arithmetic mean, total all the values to be averaged and divide the total by the number of readings. For example, a student measures the wavelength of a wave on a slinky five times: 41 cm, 43 cm, 42 cm, 46 cm, 45 cm.

The average is $\frac{(41 + 43 + 42 + 46 + 45)}{5} = \frac{217}{5} = 43.4\,\text{cm}$, or 43 cm to two significant figures, which is appropriate for the accuracy of the measurements.

Probability

Probability measures the chance of something happening in a particular time. It does not mean this will definitely happen in that time. Nuclear physicists say the probability of a nucleus decaying in unit time is a constant, called the decay constant. This constant has units of s^{-1}.

Order of magnitude

Often in physics, we are satisfied with an approximate answer. For example, we can make rough predictions knowing that the mass of an electron is about two thousand times smaller than the mass of a proton (or 10^3 times smaller to the nearest order of magnitude). For more accurate calculations we must use exact values, but order of magnitude calculations help us check calculations, rule out options and make predictions.

For an order of magnitude calculation, you should:

- express values in standard form, for example a wavelength of 700 nm could be written as 1×10^{-6} m (or as 7×10^{-7} m)
- when numbers are multiplied, powers of ten are added, for example $10^5 \times 10^9 = 10^{14}$
- when numbers are divided, powers of ten are subtracted for example $10^5/10^9 = 10^{-4}$
- some values can be approximated easily, for example π^2 is about 10.

For example, what is the energy of an electron travelling at 10% of the speed of light?

$$KE = \frac{1}{2} \times 9.11 \times 10^{-31} \times 3 \times 10^7 \times 3 \times 10^7 \text{ (exact)}$$

$$= \frac{1}{2} \times 10 \times 10^{-31} \times 10 \times 10^{14} \text{ (approximate)}$$

$= \frac{1}{2} \times 10^{(1-31+1+14)} = 0.5 \times 10^{-15}$ J, which is very close to the answer calculated using the values given: 0.40995×10^{-15} J.

Identifying uncertainties

This topic is covered in more detail in the Chapter 16.

Uncertainties are shown as an absolute value, e.g. 5.2 ± 0.2 (so its value lies between 5.0 and 5.4) or as a % uncertainty, e.g. 5.2 ± 10% (so its value lies between 4.7 and 5.7). There are some general rules to follow when combining uncertainties in data:

- to multiply or divide two quantities, add their % uncertainties
- to add or subtract quantities, add their actual uncertainties
- to square a quantity, double its % uncertainty; if it is cubed, multiply the % uncertainty by three.

TEST YOURSELF

7 Calculate the average of this data, quoting the result to the appropriate significant figures: 9.445, 9.663, 8.567, 10.345.

8 How many significant figures have these numbers got?
a) 53 **b)** 5.30 **c)** 0.00530 **d)** 53×10^5

9 Calculate the volume of a cube of side 3.4 m, quoting your answer to the appropriate number of significant figures.

10 Calculate the decay constant, λ, if a sample of radioactive material has 1000 undecayed nuclei and, then after three minutes, the number of undecayed nuclei is 960.

11 **a)** Calculate the volume of a wire, length 1.00 m ± 0.01 m and cross-sectional diameter 5.6 ± 0.1 mm (volume of a wire is $\pi r^2 l$).
b) Calculate the % error and absolute error.

12 Make order of magnitude calculations for:
a) the volume of Earth; its radius is 6378 km (hint: volume of a sphere is $\frac{4}{3}\pi r^3$)
b) the mass of Earth; its density is 5540 kg m^{-3}.

Algebra

Equations describe a physical situation using mathematical symbols. For example, by writing:

$$a = \frac{(v - u)}{t}$$

we mean acceleration equals the change in velocity (final velocity minus original velocity) divided by the time taken.

You are expected to recognise and use the different symbols in equations shown in Table 15.2.

Table 15.2 Symbols used in equations and their meanings

Symbol	Meaning	Example
=	Equal to	$x = 3$ means: x equals 3
>	Greater than	$x > 5$ means: x is greater than 5
<	Less than	$x < 8$ means: x is less than 8
>>	Much greater than	$x >> 15$ means: x is much greater than 15
<<	Much less than	$x << 6$ means: x is much less than 6
Δ	Difference between, or change in	Δx means: a change in x, or the difference between two readings of x
≈	Approximately equal to	$x \approx 7$ means: x is approximately equal to 7
∝	Proportional to	$y \propto x$ means: that x is proportional to y (or $\frac{y}{x}$ = constant)

Solving equations

To solve an equation, substitute values for each quantity using information from the question. Make sure that you use consistent units, for example convert all times into seconds. Then carry out operations in the correct order:

- complete operations inside brackets
- exponents (once these are isolated)
- multiplication and division
- addition and subtraction.

TIP
Use brackets to make clear which order to apply the operators, + – x /.

Rearranging equations

Often you must change the subject of an equation. Always do the same thing to each side of the equation in the same order. For example, to calculate the original velocity, u:

$$a = \frac{(v - u)}{t}$$

Multiply each side by t, giving: $at = v - u$
Add u to each side, giving: $at + u = v$
Subtract at from each side, giving: $u = v - at$

Rearranging non-linear equations

Some equations are non-linear, but the same rules apply: do the same thing to each side.

For example, $\mathrm{KE} = \frac{1}{2}mv^2$. To make v the subject:

Multiply each side by 2, giving: $2\mathrm{KE} = mv^2$

Divide each side by m, giving: $v^2 = \frac{2\mathrm{KE}}{m}$

Take the square root of each side, giving:

$$v = \left(\frac{2\text{KE}}{m}\right)^{\frac{1}{2}}$$

Solving quadratic equations

A quadratic equation includes x and x^2. If you have to solve a quadratic equation in the form $ax^2 + bx + c = 0$, when a, b and c are constant, then the general solution is:

$$x = \frac{-b \pm \sqrt{b^2 - 4ac}}{2a}$$

At A-level, the most commonly used quadratic equation is the equation of motion:

$$s = ut + \frac{1}{2}at^2$$

Remember that terms in the equation become zero if they involve a quantity that is zero. It often simplifies your calculation if you substitute in values from the question before using the general solution.

For example, a book falls from a shelf that is 3 m high. How long does the book fall before it reaches the ground?

The information from the question gives:

$s = 3\,\text{m}$

$u = 0\,\text{m}\,\text{s}^{-1}$

a = acceleration due to gravity, $9.81\,\text{m}\,\text{s}^{-2}$

Substituting into

$$s = ut + \frac{1}{2}at^2$$

gives: $3 = 0t + \frac{1}{2} \times 9.81t^2$

Since $0t$ is zero, the equation becomes:

$$3 = \frac{1}{2} \times 9.81t^2$$

Multiplying both sides by 2 gives:

$$6 = 9.81t^2$$

Dividing both sides by 9.81 gives:

$$0.611 = t^2$$

Taking the square root of each side gives

$$t = 0.78\,\text{m}\,\text{s}^{-1}$$

TEST YOURSELF

13 Explain in words what these equations mean:

a) $a = \frac{\Delta v}{\Delta t}$ where a is acceleration, v is velocity and t is time

b) $\sin \theta_c = \frac{n_2}{n_1}$ where n_1 and n_2 are refractive indices for materials 1 and 2, and θ_c is the critical angle.

14 Express these ideas as equations, using symbols:

a) λ, the de Broglie wavelength, is proportional to $\frac{1}{p}$, where p is momentum; h is the constant of proportionality

b) Hooke's law states force, F, equals constant, k, multiplied by change in length, l.

c) velocity, v, is the displacement, s, divided by the time, t.

15 Rearrange these equations:

a) $E = mc^2$ to make c the subject

b) $v^2 = u^2 + 2as$ to make s the subject

c) $n_1 \sin_1 = n_2 \sin_2$ to make n_2 the subject

d) $V = I(R + r)$ to make r the subject.

16 Calculate the answer using the data given:

a) Use $c = f\lambda$ to calculate the frequency, f, of light of wavelength, λ, 900 nm. The speed of light, c, is $3 \times 10^8\,\mathrm{m\,s^{-1}}$.

b) Use the equation  $\frac{1}{R} = \frac{1}{R_1} + \frac{1}{R_2}$ to calculate the combined resistance, R, for resistors of resistance 300 ohms and 500 ohms, which are connected in parallel.

c) Use $v = u + at$ to calculate the final velocity, v, of an electron travelling initially at $3 \times 10^5\,\mathrm{m\,s^{-1}}$ and accelerating for 3 seconds at a rate of $2 \times 10^4\,\mathrm{m\,s^{-2}}$.

d) Use $v^2 = u^2 + 2as$ to calculate the acceleration of a car that brakes, reducing its speed from $15\,\mathrm{m\,s^{-1}}$ to $3\,\mathrm{m\,s^{-1}}$ in a distance of 30 m.

Graphs

Graphs display useful information about the relationship between quantities at a glance, for example the behaviour of a capacitor or a moving object. The line of a graph can often be extrapolated (extended beyond the region in which data has been collected) or to make predictions about behaviour under different circumstances.

Recognising graphs

You should recognise, and be able to sketch, the shape of these graphs in Figures 15.1–15.7.

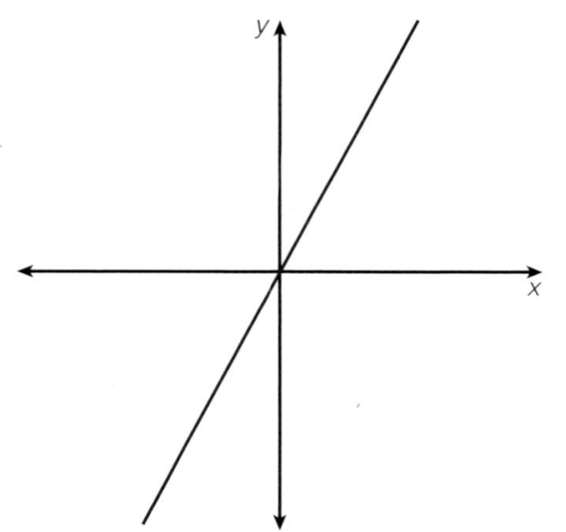

Figure 15.1 Graph of $y = kx$, where k is a constant.

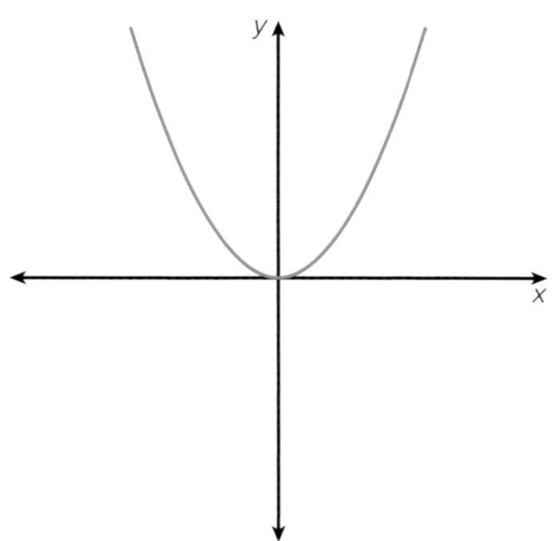

Figure 15.2 Graph of $y = kx^2$, where k is a constant.

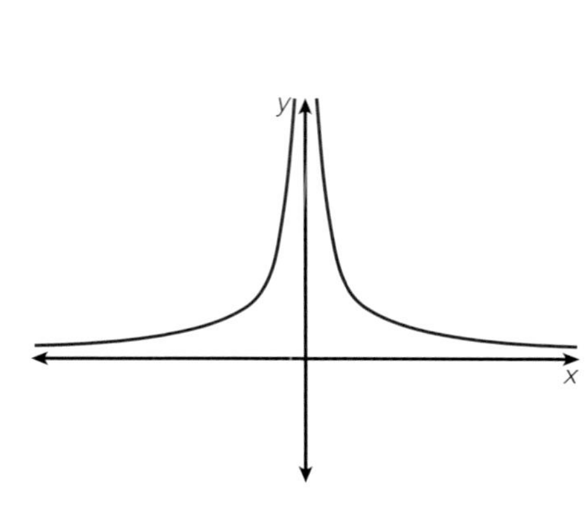

Figure 15.3 Graph of $y = 1/kx^2$, where k is a constant.

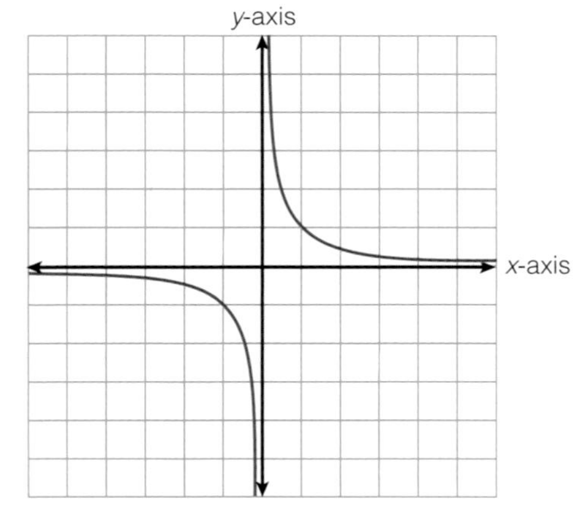

Figure 15.4 Graph of $y = 1/kx$, where k is a constant.

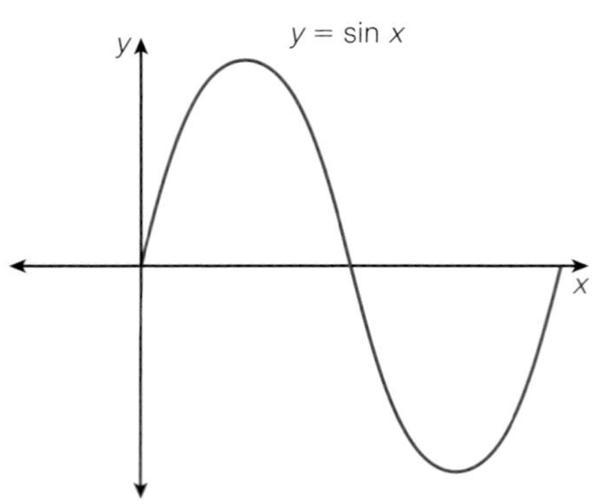

Figure 15.5 Graph of $y = \sin x$.

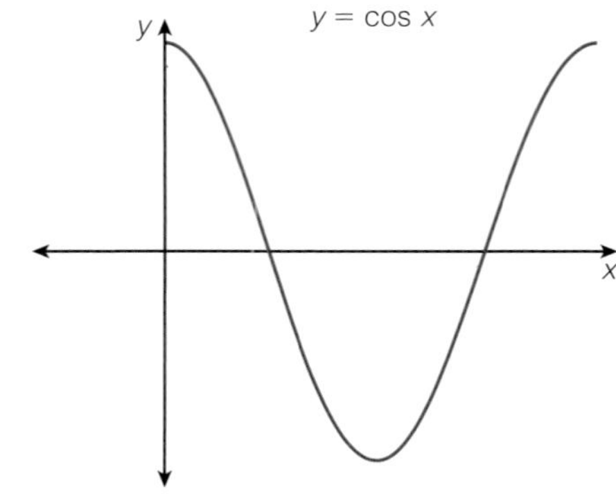

Figure 15.6 Graph of $y = \cos x$.

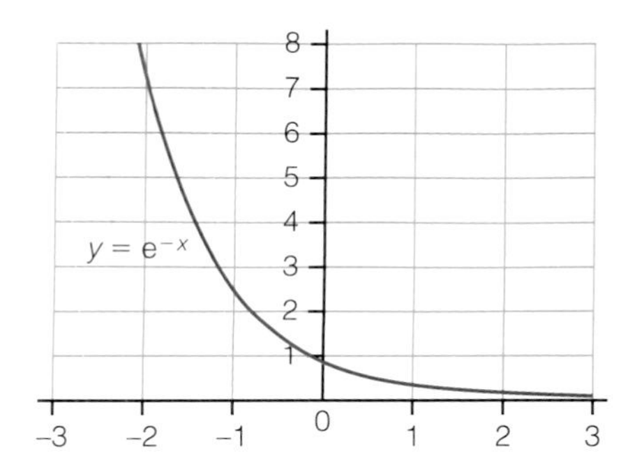

Figure 15.7 Graph of $y = e^{-x}$.

Plotting graphs

The following rules for plotting graphs are based on exam board guidance.

- You usually plot the independent variable (what you change) on the x-axis, and the dependent variable (what you measure) on the y-axis.
- Choose the scale so plotted points fill at least half of the graph grid. You need not always include the origin.
- Label axes with the quantities being plotted going from left to right.
- Choose scales that are easy to work with. For example, avoid gaps of three and label the scale reasonably frequently. Make sure the spacing and separation are regular. For example, each large square is worth two units.
- Plot points accurate to half a small square. Use a ruler to help if necessary and do not plot points in margins. Use a sharp pencil.
- The line (or curve) of best fit should be a line drawn that has approximately equal numbers of points on either side of the line. It does not need to go through the origin.

Using straight-line graphs

A straight-line graph has the form: $y = mx + c$ where m is the gradient of the graph and c is a constant. Use these rules for straight-line graphs.

- The equation must be in the form $y = mx + c$, for example $v = u + at$. Relate both equations. For example, $y = v$, $m = a$, $x = t$ and $c = u$.
- In some cases you should process data before plotting points. For example, $v^2 = u^2 + 2as$ has a straight-line form if you plot v^2 and s.
- The intercept of the line with the y-axis represents the constant, c. If the line goes through the origin, c is zero.
- Draw a large triangle when calculating the gradient of the graph.
- Use the triangle to find the change in y for a corresponding change in x. The gradient is $\frac{\Delta y}{\Delta x}$.
- Since the gradient is $\frac{\Delta y}{\Delta x}$, for graphs with t plotted on the x-axis, the gradient represents rate of change. For example the gradient of a velocity-time graph is $\frac{\Delta v}{\Delta t}$, or acceleration.

Using curved graphs

Straight-line graphs should be plotted, if possible, because anomalous points can more easily be seen. However, sometimes a non-linear function is complicated and a curved graph has to be plotted. Examples of curved graphs include motion graphs for non-uniform acceleration. Use these rules for curved graphs.

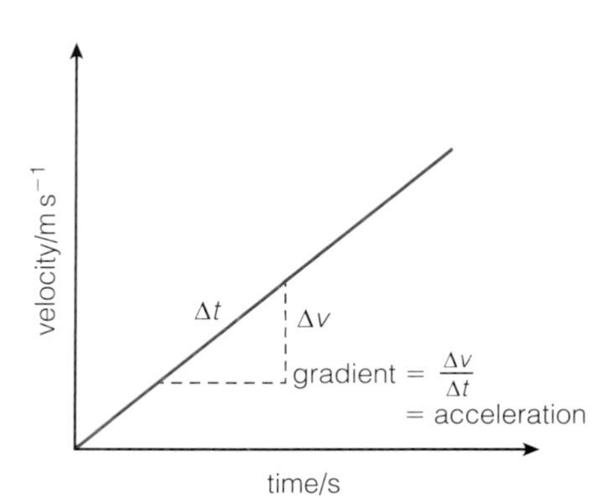

Figure 15.8 Calculating gradient.

- A curve of best fit should be smooth with equal numbers of points above and below the line.
- To calculate the gradient at a certain point, draw a large triangle with the hypotenuse as the tangent to the curve. Use the triangle to find the change in y for a corresponding change in x. The gradient is $\frac{\Delta y}{\Delta x}$.
- A curved graph shows a non-linear relationship, so the gradient changes. For example, the gradient at a particular time on a non-linear displacement–time graph shows instantaneous velocity. The average velocity is found by comparing readings over a whole section of the graph.

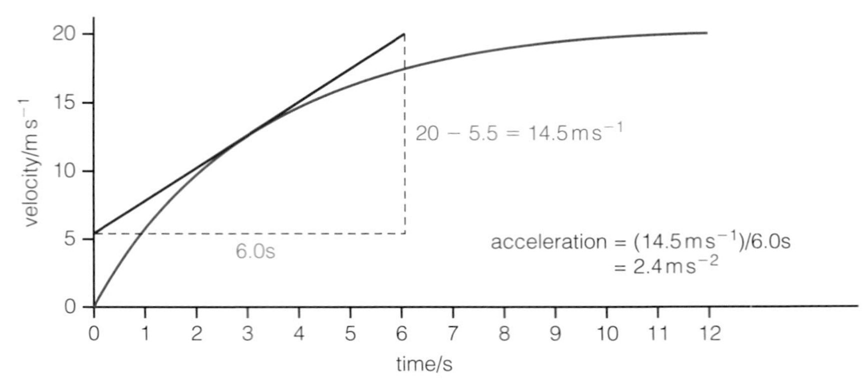

Figure 15.9 Calculating gradient on a curved graph.

Calculations using graphs

The area under certain graphs has a specific physical meaning:

- the area under a velocity–time graph represents distance travelled
- the area under the graph of force against extension represents work done by the force
- the area under the graph of potential difference across a capacitor against charge stored on the capacitor represents the energy stored in the capacitor.

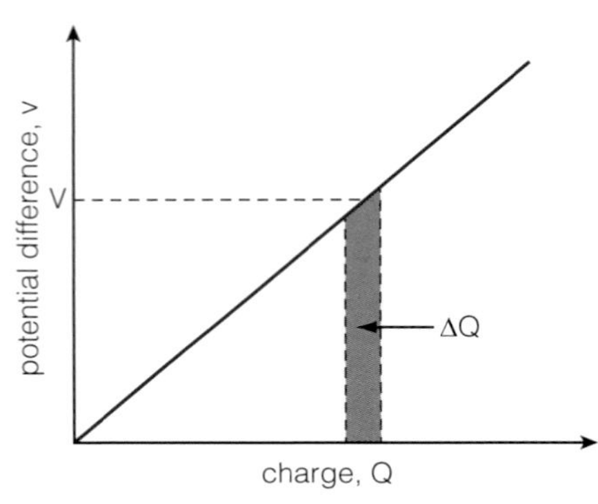

Figure 15.10 The shaded area represents the extra energy stored when the charge increases from Q to $Q + \Delta Q$.

If you are using a graph to calculate area, the axes must include $y = 0$, $x = 0$. This is needed if you are calculating a change in area, for example the additional energy stored when charge on a capacitor increases from Q to Q + ΔQ is V ΔQ.

You may need to divide the area under the graph into sections, either to make approximations if the line is curved or if you are calculating a change. The formula to calculate the area of a triangle is

$$\frac{1}{2} \times \text{base} \times \text{height}.$$

Remember to use the units shown on the axes, taking special care with prefixes to write down the correct values for the calculation.

TEST YOURSELF

17 Sketch a graph, labelling the axes, of:

a) $V = IR$ (V on the x-axis; I on the y-axis)

b) $E = \frac{1}{2}mv^2$ (v on the x-axis; E on the y-axis)

c) $E = \frac{hc}{\lambda}$ (λ on the x-axis; E on the y-axis) in this case λ can only be positive.

18 Use the data in this table to plot a correctly labelled graph of v against t.

Time (s)	Velocity (m s^{-1})
0	4.0
1	5.2
2	6.4
3	7.9
4	9.0
5	10.1
6	11.3
7	12.6
8	13.8

19 Explain what you need to plot so each of these equations produce straight-line graphs. State the physical significance of the y intercept and the gradient.

a) $s = \frac{1}{2}at^2$ (a is constant)

b) $\frac{hc}{\lambda} = \varphi + \frac{1}{2}mv^2$ (h, c, m and φ are constant)

c) $R = \frac{\rho l}{A}$ (ρ and A are constant).

20 Use the graph in Figure 15.11 to calculate the resistance when the current through a non-ohmic filament bulb is 0.4 A.

21 Use the velocity–time graph in Figure 15.12 to calculate.

a) acceleration when t = 20s

b) distance travelled between 0 seconds and 20 seconds.

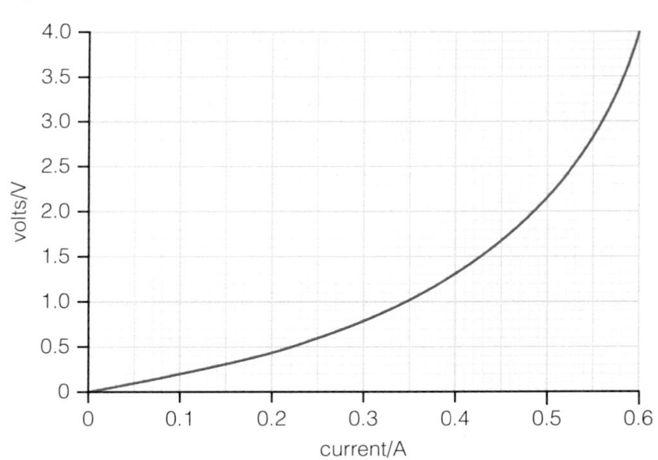

Figure 15.11 Graph for a filament bulb.

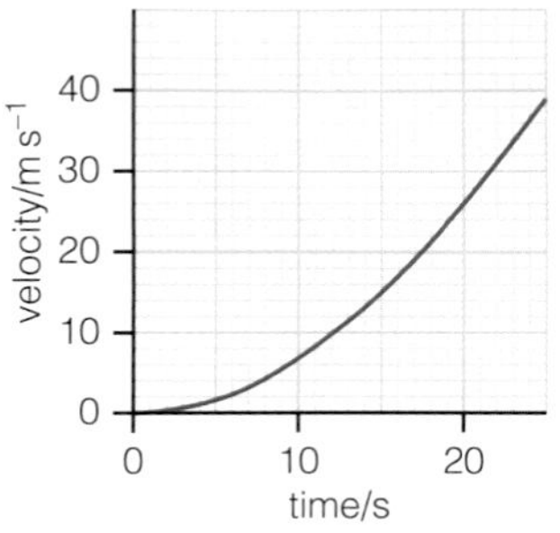

Figure 15.12 Velocity–time graph.

Geometry and trigonometry

Many physical situations are described using diagrams, for example, the interaction between nuclear particles, mechanical systems in equilibrium and the motion of waves.

Calculating areas and volumes

Often you will be given a diagram and asked to read values from it to use in a calculation. Do be aware that units may vary for different measurements. There are some basic rules to remember as shown in Tables 15.3 and 15.4.

Table 15.3 Calculating areas

Shape	Area	Example of calculations using this information
Triangle	$\frac{1}{2}$ base × height	Area under a v–t graph.
Circle of radius r	πr^2	Cross-sectional area of a wire; check if you are given the diameter or the radius
Rectangular block of sides a, b and c	$2ab + 2bc + 2ca$	
Cylinder of radius r and height h	$2\pi r^2 + 2\pi rh$	
Sphere or radius r	$4\pi r^2$	Calculate energy radiated by the Sun

Table 15.4 Calculating volumes

Shape	Volume	Example of when this is used
Rectangular block of sides a, b and c	abc	Calculate density
Cylinder of radius r and height h	$\pi r^2 h$	
Sphere of radius r	$\frac{4}{3}\pi r^3$	Calculate density

Other information that may be useful:

- the circumference of a circle of radius r is $2\pi r$; this may be used for questions involving orbits of satellites, for example
- the diameter, d, of a circle of radius r is $2r$.

Angles in triangles

You may need to calculate the angles in a triangle. These rules apply:

- the sum of the internal angles in a triangle is 180°. This may be used to interpret force diagrams, for example for objects on a ramp.
- for a right-angled triangle of sides a, b and c, Pythagoras' theorem states that: $a^2 + b^2 = c^2$. This may be used to calculate the magnitude of a force or to resolve forces into components.

Sine, cosine and tangents

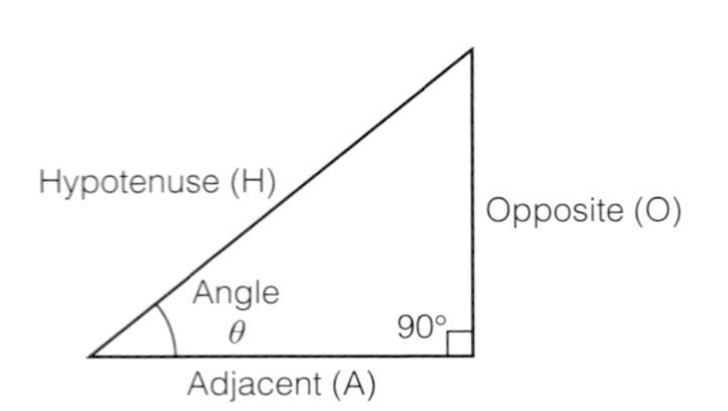

Figure 15.13 Right-angled triangle.

Right-angled triangles are used to define three functions: sine, cosine and tangent. The sides of a right-angled triangle are labelled in relation to the angle as opposite (O), adjacent (A) and hypotenuse (H).

$$\sin\theta = \text{O/H}$$

$$\cos\theta = \text{A/H}$$

$$\tan\theta = \text{O/A}$$

You can calculate an unknown value for the side of a triangle knowing the length of one other side and the angle, for example on an inclined plane. Check the diagram carefully with calculations involving an inclined plane: the triangle to use has the weight as the hypotenuse.

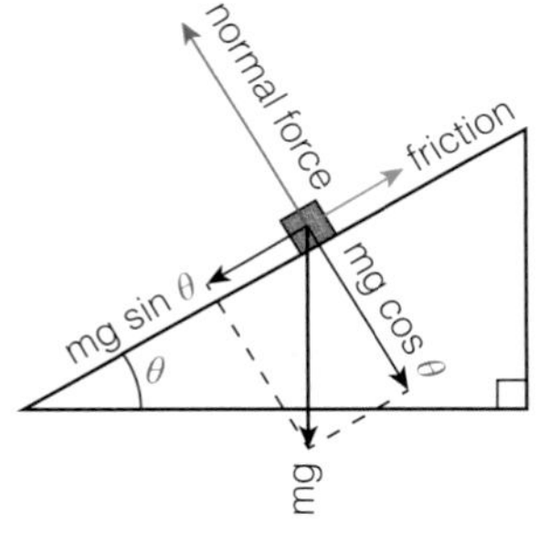

Figure 15.14 Resolving a force on an inclined plane.

A useful equation involving sine and cosine is:

$$\cos^2\theta + \sin^2\theta = 1$$

Combining this equation with Pythagoras' theorem for the vector triangle in Figure 15.14, where H is the length of the hypotenuse:

$$\text{H}^2 = \text{H}^2\cos^2\theta + \text{H}^2\sin^2\theta$$

Resultant vectors

To calculate a resultant vector, either draw a scale diagram or use calculations.

Method 1: Draw a scale diagram

- Draw each vector as an arrow whose length and direction corresponds to the vector's magnitude and direction.
- Draw each additional vector with its end touching the arrowhead of the previous vector.
- The resultant is the vector with its end touching the end of the first vector, and its arrowhead touching the arrowhead of the last vector.

For example, find the resultant of two forces if force V_1 is 10 N acting horizontally and force V_2 is 13 N and acts at 60° to the horizontal.

- Draw a scale diagram showing the two forces.
- Redraw the force V_2 vector with its tail touching the tip of vector V_1. Always include arrows to show the direction of the vector.
- The resultant force, R, joins the tail of vector V_1 to the tip of vector V_2. You can measure the vector and its angle from the scale diagram, and R = 20 N.

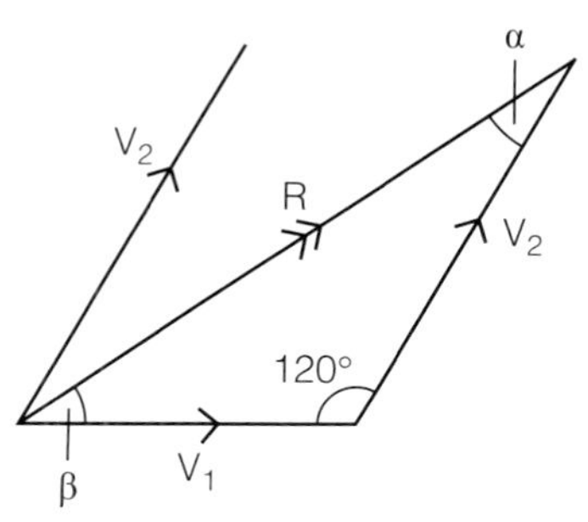

Figure 15.15 Resolving vectors.

Method 2: Calculation

The cosine rule can be used to find the resultant in the parallelogram.
It states:

$$R^2 = A^2 + B^2 - 2AB\cos\theta$$

We now substitute in to this equation, with reference to Figure 15.15 where V_1 corresponds to A, and V_2 corresponds to B:

$$= 10^2 + 13^2 - 2 \times 10 \times 13 \times \cos 120$$

$$= 399$$

$$R = \sqrt{399} = 20\,\text{N}$$

The sine rule is used to find the angle the vector makes relative to the horizontal:

$$\frac{R}{\sin\theta} = \frac{V_1}{\sin\alpha} = \frac{V_2}{\sin\beta}$$

Substituting,

$$\frac{20}{\sin 120} = \frac{10}{\sin\alpha} = \frac{13}{\sin\beta}$$

$$\sin\beta = \frac{13}{20} \times \sin 120 = 0.563$$

$$\beta = 34°$$

TIP

If you need conditions for equilibrium, draw a force diagram and find the resultant force vector. The force needed for equilibrium is equal in size and opposite in direction to this resultant force.

Degrees and radians

Degrees and radians are used to describe angles. Degrees are calculated as $\frac{1}{360}$ of the angle turned through a complete revolution. An angle in radians is the length of an arc, *s*, that the angle subtends in a circle of radius *r*. In other words, angle in radians = arc length / radius.

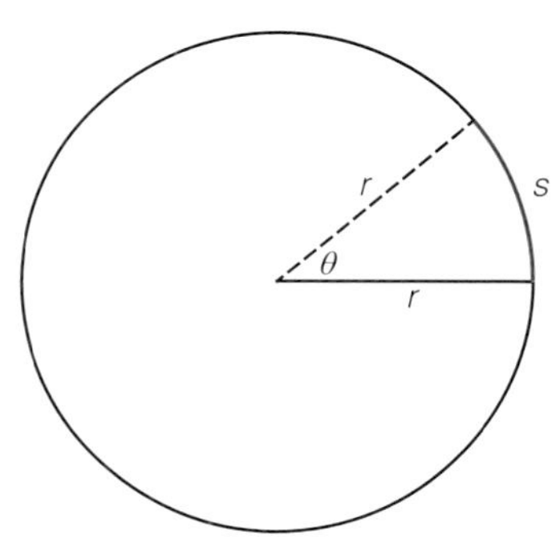

Figure 15.16 When θ is measured in radians, $\theta = \frac{s}{r}$. To go round the circle once, the arc is $2\pi r$, so the angle in radians is $\frac{2\pi r}{r} = 2\pi$.

You must be able to convert between the two units and to recognise some key values:

- 360 degrees = 2π radians, or one full revolution
- 180 degrees = π radians or half a revolution.

Small angle approximations

For very small angles, you can approximate values for the sine, cosine or tangent. This is useful when calculating the fringe separations in interference patterns. When θ is measured in radians, a small angle segment approximates to *s/r* for $\sin\theta$, $\tan\theta$ and θ where *s* is the arc length and *r* is the radius. The rules are:

- $\sin\theta \approx \tan\theta \approx \theta$
- $\cos\theta \approx 1$

To convert the angle in radians into an angle in degrees, remember that one radian is $\frac{180}{\pi} \approx 57.3$ degrees.

TEST YOURSELF

22 Calculate these quantities to the appropriate significant figures:

a) the cross-sectional area of the Earth, its radius is 6378 km

b) the cross-section area of a hair, its diameter is 90 μm

c) the surface area of a spherical light bulb of diameter 5 cm.

23 a) Work out the missing angles X, Y and Z in Figure 15.17.

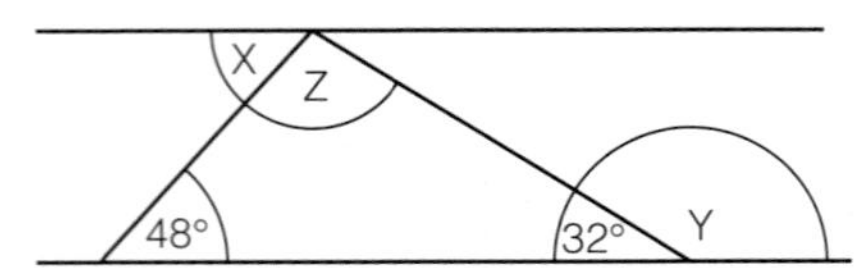

Figure 15.17 Angles between parallel lines.

b) Using Figure 15.18, explain why angle θ and angle α are the same.

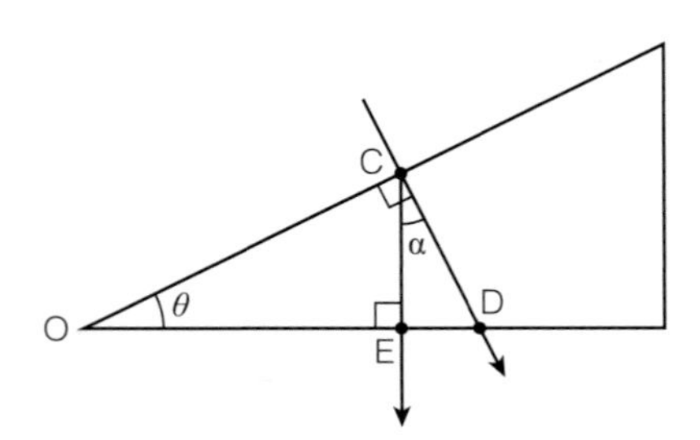

Figure 15.18 Angles on an inclined plane.

c) Using Figure 15.19, calculate the component of weight parallel and perpendicular to the plane if θ is 40° and W is 120 N.

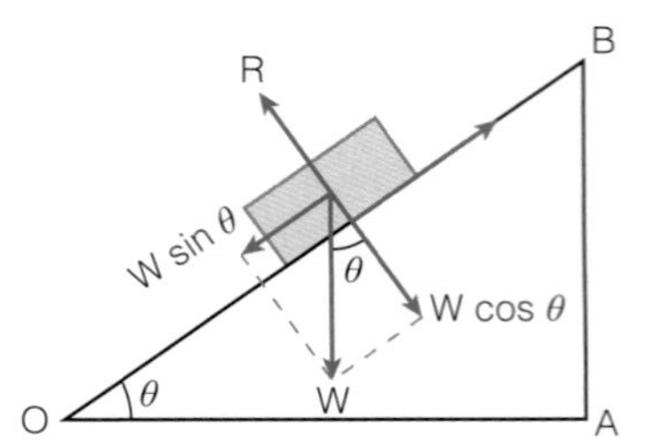

Figure 15.19 Body on an inclined plane.

d) Repeat part (c) if θ is 34° and W is 67 N.

24 Resolve these vectors into horizontal and vertical components:

a) the velocity of a ball moving at 4 m s^{-1} at an angle of 30° to the horizontal

b) the velocity of a ball moving at 12 m s^{-1} at an angle of 45° to the horizontal.

25 Convert these angles from degrees to radians:

a) 40° **b)** 175° **c)** 270°.

26 Convert these angles from radians to degrees:

a) $\pi/4$ radians **b)** 0.3 radians **c)** 1.6 radians.

27 Use the small angle rule to write down these values:

a) tan 0.01 radians

b) cos 0.05 radians

c) sin 0.03 radians.

16 Developing practical skills in physics

A successful experimental physicist must develop good practical skills and the ability to measure quantities accurately. Throughout your A-level physics course you are expected to carry out practical work that will help you to develop a range of competencies.

You will need to be able to show that you can:

- follow written procedures
- apply investigative approaches and methods during practical work
- use equipment and materials safely
- make and record observations accurately
- research, reference and report your findings.

Figure 16.1 The Mars Climate Orbiter was launched by NASA on 11 December 1998; this space mission illustrated the importance of using the correct units in practical work.

The Mars Climate Orbiter was designed to study the Martian climate and surface. The launch went well but, on 23 September 1999, as it was moving into orbit around Mars, communication with the spacecraft was lost. During the 286-day journey between Earth and Mars the thruster rockets that were being used to steer the craft were fired incorrectly. Lockheed Martin, the company carrying out the calculations, were working in imperial units (pound-force seconds) but NASA's navigation team were expecting the data in metric units (Newton seconds). This mismatch meant that the craft was flying too close to Mars. It passed through the upper atmosphere of the planet and broke up. Fortunately, the mistakes most of us make in our practical work are not so expensive.

Measurement and errors

In physics you will make measurements using a range of different instruments including rulers, stopwatches, electrical multimeters, Geiger counters, top-pan balances, Vernier calipers, oscilloscopes and thermometers. A measurement provides information about a property of an object, giving it both a magnitude and unit.

A measurement error occurs when the value that we measure is not the 'true value' of the quantity being measured. The 'true value' of measurement is the value of the measurement that would be obtained in an ideal world.

A **systematic error** is an error that affects a set of measurements in the same way each time.

Parallax occurs when the position of an object appears to be different when viewed from different positions. This can lead to measurement errors. A mirror is often placed behind the pointer of meters so that by lining up the pointer and its image you know you are avoiding parallax.

The error that caused the loss of the Mars Climate Orbiter was an example of a systematic error. A systematic error is a measurement that is consistently too small or too large. This type of error may be caused by:

- poor technique (e.g. not avoiding parallax when reading an analogue voltmeter)
- zero error on an instrument (e.g. a newton-meter that shows a value for force when there is nothing hanging from it)
- poor calibration of the instrument (the scale on a thermometer being incorrect so that one degree is too large).
- or the wrong unit being recorded.

A systematic error will not be reduced by repeating measurements. However, using different methods or instruments to obtain the same value will allow you to compare the results obtained and may identify the systematic error. For example, measuring the temperature of an object using a mercury thermometer, an infrared camera and a bimetallic strip thermometer would allow a comparison to be made between the instruments.

We can also take systematic errors into account by correcting the value of the readings taken. For example, if a newton-meter has a systematic error of + 0.2 N then we can subtract this from our results to obtain a more accurate value.

A **random error** is an error that affects a measurement in an unpredictable fashion.

A random error occurs when repeating the measurement gives an unpredictable different result. The effect of this type of error is reduced by taking repeated measurements. This also allows an average value to be calculated.

Random errors may arise due to:

- observer error (e.g. writing down a measurement incorrectly)
- the readability of the equipment (e.g. trying to measure the height to which a bouncy ball rises, or reading an ammeter when the current is changing quickly)
- external effects on the measured item (e.g. changes in ambient temperature or pressure in gas measurements).

Figure 16.2 shows the effect that systematic and random errors might have on a set of results. In this example, the 'true results' should produce a straight live graph which passes through the origin.

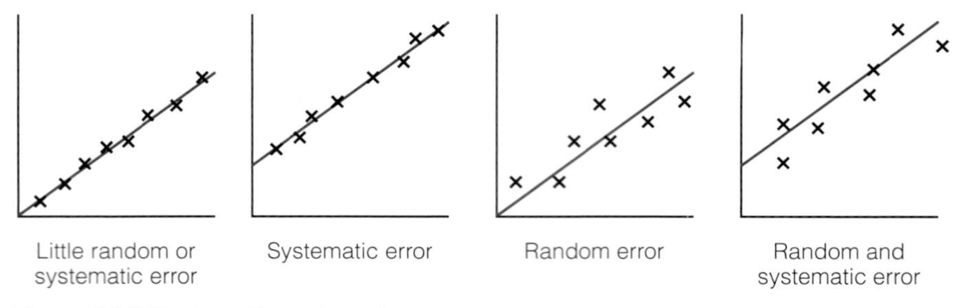

Figure 16.2 Systematic and random errors.

The quality of measurement

Once a scientist has carried out an experiment to obtain measurements, they will then consider the quality of the measurements that they have taken to enable them to draw conclusions about their work.

While the experiment is being planned, it is important that it is valid. In other words, we need to be sure that the experiment is going to measure what it is supposed to be measuring, and that all the relevant factors have been controlled.

Accuracy of a measurement describes how close it is to the 'true value'.

The accuracy of a measurement describes how closely the measurement is to the 'true value' of the quantity being measured. However, as we are unable to actually take the 'true value', the accuracy is a qualitative idea only and we cannot give it a numerical value.

Precision is how closely a set of repeated measurements are to each other.

The precision of a measurement describes how closely a number of repeated readings agree with each other. A precise measurement will have very little spread of results around the mean value. However, it does not give us any indication of how close to the 'true value' our result is.

The concepts of accuracy and precision can be illustrated using a dart board as shown in Figure 16.3 where we are aiming for the centre.

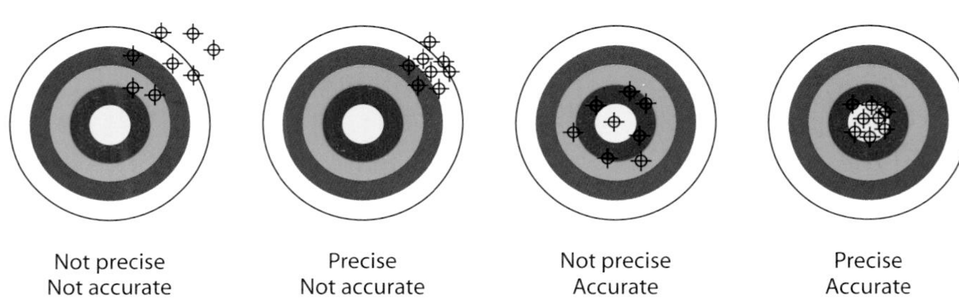

Figure 16.3 Accuracy and precision.

Repeatability is the level of consistency of a set of repeated measurements made by the same person, in the same laboratory, using the same method.

Reproducibility is the level of consistency of a set of repeated measurements made using the same method by different people in different laboratories.

The precision of a measurement is also linked the concepts of repeatability and reproducibility. Both terms describe the consistency of sets of results collected by the same method. Repeatability is the term used when the measurements are taken by the same person in the same laboratory over a short period of time. Reproducibility is the term used when the measurements are made by different people in different laboratories.

For example, if you were measuring the resistivity of a copper wire and take three measurements using the same micrometer within five minutes of each other, then you would expect all the results to be similar (repeatability). However, if someone else in your class measured the same piece of wire the following day, using a different micrometer, there might be more variation between your results and theirs (reproducibility).

In general, for the experiments you carry out during your A-level physics course, the measurements will be both repeatable and reproducible.

However, in research into a new phenomenon, it may be that measurements have not yet been shown to be reproducible, or that the exact method used means that it is hard to reproduce in another laboratory.

Sensitivity: a sensitive instrument produces a large change in output for a small change in input.

The sensitivity of an instrument is an important feature of measurement. A sensitive instrument is one that responds with a large change in output for a small change in input. For example, a sensitive analogue ammeter is one that shows a large deflection for a small change of current.

Resolution is the smallest observable change in the quantity being measured by a measuring instrument.

Measuring instruments also have a resolution, which limits the measurements being taken. The resolution is the smallest change in the quantity being measured by an instrument that gives a measurable change in the value. Using a measuring instrument with a greater resolution will increase the accuracy of the measurement.

For example, a student is measuring the diameter of a metal bar that is about 3 mm in diameter. They could use a ruler, which can measure to the nearest 0.5 mm, or they could use a micrometer, which can measure to the nearest 0.005 mm. The micrometer will give a more accurate measurement with greater resolution.

Uncertainties in measurement

In physics, error and uncertainty are not the same thing. As we have seen, error refers to the difference between the measurement of a physical quantity and the 'true value' of that quantity. In experimental work, we try to account for any known errors, for example by correcting for zero point errors.

Any error whose value we don't know is a source of uncertainty. Uncertainty is a measure of the spread of value which is likely to include the 'true value'. We can quantify the level of uncertainty for a given measurement.

For example, a top-pan balance can measure to a resolution of 0.1 g. If the value recorded for the mass of a cube of copper is 21.6 g, then the measured mass of the copper could be between 21.5 g and 21.7 g.

We would write the result as: 21.6 g ± 0.1 g.

In this example the value of uncertainty is the absolute uncertainty. It has the same units as the measurement and represents the range of possible values of the measurement.

If a set of repeated measurements are made, then the absolute uncertainty is given as half the range from the highest to the lowest value obtained.

EXAMPLE

A student takes the following measurements of length at a given value of force during an experiment to measure the Young modulus of a nylon thread.

23.1 cm, 23.0 cm, 23.1 cm, 23.2 cm, 23.0 cm

Find the measurement and absolute uncertainty for these results.

Answer

The mean value of the length of the thread

$$= \frac{23.1 + 23.0 + 23.1 + 23.2 + 23.0}{5} \text{ cm}$$

$$= 23.08 \text{ cm}$$

However, this value has four significant figures, while our measurements were only to a value of three significant figures, so we will give the value for the mean to the same number of significant figures.

Thus, mean length of thread = 23.1 cm

The absolute uncertainty

$$= \frac{23.2 - 23.0}{2} \text{ cm}$$

$$= 0.1 \text{ cm}$$

The measurement and absolute uncertainty is therefore:

23.1 cm ± 0.1 cm

TIP

It is good practice to give the calculated quantity to the same number of significant figures as the least accurate measured quantity.

When carrying out experiments in physics, we often take measurements from more than one instrument and then use these to calculate a value of a different quantity.

If we are adding or subtracting measurements, then to calculate the uncertainty in the overall measurement we add the absolute uncertainties.

In other words:

$$(a \pm \Delta a) + (b \pm \Delta b) = (a + b) \pm (\Delta a + \Delta b)$$

$$(a \pm \Delta a) - (b \pm \Delta b) = (a - b) \pm (\Delta a + \Delta b)$$

Where the calculated quantity is derived from the multiplication or division of the measured quantities then the combined percentage error of a calculated quantity is found by adding the percentage errors of the individual measurements.

$$\text{fractional uncertainty} = \frac{\text{absolute uncertainty}}{\text{mean value}}$$

$$\text{percentage uncertainty} = \frac{\text{absolute uncertainty}}{\text{mean value}} \times 100$$

For example, calculating the density of a metal cube to work out the uncertainty in the measurement of density, we can use either the fractional uncertainty or the percentage uncertainty of the measurements for mass and volume.

In general, if

$$a = bc$$

or

$$a = \frac{b}{c}$$

then

percentage error in a = (percentage error in b) + (percentage error in c)

EXAMPLE

The potential difference across a resistor is measured as 10.0 V ± 0.3 V. The current through the resistor is measured as 1.3 A ± 0.2 A. What is the percentage and absolute uncertainty in the resistance of the resistor?

Answer

$$\text{resistance} = \frac{\text{potential difference}}{\text{current}} = \frac{10.0\,\text{V}}{1.3\,\text{A}} = 7.7\,\Omega$$

$$\%\text{ uncertainty in pd} = \frac{0.3\,\text{V}}{10.0\,\text{V}} \times 100\% = 3\%$$

$$\%\text{ uncertainty in current} = \frac{0.2\,\text{A}}{1.3\,\text{A}} \times 100\% = 15\%$$

% uncertainty in resistance = % uncertainty in pd + % uncertainty in current

$$= 3\% + 15\%$$

$$= 18\%$$

$$\text{absolute uncertainty} = \frac{(\text{value} \times \%\text{ uncertainty})}{100}$$

$$\text{resistance} = 7.7\,\Omega \pm 1.4\,\Omega$$

If you go on to study physics or engineering subjects at university then you may learn about more detailed methods to estimate uncertainty in measurements. However, the treatment given in this chapter is sufficient for a first look at uncertainties.

TEST YOURSELF

1 A student is calculating the density of tin. The student measures the dimensions of a small solid tin cube as 20 mm × 20 mm × 20 mm, each dimension accurate to 0.2 mm. The mass of the cube is 58.2 g ± 0.2 g. Calculate:

a) its volume

b) the percentage uncertainty in its volume

c) its density in $kg\,m^{-3}$

d) the absolute uncertainty in its density.

2 A spring extends by 3.2 cm when a force of 10 N is applied to it. The absolute uncertainty in the extension is 0.1 cm and the uncertainty in the force is 0.1 N. The spring constant, *k*, of the spring is given by the formula spring constant = force/extension. Calculate the spring constant including the percentage uncertainty for the spring.

3 The time it takes for an athlete to run 100.00 m is measured as 9.63 s. The uncertainty in the distance is 0.01 m and in the time is 0.01 s. Calculate the absolute uncertainty in the athlete's speed.

Uncertainties in graphical data

Often in physics we will want to plot a graph using experimental measurements. We represent the uncertainties in the data using 'error bars' on the graph.

Plotting graphs

1 Choose your axis scales so that your plotted points will cover at least half of your graph paper. Using a larger area means that you can plot the points more accurately.

2 Use sensible divisions that can be plotted and read easily (e.g. multiples of two or five are often most straightforward to plot accurately).

3 Label your axes with the plotted quantity and unit in the format 'quantity/unit'

4 Plot your points as either a small horizontal cross '+' or a diagonal cross 'x'.

5 Add error bars to represent the uncertainty. The length of each bar should be the length of the absolute uncertainty for the point.

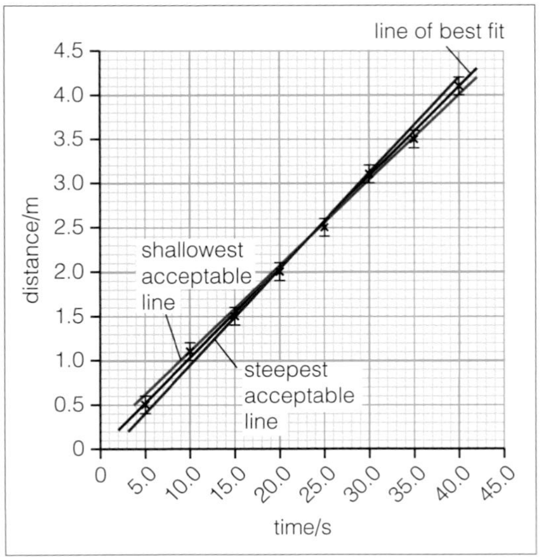

Estimating uncertainty in gradient

To estimate the uncertainty in the gradient we draw two additional lines of fit on to our data points. These are shown in Figure 16.4.

The points suggest a straight line so the line of best fit is drawn such that there are an approximately equal number of points on either side of the line.

The point (0, 0) is not usually plotted because this is not a point that was measured. Also, if there is a systematic error in the experiment the line of best fit may not pass through (0, 0), and this allows possible systematic errors to be identified.

To calculate the uncertainty in the gradient, we draw two more lines of fit; one representing the shallowest acceptable line of fit from the bottom of the upper error bar to the top of the lowest error bar (red line in Figure 16.4), and one representing the steepest acceptable line of fit from the top of the upper error bar to the bottom of the lowest error bar (blue line in Figure 16.4).

line of best fit
shallowest acceptable line
steepest acceptable line
distance/m
time/s

Figure 16.4 Estimating the uncertainty in the gradient.

The gradient of both of these lines is calculated and the uncertainty is given by:

$$\frac{1}{2} \times \text{difference between highest and lowest possible gradient values.}$$

ACTIVITY

Dropping a bouncy ball

You can either carry out this experiment or use the data given in the table.

Drop a small bouncy ball from a known height and measure the height to which the ball bounces back up.

1 Before you take any measurements, write down what the potential sources of error in your readings might be and how you will minimise or account for them.
Repeat each measurement three times and use six different heights.
Record your results in a table similar to the one shown:

Drop height (cm)	10.0	20.0	30.0	40.0	50.0	60.0
Bounce 1 (cm)	9.0	18.0	23.2	32.0	40.0	50.5
Bounce 2 (cm)	9.5	17.5	23.0	33.5	40.5	52.0
Bounce 3 (cm)	8.8	20.0	24.0	32.8	40.0	51.5
Average bounce height (cm)						

2 From your experiment, suggest a suitable value for uncertainty in the measurement of the drop height.
3 Calculate the average bounce height and the uncertainty for each drop height.
4 The height that the ball bounces to will depend in part on the efficiency of energy transfer of the ball. The ball starts at the drop height, h_d, with a gravitational potential energy of mgh_d. It then drops to the ground and rebounds to a bounce height, h_b, with a potential energy of mgh_b at the top of its bounce. As you can see from the data in the table, $h_b < h_d$.

Figure 16.5 A simple experiment to investigate the link between drop height and bounce height for a bouncy ball.

We can calculate the efficiency of energy transfer during the bounce using:

$$\text{efficiency} = \frac{\text{bounce height}}{\text{drop height}}$$

We rearrange this equation so that it is of the form

$$y = mx + c$$

bounce height = drop height × efficiency

There should be no y-intercept. If we plot a graph of drop height (on the x-axis) against the bounce height (on the y-axis), then the gradient of the graph will be equal to the efficiency of the bouncy ball.

5 Plot a graph of drop height against bounce height, including error bars on your graph.
6 Calculate the gradient of your graph and use your error bars to find a value for the uncertainty in your value.

Extension

Repeat the experiment using different balls or different surfaces.

Investigate whether the relationship you found is valid for large values of drop height.

TEST YOURSELF

4 A student measures the diameter of a piece of wire at different places along its length using a micrometer. She obtains the following measurements:

Diameter (mm)	0.19	0.21	0.18	0.81	0.22

a) Use her readings to obtain the mean value for the diameter of the wire.
b) Estimate the uncertainty in the mean diameter of the wire.

5 A student has made a sample of conductive modelling dough using flour, water and cream of tartar (potassium bitartrate, $KC_4H_5O_6$). The modelling dough can easily be made into different shapes and used to build simple circuits using LEDs and batteries. The student wanted to investigate how the resistance, R, of a fixed amount (volume) of modelling dough varied with length, l. Figure 16.6 shows the experimental set-up used. She used a multimeter set to the ohms range and connected the leads to two metal plates pressed on to each end of the cylinder of the conducting putty.

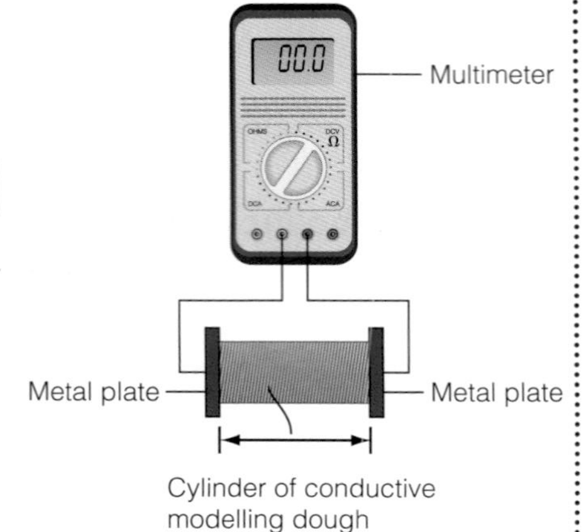

Figure 16.6 Measuring the resistance of conductive modelling dough.

The student suggests that the resistivity, ρ, of the conducting putty is given by the formula:

$$\rho = \frac{RV}{l^2}$$

where V is the volume of the conductive modelling dough.

a) Explain why plotting a graph of R against l^2 would allow you to show if the suggested relationship between R and l is correct.
b) The student took the following results shown in the table.

l (cm)	R (Ω)
4 ± 0.2	20 ± 10
8 ± 0.2	60 ± 10
12 ± 0.2	140 ± 10
16 ± 0.2	250 ± 10
20 ± 0.2	390 ± 10
24 ± 0.2	560 ± 10

i) Plot a graph of R against l^2, including error bars for both quantities.
ii) Draw a line of best fit and determine the gradient of this line.
iii) By drawing the highest and lowest acceptable lines of fit, determine the uncertainty in your value of gradient.

c) The student used 26.8 cm^3 of conductive modelling dough in her experiment. Calculate the resistivity, ρ, of the dough.

Extension

If you wish to carry out this experiment the recipe for conductive modelling dough can be found at **http://courseweb.stthomas.edu/apthomas/SquishyCircuits/conductiveDough.htm**

6 It has been suggested that there is a connection between the length of a chain of paper clips and the time it takes to swing back and forwards (periodic time). Figure 16.7 illustrates the periodic time of the chain.

It is suggested that the periodic time is given by the following relationship:

periodic time = $k \times \sqrt{\text{length}}$

Describe a possible laboratory experiment to investigate this relationship. You should include:

- the measurements to be taken
- the procedure you would use
- how to control other variables and reduce error in your measurements
- how you would analyse the data.

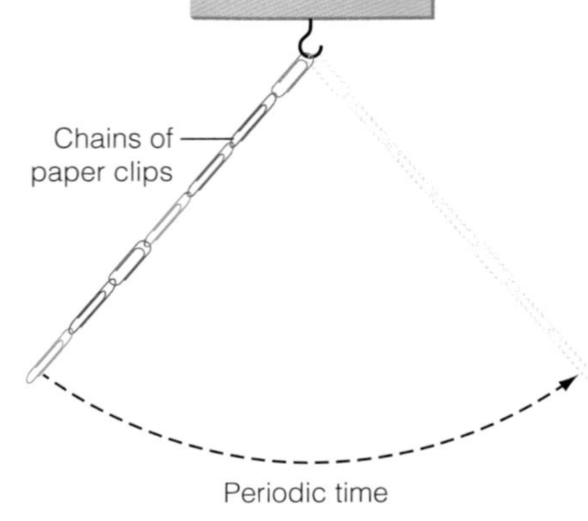

Figure 16.7 Periodic swing of a paper clip chain.

Extension

You may wish to carry out this experiment and collect the data to test the relationship.

17 Preparing for written assessments

At the end of any course of study you must demonstrate what you have learnt. That often means taking a written assessment, otherwise known as an exam. It is important to realise, however, that preparing for written assessments starts a long time before the exam itself.

Your physics lessons, and what you do after them, are a key part of learning the subject. The flowchart in Figure 17.1 shows one way to approach your studies.

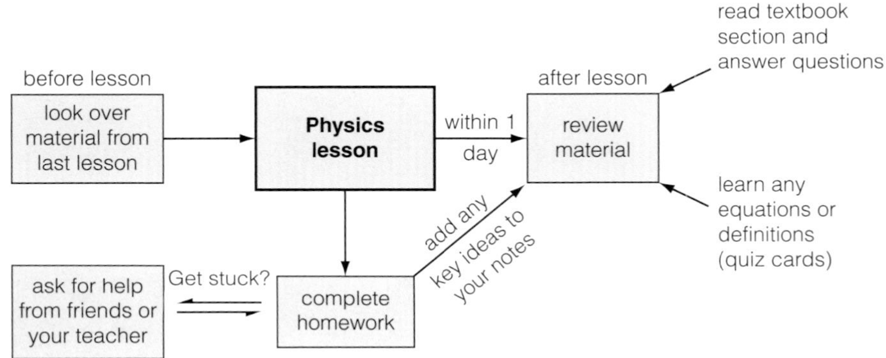

Figure 17.1 How to spend your study time.

Before each lesson you should look back at what you did the previous lesson. Ask yourself questions about what you learnt, remind yourself about any definitions or equations that were used and recall what were the main aims of the lesson.

After the lesson you should complete any homework well before any deadline that has been set. Remember that your teacher will be happy to help you if you need it, but asking ten minutes before the lesson is not a good idea.

You should also spend time soon after the lesson (ideally within a day) reviewing the material you covered. Read through the textbook section that is linked to the lesson and try to add extra information to your notes. Try the 'Test yourself' questions if you haven't done them in class. If you used any equations or were introduced to keyword definitions in the lesson, you should learn them. Although you will be provided with a formula sheet in the exam, it saves valuable thinking time if you can quickly recall the correct equation when answering questions.

Making memory quiz cards

This activity is based on 'memory pair' games where you have a collection of cards with pairs of matching pictures. The cards are shuffled and laid out face down. Each player takes it in turn to turn over two cards. If they match, the player keeps them. If they aren't a pair then they are turned over again, and the next player has a go.

We can use this principle to help learn definitions and equations. Testing yourself regularly helps you to recall definitions and equations, and helps you to store information in your long-term memory.

Use sheets of card and split each one into eight equal sections. On one small piece of card write the title of a concept and on another write the definition of that concept. Do this for each definition you need to learn, as well as for the equations that you use during the course. You could use different coloured card for the different topics in the course.

Once you have a set of cards, you can test yourself on the definitions and equations. Figure 17.2 shows an example of a set of memory pair quiz cards.

Figure 17.2 Using memory pair cards to improve recall of important information. One matching pair has just been turned over.

Before the exams

Once you know when your exams will be, you will be able to plan out your revision. If you are taking more than one subject you will need to make sure that your revision timetable allows you to spend time working on each subject regularly. Don't try to revise everything for one subject at once.

The specification for the subject is a very useful document to use during your revision. You can obtain an electronic copy of any specification from the exam board website. The specification will give you information about the structure of the exams, what you'll be examined on in each exam and the topics that you have learnt during your course.

Each exam will contain a variety of questions that aim to measure your knowledge of physics through three Assessment Objectives (AOs). These AOs are:

AO1	Demonstrate knowledge and understanding of scientific ideas, processes, techniques and procedures.
AO2	Apply knowledge and understanding of scientific ideas, processes, techniques and procedures: • in a theoretical context • in a practical context • when handling qualitative data • when handling quantitative data.
AO3	Analyse, interpret and evaluate scientific information, ideas and evidence, including in relation to issues, to: • make judgements and reach conclusions • develop and refine practical design and procedures.

In other words, you must be able to recall physics concepts (AO1), apply your knowledge to new situations (AO2), and interpret or evaluate new information (AO3).

You must always think about the number of marks being given for a particular question. For example, if a question has two marks you are not going to get both marks for a one-word answer; the examiner will be looking for two points. If a question is worth six marks, you will be expected to write at length and to include some sophisticated analysis or comment; trying to include six relevant points is a good start.

Another useful document to use, especially with past papers, is the examiner report. This is summary of how students answered the exam papers and often includes common mistakes that students made. It is written by the person in charge of the exam (the senior examiner). Reading it can help you spot what to avoid, or help explain where the marks were awarded.

For example, on a recent exam question about the horizontal motion of a projectile, the senior examiner wrote: 'It is surprising how many students do not understand that the horizontal velocity is constant and that the "suvat" equations are not necessary.'

Structuring your revision

In recent years psychologists have been researching successful ways of remembering and recalling information. Three useful findings that are particularly relevant to revision are distributed practice, testing and interleaving.

Distributed practice

Most students have, at some point in their education, spent the night before a test cramming to try and remember the information. Unsurprisingly, this is not a good way to learn. They might pass the test but the information will probably not 'stick' in their minds for the longer term.

Psychologists have found that students remember information better if they spread their learning over a longer time period, even if they spend the same amount of time learning overall. So, rather than revising for three hours in one evening, it would be better to spend 30 minutes revising per evening over six evenings.

Part of the reason for this is that, as soon as we have learnt something, we start to forget it. Using distributed practice means that we have to recall a topic a number of times, thus strengthening our memory and understanding of it.

TIP

Revise the same topic a number of times over the space of a few weeks. Don't try to learn it all in one sitting.

Test, test, test

The word 'test' often holds unpleasant memories for students. However, taking practice tests has been shown to be one of the most effective methods of learning material. The memory quiz cards described earlier in this chapter are one way of testing yourself on physics topics.

Other possible ways to include tests in your revision:

- Answer all the questions in the chapters of this textbook.
- As you read through your notes or the textbook, jot down two or three possible questions for each section. Then, without your notes, answer the questions. If they are fairly simple questions, you may be able to answer them without writing anything down. However, if you pose more difficult questions, or ones that require more explanation, then writing down the answer is a good idea.
- Use the specification to help you identify key ideas and then ask yourself questions related to that.

  For example, the specification has the following phrase as part of topic 3.2.1.1. Constituents of the atom: 'Proton number Z, nucleon number A, nuclide notation'.

  You could ask:

  - What is the proton number?
  - What is the nucleon number?
  - How are these two numbers linked?
  - If an element has the symbol $^{239}_{94}Pu$, how many protons and neutrons will it have?
  - What is the importance of the relative number of protons and neutrons in an atom?
  - Use past exam papers for question practice. As well as using questions from the current specification, exam questions from earlier specifications and other exam boards will give you a wide range of options (though do bear in mind that there will be differences in some of the content and style).

TIP

Test yourself regularly on the topic. Use different methods to make yourself recall the information you need to learn.

Interleaving

Interleaving involves creating a revision schedule that allows you to mix different topics or even subjects during each session.

When revising, many students will spend the whole of their revision session on one topic, for example gravitational fields. However, it can be more helpful for longer-term memory to 'mix up' your revision. For example, work on particles for an hour, then have a short break and work on longitudinal and progressive waves.

When doing interleaved revision, it can sometimes feel like you are not learning the material as well and that you are 'chopping and changing'. However, research suggests that interleaving improves students' performance on tests.

Interleaving also prepares you for the structure of A-level exams, where examiners will require you to draw on knowledge from different topics in physics in the same question.

Index

A

absorption, optical fibres 89
absorption spectra 49
acceleration 133
  equations of motion 137–8
  Newton's second law of motion 151, 157
  practical investigation of 159–60
  from velocity–time graphs 135
acceleration due to gravity (g) 139–40
accuracy of measurements 291
aerogels 202
algebra 280–1
alpha decay 6
ammeters 254
amplitude 71
angles
  degrees and radians 287
  of a triangle 286
annihilation 12
antinodes
  on a guitar string 102, 103
  in open-ended pipes 104–5
antiparticles 11–13, 19, 35
  antineutrinos 6, 7
antiphase 73
areas, calculation of 285
arithmetic means 278
atomic radius 18
atoms 1, 2
  charges and masses of particles 3–4
  Rutherford–Bohr model 3
Avogadro Number (N_A) 18

B

baryons 29, 30–1, 35
batteries 250
  emf and internal resistance 250–1, 253–4
Baumgartner, Felix 134
beta decay 6–7
  Feynman diagrams 25, 31
  nuclear equations 10
brittle materials 207–8

C

calculators 276–7
car safety 186–7
cells 250
  emf and internal resistance 250–1, 253–4
centre of mass 124
charge
  from current–time graphs 224–5
  of sub-atomic particles 3–4, 19, 28–30, 32, 36
charge coupled devices (CCDs) 58
charge–time graphs 225
circuits *see* electrical circuits
cloud chambers 7
coherent waves 97
collision experiments 189–90
collisions
  elastic and inelastic 190, 191–2
  in two dimensions 194
compressions, longitudinal waves 77
compressive forces 203
conductors
  ohmic and non-ohmic 231
  semiconductors 63
  superconductors 229
conservation of energy 172–3
conservation of linear momentum 188–9
continuous spectra 48
cosine (cos) 286, 287
cosmic rays 29
couples 126
creep 213
critical angle 86
crossing polarisers 80
crumple zones 186–7
current–time graphs 224–5
curved graphs 283–4

D

data handling 277–9
de Broglie wavelength 63–4, 65
deep-inelastic scattering 21
degrees and radians 286
density (ρ) 201–2
diffraction 63–4, 95, 108
  of light 108–11
diffraction gratings 48, 109–10
  applications of 111
diffraction rings experiment 64
diodes 231–2
Dirac, Paul 11
dispersion 88
displacement, equations of motion 137–8
displacement–time graphs 134
distance travelled, from velocity–time graphs 135–6
distributed practice 300
drag 141
ductility 207–8

E

efficiency 173–4
Einstein, Albert
  energy–mass equation 19, 33
  photoelectric equation 61–2
elastic collisions 190, 191–2
elasticity, Hooke's law 203–5
elastic limit 204, 212
elastic materials 204
elastic potential energy 171
  transfer to kinetic energy 175
elastic strain energy 208–11
electrical characteristics 230–2
  thermistors 233
electrical circuits
  Kirchoff's laws 248–9
  resistor networks 255–7
electrical energy 227–8
electrical power 244, 246–8
electric current 221–4
  from charge–time graphs 225
  Kirchoff's first circuit law 248–9
  variations with time 224–5
electrolytes 223
electromagnetic force 22–3
electromagnetic radiation 42, 76
  communication uses 78–9
  effect on living cells 78
  intensity of 44
  wave–particle duality 65
electromagnetic spectrum 77
electromotive force (emf) 227–8, 250–1
  investigation of 253–4
electron diffraction 63–5
electrons 3–4
  energy levels 46–7
  wave–particle duality 65
electron volts (eV) 45
elements 1
emission spectra 49
energy 168, 170–2
  conservation of 172–3
  and momentum 190–1
  practical investigations 175–6
energy levels 46–7
equilibrium
  forces 121
  turning moments 123
errors 289–90
estimating results 276
exams, preparation for 298–301
exchange particles 19, 22, 27
excitation of electrons 47
extension 204

F

Feynman diagrams 25–6
fluorescent tubes 51
force–displacement graphs 170, 171
force–extension graphs 204, 207
force plate investigations 187–8
forces 116, 151
  compressive and tensile 203
  couples 126
  equilibrium 121
  fundamental 21–4, 35
  Newton's first law of motion 152–3

Newton's second law of motion 153–4
Newton's third law of motion 155–6
practical investigation of 159–60
force–time graphs 186
fractions 276
Fraunhofer lines 49
free body diagrams 152–3, 157
frequency 72
fringes, Young's double slit experiment 99–100
fundamental forces 21–4, 35
fundamental particles 19–20

G

Galileo Galilei 139
gamma radiation 6, 78, 79
Geiger-Muller counters 7–8
Gell-Mann, Murray 24
gluons 20
gradients
of curved graphs 283–4
estimating uncertainty 294–5
graphs 282–4
charge–time 225
current–time 224–5
displacement–time 134
electrical characteristics 230–2
force–displacement 170, 171
force–extension 204, 207
force–time 186
stress–strain 212, 214–15
uncertainties 294–5
velocity–time 135–6, 137
gravitational field strength (g) 133
gravitational potential energy 171
transfer to kinetic energy 176
gravity 24
acceleration due to (*g*) 139–40
ground state 47

H

hadrons 27, 29–32, 35, 36
harmonics 103–4
hertz 72
Higgs Boson 2, 19
Hooke's law 203–7
hydrogen atom, spectral lines 50

I

impulse 184–7
incidence, angle of 83
inclined planes 120, 121
inelastic collisions 190, 191–2
interference 97–8
microwaves 107
Young's double slit experiment 99–100
interleaved revision 301
internal resistance 250–1, 253–4
intrinsic properties 236
inverse square law 44
ionisation 47
ionisation energy 47
ions 1
isotopes 5

K

kaons 32, 36
kinetic energy 171–2
Kirchoff's laws 248–9

L

Large Hadron Collider 11, 19
lasers 96, 100
leptons 27, 28, 36
light
diffraction 108–11
polarisation 80–1
see also electromagnetic radiation; photons
light-dependent resistors (LDRs) 262
light-emitting diodes (LEDs) 48
limit of proportionality 204
linear momentum, conservation of 188–9
line spectra 48
longitudinal waves 75, 77

M

material dispersion 88
materials, properties of
density 201–2
ductile and brittle materials 207–8
elastic and plastic materials 204
intrinsic properties 236
Young modulus 211–14
maths skills 274–7
maxima, diffraction gratings 110
measurements
errors 289–90
quality of 291–2
uncertainties 292–3
mechanical waves 76
memory pair cards 299
mesons 29, 31–2, 35, 36
microwaves 78, 79
interference patterns 107
stationary wave formation 106
modal dispersion 88
models 2–3
moments 123, 125
momentum 183–4
collision experiments 189–90
collisions in two dimensions 194
conservation of 188–9
elastic and inelastic collisions 191–2
and energy 190–1
and impulse 184–7
problem-solving 193
motion
displacement–time graphs 134
equations of 136–8
Newton's first law 152–3
Newton's second law 153–4, 184–5, 193
Newton's third law 155–6
projectile 142–5
velocity–time graphs 135–6
muons 28, 36
musical instruments 102–4

N

neutrinos 6
neutrons 1, 3–4
Newton's laws of motion 151, 156–7
first law 152–3
practical investigations 159–61
second law 153–4, 184–5, 193
third law 155–6
nodes
on a guitar string 102, 103
in open-ended pipes 104–5
noise-cancellation 96
non-ohmic conductors 231
nuclear equations 9–10
nuclear radius 18
nucleon number (A) 4
nucleons 4
see also neutrons; protons

O

ohmic conductors 231
Ohm's law 230–1
optical fibres 71, 88–9
orders of magnitude 279

P

paired forces, Newton's third law of motion 155–6
pair production 13
parallax 290
parallel connection 244–5
resistor networks 256–7
particle–antiparticle interactions 12
particles, sub-atomic 1, 3–4, 18–21, 35–6
classification of 26–7
de Broglie wavelength 63–4, 65
hadrons 29–32
leptons 28
quantum number conservation laws 33–4
wave–particle duality 65
path difference 97–8
percentages 276
period of a wave 72
phase change on reflection 75
phase difference 73–4, 95
photoelectric effect 58–9
Einstein's equation 61–2
explanatory model 60–1
photoelectrons 58–9
photons 10–11, 43
energy of 43–4, 60–1
Pierre Auger observatory 29
pions 32, 36
Planck, Max 59

Planck's constant 10, 43, 47–8
plasmas 51
plastic materials 204
  ductility 207–8
polarisation 79–81
  of glucose solution 82
positrons 11
potential difference (voltage) 221, 227–8
  Kirchoff's second circuit law 249
potential dividers 259
  as sensors 261–2
potential energy 171
potential wells 60, 61
potentiometers 259–60
power 168, 174–5
  electrical 244, 246–8
precision of measurements 291
probability 278
progressive waves 71
projectile motion 142–5
proton number (Z) 4
protons 1, 3–4
pulse broadening 88

Q

quadratic equations 281
quanta of energy 43, 59
quantised energy levels 46
quantum numbers
  baryons 30
  conservation laws 33–4
  leptons 28
  strangeness (S) 32
quarks 20–1, 36

R

radians 73, 286
radioactive decay 6–7
  detection of 7–8
  nuclear equations 9–10
radio waves 78
random errors 290
rarefactions, longitudinal waves 77
ratios 275
rearranging equations 280–1
reflection, phase change 75
refraction 83
  law of 84–5
  total internal reflection 86, 88
refractive index 83–4
relative atomic mass 1
repeatability of measurements 291
reproducibility of measurements 291
resistance 221, 229
  ohmic and non-ohmic conductors 231
  thermistors 233–4
resistivity (ρ) 236–7
resistors, parallel and series connection 255–7
resolution of instruments 292
resolution of vectors 120
rest mass-energy 11
resultant vectors 117, 118–19, 286–7
revision 299–301
Rutherford–Bohr model of the atom 3

S

Sankey diagrams 173–4
scalar quantities 117
semiconductors 63
sensitivity of instruments 291
sensor circuits 261–2
series connection 244
  resistor networks 255–6, 257
significant figures 277–8
sine (sin) 286, 287
Snell's law of refraction 84–5
sound waves 77
  interference 97–8
  measurement of wavelength 106
  in open-ended pipes 104–5
spark counters 7
specific charge of a particle 4
specific heat capacity 228
spectra 42–3, 48–9
  electromagnetic spectrum 77
spectral lines, energy of 50
speed 133
spring constant (k) 204
springs
  elastic strain energy 208–10
  Hooke's law 203–4
  practical investigation 205
sprites 226, 234
standard form 275
Standard Model of matter 19
stationary (standing) waves 101–3
  Kundt's tube apparatus 106
  microwaves 106–7
  practical investigation of 105
  sound waves and musical instruments 104–5
step index optical fibres 88
straight-line graphs 283
strain energy density 215–16
strangeness (S) 32, 36
stress–strain graphs 212
  interpretation of 214–15
strong nuclear force 6, 23–4
superconductors 229
superposition of waves 96–7
systematic errors 290

T

tangent (tan) 286, 287
tensile forces 203
tensile strain 212
tension, practical investigation of 159
terminal speed 141
thermionic emission 51
thermistors 233–4, 261–2
threshold energy, photoelectric effect 60
threshold frequency, photoelectric effect 59
torque *see* turning moments
total internal reflection 86, 88
transverse waves 76
triangles, angles of 286
trigonometry 286
turning moments (torque) 123, 125

U

ultimate tensile stress (UTS) 213, 214–15
ultraviolet radiation 78, 79
uncertainties 279, 292–5
units 274
  conversion between 274–5

V

vector quantities 117
  addition of 118–19, 286–7
  resolution of 120
velocity 133
  from displacement–time graphs 134
  equations of motion 137–8
velocity–time graphs 135–6, 137, 284
virtual photons 23, 35
voltmeters 254
volumes, calculation of 285

W

wave equation 72, 95
wavelength 71
  see also de Broglie wavelength
wave–particle duality 65
waves
  diffraction 108–11
  interference 97–100
  longitudinal 75, 77
  mechanical 76
  polarisation 79–81
  progressive 71–5
  stationary (standing) 101–5
  superposition 96–7
  transverse 76
weak force 24, 35
weight 200
wires, properties of 206–7
work 168, 169–70, 200
work function (φ) 62–3

X

X-ray diffraction 64
X-rays 78, 79

Y

yield point 212
Young modulus (E) 212–4
Young's double slit experiment 99–100

Z

zero errors 290

Free online resources

Answers for the following features found in this book are available online:

- Prior knowledge questions
- Test yourself questions
- Activities

You'll also find an Extended glossary to help you learn the key terms and formulae you'll need in your exam.

Scan the QR codes below for each chapter.

Alternatively, you can browse through all chapters at:
www.hoddereducation.co.uk/AQAPhysics1

How to use the QR codes

To use the QR codes you will need a QR code reader for your smartphone/tablet. There are many free readers available, depending on the smartphone/tablet you are using. We have supplied some suggestions below, but this is not an exhaustive list and you should only download software compatible with your device and operating system. We do not endorse any of the third-party products listed below and downloading them is at your own risk.

- for iPhone/iPad, search the App store for Qrafter
- for Android, search the Play store for QR Droid
- for Blackberry, search Blackberry World for QR Scanner Pro
- for Windows/Symbian, search the Store for Upcode

Once you have downloaded a QR code reader, simply open the reader app and use it to take a photo of the code. You will then see a menu of the free resources available for that topic.

1 Particles and nuclides

3 Electrons and energy levels

2 Fundamental particles

4 Particles of light

5 Waves

6 Combining Waves

7 Introduction to Mechanics

8 Motion and its measurement

9 Newton's laws of motion

10 Work, energy and power

11 Momentum

12 Properties of materials

13 Current electricity

14 Electrical circuits

15 Maths in physics

16 Developing practical skills in physics